Burkholderia

From Genomes to Function

Edited by

Tom Coenye

Laboratory of Pharmaceutical Microbiology
Ghent University
Ghent
Belgium

and

Eshwar Mahenthiralingam

Cardiff School of Biosciences
Cardiff University
Cardiff
UK

Caister Academic Press

Caister Academic Press
Norfolk, UK

www.caister.com

British Library Cataloguing-in-Publication Data
A catalogue record for this book is available from the British Library

ISBN: 978-1-908230-35-5 (hardback)
ISBN: 978-1-908230-97-3 (ebook)

Cover design adapted from an image courtesy of Dr Jan Mast and Figure 2.5

Printed and bound in Great Britain

Contents

Contributors

Claudio Aguilar
Department of Microbiology
University of Zürich
Zürich
Switzerland

claudio_aguilar@access.uzh.ch

Silvia Bazzini
Dipartimento di Biologia e Biotecnologie 'L. Spallanzani'
Università degli Studi di Pavia
Pavia
Italy

silvia.bazzini@unipv.it

Silvia Buroni
Dipartimento di Biologia e Biotecnologie 'L. Spallanzani'
Università degli Studi di Pavia
Pavia
Italy

silvia.buroni@unipv.it

Gregory S. Buzard
Booz Allen Hamilton
Frederick, MD
USA

buzard_gregory@bah.com

Tom Coenye
Laboratory of Pharmaceutical Microbiology
Ghent University
Ghent
Belgium

tom.coenye@ugent.be

Bruna G. Coutinho
Bacteriology Group
International Centre for Genetic Engineering and Biotechnology
Trieste
Italy

bruna.coutinho@icgeb.org

Mayukh Das
Department of Plant Pathology and Microbiology
Texas A&M University
College Station, TX
USA

daysmayukh77@gmail.com

Peter Dawyndt
Department of Applied Mathematics, Computer Science and Statistics
Ghent University
Ghent
Belgium

peter.dawyndt@ugent.be

Jonathan J. Dennis
Centennial Centre for Interdisciplinary Science
Department of Biological Sciences
University of Alberta
Edmonton, AB
Canada

jon.dennis@ualberta.ca

Pavel Drevinek
Department of Medical Microbiology and Department of Paediatrics
2nd Faculty of Medicine
Charles University
Prague
Czech Republic

pavel.drevinek@lfmotol.cuni.cz

Leo Eberl
Department of Microbiology
University of Zürich
Zürich
Switzerland

leberl@botinst.uzh.ch

Carlos F. Gonzalez
Department of Plant Pathology and Microbiology
Texas A&M University
College Station, TX
USA

cf-gonzalez@tamu.edu

Pimlapas Leekitcharoenphon
National Food Institute
The Technical University of Denmark
Lyngby
Denmark

pile@food.dtu.dk

John J. LiPuma
Department of Pediatrics and Communicable Diseases
University of Michigan
Ann Arbor, MI
USA

jlipuma@umich.edu

Karlene H. Lynch
Centennial Centre for Interdisciplinary Science
Department of Biological Sciences
University of Alberta
Edmonton, AB
Canada

khlynch@ualberta.ca

Andreia Madeira
Department of Bioengineering
IBB-Institute for Biotechnology and Bioengineering
Centre for Biological and Chemical Engineering
Instituto Superior Técnico
Universidade de Lisboa
Lisboa
Portugal

andreia.madeira@ist.utl.pt

Eshwar Mahenthiralingam
Cardiff School of Biosciences
Cardiff University
Cardiff
UK

MahenthiralingamE@cardiff.ac.uk

Catherine Ong
DSO National Laboratories
Singapore
Republic of Singapore

catong@dso.org.sg

Hanna Ostapska
Centre for Human Immunology
Department of Microbiology and Immunology
University of Western Ontario
London, ON
Canada

hostapsk@uwo.ca

Maria R. Pasca
Dipartimento di Biologia e Biotecnologie 'L. Spallanzani'
Università degli Studi di Pavia
Pavia
Italy

mariarosalia.pasca@unipv.it

Charlotte Peeters
Laboratory of Microbiology
Ghent University
Ghent
Belgium

charlotte.peeters@ugent.be

Gabriella Pessi
Department of Microbiology
University of Zürich
Zürich
Switzerland

gabriella.pessi@botinst.uzh.ch

Giovanna Riccardi
Dipartimento di Biologia e Biotecnologie 'L. Spallanzani'
Università degli Studi di Pavia
Pavia
Italy

giovanna.riccardi@unipv.it

Zulma R. Suárez-Moreno
Instituto de Biotecnología
Universidad Nacional de Colombia
Bogotá
Colombia

zrsuarezm@gmail.com

Roberto Rosales-Reyes
Departamento de Medicina Experimental
Facultad de Medicina
Universidad Nacional Autónoma de México
México

rosales.reyes@gmail.com

Isabel Sá-Correia
Department of Bioengineering
IBB-Institute for Biotechnology and Bioengineering
Centre for Biological and Chemical Engineering
Instituto Superior Técnico
Universidade de Lisboa
Lisboa
Portugal

isacorreia@ist.utl.pt

Sandra C. dos Santos
Department of Bioengineering
IBB-Institute for Biotechnology and Bioengineering
Centre for Biological and Chemical Engineering
Instituto Superior Técnico
Universidade de Lisboa
Lisboa
Portugal

ss.santos@ist.utl.pt

Andrea Sass
Laboratory of Pharmaceutical Microbiology
Ghent University
Ghent
Belgium

andrea.sass@ugent.be

Crystal L. Schmerk
Centre for Human Immunology
Department of Microbiology and Immunology
University of Western Ontario
London, ON
Canada

cschmerk@uwo.ca

Nadine Schmid
Department of Microbiology
University of Zürich
Zürich
Switzerland

nschmid@access.uzh.ch

Pamela A. Sokol
Department of Microbiology
Immunology and Infectious Diseases
University of Calgary
Calgary, AB
Canada

psokol@ucalgary.ca

Patrick Tan
Genome Institute of Singapore
Singapore
Republic of Singapore

tanbop@gis.a-star.edu.sg

Claudia Udine
Dipartimento di Biologia e Biotecnologie 'L. Spallanzani'
Università degli Studi di Pavia
Pavia
Italy

claudia.udine@unipv.it

David W. Ussery
Comparative Genomics Group
Division of Life Sciences
Oak Ridge National Laboratory
Oak Ridge, TN
USA

dave@cbs.dtu.dk

Miguel A. Valvano
Centre for Infection and Immunity
Queen's University Belfast
Belfast
UK

m.valvano@qub.ac.uk

Peter Vandamme
Laboratory of Microbiology
Ghent University
Ghent
Belgium

peter.vandamme@ugent.be

Vittorio Venturi
Bacteriology Group
International Centre for Genetic Engineering and Biotechnology
Trieste
Italy

venturi@icgeb.org

Bart Verheyde
Laboratory of Microbiology
Ghent University
Ghent
Belgium

bart.verheyde@ugent.be

Introduction

Eshwar Mahenthiralingam and Tom Coenye

Abstract

Burkholderia is an enigmatic Gram-negative bacterial genus that has attracted researchers from multiple disciplines. *Burkholderia* species possess considerable genetic and metabolic versatility allowing them to exist in a wide range of environmental settings as free-living or host-associated microorganisms. From a research standpoint, their interactions can be divided into either those that are beneficial or those that are problematic, often with the same *Burkholderia* species capable of both functions. The beneficial traits of *Burkholderia* are largely environmental and include their ability to promote plant growth, kill pest organisms, fix nitrogen, and degrade man-made pollutants. Pathogenicity towards multiple host organisms is the fundamental negative attribute for which *Burkholderia* have been widely researched, however, new problematic areas such as industrial contamination are emerging. We have summarized key advances in the biology of *Burkholderia* bacteria, many of which have been characterized as a result of greatly improved genomic and molecular resources for these bacteria.

Background to the book

The first book on *Burkholderia* in this series was published in 2007 (Coenye and Vandamme, 2007) as an initial attempt to capture the massive progression made on understanding the biology and diversity of these fascinating bacteria. That book, entitled *Burkholderia: Molecular Microbiology and Genomics*, focused predominantly on what was known about the *Burkholderia cepacia* complex (Bcc) group of species. It had chapters covering their diversity, epidemiology, comparative genomics, antimicrobial resistance, polysaccharides secretion, interaction with phagocytes, beneficial interactions with plants as well as their role as plant pathogens, their ability to degrade organic pollutants, modelling Bcc infection, quorum sensing and biofilms and the intracellular pathogenesis of Bcc bacteria. The virulence of *Burkholderia pseudomallei* and *Burkholderia mallei*, and a chapter on non-Bcc environmental species, completed the overview of other *Burkholderia* and the molecular biology presented in the first book.

Since 2007, multiple *Burkholderia* genome sequences have been obtained and analysis of these bacteria has truly moved into the post-genomic era, prompting the focus of this book on genomic, gene and protein functions. The first genome sequence that was publicly available for researchers to access was that of the epidemic cystic fibrosis (CF) pathogen *Burkholderia cenocepacia* J2315 (http://www.sanger.ac.uk). The sequencing project carried out by the Wellcome Trust Sanger Institute, was initiated in 2000, with partial sequence data released on their website shortly thereafter, enabling researchers to begin genomic analysis of this pathogen. The *B. cenocepacia* J2315 genome was completed in 2003 facilitating the design of a microarray to enable global gene expression analysis (Drevinek *et al.*, 2008b; Leiske *et al.*, 2006), and finally published in 2009 (Holden *et al.*, 2009). The first published genome within the genus was that of *Burkholderia xenovorans* LB400, a strain that was capable of degrading polychlorinated biphenyl pollutants (Chain *et al.*, 2006). Over 100 *Burkholderia* genome sequences have now been determined and multiple complete genomes can be accessed at online databases such as The *Burkholderia* Genomes (www.burkholderia.com) and Integrated Microbial Genomes (http://img.jgi.doe.gov/) databases.

With contributions from multiple leading investigators and as a result of the vast genomic resources available, we have been able to assemble a series of 11 chapters covering the following aspects of *Burkholderia* molecular biology: genomic taxonomy and biodiversity, comparative genomics, molecular epidemiology, transcriptomics, proteomics, molecular pathogenesis of virulence in *B. mallei/B. pseudomallei* and Bcc, molecular basis of phytopathogenicity and antibiotic resistance, signalling and biofilms, and the genomic biology of *Burkholderia* phages. The theme underpinning each chapter is the use of DNA protein sequence data and post-genomic technologies to understand *Burkholderia* biology.

Burkholderia: magnets for multiple researchers from various disciplines

As readers of the first book have already discovered, *Burkholderia* are multifaceted bacteria with considerable genetic diversity and very versatile lifestyles. As a result they make very compelling model species to study in relation to various microbial, molecular and biochemical functions (Fig. 0.1). The ability to survive in harsh, competitive, environments is a key aspect of *Burkholderia* lifestyles. They are widely distributed in terrestrial habitats including soil, especially around or on plant roots at the rhizosphere, where they interact with many biotic and abiotic factors (Suarez-Moreno *et al.*, 2012). They may form very close interactions with higher organisms such as fungi, plants and insects, ranging from surface or internal associations, through to fully endosymbiotic relationships. For the most part their environmental lifestyles play beneficial roles, prompting researchers to examine their biopesticidal, bioremedial and plant growth promotion activities (Fig. 0.1). However, plant pathogenic *Burkholderia* and species that cause human or animal infections represent major areas of study, and other traits, such as their ability to cause industrial contamination (Rushton *et al.*, 2013) or mediate insecticide resistance (Kikuchi *et al.*, 2012), represent emerging fields where they play a problematic role (Fig. 0.1). With such a polarity of attributes, researchers studying *Burkholderia* may move between multiple disciplines ranging from their natural ecology through to molecular pathogenesis. This breadth of investigation allows researchers to gain unique insights into the microbiology of these Gram-negative bacteria.

With each of the chapters covering specific aspects of molecular genomics and gene/protein function, here we will briefly emphasize key advances in *Burkholderia* biology, as well as introduce interactions not covered in detail within the book chapters. The topics reviewed below also represent (i) areas of rapid progression in terms of research; (ii) highly cited papers; (iii) recent or seminal papers that open up areas of new *Burkholderia* functions, (iv) studies where novel post-genomic techniques have been pioneered with these bacteria, and (v) topics highlighted by participants in the International *Burkholderia* cepacia Working Group (IBCWG) meetings of current and future importance. These advances in *Burkholderia* biology have been summarized as a one-stop knowledge pit stop for microbiologists entering the field of *Burkholderia* at the current time and readers may then probe many of these areas in more detail within specific chapters (Table 0.1).

Progress in taxonomy, identification and typing

The systematic classification and re-classification of *Burkholderia* species continues, but now with well developed molecular and genomic tools to assist in the accurate taxonomic identification of these bacteria (Vandamme and Dawyndt, 2011). Major advances in taxonomy to date are (Table 0.1) (i) the formal naming of 17 species within the *B. cepacia* complex; (ii) the formal naming of over 74 validly described *Burkholderia* species; and (iii) the observation that at least two major phylogenetic groups are present within

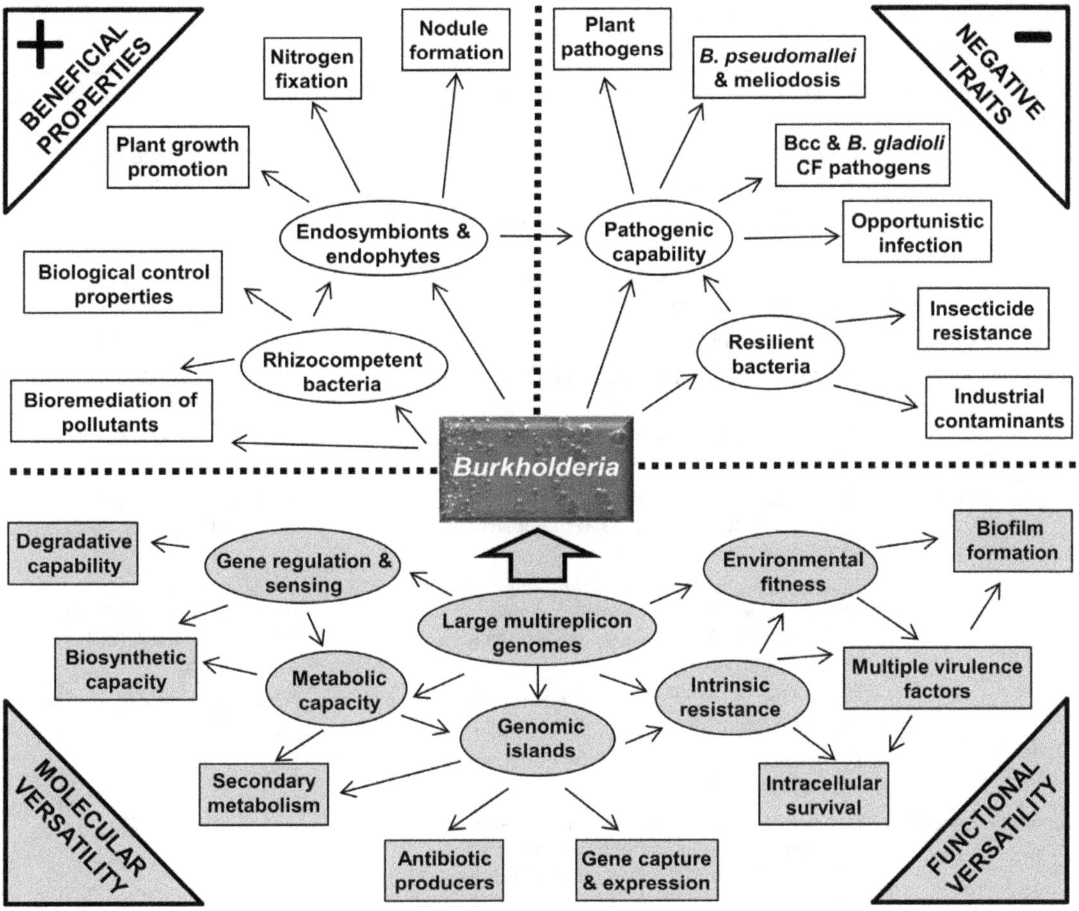

Figure 0.1 The biology of *Burkholderia* bacteria. An illustration of the beneficial and negative traits of these bacteria is shown in the top left and right panels, respectively. These properties are underpinned by a range of molecular and functional attributes that are shown in grey in the lower panel. Core interaction/functions are shown within the ovals, while additional activities are shown in the rectangles. Linkages between the various functions of *Burkholderia* are indicated by the arrows.

the genus. Regarding this latter observation, it has been suggested that there are considerable ecological differences between both major lineages (with one including human, animal and plant pathogenic species while the other cluster is mainly composed of saprophytic and plant-associated species) and it has been suggested that a subdivision of the genus *Burkholderia* would be appropriate (Estrada-de Los Santos *et al.*, 2013; Suarez-Moreno *et al.*, 2012). However, sequence-based phylogenetic analyses have indicated that several *Burkholderia* species occupy intermediate positions between the two major phylogenetic groups. In addition, considering the opportunistic nature of many of the 'pathogenic' *Burkholderia* species involved (including members of the Bcc) it seems ill-advised at this moment to put too much emphasis on the ecological differences to distinguish both groups and it is clear that more work is required before major taxonomic changes are made.

Below the species level the characterization of *Burkholderia* strains has benefited considerably from a well-developed multilocus sequence typing (MLST) scheme (Baldwin *et al.*, 2005; Spilker *et al.*, 2009) and a searchable online database (http://pubmlst.org/bcc); a number of other typing methods can also be efficiently applied to

Table 0.1 Key advances in the biology of *Burkholderia* bacteria

Feature or function	Advance	Chapter(s)	References/resource
Taxonomy/molecular epidemiology	17 species have been validly described and named within the *B. cepacia* complex	1 and 2	Vanlaere *et al.* (2009), LiPuma (2010)
	Over 74 validly described and named Burkholderia species	1 and 2	http://www.bacterio.cict.fr/b/burkholderia.html
	At least two major phylogenomic subgroups are present within the genus	1 and 2	
	MLST analysis and database; multiple strain typing methods	3	http://pubmlst.org/bcc
	Genomic evolution of *B. dolosa*	3	Lieberman *et al.* (2011)
	MALDI-TOF as a rapid species identification tool	–	Vanlaere *et al.* (2008)
	B. pseudomallei serodiagnostic protein microarray	–	Felgner *et al.* (2009)
Virulence/infection	*Burkholderia* species are capable of intracellular survival and have the ability to block host cell degradation	7	Lamothe *et al.* (2007), Schellenberg *et al.* (2007)
	Multiple infection models: *Caenorhabditis elegans*, *Galleria mellonella*, *Drosophila melanogaster*, Alfalfa plants, mice or rats, and zebra fish embryos	6 and 7	Uehlinger *et al.* (2009), Vergunst *et al.* (2009, 2010), Hasselbring *et al.* (2011), Vial *et al.* (2010)
	Type VI secretion characterized as a major virulence factor	6 and 7	Aubert *et al.* (2010), Burtnick *et al.* (2011)
	B. multivorans has become the dominant Bcc species in CF	3	LiPuma (2010)
Genomics/genetics	Bcc tolerate loss of the third chromosomal replicon, resulting in strains with reduced virulence, antifungal activity and secondary metabolism	–	Agnoli *et al.* (2012)
	Biofilm formation; the evolution of diversity being underpinned by a common set of mutations as revealed by metagenomics	10	Traverse *et al.* (2013)
	Tools to create unmarked gene mutations	7	Flannagan *et al.* (2008), Hamad *et al.* (2010)
	B. rhizoxinica genome determined as the first endofungal bacterium to be analysed	–	Lackner *et al.* (2011)
	Multiple genomes for plant-associated Burkholderia determined such as the endophyte *B. phytofirmans* and the nodule-former *B. phymatum*	–	http://burkholderia.com
Transcriptomics	Reference global gene expression dataset for *B. cenocepacia* J2315 and identification of a unique low-oxygen activated (*lxa*) locus	4	Sass *et al.* (2013)
Novel therapies	Quorum sensing inhibitors improve antibiotic efficacy and reduce biofilm formation	9	Brackman *et al.* (2011)
	Peptide linked morpholino oligos can be used to block essential gene expression and improve killing of *Burkholderia* during infection	–	Greenberg *et al.* (2010)

characterize isolates and trace outbreaks (Table 0.1). *Burkholderia* strain analysis has now also been taken to the level of mapping genomic evolution using next-generation sequencing platforms with the analysis of the *Burkholderia dolosa* outbreak of CF infection that occurred in Boston, USA (Lieberman *et al.*, 2011) (Table 0.1).

In terms of emerging non-genetic tools for bacterial species identification, matrix-assisted laser desorption ionization time-of-flight mass spectrometry (MALDI-TOF) has been successfully applied to assist with *Burkholderia* species identification (Vanlaere *et al.*, 2008). MALDI-TOF was able to distinguish Bcc from non-Bcc species, as well as resolved the majority of Bcc species except *Burkholderia anthina* and *Burkholderia pyrrocinia* (Vanlaere *et al.*, 2008). A different phenotypic identification system based on a protein microarray has also been developed for *B. pseudomallei* to enable rapid serodiagnosis of this pathogen (Felgner *et al.*, 2009). A panel of genes encoding predicted immunogenic and surface proteins were amplified by PCR, cloned by recombination into plasmid vectors and expressed *in vitro*, prior to being printed onto glass microarray slides. From a starting array containing 1205 *B. pseudomallei* proteins, 49 were identified as being most predictive of patients with active meliodosis infection as a result of serum cross-reactivity. The microarray also identified 59 candidate antigens that were equally cross-reactive across both infected and healthy individuals, and may serve as proteins suitable for a *B. pseudomallei* vaccine (Felgner *et al.*, 2009).

Key *Burkholderia* virulence traits

Virulence has arguably been the leading *Burkholderia* research area for the last three decades with these organisms possessing a diverse range of pathogenicity functions (Fig. 0.1 and Table 0.1). In particular, there is now an overwhelming body of experimental evidence that supports the intracellular survival of *Burkholderia* bacteria in single cell organisms such as amoebae (Lamothe *et al.*, 2004) and cell types such as macrophages (Lamothe *et al.*, 2007) (Table 0.1). Moreover, the development of a zebra fish embryo model of infection also showed that *B. cenocepacia* was able to survive and multiply within macrophages of a multicellular vertebrate host for the first time (Vergunst *et al.*, 2010). *Burkholderia* have evolved mechanisms that can delay acidification of mammalian macrophage vacuoles (Lamothe *et al.*, 2007) and also block intracellular degradation pathways in plant host cells. For example, *Burkholderia* possesses proteasome inhibitors such as glidobactin (Schellenberg *et al.*, 2007) which are part of a family of well-characterized bacterial syringolin toxins produced by hybrid non-ribosomal peptide–polyketide synthetase (NRPS) genes in other well-characterized plant pathogens such as *Pseudomonas syringae* (Groll *et al.*, 2008). In addition to zebra fish, a wide range of surrogate infection models have now been applied and tested on *Burkholderia* bacteria (Table 0.1). Comparative analysis infection models using the nematode *Caenorhabditis elegans*, wax moth larvae, *Galleria mellonella*, alfalfa plants, and mice or rats has also demonstrated that many of the previously characterized *Burkholderia* virulence factors are model specific in terms of active function (Uehlinger *et al.*, 2009). Of the virulence determinants screened, quorum sensing, iron acquisition via siderophores and lipopolysaccharide biosynthesis were the only factors required for full pathogenicity in all models (Uehlinger *et al.*, 2009). The soil-living amoeba *Dictyostelium discoideum* has been used to identify *B. pseudomallei* transposon mutants with diminished virulence (Hasselbring *et al.*, 2011). Vial and colleagues (Vial *et al.*, 2010) used this amoebae model as well as the model fly, *Drosophila melanogaster*, to demonstrate that a spontaneously emerging *Burkholderia ambifaria* phase variant colony type was less virulent than the wild type.

While full *Burkholderia* virulence is clearly multifactorial (Fig. 0.1), type VI secretion has recently been identified as a major *Burkholderia* virulence factor. It has been shown to be critically important for full pathogenesis of *B. pseudomallei* (Burtnick *et al.*, 2011) and of primary importance the control of biofilm formation in *B. cenocepacia* (Aubert *et al.*, 2008). Overall, the availability of multiple genome sequences and molecular strategies to pinpoint genes of interest, excellent tools for gene mutation (Flannagan *et al.*, 2008;

Hamad *et al.,* 2010), and several easy to use infection models, will greatly enhance the rate of *Burkholderia* virulence factor identification and characterization.

Dissecting functionality and evolution in biofilms

The ability to form biofilms is common trait possessed by all *Burkholderia* species (Table 0.1) and recent studies have begun to take a much more detailed look at the evolution that occurs during this sessile lifestyle. Traverse *et al.* (Traverse *et al.,* 2013) examined how clonal diversity evolves within a *B. cenocepacia* biofilm model. Using polystyrene beads to grow and then passage biofilms through multiple generations, the emergence of three colony morphotypes was followed. The functional role of each colony type, studded, ruffled spreader and wrinkly, in biofilm formation was also characterized. Although the wrinkly colony type was the best biofilm former when tested in isolation, biofilms that evolved naturally from the wild type were composed predominantly of the studded morphotype, which, in contrast was the least efficient at biofilm formation. The ruffled colony type was second in dominance among the sessile cells and the wrinkly colony type only made up a minority of the evolved biofilm.

To characterize the molecular basis for the emergence of the colony morphotypes, next generation sequencing and metagenomic analysis was used to identify the mutations that emerged during biofilm formation and growth passage over 1050 generations (Traverse *et al.,* 2013). The majority of mutations fell into five functional classes that altered metabolism of cyclic diguanosine monophosphate, lipolysaccharide production, central metabolism within the tricarboxylic acid cycle, global transcription, and the ability to acquire iron (Traverse *et al.,* 2013). Interestingly, many of the mutations overlapped with those identified in *B. dolosa* (Lieberman *et al.,* 2011) (Table 0.1) and *P. aeruginosa* (Smith *et al.,* 2006) isolates that have adapted during chronic CF lung infection, showing that the abiotic polystyrene bead model of biofilm formation enabled the modelling of evolution during infection.

Loss of replicons and insights derived from complete genome sequences from endosymbionts

The multireplicon structure of *Burkholderia* genomes has long been noted as an unusual feature of their genomes (Lessie *et al.,* 1996). Since each of the larger replicons encode ribosomal RNA (*rrn*) operons and other essential genes, they have been referred to as chromosomes (Chain *et al.,* 2006; Holden *et al.,* 2009). Genome sequence analysis revealed that the smaller replicons of *B. xenovorans* (Chain *et al.,* 2006), *B. pseudomallei* (Holden *et al.,* 2004) and *B. cenocepacia* (Holden *et al.,* 2009) had partition (*parA* and *parB*) and replication (*repA*) genes that were plasmid-like. Subsequent phylogenetic comparison also demonstrated that these regions were indeed distinct from those encoded on the largest replicon (Drevinek *et al.,* 2008a). These data suggest that the smaller *Burkholderia* replicons were acquired as megaplasmids and had become genomically stabilized as a result of the transfer of essential genes. Building on these observations, Agnoli and colleagues (Agnoli *et al.,* 2012) were recently able to prove that the third replicon in Bcc bacteria was not an essential chromosomal element and could actually be cured from strains.

Loss of the third chromosome (c3) resulted in viable cells despite the loss of over 1 Mb of DNA and the one *rrn* operon encoded on this replicon. The major functions lost by the c3-minus derivatives were virulence, antifungal activity and several secondary metabolism pathways that were not essential for primary growth (Agnoli *et al.,* 2012). It is quite amazing that a microorganism can tolerate such large loss of its genomic DNA content. This seminal finding of replicon loss (Agnoli *et al.,* 2012) adds weight to the hypothesis that *Burkholderia* species follow two major evolutionary paths. One path is based on the acquisition of new functionality associated with accessory DNA and the parallel increase in genome size that results. The other evolutionary route is based on rearrangement and loss of DNA as seen in the highly specific horse pathogen *Burkholderia mallei* (< 6.0 Mb genome) which has clonally emerged from *Burkholderia pseudomallei* that has a larger genome (> 7 Mb) (Nierman *et al.*, 2004). Within the *Burkholderia* genus, the Bcc group seems to be unique in having a third replicon encoding an *rrn* operon. Other *Burkholderia* species have two major replicons, with the second largest often encoding up to three *rrn* and multiple tRNAs, as seen in the *Burkholderia xenovorans* genome (Chain *et al.*, 2006). It will be interesting to see whether in these *Burkholderia* species as well as the Bcc, the second replicon can be cured.

In addition to adaptation to pathogenic niches as seen with *B. mallei* (Nierman *et al.*, 2004), genome size may frequently become reduced when organisms adopt an endosymbiotic lifestyle. Many *Burkholderia* species form very close interactions with a variety of multicellular host organisms and one of the most intriguing to be recently characterized was that of *Burkholderia rhizoxinica* (Lackner *et al.*, 2011). *B. rhizoxinica* is an endofungal bacterium that was originally identified within the plant pathogenic fungus, *Rhizopus*, which causes rice seedling blight disease (Partida-Martinez and Hertweck, 2005). *Rhizopus* infection causes swelling of plant tissue due to the production of a polyketide phytotoxin, rhizoxin, which was originally thought to be produced by the fungus. However, detailed characterization of *Rhizopus* demonstrated that it harboured an intracellular *Burkholderia* species, and that it was these bacterial cells that were producing the rhizoxin toxin (Partida-Martinez and Hertweck, 2005).

Full characterization of the *B. rhizoxinica* genome has shown that it has all the hallmarks of an endosymbiotic bacterial species; it is one of the smallest *Burkholderia* genomes to be characterized to date, with a total genome size of only 3.75 Mb (Lackner *et al.*, 2011). Interestingly it only has one large chromosomal replicon of 2.75 Mb that encodes all three *rrn* operons. Its GC content, 60.7%, is also considerably lower than the average within the *Burkholderia* genus (approximately 67%) (Lackner *et al.*, 2011); this low GC content may reflect *B. rhizoxinica's* evolution towards an intracellular endofungal lifestyle. In addition to the rhizoxin biosynthesis gene cluster, which constitutes a hybrid NRPS and polyketide synthase (PKS) system encoded on the large chromosome, the entire *B. rhizoxinica* genome was found to contain an additional 14 NRPS gene clusters, devoting nearly 9% of its genome to this form of secondary metabolite production. Since *B. rhizoxinica* lacks a second chromosomal replicon it is provocative to postulate that it has evolved through replicon loss (Agnoli *et al.*, 2012), rather than the small genome reflecting the ancestral state of this species.

Another major group of symbiotic *Burkholderia* are the nodulating species (*Burkholderia mimosarum*, *Burkholderia nodosa*, *Burkholderia sabiae* and *Burkholderia phymatum*), which are the first diazotropic non-alpha-proteobacteria capable of nitrogen fixation within leguminous plants (Moulin *et al.*, 2001; Suarez-Moreno *et al.*, 2012). These *Burkholderia* species have very large genomes (8.64 Mb and 8.67 Mb, for *B. mimosarum* and *B. phymatum*, respectively) and they show no evidence of genome decay as a result of their close, endosymbiotic interactions with leguminous plants (Fig. 0.1). This is probably because they are facultatively endosymbiotic in terms of nodule formation, and several of these species may also be found as free-living bacteria within bulk soil (Suarez-Moreno *et al.*, 2012). Endophytic *Burkholderia* species such as *Burkholderia phytofirmans* that live within plant tissues without causing any negative effects on their hosts, also possess large genomes (*B. phytofirmans* PsJN, 8.21 Mb). Overall, this suggests that for close interactions with plants, a large genome size can be tolerated and is actually beneficial for both the bacterium and host plant in terms of the additional functionality it brings.

Burkholderia and cystic fibrosis: what's new?

The last decade has seen a number of changes in the epidemiology of *Burkholderia* in CF infection (LiPuma, 2010). The major switch has been the reduction in the prevalence of *B. cenocepacia* infection, probably as a result of the greatly improved surveillance and infection control that has been widely implemented across CF populations (see Chapter 3). *B. multivorans* is now the most dominant Bcc species seen in several CF populations (Table 0.1), with *B. cenocepacia* generally second in dominance (LiPuma, 2010). Although this is the major trend seen in the molecular epidemiology of Bcc bacteria, regional and centre specific differences in the prevalence of certain *Burkholderia* species are still apparent. For example, *B. cepacia* has dominated at a large Portuguese CF centre in Lisbon (85% of Bcc infected patients); however, this is due to the presence of two major strains which were also found as contaminants of non-sterile saline solutions and were the primary cause of this Bcc infection outbreak (Cunha *et al.*, 2007). *Burkholderia contaminans* is the dominant Bcc species seen in Argentinian CF individuals treated at multiple centres (Martina *et al.*, 2013). While clonal strains and potential patient-to-patient transmission were observed in this study, there was also evidence of genetically distinct *B. contaminans* strains, suggesting that the dominance of this species may also reflect differences in environmental prevalence of Bcc in Argentina (Martina *et al.*, 2013). Outside the Bcc, *Burkholderia gladioli*, has also emerged as an important cause of CF infection in the US, where it is now the third most prevalent species (accounting for 15% of *Burkholderia* infections) (LiPuma, 2010). With *B. gladioli* infecting over 300 CF patients in the US, it will become increasingly important to begin to characterize the molecular pathogenicity of this species and why it is able to cause chronic lung infection.

The application of cultivation independent techniques to examine microbial diversity in CF infection also represents a major advance in this disease area that has occurred over the last 10 years. Molecular studies have shown that (i) the infected CF lung has a diverse bacterial community and high prevalence of anaerobic species (Rogers *et al.*, 2003), which are distinct from those present in the oral cavity (Tunney *et al.*, 2008); (ii) microaerophillic species, such as the *Streptococcus milleri* group (SMG), constitute novel pathogens in CF (Sibley *et al.*, 2008); (iii) cultivation-based diagnostic microbiology is limited when compared with cultivation-independent analysis (Rogers *et al.*, 2009); and (iv) increased diversity in the CF microbiota is associated with patient stability (Filkins *et al.*, 2012), while low diversity is associated with increased severity of infection (Blainey *et al.*, 2012). Although these studies have detected *Burkholderia* species as key components of the CF microbiota in parallel with what has been seen in conventional cultivation-based analysis (Rogers *et al.*, 2003), no study has yet focused on the effect of this genus on the composition of the microbiota in CF infection. Since Bcc bacteria are known to produce a wide range of antagonistic metabolites that can inhibit other CF pathogens (Mahenthiralingam *et al.*, 2011), it will be very interesting to see if they reduce or specifically alter the diversity of infection when they are present.

Novel therapies for CF infection

A number of potential therapies for the treatment of highly antibiotic-resistant *Burkholderia* infections have also emerged in the last few years. Brackman and colleagues (Brackman *et al.*, 2011) demonstrated that quorum-sensing inhibitors such as baicalin hydrate and cinnamaldehyde were effective in reducing biofilm formation by Bcc bacteria when used in combination with tobramycin (Table 0.1). The combination of tobramycin and baicalin hydrate also greatly improved the clearance of *B. cenocepacia* in a mouse model of infection (Brackman *et al.*, 2011). Another promising new small molecule therapy for infection is the use of antisense phosphorodiamidate morpholino oligomers (PMO) to block the expression of essential *Burkholderia* genes (Greenberg *et al.*, 2010) (Table 0.1). A PMO directed against the start site of *acpP* gene was selected as this gene encodes an essential component of bacterial fatty acid biosynthesis. When conjugated to a cell-penetrating peptide to form a peptide-PMO (PPMO), enhanced neutrophil killing of *B. mutlivorans* was seen *in vitro*. Treatment of *B. multivorans*-infected mice also resulted in increased survival and reduced inflammation within the spleen and lymph nodes indicating that successful modelling of morpholino therapy could be performed *in vivo*.

Phage therapy is now being widely investigated as an alternative to the use of antibiotics especially for multidrug resistant Gram-negative pathogens.

The successful clinical trials of a phage formulation for chronic ear infections caused by *Pseudomonas aeruginosa* (Wright *et al.*, 2009) heralds a promising future for these agents in the topical treatment of infections. The wider genomic biology of *Burkholderia* phages in discussed in detail within the book (chapter 11), however, it is also promising to see that inhaled phage formulations have been successfully tested in animal models of infection (Golshahi *et al.*, 2008) paving the way for the development of phage therapies for CF respiratory infection.

Two other recent findings point to simple therapies for *Burkholderia* infection. The drug glyburide is a potassium ATP channel and ATP-binding cassette (ABC) transporter inhibitor that is used to treat type 2 diabetes. It has a wide range of anti-inflammatory effects and when it was administered to *B. pseudomallei* infected individuals, it reduced the mortality associated with diabetic individuals with *B. pseudomallei* infection (Koh *et al.* 2011). When white blood cell gene expression was examined in this group of patients, it was clear that the glyburide had reduced the expression of multiple inflammatory genes, consistent with its general anti-inflammatory properties.

Another potentially very simple therapy that could reduce the minimal inhibitory concentration of antibiotics in *B. cenocepacia* was noted after a global gene expression analysis of spontaneous antibiotic resistant mutants in this pathogen (Sass *et al.*, 2011). It was observed that mutants with high-level resistance to beta-lactam antibiotics also showed increased expression of genes involved in phenylacetic acid degradation. It is well known that adding glucose often leads to repression of such alternate metabolic pathways, and indeed when the glucose concentration was increased to 20 mM, the expression of genes involved in the phenylacetic acid pathway was decreased and the minimal inhibitory concentration of several antibiotics (meropenem, ceftazidime, tetracycline and chloramphenicol) were considerably reduced in the pan resistant *B. cenocepacia* J2315 strain (Sass *et al.*, 2011). Glucose (dextrose) is a widely used clinical agent; however, whether it could be incorporated into antibiotic formulations and delivered as an inhaled therapy remains to be determined.

Novel *Burkholderia* niches: insects

In addition to the expansion of research on plant (Suarez-Moreno *et al.*, 2012) and fungal interactions (Lackner *et al.*, 2011), a study characterizing insect symbionts found that *Burkholderia* were carried at high density in the mid-guts of stink bugs, *Riptortus clavatus* (Kikuchi *et al.*, 2011). When the researchers examined the phylogeny of the *Burkholderia* endosymbionts they found that it did not map onto that of the host insects. In addition, when they examined the stink bug eggs from insects reared in the laboratory they did not find any evidence of the bacteria, suggesting the *Burkholderia* were not transmitted vertically. However, rearing of the sterile insects in soil led to successful colonization suggesting the *Burkholderia* was environmentally acquired at an early stage in the insects development (Kikuchi *et al.*, 2011).

Follow up to this work led to the intriguing finding that insecticide resistance was one of the functions the *Burkholderia* endosymbionts was able to confer on the host insects (Kikuchi *et al.*, 2012) (Fig. 0.1). The ability to degrade a wide variety of man-made organic compounds and pollutants such as polychlorinated biphenyls is a well known *Burkholderia* trait (Chain *et al.*, 2006). Kikuchi and colleagues (Kikuchi *et al.*, 2012) demonstrated that the organophosphate insecticide, fenitrothion, could be degraded by *Burkholderia* that were carried within the stink bug pest, *Riptortus pedestris*. In agricultural fields where fenitrothion had been intensively used to protect sugarcane crops, resident stink bugs were shown to have a gut community that was highly enriched for *Burkholderia* species that could degrade this insecticide (Kikuchi *et al.*, 2012). Although other functions of the *Burkholderia* symbionts in these insects have not been well characterized, it is amazing that an anthropogenic activity such as insecticide use can drive the evolution of a novel interaction between *Burkholderia* bacteria that is beneficial to its insect host.

Novel *Burkholderia* niches: an ability to survive in microaerophilic environments

Resilience and broad environmental fitness traits enable *Burkholderia* to grow in a number of harsh environments (Fig. 0.1). Classical traits such as metabolic versatility, antibiotic resistance, production of multiple virulence factors and the ability to form biofilms which enable *Burkholderia* to colonize various environments have already been investigated in some depth as is illustrated in the various chapters in this book. However, one attribute that is not widely studied in the context of aerobic bacteria is the influence of oxygen, and in particular what happens when its concentration is reduced. The majority of *Burkholderia* species are considered to be aerobic and unable to grow without oxygen. Those species that fix nitrogen, an oxygen-sensitive process, must do so in an anoxic environment such as within the nodules of leguminous plants, suggesting that these species must have the capability to survive without oxygen (Suarez-Moreno *et al.*, 2012). Microbial diversity analysis of CF sputum has shown that anaerobic bacteria are present in high numbers (Rogers *et al.*, 2003; Tunney *et al.*, 2008), confirming that the concentration of oxygen in sputum can be close to zero. Therefore, for Bcc bacteria to be such successful CF pathogens, they must have the ability to tolerate low levels of oxygen and grow microaerophilically.

Other than the ability of *Burkholderia* to grow in biofilms (Peeters *et al.*, 2010) where oxygen can become limited, little is known about how they grow microaerophilically. Sass *et al.* (2013) examined *B. cenocepacia* global gene expression under a variety of environmental conditions pertinent to the CF lung, with one of the conditions being a reduction in oxygen concentration to 6% compared with the atmospheric level of 21%. This was a convenient level to produce experimentally using a commercial gas-pack sachet optimized for growth of facultatively anaerobic *Campylobacter* species (Sass *et al.*, 2013). The effect of reduced oxygen concentration on gene expression was staggering, with 1987 of the 7251 (27%) of the coding sequences (CDS) on the *B. cenocepacia* J2315 microarray altered in expression. The only other growth condition to alter expression of a larger number of genes was growth at stationary phase in a minimal medium, where 29% of the CDS altered in expression. The low oxygen level also resulted in the induction of a novel regulon of genes, the majority of which clustered in one unique 50 Kb gene cluster that was designated the low oxygen activated locus (*lxa*). The intact *lxa* region was only found in *B. cenocepacia* with other *Burkholderia* species possessing only fragments of this region as small gene clusters or individual homologous genes. The gene content of the *lxa* cluster was also highly unusual with poorly characterized predicted functions in metabolism, transport, stress (six universal stress proteins) and gene regulation (Sass *et al.*, 2013). Overall, the discovery of the *lxa* locus suggests that *Burkholderia* have evolved mechanisms to cope with anoxia and that because microaerophilic environments probably reflect 'reality' for most aerobic bacteria, we should be doing more to model this growth parameter.

Undesirable presence of *Burkholderia*: contamination

In addition to infection, an increasing number of studies are reporting the presence of *Burkholderia* bacteria as undesirable contaminants of man-made products. Historical outbreaks of infection due to the contamination of disinfectants and medical solutions have been well characterized, but the wider prevalence of *Burkholderia* as contaminants was not revealed until Jimenez (2007) recently reported that '*Burkholderia cepacia*' was the most frequently isolated microbial species in product recalls is logged by the US Food and Drug Agency. It was present in 22% of non-sterile product recalls and 2.5% of sterile product contaminations (Jimenez, 2007). A major problem with many of these reports of contamination was the fact that accurate identification of the contaminating *Burkholderia* species was often not performed. Recent studies have begun to systematically identify the species involved in microbial fouling and have shown that a wide range of *Burkholderia* have the capability to cause industrial contamination (Fig. 0.1).

MLST analysis of clonal sequence types associated with infection and the environment had demonstrated that *Burkholderia stabilis* was associated with contamination of shampoo and albuterol inhaler solutions (Baldwin *et al.*, 2007). The formal name given to *B. contaminans* (Vanlaere *et al.*, 2009) had reflected the fact that the organism was linked to multiple, globally distributed instances of contamination and opportunistic infection (Mahenthiralingam *et al.*, 2006). *Burkholderia* species may even contaminate fuels with *Burkholderia vietnamiensis*, *B. lata* and novel Bcc species found at a single oil refinery site (White *et al.*, 2011). A recent systematic survey of 60 *Burkholderia* industrial process contaminants demonstrated that *Burkholderia lata* and *B. cenocepacia* were the two most common species in this collection (25% and 18%, respectively); *Burkholderia vietnamiensis*, *Burkholderia arboris*, *B. stabilis*, *B. cepacia*, *B. multivorans*, *B. contaminans*, *B. ambifaria* and potentially novel Bcc species made up the remaining contaminating isolates (Rushton *et al.*, 2013). The key features which allow *Burkholderia* to survive in such harsh anthropogenic environments include (i) their innate resistance to antimicrobials, with efflux shown to mediate preservative resistance (Rushton *et al.*, 2013), and (ii) their metabolic diversity to grow on a range of substrates including complex hydrocarbons found in fuels (White *et al.*, 2011).

Conclusions and future perspectives

With such wide diversity in terms of biological functionality it is not hard to see why there has been increased interest in the study of *Burkholderia*. Not many bacteria can be used to study both infection and positive environmental interactions at the same time. With the increased availability and reduced costs of next generation sequencing technologies, it will become technically feasible to dissect the complex functions encoded by large *Burkholderia* genomes. But what are the gaps in our understanding of these bacteria to which these tools can be applied? We can certainly continue to probe their disease-causing ability in more detail to developing new therapies for infection. However, it is probably their vast and varied environmental interactions that represent the greatest challenge. There is no doubt that *Burkholderia* bacteria bring considerable benefit to many natural interactions especially in terms of their plant growth promotion capabilities. Understanding how to harness these positive environmental properties without tipping the balance towards *Burkholderia* pathogenicity will be a major future challenge.

Acknowledgements

E.M. and T.C. thank the European Cooperation in Science and Technology (COST) action BM1003 for enabling networking and collaborative interactions with many researchers who have contributed towards this book. We would also like to thank members of the International *Burkholderia cepacia* Working Group for their input and valuable suggestions.

References

Agnoli, K., Schwager, S., Uehlinger, S., Vergunst, A., Viteri, D.F., Nguyen, D.T., Sokol, P.A., Carlier, A., and Eberl, L. (2012). Exposing the third chromosome of *Burkholderia cepacia* complex strains as a virulence plasmid. Mol. Microbiol. *83*, 362–378.

Aubert, D.F., Flannagan, R.S., and Valvano, M.A. (2008). A novel sensor kinase-response regulator hybrid controls biofilm formation and type VI secretion system activity in *Burkholderia cenocepacia*. Infect. Immun. *76*, 1979–1991.

Aubert, D., Macdonald, D.K., and Valvano, M.A. (2010). BcsKC Is an Essential Protein for the Type VI Secretion System Activity in *Burkholderia cenocepacia* that forms an outer membrane complex with BcsLB. J. Biol. Chem. *285*, 35988–35998.

Baldwin, A., Mahenthiralingam, E., Thickett, K.M., Honeybourne, D., Maiden, M.C., Govan, J.R., Speert, D.P., LiPuma, J.J., Vandamme, P., and Dowson, C.G. (2005). Multilocus sequence typing scheme that provides both species and strain differentiation for the *Burkholderia cepacia* complex. J. Clin. Microbiol. *43*, 4665–4673.

Baldwin, A., Mahenthiralingam, E., Drevinek, P., Vandamme, P., Govan, J.R., Waine, D.J., LiPuma, J.J., Chiarini, L., Dalmastri, C., Henry, D.A., *et al.* (2007). Environmental *Burkholderia cepacia* complex isolates in human infections. Emerg Infect. Dis. *13*, 458–461.

Blainey, P.C., Milla, C.E., Cornfield, D.N., and Quake, S.R. (2012). Quantitative analysis of the human airway microbial ecology reveals a pervasive signature for cystic fibrosis. Sci. Transl. Med. *4*, 153ra130.

Brackman, G., Cos, P., Maes, L., Nelis, H.J., and Coenye, T. (2011). Quorum sensing inhibitors increase the susceptibility of bacterial biofilms to antibiotics *in vitro* and *in vivo*. Antimicrob. Agents Chemother. *55*, 2655–2661.

Burtnick, M.N., Brett, P.J., Harding, S.V., Ngugi, S.A., Ribot, W.J., Chantratita, N., Scorpio, A., Milne, T.S., Dean, R.E., Fritz, D.L., *et al.* (2011). The cluster 1 type

VI secretion system is a major virulence determinant in *Burkholderia pseudomallei*. Infect. Immun. *79*, 1512–1525.

Chain, P.S., Denef, V.J., Konstantinidis, K.T., Vergez, L.M., Agullo, L., Reyes, V.L., Hauser, L., Cordova, M., Gomez, L., Gonzalez, M., *et al.* (2006). *Burkholderia xenovorans* LB400 harbors a multi-replicon, 9.73-Mbp genome shaped for versatility. Proc. Natl. Acad. Sci. U.S.A. *103*, 15280–15287.

Coenye, T., Vandamme, P., Eds. (2007). *Burkholderia*: molecular microbiology and genomics. Horizon Bioscience, Wymondham, Norfolk, United Kingdom.

Cunha, M.V., Pinto-de-Oliveira, A., Meirinhos-Soares, L., Salgado, M.J., Melo-Cristino, J., Correia, S., Barreto, C., and Sa-Correia, I. (2007). Exceptionally high representation of *Burkholderia cepacia* among *B. cepacia* Complex isolates recovered from the major Portuguese cystic fibrosis Center. J. Clin. Microbiol. *45*, 1628–1633.

Drevinek, P., Baldwin, A., Dowson, C.G., and Mahenthiralingam, E. (2008a). Diversity of the *parB* and *repA* genes of the *Burkholderia cepacia* complex and their utility for rapid identification of *Burkholderia cenocepacia*. BMC Microbiol. *8*, 44.

Drevinek, P., Holden, M.T., Ge, Z., Jones, A.M., Ketchell, I., Gill, R.T., and Mahenthiralingam, E. (2008b). Gene expression changes linked to antimicrobial resistance, oxidative stress, iron depletion and retained motility are observed when *Burkholderia cenocepacia* grows in cystic fibrosis sputum. BMC Infect. Dis. *8*, 121.

Estrada-de Los Santos, P., Vinuesa, P., Martinez-Aguilar, L., Hirsch, A.M., and Caballero-Mellado, J. (2013). Phylogenetic analysis of *Burkholderia* species by multilocus sequence analysis. Curr. Microbiol. DOI 10.1007/s00284-013-0330-9.

Felgner, P.L., Kayala, M.A., Vigil, A., Burk, C., Nakajima-Sasaki, R., Pablo, J., Molina, D.M., Hirst, S., Chew, J.S., Wang, D., *et al.* (2009). A *Burkholderia pseudomallei* protein microarray reveals serodiagnostic and cross-reactive antigens. Proc. Natl. Acad. Sci. U.S.A. *106*, 13499–13504.

Filkins, L.M., Hampton, T.H., Gifford, A.H., Gross, M.J., Hogan, D.A., Sogin, M.L., Morrison, H.G., Paster, B.J., and O'Toole, G.A. (2012). Prevalence of streptococci and increased polymicrobial diversity associated with cystic fibrosis patient stability. J. Bacteriol. *194*, 4709–4717.

Flannagan, R.S., Linn, T., and Valvano, M.A. (2008). A system for the construction of targeted unmarked gene deletions in the genus *Burkholderia*. Environ. Microbiol. *10*, 1652–1660.

Golshahi, L., Seed, K.D., Dennis, J.J., and Finlay, W.H. (2008). Toward modern inhalational bacteriophage therapy: nebulization ofbacteriophages of *Burkholderia cepacia* complex. J. Aerosol. Med. Pulm. Drug. Deliv. *21*, 351–360.

Greenberg, D.E., Marshall-Batty, K.R., Brinster, L.R., Zarember, K.A., Shaw, P.A., Mellbye, B.L., Iversen, P.L., Holland, S.M., and Geller, B.L. (2010). Antisense phosphorodiamidate morpholino oligomers targeted to an essential gene inhibit *Burkholderia cepacia* complex. J. Infect. Dis. *201*, 1822–1830.

Groll, M., Schellenberg, B., Bachmann, A.S., Archer, C.R., Huber, R., Powell, T.K., Lindow, S., Kaiser, M., and Dudler, R. (2008). A plant pathogen virulence factor inhibits the eukaryotic proteasome by a novel mechanism. Nature *452*, 755–758.

Hamad, M.A., Skeldon, A.M., and Valvano, M.A. (2010). Construction of aminoglycoside-sensitive *Burkholderia cenocepacia* strains for use in studies of intracellular bacteria with the gentamicin protection assay. Appl. Environ. Microbiol. *76*, 3170–3176.

Hasselbring, B.M., Patel, M.K., and Schell, M.A. (2011). Dictyostelium discoideum as a model system for identification of *Burkholderia pseudomallei* virulence factors. Infect. Immun. *79*, 2079–2088.

Holden, M.T., Titball, R.W., Peacock, S.J., Cerdeno-Tarraga, A.M., Atkins, T., Crossman, L.C., Pitt, T., Churcher, C., Mungall, K., Bentley, S.D., *et al.* (2004). Genomic plasticity of the causative agent of melioidosis, *Burkholderia pseudomallei*. Proc. Natl. Acad. Sci. U.S.A. *101*, 14240–14245.

Holden, M.T., Seth-Smith, H.M., Crossman, L.C., Sebaihia, M., Bentley, S.D., Cerdeno-Tarraga, A.M., Thomson, N.R., Bason, N., Quail, M.A., Sharp, S., *et al.* (2009). The genome of *Burkholderia cenocepacia* J2315, an epidemic pathogen of cystic fibrosis patients. J. Bacteriol. *191*, 261–277.

Jimenez, L. (2007). Microbial diversity in pharmaceutical product recalls and environments. PDA J. Pharm. Sci. Technol. *61*, 383–399.

Kikuchi, Y., Hosokawa, T., and Fukatsu, T. (2011). An ancient but promiscuous host–symbiont association between *Burkholderia* gut symbionts and their heteropteran hosts. ISME J. *5*, 446–460.

Kikuchi, Y., Hayatsu, M., Hosokawa, T., Nagayama, A., Tago, K., and Fukatsu, T. (2012). Symbiont-mediated insecticide resistance. Proc. Natl. Acad. Sci. U.S.A. *109*, 8618–8622.

Lackner, G., Moebius, N., Partida-Martinez, L.P., Boland, S., and Hertweck, C. (2011). Evolution of an endofungal lifestyle: Deductions from the *Burkholderia rhizoxinica* genome. BMC Genomics *12*, 210.

Lamothe, J., Thyssen, S., and Valvano, M.A. (2004). *Burkholderia cepacia* complex isolates survive intracellularly without replication within acidic vacuoles of *Acanthamoeba polyphaga*. Cell Microbiol *6*, 1127–1138.

Lamothe, J., Huynh, K.K., Grinstein, S., and Valvano, M.A. (2007). Intracellular survival of *Burkholderia cenocepacia* in macrophages is associated with a delay in the maturation of bacteria-containing vacuoles. Cell. Microbiol. *9*, 40–53.

Leiske, D.L., Karimpour-Fard, A., Hume, P.S., Fairbanks, B.D., and Gill, R.T. (2006). A comparison of alternative 60-mer probe designs in an in-situ synthesized oligonucleotide microarray. BMC Genomics *7*, 72.

Lessie, T.G., Hendrickson, W., Manning, B.D., and Devereux, R. (1996). Genomic complexity and plasticity of *Burkholderia cepacia*. FEMS Microbiol. Lett. *144*, 117–128.

Lieberman, T.D., Michel, J.B., Aingaran, M., Potter-Bynoe, G., Roux, D., Davis Jr, M.R., Skurnik, D., Leiby, N., LiPuma, J.J., Goldberg, J.B., *et al.* (2011). Parallel bacterial evolution within multiple patients identifies candidate pathogenicity genes. Nature Genet. *43*, 1275–1280.

LiPuma, J.J. (2010). The changing microbial epidemiology in cystic fibrosis. Clin. Microbiol. Rev. *23*, 299–323.

Mahenthiralingam, E., Baldwin, A., Drevinek, P., Vanlaere, E., Vandamme, P., LiPuma, J.J., and Dowson, C.G. (2006). Multilocus sequence typing breathes life into a microbial metagenome. PLoS One *1*, e17.

Mahenthiralingam, E., Song, L., Sass, A., White, J., Wilmot, C., Marchbank, A., Boaisha, O., Paine, J., Knight, D., and Challis, G.L. (2011). Enacyloxins are products of an unusual hybrid modular polyketide synthase encoded by a cryptic *Burkholderia ambifaria* genomic island. Chem. Biol. *18*, 665–677.

Martina, P., Bettiol, M., Vescina, C., Montanaro, P., Mannino, M.C., Prieto, C.I., Vay, C., Naumann, D., Schmitt, J., Yantorno, O., *et al.* (2013). Genetic diversity of *Burkholderia contaminans* isolates from cystic fibrosis patients in Argentina. J. Clin. Microbiol. *51*, 339–344.

Moulin, L., Munive, A., Dreyfus, B., and Boivin-Masson, C. (2001). Nodulation of legumes by members of the β-subclass of Proteobacteria. Nature *411*, 948–950.

Nierman, W.C., DeShazer, D., Kim, H.S., Tettelin, H., Nelson, K.E., Feldblyum, T., Ulrich, R.L., Ronning, C.M., Brinkac, L.M., Daugherty, S.C., *et al.* (2004). From the Cover: Structural flexibility in the *Burkholderia mallei* genome. Proc. Natl. Acad. Sci. U.S.A. *101*, 14246–14251.

Partida-Martinez, L.P., and Hertweck, C. (2005). Pathogenic fungus harbours endosymbiotic bacteria for toxin production. Nature *437*, 884–888.

Peeters, E., Sass, A., Mahenthiralingam, E., Nelis, H., and Coenye, T. (2010). Transcriptional response of *Burkholderia cenocepacia* J2315 sessile cells to treatments with high doses of hydrogen peroxide and sodium hypochlorite. BMC Genom. *11*, 90.

Rogers, G.B., Hart, C.A., Mason, J.R., Hughes, M., Walshaw, M.J., and Bruce, K.D. (2003). Bacterial diversity in cases of lung infection in cystic fibrosis patients: 16S ribosomal DNA (rDNA) length heterogeneity PCR and 16S rDNA terminal restriction fragment length polymorphism profiling. J. Clin. Microbiol. *41*, 3548–3558.

Rogers, G.B., Daniels, T.W., Tuck, A., Carroll, M.P., Connett, G.J., David, G.J., and Bruce, K.D. (2009). Studying bacteria in respiratory specimens by using conventional and molecular microbiological approaches. BMC Pulm. Med. *9*, 14.

Rushton, L., Sass, A., Baldwin, A., Dowson, C.G., Donoghue, D., and Mahenthiralingam, E. (2013). A key role for efflux in the preservative susceptibility and adaptive resistance of *Burkholderia cepacia* complex bacteria. Antimicrob. Agents Chemother. *57*, 2972–2980

Sass, A., Marchbank, A., Tullis, E., LiPuma, J.J., and Mahenthiralingam, E. (2011). Spontaneous and evolutionary changes in the antibiotic resistance of *Burkholderia cenocepacia* observed by global gene expression analysis. BMC Genom. *12*, 373.

Sass, A.M., Schmerk, C., Agnoli, K., Norville, P.J., Eberl, L., Valvano, M.A., and Mahenthiralingam, E. (2013). The unexpected discovery of a novel low-oxygen-activated locus for the anoxic persistence of *Burkholderia cenocepacia*. ISME J. DOI 10.1038/ismej.2013.1036.

Schellenberg, B., Bigler, L., and Dudler, R. (2007). Identification of genes involved in the biosynthesis of the cytotoxic compound glidobactin from a soil bacterium. Environ. Microbiol. *9*, 1640–1650.

Sibley, C.D., Parkins, M.D., Rabin, H.R., Duan, K., Norgaard, J.C., and Surette, M.G. (2008). A polymicrobial perspective of pulmonary infections exposes an enigmatic pathogen in cystic fibrosis patients. Proc. Natl. Acad. Sci. U.S.A. *105*, 15070–15075.

Smith, E.E., Buckley, D.G., Wu, Z., Saenphimmachak, C., Hoffman, L.R., D'Argenio, D.A., Miller, S.I., Ramsey, B.W., Speert, D.P., Moskowitz, S.M., *et al.* (2006). Genetic adaptation by *Pseudomonas aeruginosa* to the airways of cystic fibrosis patients. Proc. Natl Acad Sci U.S.A. *103*, 8487–8492.

Spilker, T., Baldwin, A., Bumford, A., Dowson, C.G., Mahenthiralingam, E., and LiPuma, J.J. (2009). Expanded multilocus sequence typing for *Burkholderia* species. J. Clin. Microbiol. *47*, 2607–2610.

Suarez-Moreno, Z.R., Caballero-Mellado, J., Coutinho, B.G., Mendonca-Previato, L., James, E.K., and Venturi, V. (2012). Common features of environmental and potentially beneficial plant-associated *Burkholderia*. Microb. Ecol. *63*, 249–266.

Traverse, C.C., Mayo-Smith, L.M., Poltak, S.R., and Cooper, V.S. (2013). Tangled bank of experimentally evolved *Burkholderia* biofilms reflects selection during chronic infections. Proc. Natl. Acad. Sci. U.S.A. *110*, E250–259.

Tunney, M.M., Field, T.R., Moriarty, T.F., Patrick, S., Doering, G., Muhlebach, M.S., Wolfgang, M.C., Boucher, R., Gilpin, D.F., McDowell, A., *et al.* (2008). Detection of anaerobic bacteria in high numbers in sputum from patients with cystic fibrosis. Am. J. Respir. Crit. Care Med. *177*, 995–1001.

Uehlinger, S., Schwager, S., Bernier, S.P., Riedel, K., Nguyen, D.T., Sokol, P.A., and Eberl, L. (2009). Identification of specific and universal virulence factors in *Burkholderia cenocepacia* strains by using multiple infection hosts. Infect. Immun. *77*, 4102–4110.

Vandamme, P., and Dawyndt, P. (2011). Classification and identification of the *Burkholderia cepacia* complex: Past, present and future. Syst. Appl. Microbiol. *34*, 87–95.

Vanlaere, E., Sergeant, K., Dawyndt, P., Kallow, W., Erhard, M., Sutton, H., Dare, D., Devreese, B., Samyn, B., and Vandamme, P. (2008). Matrix-assisted laser desorption ionisation-time-of of-flight mass spectrometry of intact cells allows rapid identification of *Burkholderia cepacia* complex. J. Microbiol. Meth. *75*, 279–286.

Vanlaere, E., Baldwin, A., Gevers, D., Henry, D., De Brandt, E., LiPuma, J.J., Mahenthiralingam, E., Speert,

D.P., Dowson, C., and Vandamme, P. (2009). Taxon K, a complex within the *Burkholderia cepacia* complex, comprises at least two novel species, *Burkholderia contaminans* sp. nov. and *Burkholderia lata* sp. nov. Int. J. Syst. Evol. Microbiol. 59, 102–111.

Vergunst, A.C., Meijer, A.H., Renshaw, S.A., and O'Callaghan, D. (2010). *Burkholderia cenocepacia* creates an intramacrophage replication niche in zebrafish embryos, followed by bacterial dissemination and establishment of systemic infection. Infect. Immun. 78, 1495–1508.

Vial, L., Groleau, M.C., Lamarche, M.G., Filion, G., Castonguay-Vanier, J., Dekimpe, V., Daigle, F., Charette, S.J., and Deziel, E. (2010). Phase variation has a role in *Burkholderia ambifaria* niche adaptation. ISME J. 4, 49–60.

White, J., Gilbert, J., Hill, G., Hill, E., Huse, S.M., Weightman, A.J., and Mahenthiralingam, E. (2011). Culture-independent analysis of bacterial fuel contamination provides insight into the level of concordance with the standard industry practice of aerobic cultivation. Appl. Environ. Microbiol. 77, 4527–4538.

Wright, A., Hawkins, C.H., Anggard, E.E., and Harper, D.R. (2009). A controlled clinical trial of a therapeutic bacteriophage preparation in chronic otitis due to antibiotic-resistant *Pseudomonas aeruginosa*; a preliminary report of efficacy. Clin. Otolaryngology. 34, 349–357.

Genomic Taxonomy and Biodiversity of the *Burkholderia cepacia* Complex

1

Peter Vandamme, Bart Verheyde, Charlotte Peeters and Peter Dawyndt

Abstract

The *Burkholderia cepacia* complex is a group of closely related bacteria with conflicting biological properties and for which an unusually large amount of taxonomic data is available. These bacteria therefore represent a unique opportunity to evaluate traditional polyphasic taxonomic practices by means of novel whole-genome sequence based parameters. The present state-of-the-art demonstrates that taxonomists can replace DNA–DNA hybridization experiments as the pivotal instrument for species level demarcation and 16S rRNA sequence analysis for studying phylogeny by superior whole-genome sequence based alternatives in the frame of polyphasic taxonomic studies.

Introduction

For more than two decades the percentage of whole-genome DNA–DNA hybridization has been the best approach for estimating the degree of whole-genome sequence similarity and DNA–DNA hybridization was naturally incorporated in the practice of polyphasic taxonomy as the most decisive criterion for the delineation of novel species (Vandamme *et al.*, 1996). Polyphasic taxonomy refers to the integration of all available genotypic, phenotypic, and phylogenetic information into a consensus type of general-purpose classification. It departs from the assumption that the overall biological diversity cannot be encoded in a single molecule and that the variability of characters is group dependent. Furthermore, polyphasic taxonomy is phylogeny-based and traditionally uses sequence analysis and signature features of rRNA for the deduction of a phylogenetic framework for the classification of bacteria (Vandamme *et al.*, 1996). Nowadays a growing number of whole-genome sequences are becoming available and there is a general interest in evaluating the practice of polyphasic taxonomy using these novel data (Richter and Rossello-Mora, 2009). To this end a considerable number of whole-genome sequences of strains representing the same, closely related and distantly related species will be required to determine if whole-genome sequence based parameters can be identified that are equivalent or superior alternatives to the traditional methods.

The genus *Burkholderia* could be a valuable model for doing this exercise as, because of its general importance, an unusually large number of whole-genome sequences have become available. This genus contains over 70 species and a rapidly growing number of unnamed species, which occupy diverse ecological niches, including contaminated soils, plant rhizospheres and phytospheres, invertebrate intestinal tracts and the respiratory tract of humans (Coenye and Vandamme, 2003; Compant *et al.*, 2008; Suarez-Moreno *et al.*, 2012). Within this genus the *Burkholderia cepacia* complex (Bcc) is ubiquitous in nature and can be found in soil, water, the plant rhizosphere and phytosphere, in humans and various animal species and in the hospital environment. Bcc isolates have been exploited for various purposes, including biological control of plant pathogens, bioremediation of polluted environments and plant growth promotion. An increasing number of *Burkholderia* species, which includes all Bcc species, have been involved in human infections, and safety issues regarding these human infections are hampering widespread biotechnological applications. Taxonomically, Bcc bacteria are closely related species that share a high level of 16S rRNA gene sequence similarity and moderate DNA–DNA hybridization values (Coenye *et al.*, 2001; Vandamme *et al.*, 1997). Their large genomes (7.5–8.5 Mb) provide them with an unusual array of metabolic capacities. This review presents the traditional polyphasic, multilocus sequence and genomic data used for the classification and identification of these bacteria and updates an overview presented earlier (Vandamme and Dawyndt, 2011). A phylogenetic tree based on 16S rRNA gene sequences, showing the positions of all present *Burkholderia* species and representatives of related genera, is shown in Fig. 1.1.

Historical background

Bcc bacteria were first described in the 1940s by Walter Burkholder, who reported on 'sour skin' or 'slippery skin', a disease of onion bulbs, and who proposed to name them *Pseudomonas cepacia* (Burkholder, 1950). *Pseudomonas kingii*, a human opportunistic pathogen, and *Pseudomonas multivorans*, an environmental organism isolated from soil and water in Trinidad (Jonsson, 1970; Morris and Roberts, 1959; Stanier *et al.*, 1966), were later considered to be synonyms of *P. cepacia* (Ballard *et al.*, 1970; Samuels *et al.*, 1973; Sinsabaugh and Howard, 1975; Snell *et al.*, 1972). Of these species names, *P. cepacia* had nomenclatural priority (Palleroni and Holmes, 1981). With the advent of the first rRNA-based techniques for studying bacterial phylogeny in the 1970s, taxonomists generated a phylogenetic framework in which they revised the classification and nomenclature of many established genera including that of the genus *Pseudomonas*. It became apparent that the genus *Pseudomonas* was phylogenetically diverse and formed a polyphyletic part of a major group of bacteria, now known as the *Proteobacteria* (De Vos and De Ley, 1983). In 1992, the new genus *Burkholderia* (Yabuuchi *et al.*, 1992) was proposed to accommodate *P. cepacia* and six other species of the so-called *Pseudomonas* rRNA group II (Palleroni *et al.*, 1973). In the years that followed, the interest in burkholderias increased exponentially and many novel species were isolated, described and named. They appeared to be most versatile bacteria and occupied a surprisingly wide array of ecological niches (Coenye and Vandamme, 2003; Compant *et al.*, 2008). Their characteristics were exploited for biocontrol, bioremediation and plant growth promotion purposes, but safety issues regarding human infections, especially in cystic fibrosis patients, were not solved. By now, the number of *Burkholderia* species has risen to more than 60 validly named species (see http://www.bacterio.cict.fr/) and a large number of candidate species (Lemaire *et al.*, 2011, 2012; Van Oevelen *et al.*, 2004; Verstraete *et al.*, 2011).

The confounding Bcc taxonomy

From the early 1990s a marked heterogeneity among strains identified as *B. cepacia* using traditional and molecular identification approaches was noted (Bevivino *et al.*, 1994; Butler *et al.*, 1995; Gillis *et al.*, 1995; Simpson *et al.*, 1994; Tabacchioni *et al.*, 1995; Yohalem and Lorbeer, 1994). This heterogeneity rendered correct identification problematic and evaluation of many commonly used identification techniques showed that they were either not very sensitive, not very

Figure 1.1 (opposite page) Phylogenetic tree based on partial 16S rRNA gene sequences of *Burkholderia* species. Sequences were aligned against the Silva reference database (www.arb-silva.de) using the Mothur software (Schloss *et al.*, 2009). Phylogenetic analysis was conducted using MEGA5 (Tamura *et al.*, 2011). All positions with less than 95% site coverage were eliminated, resulting in a total of 1117 positions in the final dataset. The bootstrap consensus tree, inferred from 1000 replicates, was constructed using the maximum likelihood method based on the Tamura–Nei model (Tamura and Nei, 1993). Branches corresponding to partitions reproduced in less than 50% bootstrap replicates were collapsed. The percentage of replicate trees in which the associated taxa clustered together in the bootstrap test are shown next to the branches if greater than 50%. The sequence of *Pandoraea apista* CCUG 38412^T was used as out group. The scale bar indicates the number of substitutions per site. The main species groups are shown.

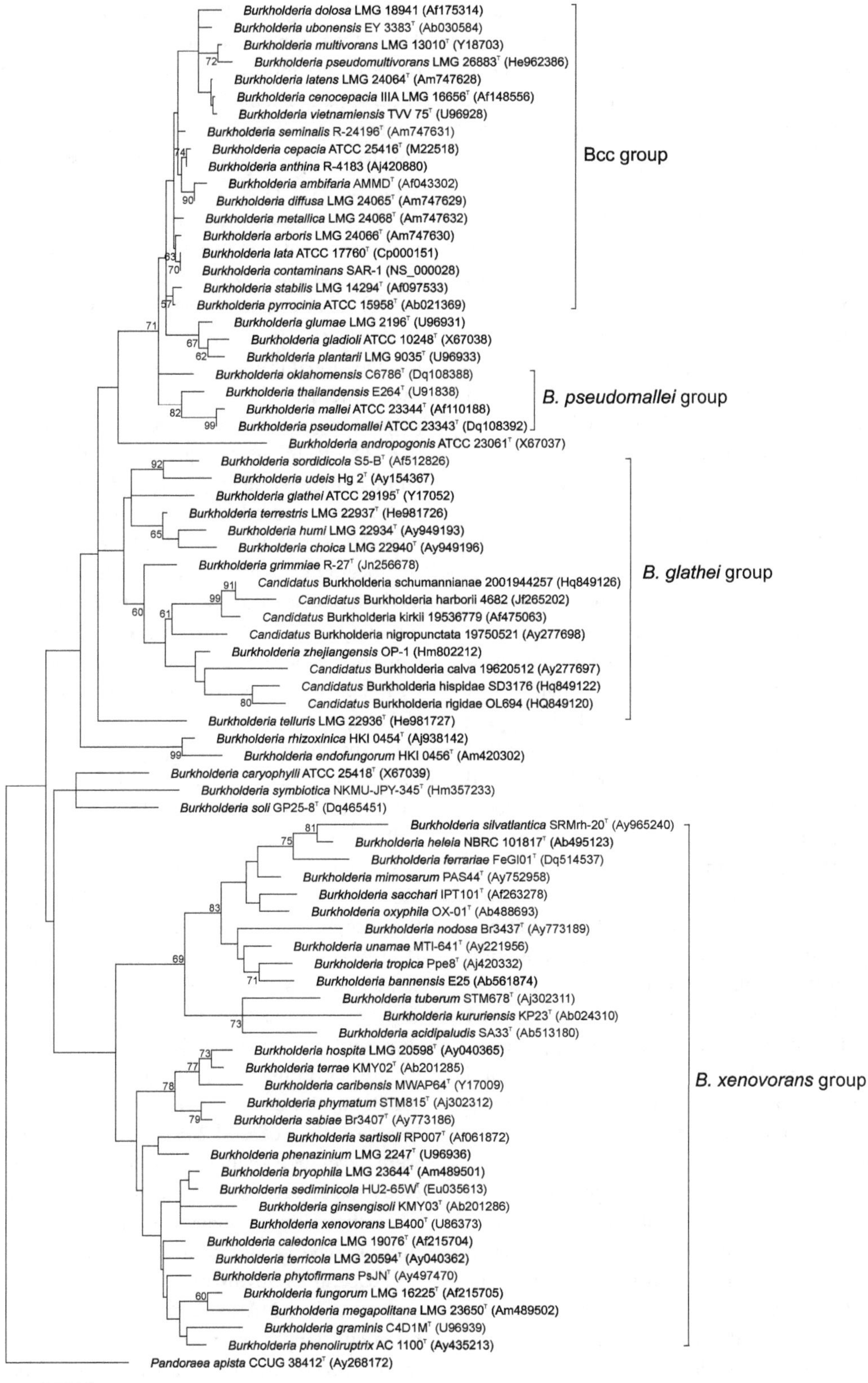
Burkholderia dolosa LMG 18941 (Af175314)
Burkholderia ubonensis EY 3383^{T} (Ab030584)
Burkholderia multivorans LMG 13010^{T} (Y18703)
Burkholderia pseudomultivorans LMG 26883^{T} (He962386)
Burkholderia latens LMG 24064^{T} (Am747628)
Burkholderia cenocepacia IIIA LMG 16656^{T} (Af148556)
Burkholderia vietnamiensis TVV 75^{T} (U96928)
Burkholderia seminalis R-24196^{T} (Am747631)
Burkholderia cepacia ATCC 25416^{T} (M22518)
Burkholderia anthina R-4183 (Aj420880)
Burkholderia ambifaria AMMD^{T} (Af043302)
Burkholderia diffusa LMG 24065^{T} (Am747629)
Burkholderia metallica LMG 24068^{T} (Am747632)
Burkholderia arboris LMG 24066^{T} (Am747630)
Burkholderia lata ATCC 17760^{T} (Cp000151)
Burkholderia contaminans SAR-1 (NS_000028)
Burkholderia stabilis LMG 14294^{T} (Af097533)
Burkholderia pyrrocinia ATCC 15958^{T} (Ab021369)
Bcc group
Burkholderia glumae LMG 2196^{T} (U96931)
Burkholderia gladioli ATCC 10248^{T} (X67038)
Burkholderia plantarii LMG 9035^{T} (U96933)
Burkholderia oklahomensis C6786^{T} (Dq108388)
Burkholderia thailandensis E264^{T} (U91838)
Burkholderia mallei ATCC 23344^{T} (Af110188)
Burkholderia pseudomallei ATCC 23343^{T} (Dq108392)
B. pseudomallei group
Burkholderia andropogonis ATCC 23061^{T} (X67037)
Burkholderia sordidicola S5-B^{T} (Af512826)
Burkholderia udeis Hg 2^{T} (Ay154367)
Burkholderia glathei ATCC 29195^{T} (Y17052)
Burkholderia terrestris LMG 22937^{T} (He981726)
Burkholderia humi LMG 22934^{T} (Ay949193)
Burkholderia choica LMG 22940^{T} (Ay949196)
Burkholderia grimmiae R-27^{T} (Jn256678)
Candidatus Burkholderia schumannianae 2001944257 (Hq849126)
Candidatus Burkholderia harborii 4682 (Jf265202)
Candidatus Burkholderia kirkii 19536779 (Af475063)
Candidatus Burkholderia nigropunctata 19750521 (Ay277698)
Burkholderia zhejiangensis OP-1 (Hm802212)
Candidatus Burkholderia calva 19620512 (Ay277697)
Candidatus Burkholderia hispidae SD3176 (Hq849122)
Candidatus Burkholderia rigidae OL694 (HQ849120)
B. glathei group
Burkholderia telluris LMG 22936^{T} (He981727)
Burkholderia rhizoxinica HKI 0454^{T} (Aj938142)
Burkholderia endofungorum HKI 0456^{T} (Am420302)
Burkholderia caryophylli ATCC 25418^{T} (X67039)
Burkholderia symbiotica NKMU-JPY-345^{T} (Hm357233)
Burkholderia soli GP25-8^{T} (Dq465451)
Burkholderia silvatlantica SRMrh-20^{T} (Ay965240)
Burkholderia heleia NBRC 101817^{T} (Ab495123)
Burkholderia ferrariae FeGl01^{T} (Dq514537)
Burkholderia mimosarum PAS44^{T} (Ay752958)
Burkholderia sacchari IPT101^{T} (Af263278)
Burkholderia oxyphila OX-01^{T} (Ab488693)
Burkholderia nodosa Br3437^{T} (Ay773189)
Burkholderia unamae MTI-641^{T} (Ay221956)
Burkholderia tropica Ppe8^{T} (Aj420332)
Burkholderia bannensis E25 (Ab561874)
Burkholderia tuberum STM678^{T} (Aj302311)
Burkholderia kururiensis KP23^{T} (Ab024310)
Burkholderia acidipaludis SA33^{T} (Ab513180)
Burkholderia hospita LMG 20598^{T} (Ay040365)
Burkholderia terrae KMY02^{T} (Ab201285)
Burkholderia caribensis MWAP64^{T} (Y17009)
Burkholderia phymatum STM815^{T} (Aj302312)
Burkholderia sabiae Br3407^{T} (Ay773186)
Burkholderia sartisoli RP007^{T} (Af061872)
Burkholderia phenazinium LMG 2247^{T} (U96936)
Burkholderia bryophila LMG 23644^{T} (Am489501)
Burkholderia sediminicola HU2-65W^{T} (Eu035613)
Burkholderia ginsengisoli KMY03^{T} (Ab201286)
Burkholderia xenovorans LB400^{T} (U86373)
Burkholderia caledonica LMG 19076^{T} (Af215704)
Burkholderia terricola LMG 20594^{T} (Ay040362)
Burkholderia phytofirmans PsJN^{T} (Ay497470)
Burkholderia fungorum LMG 16225^{T} (Af215705)
Burkholderia megapolitana LMG 23650^{T} (Am489502)
Burkholderia graminis C4D1M^{T} (U96939)
Burkholderia phenoliruptrix AC 1100^{T} (Ay435213)
B. xenovorans group
Pandoraea apista CCUG 38412^{T} (Ay268172)
0,01

specific, or neither sensitive nor specific (Kiska *et al.*, 1996; Larsen *et al.*, 1993; Leff *et al.*, 1995; Liu *et al.*, 1995; Simpson *et al.*, 1994; Stead, 1992). The diversity among these presumed *B. cepacia* strains and the lack of reliable identification schemes led Vandamme *et al.* (1997) to perform a polyphasic taxonomic study which revealed that *B. cepacia* isolates from cystic fibrosis patients and other human clinical and environmental samples represented at least five genomic species or genomovars (Ursing *et al.*, 1995), collectively referred to as the *B. cepacia* complex. In this polyphasic study, whole cell protein electrophoresis, whole cell fatty acid methyl ester analyses and DNA–DNA and DNA–rRNA hybridizations were the key techniques used. Driven by the devastating effect of these infections in cystic fibrosis patients, microbiologists developed a range of alternative identification approaches to reassess the clinical role of each *B. cepacia* complex species, to study their epidemiology and to improve patient management. These methods included species specific PCR tests, amplified fragment length polymorphism analysis of genomic DNA (Coenye *et al.*, 1999), ribotyping (Brisse *et al.*, 2000, 2004), restriction fragment length polymorphism analysis of PCR amplified gene fragments (Mahenthiralingam *et al.*, 2000a; Segonds *et al.*, 1999; Vermis *et al.*, 2002b) and tRNA profiling (Storms *et al.*, 2002). Although several of these approaches proved useful for species identification of Bcc bacteria, their sensitivity and specificity needed constant re-evaluation because of the regular description of novel members of this species complex. At present, *recA* gene and multilocus sequence based approaches have been established as powerful, objective and portable taxonomic tools that are very useful for both identification and classification purposes. The need to have well characterized strains representative of these *B. cepacia* complex species to assist with accurate identification, epidemiological analysis and systematic studies of virulence on this important group of opportunistic pathogens led to the description of experimental strain panels that were made publicly available (Coenye *et al.*, 2003; Mahenthiralingam *et al.*, 2000b).

16S rRNA gene-based analyses

16S rRNA gene sequences are used to organize bacterial taxonomy at various hierarchical ranks (Ludwig and Klenk, 2005). In a 16S rRNA based phylogenetic analysis, the genus *Burkholderia* consists of multiple main clades of species (Fig. 1.1) (Lemaire *et al.*, 2012; Suarez-Moreno *et al.*, 2012) with human pathogens like *Burkholderia mallei*, *Burkholderia pseudomallei* and Bcc species, and several plant pathogens clustering together. Beneficial or putatively beneficial plant associated *Burkholderia* species however, occur in multiple lineages and it is unclear if there are characteristics that systematically differentiate members of the major lineages within this genus. When comparing 16S rRNA gene sequences of Bcc reference strains, high similarity values (typically above 98%) are measured (Coenye *et al.*, 2001; Vanlaere *et al.*, 2008a, 2009), indicating that Bcc species are phylogenetically very closely related. In addition, as these species also exhibit up to 2% intraspecies diversity in their 16S rRNA gene sequences, they cannot reliably be identified at the species level by means of simple comparison of complete 16S rRNA gene sequences.

Early efforts to develop genotypic methods for identifying *B. cepacia* yielded a number of candidate rRNA gene based PCR assays (Campbell *et al.*, 1995; Karpati and Jonasson, 1996; O'Callaghan *et al.*, 1994; Tyler *et al.*, 1995). However, these studies were conducted before the recognition that several species comprise the *B. cepacia* complex and most relied on published sequence data derived from analyses of culture collection strains that, in retrospect, were poorly representative of the majority of human clinical isolates (LiPuma *et al.*, 1999). Specific PCR tests for *B. multivorans*, *B. vietnamiensis* and *B. dolosa*, based on the 16S and 23S rRNA gene signatures were developed (Bauernfeind *et al.*, 1999; LiPuma *et al.*, 1999; Vermis *et al.*, 2004) but rRNA signatures that allowed species level identification of other *B. cepacia* complex bacteria were not found.

Procedures involving PCR based amplification of 16S rRNA genes followed by restriction fragment length polymorphism analysis using multiple restriction enzymes have been elaborated but are most likely only rarely used today as they are complex, time consuming, expensive and poorly discriminatory compared to currently available methods (Fiore *et al.*, 2001; Segonds *et al.*, 1999; Vermis *et al.*, 2002b).

recA gene-based analyses

Since the rRNA genes have limited taxonomic resolution, Mahenthiralingam *et al.* (2000a) explored the utility of the *recA* gene for species identification in the *B. cepacia* complex. They demonstrated that the *recA* genes of these bacteria typically show 94–95% similarity between different Bcc species and 98–99% similarity within Bcc species (Mahenthiralingam *et al.*, 2000a; Vanlaere *et al.*, 2008a, 2009). Remarkably, *B. cenocepacia* comprises four subpopulations (referred to as *B. cenocepacia* IIIA, IIIB, IIIC and IIID) with different *recA* alleles. In addition, *recA* sequence based identification of the so-called taxon K bacteria which currently comprise two named species (*B. contaminans* and *B. lata*) is problematic possibly because the species complexity in this lineage is only partly resolved (Vanlaere *et al.*, 2009).

Species-specific *recA* gene-derived primer pairs have been developed for the identification of several Bcc species, but subsequent studies revealed that some of these assays were poorly sensitive or specific (Vermis *et al.*, 2002a). Initially, *recA* restriction fragment length polymorphism studies appeared very promising as a simple, rapid, cheap and accurate identification approach for Bcc bacteria (Mahenthiralingam *et al.*, 2000a). However, the traditional agarose electrophoresis used to separate the restriction fragments proved poorly discriminatory between a number of similar restriction patterns and was therefore replaced by polyacrylamide electrophoresis (Vanlaere *et al.*, 2005). By that time, more than 70 different *Hae*III-*recA* restriction profiles among Bcc bacteria could be distinguished (P. Vandamme, unpublished data). Most species include multiple restriction profiles (Coenye *et al.*, 2001; Mahenthiralingam *et al.*, 2000a) and overlap of restriction profiles is observed in multiple species (Mahenthiralingam *et al.*, 2000b; McDowell *et al.*, 2001; Vandamme *et al.*, 2002); together, these problems render this approach not longer efficient or effective.

Multilocus sequence analysis

One of the particularly interesting developments in bacterial taxonomy is multilocus sequence analysis (MLSA) (Gevers *et al.*, 2005) (see also Chapter 2). Initially MLSA methods were developed for strain genotyping below the species level and designated as multilocus sequence typing (MLST) (Maiden *et al.*, 1998); we will use the term MLSA going forward as these methods have also proved useful for bacterial classification above the strain level. MLSA employs phylogenetic procedures based on the nucleotide sequences of multiple gene fragments to reveal similarities between strains representing different species and genera. Especially for depicting relationships within and between closely related species like the Bcc bacteria, this approach has a taxonomic resolution superior to that of the traditional 16S rRNA gene sequence analysis. The deduced phylogenetic trees not only provide a phylogenetic backbone but also reveal intraspecies relationships at a level where comparative 16S rRNA sequence analysis is not longer discriminatory (cfr. chapter 2).

Within the genus *Burkholderia*, MLST schemes have been developed for *B. pseudomallei* and allied species (including *B. mallei* and *Burkholderia thailandensis*) (Godoy *et al.*, 2003) and Bcc bacteria (Baldwin *et al.*, 2005, 2007, 2008;; Dalmastri *et al.*, 2007; Vanlaere *et al.*, 2008a, 2009; Waine *et al.*, 2007). More recently, Spilker *et al.* (2009) redesigned PCR primers targeting loci in the current Bcc MLST scheme to more reliably amplify these loci from Bcc species, to amplify these same loci from additional *Burkholderia* species, and to enable the use of a single primer set per locus for both amplification and DNA sequencing. Interestingly from a taxonomic perspective, Vanlaere *et al.* (2009) compared the average nucleotide divergence of concatenated allele sequences between Bcc species and demonstrated that an average concatenated sequence divergence of 3% can be used as a threshold value to distinguish between Bcc species. The diversity within species varied from 0.4% among *B. vietnamiensis* strains to 2.9% within *B. anthina* strains. The inter-species nucleotide divergence values of *B. cepacia* complex species to other *Burkholderia* species were at least 7.5% (Vanlaere *et al.*, 2009).

Phenotypic methods for the classification and identification of *B. cepacia* complex species

Biochemical identification

The progress made in recent years in refining the taxonomy of these bacteria has provided the basis for improved identification algorithms (Vanlaere *et al.*, 2009). Nevertheless, reliable differentiation of these species from other related taxa such as *Ralstonia, Cupriavidus, Pandoraea, Achromobacter, Brevundimonas, Comamonas* and *Delftia* species is difficult. Differentiation of species within the *B. cepacia* complex can be particularly problematic, even with an extended panel of biochemical tests (Henry *et al.*, 2001), as they are phenotypically very similar and most currently available commercial bacterial identification systems cannot reliably distinguish between them.

Protein analysis

Analysis of whole-cell protein content has been used in taxonomic studies of a large number of bacteria, initially through whole-cell protein electrophoresis, nowadays more commonly via matrix assisted laser desorption ionization time of flight mass spectroscopy (MALDI-TOF MS) (Degand *et al.*, 2008; Vanlaere *et al.*, 2008b). The rationale for using whole-cell protein electrophoresis in bacterial systematics is that bacteria that are closely related are likely to have a highly similar protein content when grown in standardized conditions (Pot *et al.*, 1994). The subset of proteins analysed in MALDI-TOF MS is much smaller than that in traditional whole-cell protein electrophoresis, as it are primarily peptides from high abundance proteins like ribosomal proteins, of a molecular length less than 20,000 dalton (Mellmann *et al.*, 2008) that are detected. Vandamme *et al.* (1997) reported that SDS-PAGE of whole-cell proteins was a suitable technique for the identification of members of the *B. cepacia* complex. However, subsequent comparisons of whole cell protein profile based identification results with those of other approaches revealed several discrepancies and a poor discrimination among several Bcc species was noted. Although this method was therefore abandoned for the identification of Bcc isolates rather quickly it remained instrumental and was used as a first line screening method for many years. Indeed, the availability of a comprehensive database of highly standardized whole cell protein profiles of thousands of human and environmental isolates (including established and new cystic fibrosis pathogens and a range of bacteria that are occasionally isolated from cystic fibrosis specimens) allowed a straightforward and accurate species level identification of many non-*B. cepacia* complex isolates. Even when no identification was achieved, the profiles of new isolates could be compared with those of previous unknowns in the database and clusters of isolates representing putative new species (or simply organisms not yet present in the protein profile database) were delineated; these could be further examined by polyphasic studies pending the availability of a minimal number of isolates and sufficient time for analyses. Current commercial MALDI-TOF MS systems use this same working rationale (Degand *et al.*, 2008; Desai *et al.*, 2012; Lau *et al.*, 2012; Marko *et al.*, 2012; Minan *et al.*, 2009; Vanlaere *et al.*, 2008b). Specifically with regards to Bcc bacteria, MALDI-TOF MS identifies most isolates accurately at the species level and all isolates at the genus level (Degand *et al.*, 2008; Desai *et al.*, 2012; Vanlaere *et al.*, 2008b).

Fatty acid analysis

The high degree of automation, the relative simplicity and the low cost of whole-cell fatty acid analysis made it a valuable technique for rapid identification of isolates in clinical laboratories (Welch, 1991). However, cellular fatty acid methyl ester analysis appeared useful for identification of *Burkholderia* strains at the genus level, but was of limited utility for identification of individual *B. cepacia* complex species and did not differentiate *B. gladioli* (Clode *et al.*, 1999; Vandamme *et al.*, 1997).

Whole-genome studies

The number of whole-genome sequences is increasing rapidly and allows assessment of genome level variation within and between species. A range of novel approaches for assessing these taxonomic relationships within and between species has become available (Richter and Rossello-Mora, 2009). As mentioned above, in the present taxonomic practice the level of DNA–DNA hybridization is still used as the most decisive criterion for species delineation (Tindall *et al.*, 2010). Two distinct approaches are primarily useful to assess genome level variation within species and can be used as *in silico* imitations of DNA–DNA hybridization experiments: the calculation of the percentage of conserved DNA (pcDNA) (Goris *et al.*, 2007) and of MUM-indices (Deloger *et al.*, 2009). The former computes a pairwise whole-genome similarity by cutting the sequence of one genome (the query genome) into consecutive fragments of 1020 nucleotides, which are then blasted against the complete genome sequence of the other strain (the reference genome). This computation can be made reflexive by repeating the blast approach with the roles of the query and reference genomes reversed, so that the percentage of conserved DNA can be computed as the sum of the lengths of the alignable regions of both genomes divided by the sum of the total length of both genomes (Vandamme and Dawyndt, 2011).

The MUM-index (Deloger *et al.*, 2009) is an alternative method to compute the percentage of conserved DNA that makes use of the mummer software (Kurtz *et al.*, 2004) instead of BLAST. This software package determines the maximal unique exact matches shared between two genomes using suffix arrays, a process that works considerably faster than finding reciprocal matches using BLAST. In a previous comparison of *Burkholderia* whole-genome sequences, we demonstrated that the pcDNA method showed a high correlation with the MUM-index (Vandamme and Dawyndt, 2011). An analysis of 36 whole-genome sequences using the MUM-index was performed as reported previously (Vandamme and Dawyndt, 2011) and confirms and extends previous findings (Fig. 1.2). The main conclusions are that (i) strain *B. cenocepacia* J2315^{T} which represents the *recA* IIIA allele is rather different from the three *B. cenocepacia* IIIB strains as first revealed by DNA–DNA hybridization data (Vandamme *et al.*, 2003), (ii) *B. lata* and *B. contaminans* are closely related species (Vanlaere *et al.*, 2009), (iii) given the variance in MUM-indices observed among multiple *B. cenocepacia* strains, the rice endophyte with antifungal activity, strain KJ006 (Kwak *et al.*, 2012), is a *B. vietnamiensis* strain (this is confirmed by a *recA* sequence similarity of 99% towards the *B. vietnamiensis* type strain [data taken from (Kwak *et al.*, 2012)]), and (iv) *B. mallei* and *B. pseudomallei* represent the same genospecies in which the former represents a single line of descent (again confirming DNA–DNA hybridization and also MLSA data) (Godoy *et al.*, 2003). Although analyses of pcDNA and of MUM-indices poorly resolve evolutionary relationships among the more distantly related *Burkholderia* species, they are still applicable for inferring the phylogeny of more closely related *Burkholderia* species, such as members of the *B. cepacia* complex or of the *B. pseudomallei* group (Vandamme and Dawyndt, 2011).

Complementary to these approaches, the average nucleotide identity (ANI) (Konstantinidis and Tiedje, 2005; Richter and Rossello-Mora, 2009) and core gene identity (CGI) methods (Vanlaere *et al.*, 2009) are both natural extensions of MLSA in case complete or draft genome sequences are available. Whereas MLSA computes similarity values as the average similarity between a limited set of orthologous protein coding genes shared between all organisms studied, ANI and CGI maximize the information content by taking all orthologous protein coding genes of the genomes into consideration. The main difference between both approaches is that ANI computes a pairwise genome similarity based on all orthologous genes shared between two genomes, while CGI restricts the computation of pairwise genome similarity to orthologous genes shared by all genomes under study. ANI may thus induce a bias in the pairwise similarities when computed for a heterogeneous set of genome sequences or for draft genomes with random sets of protein coding genes missing. When working with more dispersed organisms, the set of single-copy core genes might become too small to build a reliable whole-genome based similarity parameter so that in such cases the ANI similarity parameter would be preferable as it computes the pairwise similarity based on the maximal set of information shared by every two genomes. The computation of ANI and CGI depends on the annotation of protein coding genes, but can be applied to both

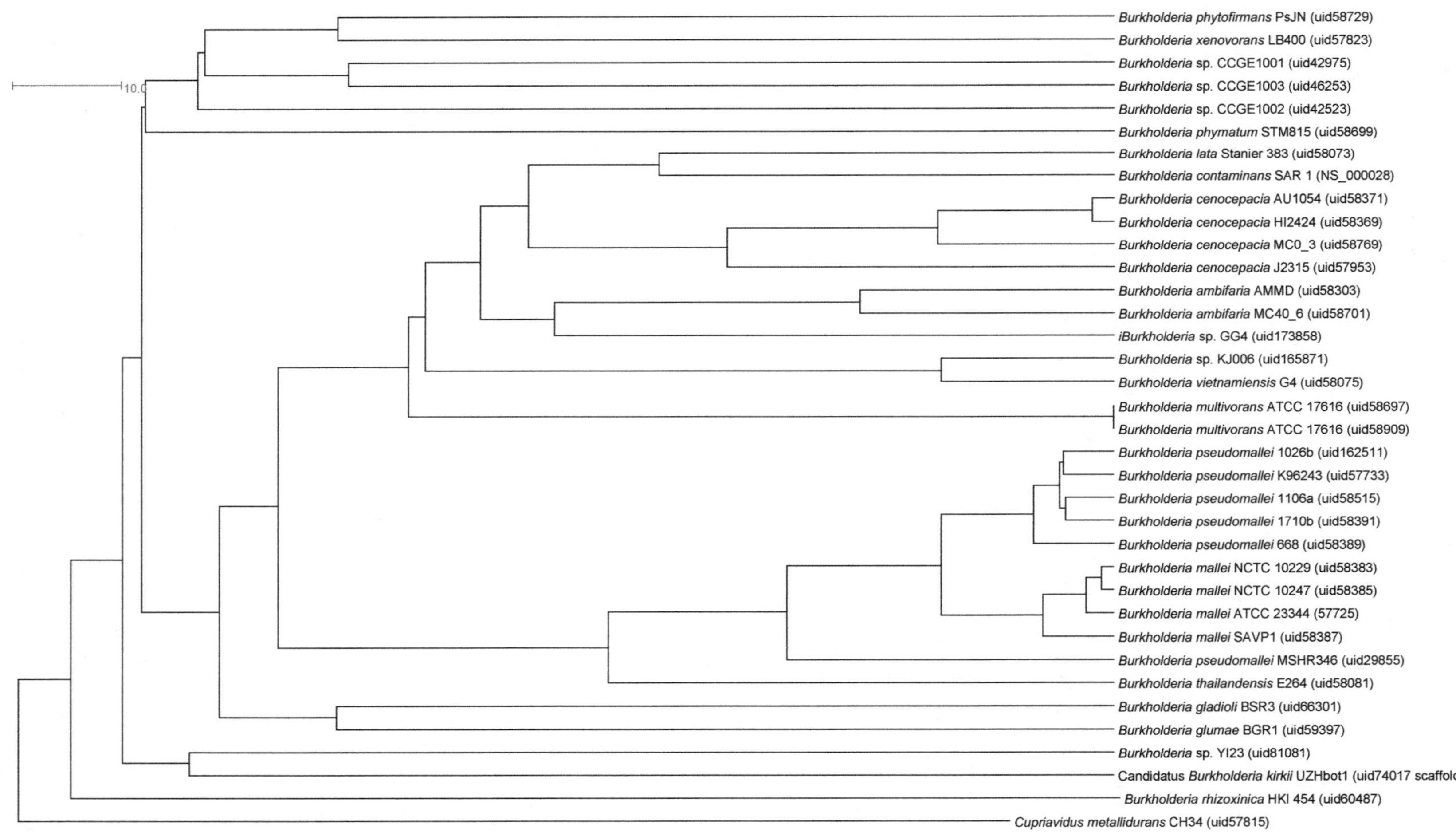

Figure 1.2 UPGMA tree based on maximal unique matches (MUMi) similarity values calculated between *Burkholderia* whole-genome sequences. Thirty six whole-genome sequences and their unique identifiers (uid numbers) were downloaded from the National Center for Biotechnology Information's (NCBI) ftp site (ftp://ftp.ncbi.nih.gov). *Cupriavidus metallidurans* CH34 was chosen as an out group.

complete and draft genomes when a sufficient number of shared orthologous genes can be taken into consideration. Analysis of ANI or CGI values therefore represent approaches that primarily assess genome level variation between, rather than within species or between closely related species. For a group of closely related organisms both approaches result in highly correlated similarity matrices (Vandamme and Dawyndt, 2011).

When applied to the *Burkholderia* genomes, OrthoMCL, an algorithm designed to identify homologous proteins based on sequence similarity and to distinguish orthologous from paralogous relationships (Li *et al.*, 2003; Vandamme and Dawyndt, 2011) showed that they share a set of about 700 single-copy core genes (draft genomes are included which reduces the number of remaining single-copy core genes considerably). Fig. 1.3 shows a maximum-likelihood phylogenetic tree with 100 bootstrap replicates obtained for a subset of 100,000 randomly selected amino acid positions from the concatenated single copy core gene alignment. In contrast with the 16S rRNA based phylogenetic tree (Fig. 1.1), bootstrap values showed that the majority of branches of the tree were highly supported, although phylogenetic inference was based on both complete and draft genomes. The tree thus provides a much better resolved picture of the phylogenetic relationships among the more distantly and more closely related *Burkholderia* species than the 16S rRNA-based tree. In this CGI tree, all Bcc genomes formed a distinct line of descent within the genus *Burkholderia*, well separated from *B. gladioli*, *B. glumae* and the *B. pseudomallei* group, which are their nearest neighbours in terms of 16S rRNA sequence similarity (Fig. 1.1). The *B. pseudomallei* group also includes *B. thailandensis*, *B. oklahomensis* and strain MSMB 43, an unusual *B. thailandensis*-like strain from an Australian water source (Gee *et al.*, 2008), confirming MLSA data (Godoy *et al.*, 2003). On the basis of their minimal divergence from the respective type strains, strains KJ006 and Ch1-1 represent *B. vietnamiensis* (confirming results from the MUM-index and *recA* analyses) and *B. xenovorans* strains, respectively. The sequence of strain YI23, a fenitrothion-degrading bacterium (Lim *et al.*, 2012), represents a whole genome sequence of a member of the *B. glathei* group (as revealed by analysis of the 16S rRNA sequence within that genome sequence [data not shown]) and confirms that the *B. glathei* group represents a unique, deep branch within the genus *Burkholderia*. The phylogenetic positions of the remaining unnamed *Burkholderia* strains (CCGE 1001, CCGE 1002, CCGE 1003 and H160) indicate that they represent species of the *B. xenovorans* group for which there are currently no annotated reference whole genome sequences publicly available (Fig. 1.3). Finally, the CGI analysis also confirms that *B. rhizoxinica* represents a very distinct *Burkholderia* lineage altogether, again supporting its position in the 16S rRNA-based tree (Fig. 1.1).

Conclusion

During the past two decades *B. cepacia* has been transformed into a species complex that now consists of 17 validly named species. Published and unpublished results from *recA* sequencing and MLSA studies demonstrate that there is still a considerable number of unnamed *B. cepacia* complex and other *Burkholderia* species waiting to be cultured, described and formally named. The devastating infections in cystic fibrosis patients triggered an unusual number of studies and variety of data. In about 20 years, thousands of strains have been included in polyphasic taxonomic analyses by members of an international working group (http://go.to/cepacia/); more than 1200 of these have been examined in MLST studies (http://pubmlst.org/bcc/) and the whole-genome sequences of several dozens of Bcc and other *Burkholderia* strains are publicly available or are being determined (http://www.genomesonline.org/cgi-bin/GOLD/index.cgi). The high interspecies DNA–DNA hybridization values, the presence of many intermediate DNA–DNA hybridization values, the MLSA and whole-genome data for *B. cenocepacia* and the taxon K strains in particular, and their biochemical similarity suggest that Bcc bacteria constitute a continuum in which separated entities which we now call species have only relatively recently developed. It is unlikely that large populations of related bacteria can invariably be divided into discrete clusters. They would rather represent a genetic continuum where clear

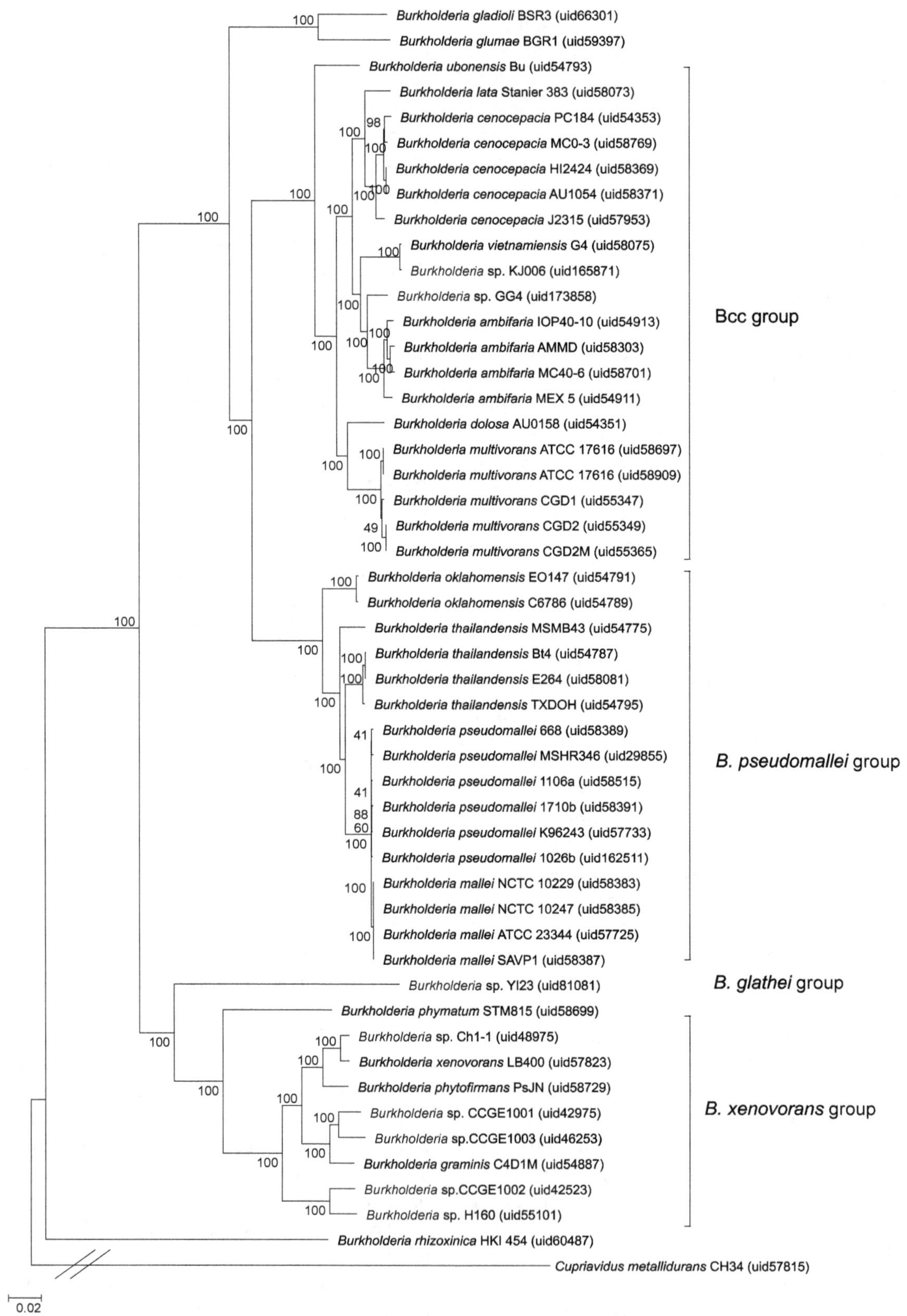
Burkholderia gladioli BSR3 (uid66301)
Burkholderia glumae BGR1 (uid59397)
Burkholderia ubonensis Bu (uid54793)
Burkholderia lata Stanier 383 (uid58073)
Burkholderia cenocepacia PC184 (uid54353)
Burkholderia cenocepacia MC0-3 (uid58769)
Burkholderia cenocepacia HI2424 (uid58369)
Burkholderia cenocepacia AU1054 (uid58371)
Burkholderia cenocepacia J2315 (uid57953)
Burkholderia vietnamiensis G4 (uid58075)
Burkholderia sp. KJ006 (uid165871)
Burkholderia sp. GG4 (uid173858)
Burkholderia ambifaria IOP40-10 (uid54913)
Burkholderia ambifaria AMMD (uid58303)
Burkholderia ambifaria MC40-6 (uid58701)
Burkholderia ambifaria MEX 5 (uid54911)
Burkholderia dolosa AU0158 (uid54351)
Burkholderia multivorans ATCC 17616 (uid58697)
Burkholderia multivorans ATCC 17616 (uid58909)
Burkholderia multivorans CGD1 (uid55347)
Burkholderia multivorans CGD2 (uid55349)
Burkholderia multivorans CGD2M (uid55365)
Bcc group
Burkholderia oklahomensis EO147 (uid54791)
Burkholderia oklahomensis C6786 (uid54789)
Burkholderia thailandensis MSMB43 (uid54775)
Burkholderia thailandensis Bt4 (uid54787)
Burkholderia thailandensis E264 (uid58081)
Burkholderia thailandensis TXDOH (uid54795)
Burkholderia pseudomallei 668 (uid58389)
Burkholderia pseudomallei MSHR346 (uid29855)
Burkholderia pseudomallei 1106a (uid58515)
Burkholderia pseudomallei 1710b (uid58391)
Burkholderia pseudomallei K96243 (uid57733)
Burkholderia pseudomallei 1026b (uid162511)
Burkholderia mallei NCTC 10229 (uid58383)
Burkholderia mallei NCTC 10247 (uid58385)
Burkholderia mallei ATCC 23344 (uid57725)
Burkholderia mallei SAVP1 (uid58387)
B. pseudomallei group
Burkholderia sp. YI23 (uid81081)
B. glathei group
Burkholderia phymatum STM815 (uid58699)
Burkholderia sp. Ch1-1 (uid48975)
Burkholderia xenovorans LB400 (uid57823)
Burkholderia phytofirmans PsJN (uid58729)
Burkholderia sp. CCGE1001 (uid42975)
Burkholderia sp.CCGE1003 (uid46253)
Burkholderia graminis C4D1M (uid54887)
Burkholderia sp.CCGE1002 (uid42523)
Burkholderia sp. H160 (uid55101)
B. xenovorans group
Burkholderia rhizoxinica HKI 454 (uid60487)
Cupriavidus metallidurans CH34 (uid57815)
0.02

Figure 1.3 (opposite page) Maximum-likelihood phylogenetic tree with 100 bootstrap replicates obtained for a subset of 100.000 randomly selected amino acid positions from the concatenated single copy core gene alignment. The genome sequences, their annotations and unique identifiers (uid numbers) were downloaded from the National Center for Biotechnology Information's (NCBI) ftp site (ftp://ftp.ncbi.nih.gov). *Cupriavidus metallidurans* CH34 was chosen as an out group (non-proportional branch length).

divisions not always exist. There is a pragmatic need to classify the large majority of bacterial isolates in practical units and in that respect it should not be considered a problem if we regularly encounter intermediate strains. The fact that we recognize them as intermediate is more important than our current failure to classify them formally.

The present state-of-the-art demonstrates that both pcDNA and MUM-index parameters are equivalent (because they have the same taxonomic resolution) but superior (because they are sequence based and thus cumulative) approaches to the traditional DNA–DNA hybridizations for the species-level classification of these bacteria. Similarly, the ANI and CGI parameters are equivalent but superior tools compared with traditional 16S rRNA for studying *Burkholderia* phylogeny because they are based on a much larger part of the genome and because they have a better resolution for discriminating distantly and closely related bacteria. It is timely that the international taxonomic community embraces these novel methods in their daily practices and promotes modern genomic and polyphasic taxonomy in bacterial systematics.

References

Baldwin, A., Mahenthiralingam, E., Thickett, K.M., Honeybourne, D., Maiden, M.C., Govan, J.R., Speert, D.P., LiPuma, J.J., Vandamme, P., and Dowson, C.G. (2005). Multilocus sequence typing scheme that provides both species and strain differentiation for the *Burkholderia cepacia* complex. J. Clin. Microbiol. *43*, 4665–4673.

Baldwin, A., Mahenthiralingam, E., Drevinek, P., Vandamme, P., Govan, J.R., Waine, D.J., LiPuma, J.J., Chiarini, L., Dalmastri, C., Henry, D.A., Speert, D.P., Honeybourne, D., Maiden, M.C.J., and Dowson, C.G. (2007). Environmental *Burkholderia cepacia* complex isolates in human infections. Emerg. Infect. Dis. *13*, 458–461.

Baldwin, A., Mahenthiralingam, E., Drevinek, P., Pope, C., Waine, D.J., Henry, D.A., Speert, D.P., Carter, P., Vandamme, P., LiPuma, J.J., and Dowson, C.G. (2008). Elucidating global epidemiology of *Burkholderia multivorans* in cases of cystic fibrosis by multilocus sequence typing. J. Clin. Microbiol. *46*, 290–295.

Ballard, R.W., Palleroni, N.J., Doudoroff, M., Stanier, R.Y., and Mandel, M. (1970). Taxonomy of aerobic pseudomonads – *Pseudomonas cepacia, Pseudomonas marginata, Pseudomonas alliicola* and *Pseudomonas caryophylli*. J. Gen. Microbiol. *60*, 199–214.

Bauernfeind, A., Schneider, I., Jungwirth, R., and Roller, C. (1999). Discrimination of *Burkholderia multivorans* and *Burkholderia vietnamiensis* from *Burkholderia cepacia* genomovars I, III, and IV by PCR. J. Clin. Microbiol. *37*, 1335–1339.

Bevivino, A., Tabacchioni, S., Chiarini, L., Carusi, M.V., Delgallo, M., and Visca, P. (1994). Phenotypic comparison between rhizosphere and clinical isolates of *Burkholderia cepacia*. Microbiology-Uk *140*, 1069–1077.

Brisse, S., Verduin, C.M., Milatovic, D., Fluit, A., Verhoef, J., Laevens, S., Vandamme, P., Tummler, B., Verbrugh, H.A., and van Belkum, A. (2000). Distinguishing species of the *Burkholderia cepacia* complex and *Burkholderia gladioli* by automated ribotyping. J. Clin. Microbiol. *38*, 1876–1884.

Brisse, S., Cordevant, C., Vandamme, P., Bidet, P., Loukil, C., Chabanon, G., Lange, M., and Bingen, E. (2004). Species distribution and ribotype diversity of *Burkholderia cepacia* complex isolates from French patients with cystic fibrosis. J. Clin. Microbiol. *42*, 4824–4827.

Burkholder, W.H. (1950). Sour skin, a bacterial rot of onion bulbs. Phytopathology *40*, 115.

Butler, S.L., Doherty, C.J., Hughes, J.E., Nelson, J.W., and Govan, J.R.W. (1995). *Burkholderia cepacia* and cystic-fibrosis – do natural environments present a potential hazard? J. Clin. Microbiol. 33, 1001–1004.

Campbell, P.W., Phillips, J.A., Heidecker, G.J., Krishnamani, M.R.S., Zahorchak, R., and Stull, T.L. (1995). Detection of *Pseudomonas (Burkholderia) cepacia* using PCR. Pediatr. Pulmonol. *20*, 44–49.

Clode, F.E., Metherell, L.A., and Pitt, T.L. (1999). Nosocomial acquisition of *Burkholderia gladioli* in patients with cystic fibrosis. Am. J. Resp. Critical Care Med. *160*, 374–374.

Coenye, T., and Vandamme, P. (2003). Diversity and significance of *Burkholderia* species occupying diverse ecological niches. Environ. Microbiol. *5*, 719–729.

Coenye, T., Schouls, L.M., Govan, J.R., Kersters, K., and Vandamme, P. (1999). Identification of *Burkholderia* species and genomovars from cystic fibrosis patients by AFLP fingerprinting. Int. J. Syst. Bacteriol. *49*, 1657–1666.

Coenye, T., Vandamme, P., Govan, J.R., and LiPuma, J.J. (2001). Taxonomy and identification of the *Burkholderia cepacia* complex. J. Clin. Microbiol. *39*, 3427–3436.

Coenye, T., Vandamme, P., LiPuma, J.J., Govan, J.R., and Mahenthiralingam, E. (2003). Updated version of the *Burkholderia cepacia* complex experimental strain panel. J. Clin. Microbiol. *41*, 2797–2798.

Compant, S., Nowak, J., Coenye, T., Clement, C., and Barka, E.A. (2008). Diversity and occurrence of *Burkholderia* spp. in the natural environment. FEMS Microbiol. Rev. *32*, 607–626.

Dalmastri, C., Baldwin, A., Tabacchioni, S., Bevivino, A., Mahenthiralingam, E., Chiarini, L., and Dowson, C. (2007). Investigating *Burkholderia cepacia* complex populations recovered from Italian maize rhizosphere by multilocus sequence typing. Environ. Microbiol. *9*, 1632–1639.

De Vos, P., and De Ley, J. (1983). Intra- and intergeneric similarities of *Pseudomonas* and *Xanthomonas* ribosomal ribonucleic acid cistrons. Int. J. Syst. Bacteriol. *33*, 487–509.

Degand, N., Carbonnelle, E., Dauphin, B., Beretti, J.L., Le Bourgeois, M., Sermet-Gaudelus, I., Segonds, C., Berche, P., Nassif, X., and Ferroni, A. (2008). Matrix-assisted laser desorption ionization-time of flight mass spectrometry for identification of nonfermenting Gram-negative bacilli isolated from cystic fibrosis patients. J. Clin. Microbiol. *46*, 3361–3367.

Deloger, M., El Karoui, M., and Petit, M.A. (2009). A genomic distance based on MUM indicates discontinuity between most bacterial species and genera. J. Bacteriol. *191*, 91–99.

Desai, A.P., Stanley, T., Atuan, M., McKey, J., LiPuma, J.J., Rogers, B., and Jerris, R. (2012). Use of matrix assisted laser desorption ionisation-time of flight mass spectrometry in a paediatric clinical laboratory for identification of bacteria commonly isolated from cystic fibrosis patients. J. Clin. Pathol. *65*, 835–838.

Fiore, A., Laevens, S., Bevivino, A., Dalmastri, C., Tabacchioni, S., Vandamme, P., and Chiarini, L. (2001). *Burkholderia cepacia* complex: distribution of genomovars among isolates from the maize rhizosphere in Italy. Environ. Microbiol. 3, 137–143.

Gee, J.E., Glass, M.B., Novak, R.T., Gal, D., Mayo, M.J., Steigerwalt, A.G., Wilkins, P.P., and Currie, B.J. (2008). Recovery of a *Burkholderia thailandensis*-like isolate from an Australian water source. BMC Microbiol. *8*, 54.

Gevers, D., Cohan, F.M., Lawrence, J.G., Spratt, B.G., Coenye, T., Feil, E.J., Stackebrandt, E., Van de Peer, Y., Vandamme, P., Thompson, F.L., *et al.* (2005). Opinion: Re-evaluating prokaryotic species. Nat. Rev. Microbiol. 3, 733–739.

Gillis, M., Vanvan, T., Bardin, R., Goor, M., Hebbar, P., Willems, A., Segers, P., Kersters, K., Heulin, T., and Fernandez, M.P. (1995). Polyphasic taxonomy in the genus *Burkholderia* leading to an emended description of the genus and proposition of *Burkholderia vietnamiensis* sp. nov. for N_2-fixing isolates from rice in Vietnam. Int. J. Syst. Bacteriol. *45*, 274–289.

Godoy, D., Randle, G., Simpson, A.J., Aanensen, D.M., Pitt, T.L., Kinoshita, R., and Spratt, B.G. (2003). Multilocus sequence typing and evolutionary relationships among the causative agents of melioidosis and glanders, *Burkholderia pseudomallei* and *Burkholderia mallei*. J. Clin. Microbiol. *41*, 2068–2079.

Goris, J., Konstantinidis, K.T., Klappenbach, J.A., Coenye, T., Vandamme, P., and Tiedje, J.M. (2007). DNA–DNA hybridization values and their relationship to whole-genome sequence similarities. Int. J. Syst. Evol. Microbiol. *57*, 81–91.

Henry, D.A., Mahenthiralingam, E., Vandamme, P., Coenye, T., and Speert, D.P. (2001). Phenotypic methods for determining genomovar status of the *Burkholderia cepacia* complex. J. Clin. Microbiol. *39*, 1073–1078.

Jonsson, V. (1970). Proposal of a new species *Pseudomonas kingii*. Int. J. Syst. Bacteriol. *20*, 255–257.

Karpati, F., and Jonasson, J. (1996). Polymerase chain reaction for the detection of *Pseudomonas aeruginosa*, *Stenotrophomonas maltophilia* and *Burkholderia cepacia* in sputum of patients with cystic fibrosis. Mol. Cell. Probes *10*, 397–403.

Kiska, D.L., Kerr, A., Jones, M., Caracciolo, J.A., Eskridge, B., Jordan, M., Miller, S., Hughes, D., King, N., and Gilligan, P.H. (1996). Accuracy of four commercial systems for identification of *Burkholderia cepacia* and other Gram-negative nonfermenting bacilli recovered from patients with cystic fibrosis. J. Clin. Microbiol. *34*, 886–891.

Konstantinidis, K.T., and Tiedje, J.M. (2005). Genomic insights that advance the species definition for prokaryotes. Proc. Natl. Acad. Sci. U.S.A. *102*, 2567–2572.

Kurtz, S., Phillippy, A., Delcher, A.L., Smoot, M., Shumway, M., Antonescu, C., and Salzberg, S.L. (2004). Versatile and open software for comparing large genomes. Genome Biol. *5*, *R12*.

Kwak, M.J., Song, J.Y., Kim, S.Y., Jeong, H., Kang, S.G., Kim, B.K., Kwon, S.K., Lee, C.H., Yu, D.S., Park, S.H., and Kim, J.F. (2012). Complete genome sequence of the endophytic bacterium *Burkholderia* sp. strain KJ006. J. Bacteriol. *194*, 4432–4433.

Larsen, G.Y., Stull, T.L., and Burns, J.L. (1993). Marked phenotypic variability in *Pseudomonas cepacia* isolated from a patient with cystic fibrosis. J. Clin. Microbiol. *31*, 788–792.

Lau, S.K.P., Tang, B.S.F., Curreem, S.O.T., Chan, T.M., Martelli, P., Tse, C.W.S., Wu, A.K.L., Yuen, K.Y., and Woo, P.C.Y. (2012). Matrix-assisted laser desorption ionization-time of flight mass spectrometry for rapid identification of *Burkholderia pseudomallei*: importance of expanding databases with pathogens endemic to different localities. J. Clin. Microbiol. *50*, 3142–3143.

Leff, L.G., Kernan, R.M., McArthur, J.V., and Shimkets, L.J. (1995). Identification of aquatic *Burkholderia (Pseudomonas) cepacia* by hybridization with species-specific ribosomal RNA gene probes. Appl. Environ. Microbiol. *61*, 1634–1636.

Lemaire, B., Robbrecht, E., van Wyk, B., Van Oevelen, S., Verstraete, B., Prinsen, E., Smets, E., and Dessein, S. (2011). Identification, origin, and evolution of leaf nodulating symbionts of *Sericanthe (Rubiaceae)*. J. Microbiol. *49*, 935–941.

Lemaire, B., Van Oevelen, S., De Block, P., Verstraete, B., Smets, E., Prinsen, E., and Dessein, S. (2012). Identification of the bacterial endosymbionts in leaf nodules of *Pavetta (Rubiaceae)*. Int. J. Syst. Evol. Microbiol. *62*, 202–209.

Li, L., Stoeckert, C.J., Jr., and Roos, D.S. (2003). OrthoMCL: identification of ortholog groups for eukaryotic genomes. Genome Res. *13*, 2178–2189.

Lim, J.S., Choi, B.S., Choi, A.Y., Kim, K.D., Kim, D.I., Choi, I.Y., and Ka, J.O. (2012). Complete genome sequence of the fenitrothion-degrading *Burkholderia* sp. strain YI23. J. Bacteriol. *194*, 896–896.

LiPuma, J.J., Dulaney, B.J., McMenamin, J.D., Whitby, P.W., Stull, T.L., Coenye, T., and Vandamme, P. (1999). Development of rRNA-based PCR assays for identification of *Burkholderia cepacia* complex isolates recovered from cystic fibrosis patients. J. Clin. Microbiol. *37*, 3167–3170.

Liu, P.Y.F., Shi, Z.Y., Lau, Y.J., Hu, B.S., Shyr, J.M., Tsai, W.S., Lin, Y.H., and Tseng, C.Y. (1995). Comparison of different pcr approaches for characterization of *Burkholderia (Pseudomonas) cepacia* isolates. J. Clin. Microbiol. 33, 3304–3307.

Ludwig, W., and Klenk, H.P. (2005). Overview: a phylogenetic backbone and taxonomic framework for Prokaryotic systematics. In Bergey's Manual of Systematic Bacteriology, The Proteobacteria Part A Introductory Essays, Garrity, G.M., Bean, D.J., Krieg, N.R., and Staley, J.T., eds. (Springer), vol. *2*, 49–65.

McDowell, A., Mahenthiralingam, E., Moore, J.E., Dunbar, K.E.A., Webb, A.K., Dodd, M.E., Martin, S.L., Millar, B.C., Scott, C.J., Crowe, M., and Elborn, J.S. (2001). PCR-based detection and identification of *Burkholderia cepacia* complex pathogens in sputum from cystic fibrosis patients. J. Clin. Microbiol. *39*, 4247–4255.

Mahenthiralingam, E., Bischof, J., Byrne, S.K., Radomski, C., Davies, J.E., Av-Gay, Y., and Vandamme, P. (2000a). DNA-Based diagnostic approaches for identification of *Burkholderia cepacia* complex, *Burkholderia vietnamiensis, Burkholderia multivorans, Burkholderia stabilis*, and *Burkholderia cepacia* genomovars I and III. J. Clin. Microbiol. 38, 3165–3173.

Mahenthiralingam, E., Coenye, T., Chung, J.W., Speert, D.P., Govan, J.R., Taylor, P., and Vandamme, P. (2000b). Diagnostically and experimentally useful panel of strains from the *Burkholderia cepacia* complex. J. Clin. Microbiol. *38*, 910–913.

Maiden, M.C.J., Bygraves, J.A., Feil, E., Morelli, G., Russell, J.E., Urwin, R., Zhang, Q., Zhou, J.J., Zurth, K., Caugant, D.A., *et al.* (1998). Multilocus sequence typing: a portable approach to the identification of clones within populations of pathogenic microorganisms. Proc. Natl. Acad. Sci. U.S.A. *95*, 3140–3145.

Marko, D.C., Saffert, R.T., Cunningham, S.A., Hyman, J., Walsh, J., Arbefeville, S., Howard, W., Pruessner, J., Safwat, N., Cockerill, F.R., *et al.* (2012). Evaluation of the Bruker Biotyper and Vitek MS matrix-assisted laser desorption ionization-time of flight mass spectrometry systems for identification of nonfermenting Gram-negative bacilli isolated from cultures from cystic fibrosis patients. J. Clin. Microbiol. *50*, 2034–2039.

Mellmann, A., Cloud, J., Maier, T., Keckevoet, U., Ramminger, I., Iwen, P., Dunn, J., Hall, G., Wilson, D., LaSala, P., *et al.* (2008). Evaluation of matrix-assisted laser desorption ionization-time-of-flight mass spectrometry in comparison to 16S rRNA gene sequencing for species identification of nonfermenting bacteria. J. Clin. Microbiol. *46*, 1946–1954.

Minan, A., Bosch, A., Lasch, P., Stammler, M., Serra, D.O., Degrossi, J., Gatti, B., Vay, C., D'Aquino, M., Yantorno, O.,and Naumann, D. (2009). Rapid identification of *Burkholderia cepacia* complex species including strains of the novel Taxon K, recovered from cystic fibrosis patients by intact cell MALDI-TOF mass spectrometry. Analyst *134*, 1138–1148.

Morris, M.B., and Roberts, J.B. (1959). Group of pseudomonads able to synthesize poly-beta-hydroxybutyric acid. Nature *183*, 1538–1539.

O'Callaghan, E.M., Tanner, M.S., and Boulnois, G.J. (1994). Development of a PCR probe test for identifying *Pseudomonas aeruginosa* and *Pseudomonas (Burkholderia) cepacia*. J. Clin. Pathol. *47*, 222–226.

Palleroni, N.J., and Holmes, B. (1981). *Pseudomonas cepacia* sp. nov, nom. rev. Int. J. Syst. Bacteriol. *31*, 479–481.

Palleroni, N.J., Kunisawa, R., Contopoulos, R., and Doudoroff, M. (1973). Nucleic acid homologies in the genus *Pseudomonas*. Int. J. Syst. Bacteriol. *23*, 333–339.

Pot, B., Vandamme, P., and Kersters, K. (1994). Analysis of electrophoretic whole-organism protein fingerprinting. In Modern microbial methods: chemical methods in bacterial systematics, Goodfellow, M., and O'Donnell, A.G., eds. (Chichester, United Kingdom, Wiley), pp. 493–521.

Richter, M., and Rossello-Mora, R. (2009). Shifting the genomic gold standard for the prokaryotic species definition. Proc. Natl. Acad. Sci. U.S.A. *106*, 19126–19131.

Samuels, S.B., Moss, C.W., and Weaver, R.E. (1973). The fatty acids of *Pseudomonas multivorans* (*Pseudomonas cepacia*) and *Pseudomonas kingii*. J. Gen. Microbiol. *74*, 275–279.

Schloss, P.D., Westcott, S.L., Ryabin, T., Hall, J.R., Hartmann, M., Hollister, E.B., Lesniewski, R.A., Oakley, B.B., Parks, D.H., Robinson, C.J.,*et al.* (2009). Introducing mothur: open-source, platform-independent, community-supported software for describing and comparing microbial communities. Appl. Environ. Microbiol. 75, 7537–7541.

Segonds, C., Heulin, T., Marty, N., and Chabanon, G. (1999). Differentiation of *Burkholderia* species by PCR-restriction fragment length polymorphism analysis of the 16S rRNA gene and application to cystic fibrosis isolates. J. Clin. Microbiol. 37, 2201–2208.

Simpson, I.N., Finlay, J., Winstanley, D.J., Dewhurst, N., Nelson, J.W., Butler, S.L., and Govan, J.R.W. (1994). Multi-resistance isolates possessing characteristics of both *Burkholderia (Pseudomonas) cepacia* and

Burkholderia gladioli from patients with cystic fibrosis. J. Antimicrobial Chemother. *34*, 353–361.

Sinsabaugh, H.A., and Howard, G.W. (1975). Emendation of the description of *Pseudomonas cepacia* Burkholder (synonyms: *Pseudomonas multivorans* Stanier *et al.*, *Pseudomonas kingae* Jonsson; EO-1 Group). Int. J. Syst. Bacteriol. *25*, 187–201.

Snell, J.J.S., Hill, L.R., Lapage, S.P., and Curtis, M.A. (1972). Identification of *Pseudomonas cepacia* Burkholder and its synonymy with *Pseudomonas kingii* Jonsson. Int. J. Syst. Bacteriol. *22*, 127–138.

Spilker, T., Baldwin, A., Bumford, A., Dowson, C.G., Mahenthiralingam, E., and LiPuma, J.J. (2009). Expanded multilocus sequence typing for *Burkholderia* species. J. Clin. Microbiol. *47*, 2607–2610.

Stanier, R.Y., Palleroni, N.J., and Doudoroff, M. (1966). Aerobic pseudomonads – a taxonomic study. J. Gen. Microbiol. *43*, 159–271.

Stead, D.E. (1992). Grouping of plant-pathogenic and some other pseudomonas spp. by using cellular fatty acid profiles. Int. J. Syst. Bacteriol. *42*, 281–295.

Storms, V., van den Vreken, N., Gillis, M., and Vandamme, P. (2002). Evaluation of tRNA intergenic length polymorphism (tDNA-pCR) for the differentiation of the members of the *Burkholderia cepacia* complex. Syst. Appl. Microbiol *25*, 376–385.

Suarez-Moreno, Z.R., Caballero-Mellado, J., Coutinho, B.G., Mendonca-Previato, L., James, E.K., and Venturi, V. (2012). Common features of environmental and potentially beneficial plant-associated *Burkholderia*. Microb. Ecol. *63*, 249–266.

Tabacchioni, S., Visca, P., Chiarini, L., Bevivino, A., Diserio, C., Fancelli, S., and Fani, R. (1995). Molecular characterisation of rhizosphere and clinical isolates of *Burkholderia cepacia*. Res. Microbiol. *146*, 531–542.

Tamura, K., and Nei, M. (1993). Estimation of the number of nucleotide substitutions in the control region of mitochondrial-dna in humans and chimpanzees. Mol. Biol. Evol. *10*, 512–526.

Tamura, K., Peterson, D., Peterson, N., Stecher, G., Nei, M., and Kumar, S. (2011). MEGA5: Molecular evolutionary genetics analysis using maximum likelihood, evolutionary distance, and maximum parsimony methods. Mol. Biol. Evol. *28*, 2731–2739.

Tindall, B.J., Rossello-Mora, R., Busse, H.J., Ludwig, W., and Kampfer, P. (2010). Notes on the characterization of prokaryote strains for taxonomic purposes. Int. J. Syst. Evol. Microbiol. *60*, 249–266.

Tyler, S.D., Strathdee, C.A., Rozee, K.R., and Johnson, W.M. (1995). Oligonucleotide primers designed to differentiate pathogenic pseudomonads on the basis of the sequencing of genes coding for 16S-23S ribosomal RNA internal transcribed spacers. Clin. Diagn. Lab. Immunol. *2*, 448–453.

Ursing, J.B., Rossello-Mora, R.A., Garcia-Valdes, E., and Lalucat, J. (1995). A pragmatic approach to the nomenclature of phenotypically similar genomic groups. Int. J. Syst. Bacteriol. *45*, 604.

Van Oevelen, S., De Wachter, R., Vandamme, P., Robbrecht, E., and Prinsen, E. (2004). 'Candidatus *Burkholderia calva*' and 'Candidatus *Burkholderia nigropunctata*' as leaf gall endosymbionts of African Psychotria. Int. J. Syst. Evol. Microbiol. *54*, 2237–2239.

Vandamme, P., and Dawyndt, P. (2011). Classification and identification of the *Burkholderia cepacia* complex: past, present and future. Syst. Appl. Microbiol. *34*, 87–95.

Vandamme, P., Pot, B., Gillis, M., DeVos, P., Kersters, K., and Swings, J. (1996). Polyphasic taxonomy, a consensus approach to bacterial systematics. Microbiol. Rev. *60*, 407–438.

Vandamme, P., Holmes, B., Vancanneyt, M., Coenye, T., Hoste, B., Coopman, R., Revets, H., Lauwers, S., Gillis, M., Kersters, K., and Govan, J.R.W. (1997). Occurrence of multiple genomovars of *Burkholderia cepacia* in cystic fibrosis patients and proposal of *Burkholderia multivorans* sp. nov. Int. J. Syst. Bacteriol. *47*, 1188–1200.

Vandamme, P., Henry, D., Coenye, T., Nzula, S., Vancanneyt, M., LiPuma, J.J., Speert, D.P., Govan, J.R., and Mahenthiralingam, E. (2002). *Burkholderia anthina* sp. nov. and *Burkholderia pyrrocinia*, two additional *Burkholderia cepacia* complex bacteria, may confound results of new molecular diagnostic tools. FEMS Immunol. Med. Microbiol. *33*, 143–149.

Vandamme, P., Holmes, B., Coenye, T., Goris, J., Mahenthiralingam, E., LiPuma, J.J., and Govan, J.R. (2003). *Burkholderia cenocepacia* sp. nov.--a new twist to an old story. Res. Microbiol. *154*, 91–96.

Vanlaere, E., Coenye, T., Samyn, E., Van den Plas, C., Govan, J., De Baets, F., De Boeck, K., Knoop, C., and Vandamme, P. (2005). A novel strategy for the isolation and identification of environmental *Burkholderia cepacia* complex bacteria. FEMS Microbiol. Lett. *249*, 303–307.

Vanlaere, E., LiPuma, J.J., Baldwin, A., Henry, D., De Brandt, E., Mahenthiralingam, E., Speert, D., Dowson, C., and Vandamme, P. (2008a). *Burkholderia latens* sp. nov., *Burkholderia diffusa* sp. nov., *Burkholderia arboris* sp. nov., *Burkholderia seminalis* sp. nov. and *Burkholderia metallica* sp. nov., novel species within the *Burkholderia cepacia* complex. Int. J. Syst. Evol. Microbiol. *58*, 1580–1590.

Vanlaere, E., Sergeant, K., Dawyndt, P., Kallow, W., Erhard, M., Sutton, H., Dare, D., Devreese, B., Samyn, B., and Vandamme, P. (2008b). Matrix-assisted laser desorption ionisation-time-of of-flight mass spectrometry of intact cells allows rapid identification of *Burkholderia cepacia* complex. J. Microbiol. Methods *75*, 279–286.

Vanlaere, E., Baldwin, A., Gevers, D., Henry, D., De Brandt, E., LiPuma, J.J., Mahenthiralingam, E., Speert, D.P., Dowson, C., and Vandamme, P. (2009). Taxon K, a complex within the *Burkholderia cepacia* complex, comprises at least two novel species, *Burkholderia contaminans* sp. nov. and *Burkholderia lata* sp. nov. Int. J. Syst. Evol. Microbiol. *59*, 102–111.

Vermis, K., Coenye, T., Mahenthiralingam, E., Nelis, H.J., and Vandamme, P. (2002a). Evaluation of species-specific recA-based PCR tests for genomovar level identification within the *Burkholderia cepacia* complex. J. Med. Microbiol. *51*, 937–940.

Vermis, K., Vandekerckhove, C., Nelis, H.J., and Vandamme, P.A. (2002b). Evaluation of restriction fragment length polymorphism analysis of 16S rDNA as a tool for genomovar characterisation within the *Burkholderia cepacia* complex. FEMS Microbiol. Lett. *214*, 1–5.

Vermis, K., Coenye, T., LiPuma, J.J., Mahenthiralingam, E., Nelis, H.J., and Vandamme, P. (2004). Proposal to accommodate *Burkholderia cepacia* genomovar VI as *Burkholderia dolosa* sp. nov. Int. J. Syst. Evol. Microbiol. *54*, 689–691.

Verstraete, B., Van Elst, D., Steyn, H., Van Wyk, B., Lemaire, B., Smets, E., and Dessein, S. (2011). Endophytic bacteria in toxic south African plants: identification, phylogeny and possible involvement in Gousiekte. PLoS ONE *6*, e19265.

Waine, D.J., Henry, D.A., Baldwin, A., Speert, D.P., Honeybourne, D., Mahenthiralingam, E., and Dowson, C.G. (2007). Reliability of multilocus sequence typing of the *Burkholderia cepacia* complex in cystic fibrosis. J. Cyst. Fibros. *6*, 215–219.

Welch, D.F. (1991). Applications of cellular fatty acid analysis. Clin. Microbiol. Rev. *4*, 422–438.

Yabuuchi, E., Kosako, Y., Oyaizu, H., Yano, I., Hotta, H., Hashimoto, Y., Ezaki, T., and Arakawa, M. (1992). Proposal of *Burkholderia* gen. nov.; and transfer of seven species of the *Pseudomonas* homology group II to the new genus, with the type species *Burkholderia cepacia* (Palleroni and Holmes 1981) comb. nov. Microbiol. Immunol. *36*, 1251–1275.

Yohalem, D.S., and Lorbeer, J.W. (1994). Multilocus isoenzyme diversity among strains of *Pseudomonas cepacia* isolated from decayed onions, soils, and clinical sources. Syst. Appl. Microbiol. *17*, 116–124.

Comparative Genomics in the Genus *Burkholderia* 2

Pimlapas Leekitcharoenphon, Gregory S. Buzard and David W. Ussery

Abstract

The *Burkholderia* genus contains over 60 species that have colonized a wide diversity of biological niches, including those as animal, human and plant pathogens. We have collected the publicly available whole genome sequences of *Burkholderia* spp. to explore the genomic diversity within this genus, and we report here on the comparison of 110 *Burkholderia* genomes. We have constructed three types of phylogenetic trees, based on different sets of genetic information [16S rRNA genes, multiple locus sequence tags (MLST) and pan-genome sequences], as alternative ways to display taxonomic groupings and evolutionary relationships. We have found, among the more than 35,000 gene families that are represented in the *Burkholderia* pan-genome, a set of 717 core gene families that are conserved across all the *Burkholderia* genomes we studied. We have also identified the species-unique core genes for each species. Our findings can be used to find new signature sequences for the rapid and accurate detection of the different *Burkholderia* species. We have also constructed a BLAST atlas that compares the *B. pseudomallei* and *B. mallei* genomes which helps visualize gene homologies and shows the fraction of genes found in common in pair-wise comparison and reveals the duplication of genes within their genomes. The *B. mallei* genome is known to be significantly smaller in size than that of *B. pseudomallei*; this is also obvious from our analysis that demonstrates that almost all of the *B. mallei* proteome is present in *B. pseudomallei*.

Introduction

The *Burkholderia* genus refers to a group of virtually ubiquitous Gram-negative, motile, obligately aerobic, rod-shaped bacteria. The genus is best-known for its human-pathogenic members: *Burkholderia mallei*, responsible for glanders, a disease that occurs mostly in horses and related animals; *Burkholderia pseudomallei*, the causative agent of human melioidosis; and *Burkholderia cepacia* complex (Bcc), an increasingly important group of 17 related species of *Burkholderia* frequently responsible for human pulmonary infections and other opportunistic infections (Galyov *et al.*, 2010).

The genus *Burkholderia* contains important animal, human and plant pathogens, and *Burkholderia* species have colonized a wide diversity of other biological niches; they range from free-living to commensal soil and water organisms, and they can be obligate plant and animal symbionts or pathogens (Bontemps *et al.*, 2010). At this moment the genome sequence of over 110 different *Burkholderia* strains has been determined (or is being determined).

B. pseudomallei causes melioidosis, a naturally occurring human disease of predominantly tropical areas of Southeast Asia and north-western Australia. It is the most frequent and virulent human pathogen of the *Burkholderia* genus. Melioidosis can present as several distinct forms, ranging from a prolonged subclinical illness to an overwhelming septicaemia with a 90% mortality rate within 24–48 h of onset. Two genetically closely related species, *B. thailandensis* and *B. oklahomensis*, also live in soils, although they are far less pathogenic to humans than either *B. pseudomallei* or *B. mallei*.

B. mallei is mainly known as an equine pathogen, although it also causes rare cases of glanders in humans, usually from occupational exposures (Estes *et al.*, 2010). *B. mallei* is thought to have diverged relatively recently, roughly 3.5 million years ago, from a *B. pseudomallei*-like ancestor that had migrated to Southeast Asia from Australia. The monophyletic *B. mallei* pathogen evolved to become an obligate host-adapted species, i.e. it no longer survives well in soil, and *B. mallei* is not commonly found in the environment outside of its host, unlike *B. pseudomallei*. The genomes of these two *Burkholderia* species are unusually large (5.7 and 7.0 Mb, for *B. mallei* and *B. pseudomallei*, respectively), with that of *B. pseudomallei* being far more evolutionarily diverse than that of *B. mallei*. A remarkably high level of horizontal gene transfer contributes to the genetic diversity of the two species and this promiscuous swapping of genetic material may also be an important determinant of their virulence potential (Losada *et al.*, 2010).

Several other members of the genus *Burkholderia* also cause human infections, including species belonging to the Bcc (Lynch, 2009). This 'complex' currently comprises 17 closely related species that are widely and heterogeneously distributed in natural environments such as soil and water. Some Bcc strains are of significant environmental interest because they have proven to be very efficient in biocontrol and plant growth promotion usage (Vial *et al.*, 2011). Certain other Bcc species, particularly *B. cenocepacia* and *B. multivorans*, are opportunistic and transmissible pathogens that cause infections in people with cystic fibrosis and other patient populations at-risk (Lynch, 2009).

B. thailandensis is the most closely related species to *B. pseudomallei* and *B. mallei*, but, until recently, was considered so avirulent in humans that it was often used as a surrogate for these pathogens in proficiency tests (Deshazer, 2007). Another close relative, *B. oklahomensis*, was first isolated in 1973 from a deep wound infection resulting from a farming accident in Oklahoma, USA. More recently it was found in a clinical isolate recovered in Georgia, USA, from a person involved in an automobile accident that resulted in dirt in the ocular orbit (Glass *et al.*, 2009). Thus far, no other isolates have been reported.

B. mallei and *B. pseudomallei* are classified by the Centers for Disease Control and Prevention (CDC) as Tier I biological warfare agents because they have been previously weaponized, are naturally highly infectious, and because there currently is no approved effective vaccine available (Larsen and Johnson, 2009). Their significant health burden in endemic areas, plus the classification of two of their members as biological warfare agents, has resulted in an escalation of research interest in all the potentially pathogenic strains of the *Burkholderia* genus.

In this chapter, we have conducted a comparative genomic analysis of 110 *Burkholderia* genomes, including analysing core/pan genomes and identifying unique genes for each species and for multiple replicons. The general information for the *Burkholderia* genomes we examined is summarized in Table 2.1.

Evolutionary relationships

We have constructed the phylogenetic tree shown in Fig. 2.1 based on the 16S rRNA gene sequences in each genome; these genes are functionally conserved across all bacteria. The 16S rRNA sequences were identified using RNAmmer software (Lagesen *et al.*, 2007); the tree was computed using MEGA5 (Tamura *et al.*, 2011). In the recently sequenced genome of *Burkholderia* KJ006 (Table 2.1), the 16S rRNA gene clusters within the Bcc, suggesting KJ006 is part of this group (Fig. 2.1).

The *in silico* MLST tree in Fig. 2.2 was generated based on the sequence of seven housekeeping genes: *ace*, *gltB*, *gmhD*, *lepA*, *lipA*, *narK*, *ndh*, as commonly used in the *B. pseudomallei* MLST scheme (http://www.mlst.net). These gene sequences were extracted from the *Burkholderia* genomes, concatenated and then aligned to build a phylogenetic tree using MEGA5 (Tamura *et al.*, 2011). The 16S rRNA gene and MLST trees both show good representations of the relationships of different *Burkholderia* species. They can divide the sequenced *Burkholderia* genomes into a *B. pseudomallei* group and the Bcc. Additionally, within

Table 2.1 General information for the 110 *Burkholderia* genomes used in this study

Group	Species	Strain	No. of contigs	No. of proteins	BioProject	PID	Sequence source*	Status
Pseudomallei group	*B. pseudomallei*	K96243	2	5947	PRJNA57733	57733	Sanger Institute	Complete
	B. pseudomallei	1026b	2	5960	PRJNA162511	162511	University of Washington	Complete
	B. pseudomallei	1106a	2	5759	PRJNA58515	58515	TIGR	Complete
	B. pseudomallei	1106b	2	5786	PRJNA54361	54361	TIGR	Complete
	B. pseudomallei	1710a	2	5985	PRJNA54285	54285	TIGR	Complete
	B. pseudomallei	1710b	2	5977	PRJNA58391	58391	TIGR	Complete
	B. pseudomallei	668	2	5794	PRJNA58389	58389	TIGR	Complete
	B. pseudomallei	Bp22	2	6001	PRJNA49645	49645	Genome Institute of Singapore	Complete
	B. pseudomallei	MSHR346	2	6046	PRJNA55259	55259	LANL	complete
	B. pseudomallei	576_BUC	21	5954	PRJNA55425	55425	JCVI	draft
	B. pseudomallei	305	36	6170	PRJNA54601	54601	TIGR	draft
	B. pseudomallei	Pakistan_9	71	5933	PRJNA55843	55843	J. Craig Venter Institute	draft
	B. pseudomallei	ATCC15682	133	5788	–	–	Sanger Institute	SRA
	B. pseudomallei	EY7	154	5848	–	–	Sanger Institute	SRA
	B. pseudomallei	P6	167	5752	–	–	Sanger Institute	SRA
	B. pseudomallei	EY1	166	6186	–	–	Sanger Institute	SRA
	B. pseudomallei	S13	169	6140	PRJNA54287	54287	TIGR	draft
	B. pseudomallei	EY2	185	6001	–	–	Sanger Institute	SRA
	B. pseudomallei	EY6	185	5941	–	–	Sanger Institute	SRA
	B. pseudomallei	EY4	194	5997	–	–	Sanger Institute	SRA
	B. pseudomallei	1655	194	5842	PRJNA54283	54283	TIGR	Draft
	B. pseudomallei	Pasteur_52237	217	6089	PRJNA54289	54289	TIGR	Draft
	B. pseudomallei	406e	271	6020	PRJNA54371	54371	TIGR	Draft
	B. pseudomallei	1026a	291	5940	PRJNA73057	73057	University of Washington	Draft

Table 2.1 continued

Group	Species	Strain	No. of contigs	No. of proteins	BioProject	PID	Sequence source*	Status
	B. pseudomallei	354e	338	5863	PRJNA73153	73153	University of Washington	Draft
	B. pseudomallei	1258a	340	5588	PRJNA73147	73147	University of Washington	Draft
	B. pseudomallei	1258b	352	5840	PRJNA73149	73149	University of Washington	Draft
	B. pseudomallei	354a	907	6008	PRJNA73151	73151	University of Washington	Draft
	B. pseudomallei	BCC215	1030	5885	PRJNA54765	54765	NMRC	Draft
	B. pseudomallei	NCTC_13177	1077	6075	PRJNA54767	54767	NMRC	Draft
	B. pseudomallei	112	1274	5927	PRJNA54769	54769	NMRC	Draft
	B. pseudomallei	K11277244–293	1368	5702	–	–	Sanger Institute	SRA
	B. pseudomallei	BP20	1389	5917	–	–	Sanger Institute	SRA
	B. pseudomallei	B7210	1424	5868	PRJNA54773	54773	NMRC	Draft
	B. pseudomallei	EM10266_01	1481	5685	–	–	Sanger Institute	SRA
	B. pseudomallei	BP17	1520	5587	–	–	Sanger Institute	SRA
	B. pseudomallei	7894	1568	6002	PRJNA54771	54771	NMRC	Draft
	B. pseudomallei	15-10	1625	5573	–	–	Sanger Institute	SRA
	B. pseudomallei	P163_04	1639	5746	–	–	Sanger Institute	SRA
	B. pseudomallei	P157_06	1656	5651	–	–	Sanger Institute	SRA
	B. pseudomallei	DR08726_01	1664	5389	–	–	Sanger Institute	SRA
	B. pseudomallei	91	1690	5976	PRJNA54779	54779	NMRC	draft
	B. pseudomallei	4044-138	1710	5601	–	–	Sanger Institute	SRA
	B. pseudomallei	488	1717	5553	–	–	Sanger Institute	SRA
	B. pseudomallei	D107310-3154	1724	5551	–	–	Sanger Institute	SRA
	B. pseudomallei	9	1762	5915	PRJNA54777	54777	NMRC	Draft
	B. pseudomallei	DB	1764	5446	–	–	Sanger Institute	SRA
	B. pseudomallei	79_96	1805	5545	–	–	Sanger Institute	SRA

B. pseudomallei	P166_04	1810	5466	–	–	Sanger Institute	SRA
B. pseudomallei	EY10	1835	5657	–	–	Sanger Institute	SRA
B. pseudomallei	TRF661	1858	5446	–	–	Sanger Institute	SRA
B. pseudomallei	612	1884	5716	–	–	Sanger Institute	SRA
B. pseudomallei	14	1888	5793	PRJNA54781	54781	NMRC	Draft
B. pseudomallei	I64043-3096	1894	5464	–	–	Sanger Institute	SRA
B. mallei	ATCC-23344	2	5001	PRJNA57725	57725	TIGR	Complete
B. mallei	NCTC 10229	2	4913	PRJNA58383	58383	TIGR	Complete
B. mallei	NCTC 10247	2	5016	PRJNA58385	58385	TIGR	Complete
B. mallei	SAVP1	2	4487	PRJNA58387	58387	TIGR	Complete
B. mallei	ATCC_10399	106	5101	PRJNA54279	54279	TIGR	Draft
B. mallei	GB8_horse_4	181	5009	PRJNA54281	54281	TIGR	Draft
B. mallei	JHU	184	4959	PRJNA54293	54293	TIGR	Draft
B. mallei	FMH	205	5023	PRJNA54291	54291	TIGR	Draft
B. mallei	2002721280	208	4918	PRJNA54379	54379	TIGR	Draft
B. mallei	PRL-20	274	4923	PRJNA54653	54653	TIGR	Draft
B. thailandensis	ATCC_700388	2	5396	PRJNA54655	54655	Genome Institute of Singapore	Complete
B. thailandensis	E264	2	5652	PRJNA58081	58081	TIGR	Complete
B. thailandensis	MSMB43	176	6275	PRJNA54775	54775	NMRC	Draft
B. thailandensis	E555	521	5321	PRJNA50871	50871	Genome Institute of Singapore	Draft
B. thailandensis	Bt4	803	5713	PRJNA54787	54787	NMRC	Draft
B. thailandensis	TXDOH	810	5684	PRJNA54795	54795	NMRC	Draft
B. oklahomensis	C6786	633	6104	PRJNA54789	54789	NMRC	Draft
B. oklahomensis	EO147	886	6242	PRJNA54791	54791	NMRC	Draft

Table 2.1 continued

Group	Species	Strain	No. of contigs	No. of proteins	BioProject	PID	Sequence source*	Status
B. cepacia complex (Bcc)	*B. cenocepacia*	AU1054	3	6514	PRJNA58371	58371	DOE	Complete
	B. cenocepacia	HI2424	3	6723	PRJNA58369	58369	DOE	Complete
	B. cenocepacia	J2315	3	7207	PRJNA57953	57953	DOE	Complete
	B. cenocepacia	MC0-3	3	7116	PRJNA58769	58769	DOE	Complete
	B. cenocepacia	H111	71	6867	PRJNA69823	69823	University of Zurich	Draft
	B. cenocepacia	PC184	174	6401	PRJNA54353	54353	Broad Institute	Draft
	B. cenocepacia	K56-2Valvano	1624	6755	PRJNA62783	62783	J. Craig Venter Institute	SRA
	B. cepacia	Bu72	1667	6771	PRJNA61371	61371	DOE	SRA
	B. multivorans	ATCC 17616	3	6205	PRJNA58697	58697	DOE	Complete
	B. multivorans	CGD1	34	5870	PRJNA55347	55347	J. Craig Venter Institute	Draft
	B. multivorans	CGD2	37	5869	PRJNA55349	55349	J. Craig Venter Institute	Draft
	B. multivorans	CGD2M	27	5872	PRJNA55365	55365	J. Craig Venter Institute	Draft
Group	*Species*	Strain	No. of contigs	No. of proteins	BioProject	PID	Sequence source*	Status
	B. multivorans	CF2	1678	5441	PRJNA62779	62779	J. Craig Venter	SRA
	B. multivorans	ATCC BAA-247	1635	5151	PRJNA62777	62777	J. Craig Venter Institute	SRA
	B. ambifaria	AMMD	3	6561	PRJNA58303	58303	DOE	Complete
	B. ambifaria	MC40-6	3	6459	PRJNA58701	58701	DOE	Complete
	B. ambifaria	IOP40–10	329	6872	PRJNA54913	54913	DOE	Draft
	B. ambifaria	MEX_5	539	6003	PRJNA54911	54911	DOE	Draft
	B. dolosa	AUO158	233	5757	PRJNA54351	54351	Broad Institute	Draft
	B. vietnamiensis	G4	3	6507	PRJNA58075	58075	DOE	Complete
	B. ubonensis	Bu	1143	6325	PRJNA54793	54793	NMRC	Draft

	B. lata	383	3	7713	PRJNA58073	58073	DOE	Complete
	Burkholderia spp.	KJ006	3	5736	PRJNA165871	165871	KRIBB	Complete
None	*B. phymatum*	STM815	2	5527	PRJNA58699	58699	DOE	Complete
None	*B. phytofirmans*	PsJN	2	7174	PRJNA58729	58729	DOE	Complete
None	*B. xenovorans*	LB400	3	8703	PRJNA57823	57823	DOE	Complete
None	*B. graminis*	C4D1M	70	6707	PRJNA54887	54887	DOE	Draft
None	*B. gladioli*	BSR3	2	6910	PRJNA66301	66301	Seoul National University	Complete
None	*B. glumae*	BGR1	2	5866	PRJNA59397	59397	Seoul National University	Complete
None	*B. rhizoxinica*	HKI_454	1	2499	PRJNA60487	60487	HKI	Complete
None	*B. terrae*	BS001	330	10360	PRJNA168186	168186	RuG, NL	Draft
Burkholderia spp.	*Burkholderia* spp.	CCGE1001	2	6017	PRJNA42975	42975	DOE	Complete
	Burkholderia spp.	CCGE1002	3	6665	PRJNA42523	42523	DOE	Complete
	Burkholderia spp.	CCGE1003	2	6080	PRJNA46253	46253	DOE	Complete
	Burkholderia spp.	YI23	3	5950	PRJNA81081	81081	Seoul National University	Complete
	Burkholderia spp.	Ch1–1	35	7946	PRJNA48975	48975	DOE	Draft
	Burkholderia spp.	H160_ctg00250	310	7205	PRJNA55101	55101	DOE	Draft
	Burkholderia spp.	SJ98	79	7257	PRJNA160003	160003	Institute of Microbial Technology, India	Draft

*DOE, US Department of Energy Joint Genome Institute; JCVI, J. Craig Venter Institute; TIGR, The Institute of Genome Research; NMRC, Naval Medical Research Centre/ Defence Research Directorate, Genomics, USA; LANL, Los Alamos National Laboratory; HKI, Leibniz Institute for Natural Product Research and Infection Biology; KRIBB, Korea Research Institute of Bioscience and Biotechnology.

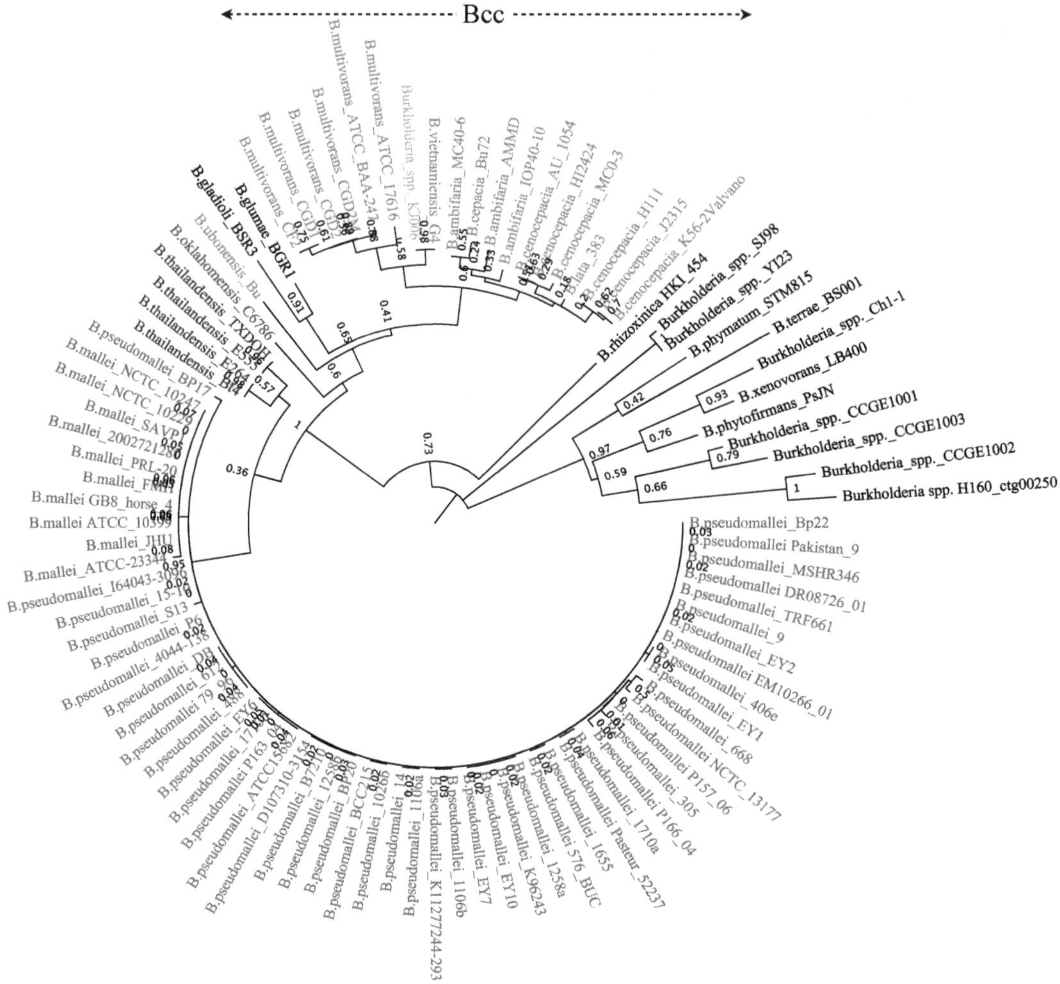

Figure 2.1 Phylogenetic analysis of the 16S rRNA gene from 96 *Burkholderia* genomes. Not all genomes listed in Table 2.1 are included here because the full length of their 16S rRNA gene could not be identified. The tree was generated by the maximum-likelihood method. The numbers next to the branches represent the bootstrap value, which was based on 1000iterations. Species names were labelled according to *Burkholderia* groups; *B. pseudomallei* (pink), *B. mallei* (green), *B. thailandensis* (blue), *B. oklahomensis* (purple), Bcc (orange), *Burkholderia* spp. and the rest are shaded with black. A colour version of this figure is located in the plate section at the back of the book.

the *B. pseudomallei* group, there was a mix-up between *B. pseudomallei* BP17 and some *B. mallei* species, as judged by 16S rRNA sequences (Fig. 2.1); however, this overlap was easily resolved in the MLST tree (Fig. 2.2). The MLST tree also placed *Burkholderia* spp. KJ006 (Table 2.1) adjacent to *B. vietnamiensis* G4 (Fig. 2.2), suggesting it belongs to the latter Bcc species.

Another means of phylogenic depiction, the pan-genome tree, is shown in Fig. 2.3. This tree was constructed from the whole-genomic content of *Burkholderia.* The pan-genome matrix consists of genes (rows) and genomes (columns). The absence and presence of genes across genomes are represented by 0s and 1s, respectively. The relative Manhattan distance between genomes was used for hierarchical clustering (Snipen and Ussery, 2009). The pan-genome tree presents itself as a good illustration of the evolutionary relationships among the *Burkholderia* species. The tree divides *Burkholderia* into the *B. pseudomallei* and Bcc groups, however, *B. gladioli* and *B. glumae,* two

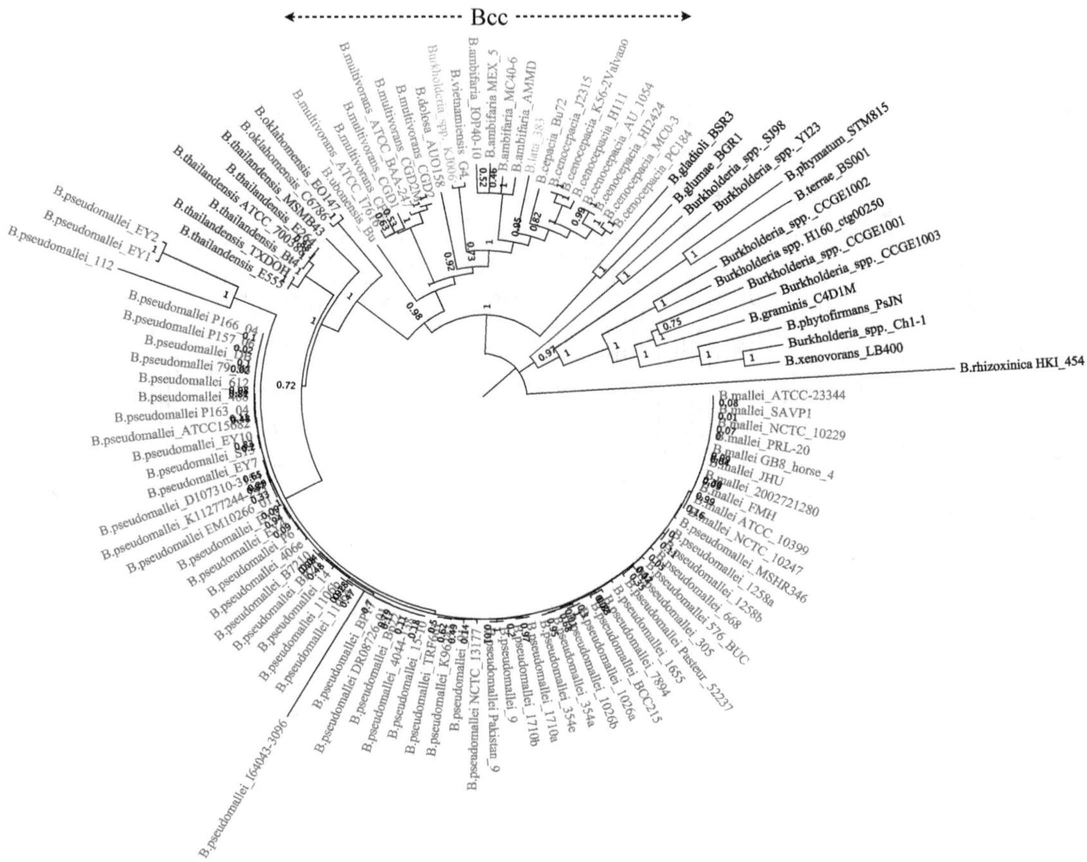

Figure 2.2 *In silico* MLST tree. This tree was constructed from the concatenated sequences of seven *housekeeping* genes used for MLST typing of *B. pseudomallei*. The tree was simulated by MEGA5, with the maximum-likelihood method using 1000 iterations of bootstrapping. The numbers next to the branches represent the bootstrap value. The species groups are labelled similarly to Fig. 2.1. A colour version of this figure is located in the plate section at the back of the book.

non-Bcc species cluster with other Bcc species. It is interesting that analysis of the seven genes in the MLST tree provides a tighter grouping of Bcc species (Fig. 2.2) compared to the pan-genome tree (Fig. 2.3). In addition, *B. pseudomallei*, *B. mallei* and other species are grouped properly according to their taxonomy.

Core/pan-genome plot

Determining gene conservation across multiple genomes is not overly difficult, but certain choices have to be made which will affect the final outcome. In one of the pioneering works to determine gene conservation across genomes, Tettelin *et al.* (2005) performed three separate Smith–Waterman alignments on each genome pair: the proteome of one versus the proteome of the other, the DNA sequence of all open reading frames (ORFs) of one versus the ORFs of the other, and a translated protein alignment against the entire nucleotide sequence of the other genome. For a gene to be considered conserved it needed to produce a hit in at least one of the three alignments, defined as having a minimum of 50% sequence identity over 50% of the sequence length. While this method worked for comparing eight genomes (as in their original paper), the speed for these comparisons unfortunately does not scale well, such that comparison of as few as 50 genomes can be quite time-consuming. To put this in proper perspective, probably thousands of

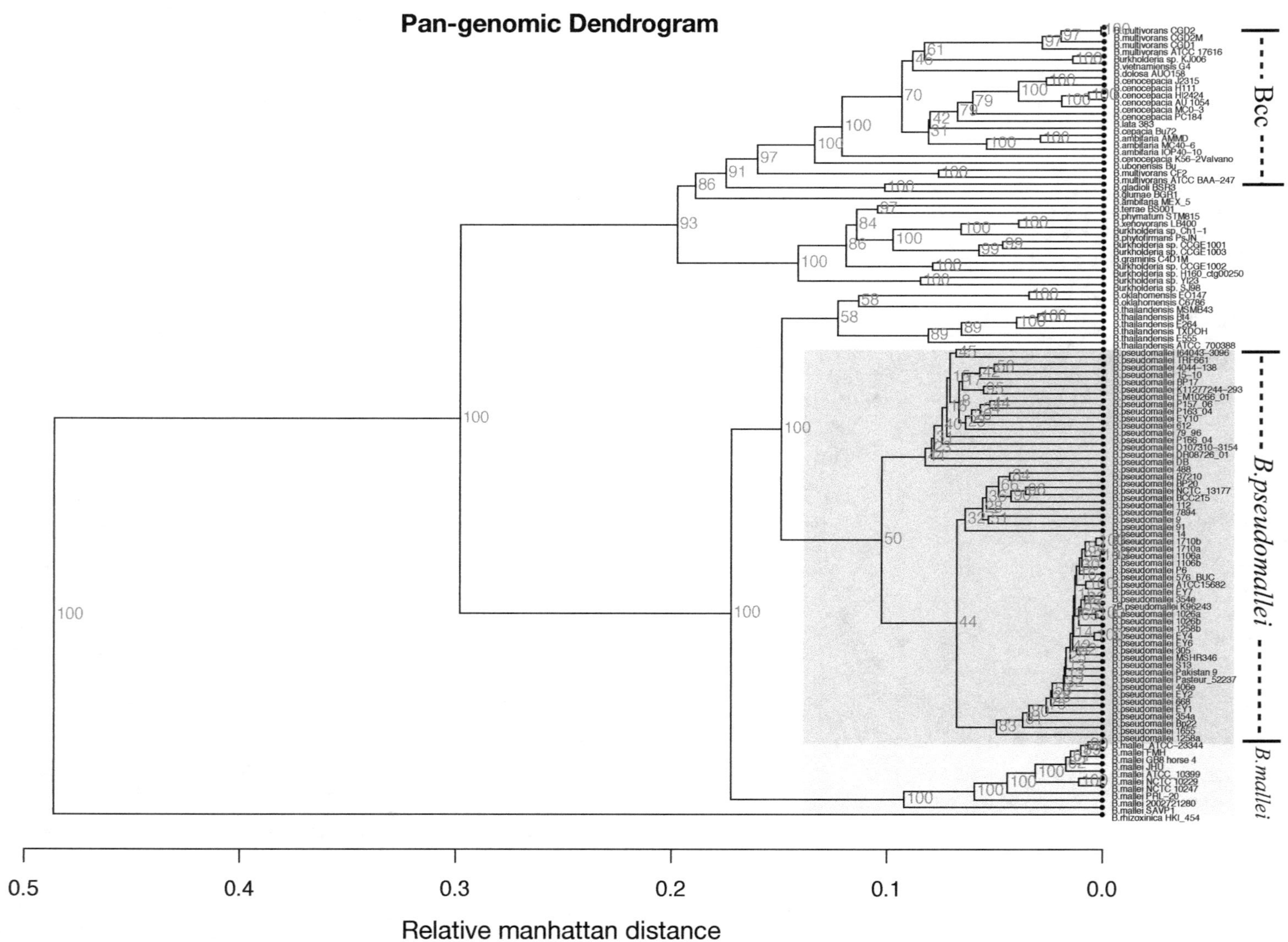
Pan-genomic Dendrogram
Bcc
B.pseudomallei
B.mallei
0.5
0.4
0.3
0.2
0.1
0.0
Relative manhattan distance

Figure 2.3 (opposite page) A pan-genome family tree of 110 *Burkholderia* genomes. This is an alternative approach to construct a phylogenetic tree based on all the gene contents in the genome (pan-genome). The tree was generated from the presence or absence of gene families across all *Burkholderia* genomes. The red values next to the branches are bootstrap values. A colour version of this figure is located in the plate section at the back of the book.

genome sequences would need to be compared in epidemiological studies, and this method, as originally stated, simply would take too much time (years of computing).

Using protein BLAST alignments (BLASTP), but still retaining the 50/50 cut-off requirement, a much faster approximation of Tettelin's approach can be achieved (Ussery *et al.*, 2008). Using this method on the 110 currently available *Burkholderia* genomes, a pan- and core-genome plot was constructed. Any two genes having a minimum of 50% sequence identity over 50% of the sequence length were considered as belonging to the same gene family (Ussery and Wassenaar, 2008). The gene families that were found in all genomes were collected as the core-genome and all others were considered part of the larger pan-genome (Friis *et al.*, 2010; Leekitcharoenphon *et al.*, 2012). The pan- versus core-genome plot of 110 *Burkholderia* genomes is shown in Fig. 2.4, where it can be seen that the number of new gene families in the 'pan-genome' gradually increases when adding more genomes, while the number of conserved gene families or 'core-genome' reciprocally decreases slightly. Based on this set of 110 *Burkholderia* genomes, there are 35,680 pan-gene families and 717 core-gene families.

The number of core gene families being unique to each *Burkholderia* species was identified and summarized in Fig. 2.5. The species *B. pseudomallei* (54 genomes sequenced) contains 10 unique core-gene families, while *B. mallei* (10 genomes) possesses 10 unique core-gene families (the unique core-gene sequences can be obtained from the corresponding author). These unique core-genes are potentially highly useful as target genes to distinguish different *Burkholderia* species.

Multiple replicon genomes

The core/pan-genome plots of the first and second replicons (chromosomes) of nine *B. pseudomallei* strains are illustrated in Fig. 2.6. The number of core genes from the same replicon decreases steadily when adding more genomes. However, the core number drops dramatically when adding the genes from different replicons, suggesting that the chromosomes share few genes in common. There are 2813 and 1913 unique core genes for the first and second replicon, respectively (Table 2.2).

A number of essential biological elements was determined from the unique core genes. Ribosomal and sigma factor proteins were identified using a pfam database (http://pfam.sanger.ac.uk/) (Punta *et al.*, 2012). The tRNA genes were determined using tRNAscan (Fichant and Burks, 1991). Membrane proteins were identified using TMHMM (http://www.cbs.dtu.dk/services/TMHMM/) (Krogh *et al.*, 2011). Interestingly, the first chromosome contains 57 ribosomal protein genes (2% of its total unique core genes); meanwhile, the second replicon holds only 4 ribosomal proteins (0.2%). In contrast, both replicons consist of an equal number of tRNA, membrane protein and sigma factor genes (Table 2.2).

The core/pan-genome plots of the three replicons/chromosomes of five Bcc members (four *B. cenocepacia* and one *B.* multivorans) are shown in Fig. 2.7. Similar to the plot in Fig. 2.6, the number of the Bcc core genes drops sharply when a different replicon genome is added. There are 1901, 1160 and 90 unique core genes for the first, second and third replicons, respectively (Table 2.2). Again, the first chromosome contains the vast majority of the ribosomal proteins (49 proteins, or 2.6% of the sequence) compared to the other chromosomes. In addition, the third replicon's lack of important elements, such as ribosomal and sigma factor proteins, suggests that the third replicon might not be biologically essential to the Bcc species. This assumption is supported by the recent study by Agnoli *et al.* (Agnoli *et al.*, 2012). They demonstrated that the third replicon of *Burkholderia cepacia* complex species could be deleted while still retaining viability, showing

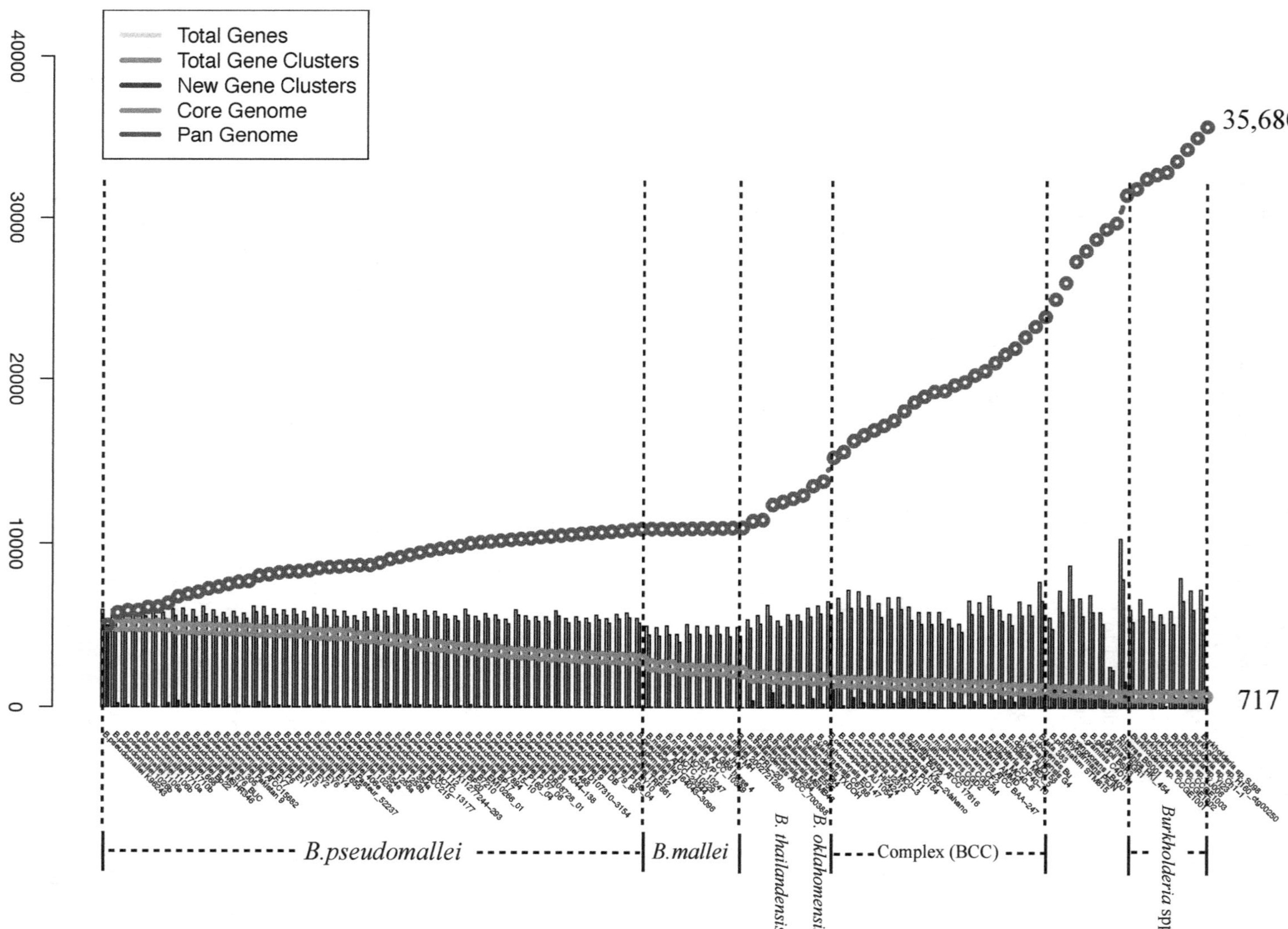

Total Genes
Total Gene Clusters
New Gene Clusters
Core Genome
Pan Genome
35,680
717
40000
30000
20000
10000
0
B.pseudomallei
B.mallei
B. thailandensis
B. oklahomensis
Complex (BCC)
Burkholderia spp.

Figure 2.4 (opposite page) Core/pan-genome plot of 110 *Burkholderia* genomes. The increasing number of genes in the pan-genome and core-genome are shown in blue and red dots, respectively. The last points express the total number of pan- (35,680 gene families) and core- genomes (717 gene families). A colour version of this figure is located in the plate section at the back of the book.

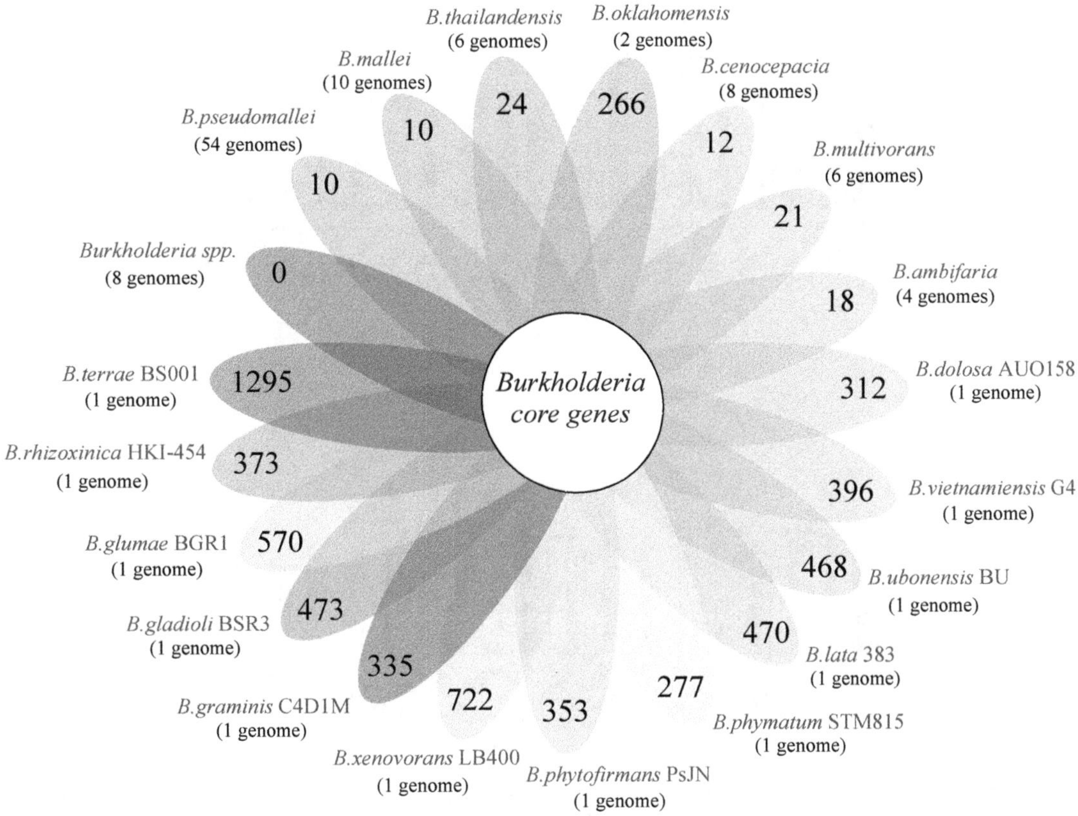

Figure 2.5 FlowerPlot of unique core gene families in each *Burkholderia* species. The figure represents the number of core gene families found in each genome as being unique to the *Burkholderia* species. The coloured petals represent different *Burkholderia* groups or species. A colour version of this figure is located in the plate section at the back of the book. The number of sequenced genomes used for each species is given in parenthesis below the name.

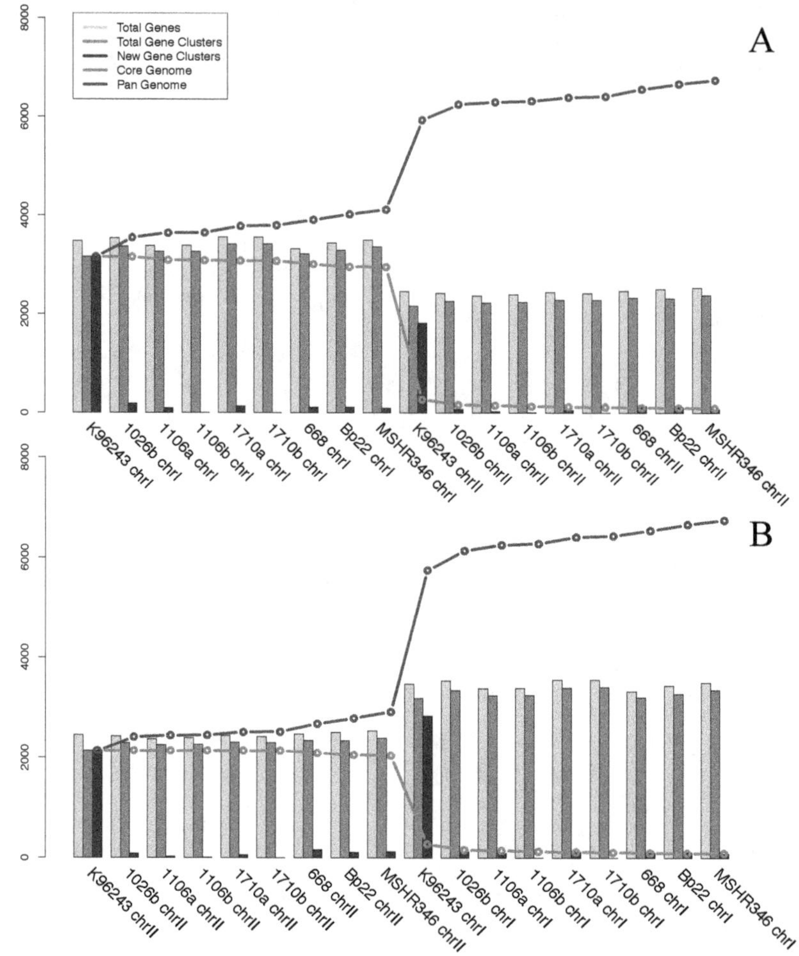

Figure 2.6 Core/pan-genome plot of nine *Burkholderia pseudomallei* genomes. (A) The plot starts with the first replicon. (B) The plot starts with the second chromosome. The increasing numbers of genes in the pan-genome and core-genome are shown in blue and red dots, respectively. A colour version of this figure is located in the plate section at the back of the book.

Table 2.2 A number of essential biological elements that has been found within the core genes from multiple replicon genomes

Genomes	Core genes	Ribosomal proteins	tRNAs	Membrane proteins	Sigma factors
B. pseudomallei chrI	2813	57	164	679	14
B. pseudomallei chrII	1913	4	135	460	11
Bcc (*B. cenocepacia* and *B. multivorans*) chrI	1901	49	99	466	15
Bcc (*B. cenocepacia* and *B. multivorans*) chrII	1160	1	62	300	4
Bcc (*B. cenocepacia* and *B. multivorans*) chrIII	90	0	4	18	0

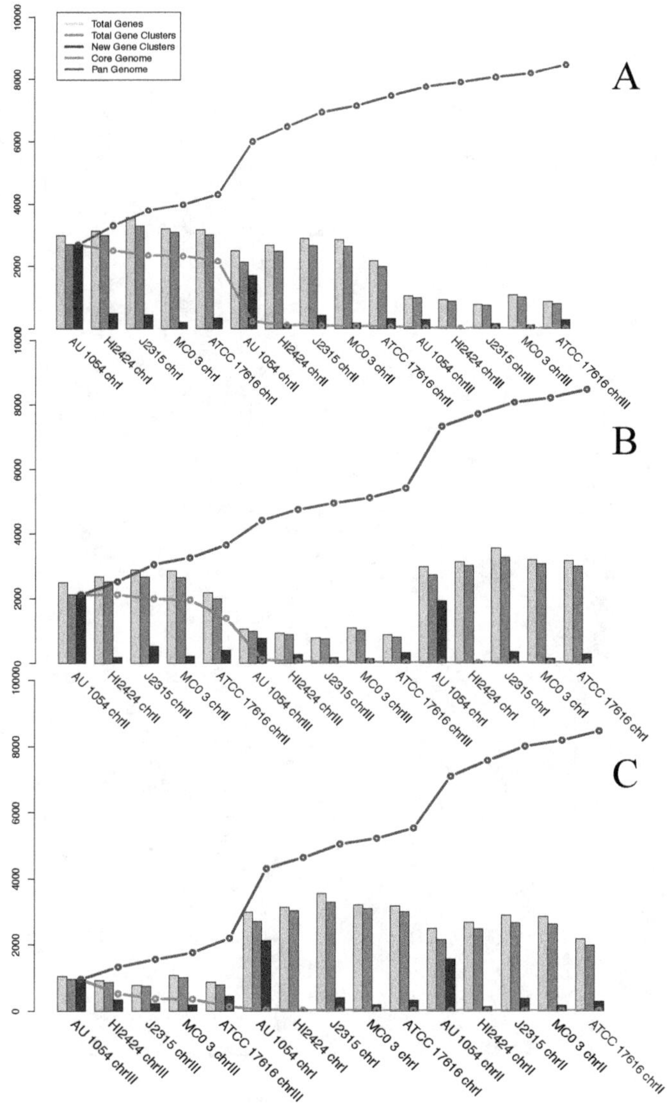

Figure 2.7 Core/pan-genome plot of four *Burkholderia cenocepacia* plus one *Burkholderia multivorans* genomes. (A) The plot starts from the first replicon, (B) the plot starts from the second chromosome, and (C) the plot starts from the third replicon. The increasing numbers of genes in the pan-genome and core-genome are shown in blue and red dots, respectively. A colour version of this figure is located in the plate section at the back of the book.

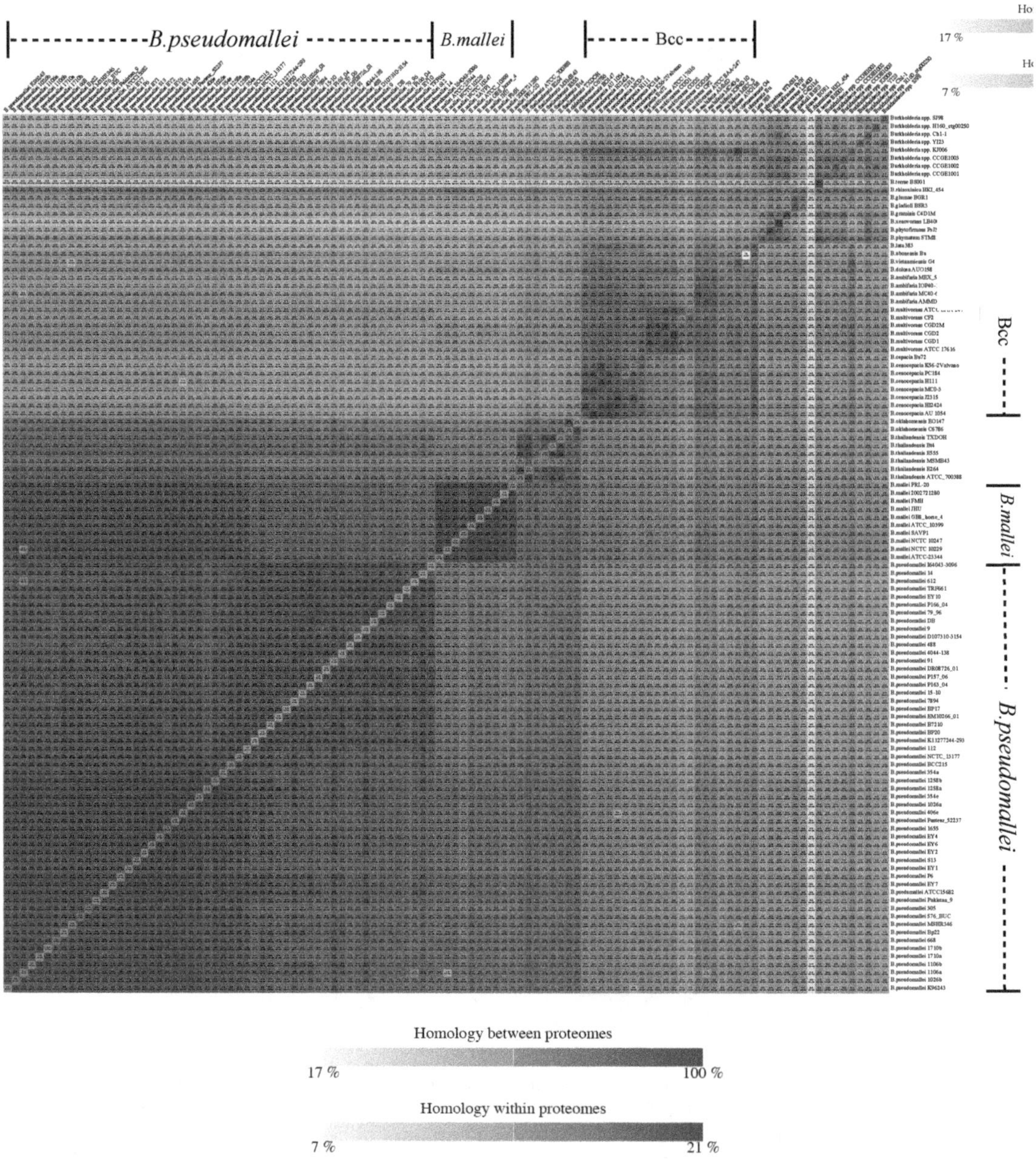

Figure 2.8 BLAST Matrix of 110 *Burkholderia* genomes. The figure shows the percentage of gene families found in common between any pair-genomes (pink). The blue colour represents the degree of gene duplication within each *Burkholderia* genome. The pairwise all-against-all alignment was performed by reciprocal BLASTP at the amino acid level. A colour version of this figure is located in the plate section at the back of the book.

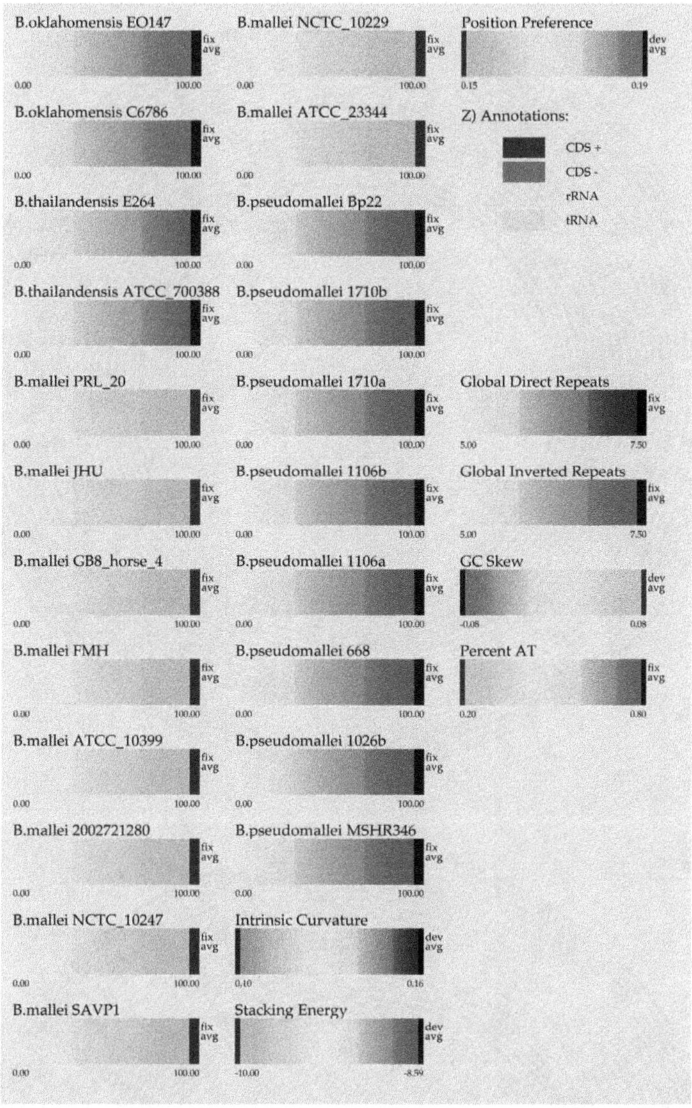

Figure 2.9 BLAST Atlas of eight *B. pseudomallei*, ten *B. mallei*, two *B. thailandensis* and two *B. oklahomensis*. The figure shows the BLAST atlas with *B. pseudomallei* K96243 chromosome I (above) and chromosome II (overleaf) as references (continued on next page). A colour version of this figure is located in the plate section at the back of the book.

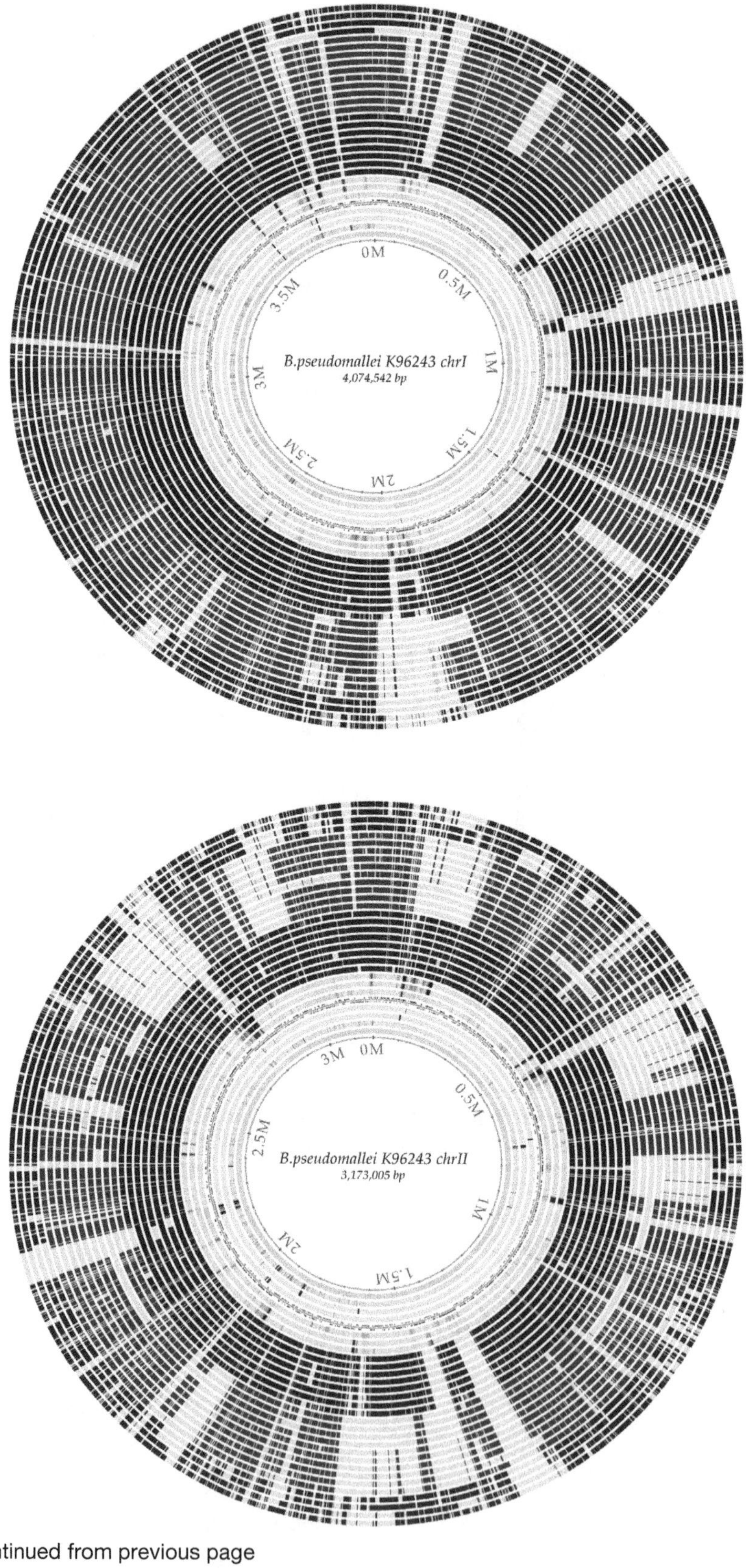

Figure 2.9 Continued from previous page

the chromosome/replicon III was an accessory genomic element that encoded several diverse, but non-vital, functions.

BLAST matrix

A BLAST matrix can show a comparison of one proteome to another. All 110 proteomes were compared, using reciprocal BLASTP and the 50/50 rule, to categorize the genes into genes families. The percentages represent the proportion of proteins shared between two proteomes, along with the corresponding fraction showing the number of gene families present in both genomes over the total amount of gene families in the two strains (Binnewies *et al.*, 2005). The BLAST matrix of 110 *Burkholderia* genomes is shown in Fig. 2.8. The similarity of gene families between any two *Burkholderia* genomes is above 17% while the similarity of the gene families within a genome (for example, duplicate genes) is less than 21%. The matrix also illustrates the high conservation of gene families within species *B. pseudomallei, B. mallei, B. thailandensis* and *B. cenocepacia* (Fig. 2.8, dark pink).

BLAST atlas of *B. pseudomallei* and *B. mallei* species

The BLAST atlases are the visualization of gene similarity between *B. pseudomallei* and *B. mallei*, including *B. thailandensis* and *B. oklahomensis*, and are illustrated in Fig. 2.9. The proteome of the *B. pseudomallei* and *B. mallei* strains were aligned against a reference genome (*B. pseudomallei* K96243). The colour intensity represents proteome conservation. The BLAST atlases for both chromosomes (especially for the chromosome II) show that all of the proteome of the test *B. mallei* is present in the reference *B. pseudomallei*. This is further evidence supporting the theory that *B. mallei* is a descendent of *B. pseudomallei* and has undergone a remarkably extensive genome reduction, accounting for a loss of as much as 20% of its genome (Losada *et al.*, 2010).

Conclusions

We have examined more than 100 *Burkholderia* genomes, and have compared them with each other and in three broad groups, which cluster together in several different types of trees. In the case of *B. pseduomallei* (with nine complete genomes sequenced), there are about 5000 core genes, with roughly 2800 genes that are consistently on chromosome I, and another 1900 genes found on chromosome II. Closer examination of the *B. mallei* versus *B. pseudomallei* chromosomes reveal several large regions of missing genes in *B. mallei*, in particular in chromosome II. For most of the *Burkholderia* species for which whole-genome sequences are available, we were able to identify taxa-specific genes.

Acknowledgement

We thank the several sequencing centres that have provided finished and unfinished genomic data. We also acknowledge the grant from the Centre for Genomic Epidemiology (09- 067103/DSF) (http://www.genomicepidemiology.org.)

References

Agnoli, K., Schwager, S., Uehlinge,r S., Vergunst, A., Viteri, D.F., Nguyen, D.T., Sokol, P.A., and Carlier, A. (2012). Exposing the third chromosome of *Burkholderia cepacia* complex strains as a virulence plasmid. Mol. Microbiol. *83*, 362–378.

Binnewies, T.T., Hallin, P.F., Staerfeldt, H.-H., and Ussery, D.W. (2005). Genome Update: proteome comparisons. Microbiology *151*, 1–4.

Bontemps, C., Elliott, G.N., Simon, M.F., Dos Reis Junior, F.B., Gross, E., Lawton, R.C., Neto, N.E., De Fatima Loureiro, M., De Faria, S.M., Sprent, J.I., *et al.* (2010). Burkholderia species are ancient symbionts of legumes. Mol. Ecol. *19*, 44–52.

Deshazer, D. (2007). Virulence of clinical and environmental isolates of *Burkholderia oklahomensis* and *Burkholderia thailandensis* in hamsters and mice. FEMS Microbiol. Lett. *277*, 64–69.

Estes, D.M., Dow, S.W., Schweizer, H.P., and Torres, A.G. (2010). Present and future therapeutic strategies for melioidosis and glanders. Expert. Rev. Anti Infect. Ther. *8*, 325–338.

Fichant, G., and Burks, C. (1991). Identifying potential tRNA genes in genomic DNA sequences. J. Mol. Biol. *220*, 659–671.

Friis, C., Wassenaar, T.M., Javed, M.A., Snipen, L., Lagesen, K., Hallin, P.F., Newell, D.G., Toszeghy,

M., Ridley, A., Manning, G., *et al.* (2010). Genomic characterization of Campylobacter jejuni strain M1. PloS One *5*, e12253.

Galyov, E.E., Brett, P.J., and DeShazer, D. (2010). Molecular insights into *Burkholderia pseudomallei* and *Burkholderia mallei* pathogenesis. Annu. Rev. Microbiol. *64*, 495–517.

Glass, M.B., Beesley, C.A., Wilkins, P.P., and Hoffmaster, A.R. (2009). Comparison of four selective media for the isolation of *Burkholderia mallei* and *Burkholderia pseudomallei*. Am. J. Trop. Med. Hyg. *80*, 1023–1028.

Krogh, A., Larsson, B., von Heijne G., and Sonnhammer, E.L. (2011). Predicting transmembrane protein topology with a hidden Markov model: application to complete genomes. J. Mol. Biol. *305*, 567–580.

Lagesen, K., Hallin, P., Rødland, E.A., Stærfeldt, H.-H., Rognes, T., and Ussery, D.W. (2007). RNAmmer: consistent and rapid annotation of ribosomal RNA genes. Nucleic Acids Res. *35*, 3100–3108.

Larsen, J., and Johnson, N. (2009). Pathogenesis of *Burkholderia pseudomallei* and *Burkholderia mallei*. Mil. Med. *174*, 647–651.

Leekitcharoenphon, P., Lukjancenko, O., Friis, C., Aarestrup, F.M., and Ussery, D.W. (2012). Genomic variation in *Salmonella enterica* core genes for epidemiological typing. BMC Genomics *13*, 88.

Losada, L., Ronning, C.M., DeShazer, D., Woods, D., Fedorova, N., Kim, H.S., Shabalina, S.A., Pearson, T.R., Brinkac, L., Tan, P., *et al.* (2010). Continuing evolution of *Burkholderia mallei* through genome reduction and large-scale rearrangements. Genome Biol. Evol. *2*, 102–116.

Lynch, J.P. (2009). *Burkholderia cepacia* complex: impact on the cystic fibrosis lung lesion. Semin. Respir. Crit. Care Med. *30*, 596–610.

Punta, M., Coggill, P.C., Eberhardt, R.Y., Mistry, J., Tate, J., Boursnell, C., Pang, N., Forslund, K., Ceric, G., Clements, J., *et al.* (2012). The Pfam protein families database. Nucleic Acids Res. *40*, 290–301.

Snipen, L., and Ussery, D.W. (2009). Standard operation procedure for computing pangenome trees. Stand. Genomics Sci. *2*, 135–141.

Tamura, K., Peterson, D., Peterson, N., Stecher, G., Nei, M., and Kumar, S. (2011). MEGA5: molecular evolutionary genetics analysis using maximum likelihood, evolutionary distance, and maximum parsimony methods. Mol. Biol. Evol. *28*, 2731–2739.

Tettelin, H., Masignani, V., Cieslewicz, M.J., Donati, C., Medini, D., Ward, N.L., Angiuoli, S.V., Crabtree, J., Jones, A.L., Durkin, A.S., *et al.* (2005). Genome analysis of multiple pathogenic isolates of *Streptococcus agalactiae*: implications for the microbial 'pan-genome'. Proc. Natl. Acad. Sci. U.S.A. *102*, 13950–13955.

Ussery, D., Wassenaar, T.M., Borini, S (2008). Computing for Comparative Microbial Genomics: Bioinformatics for Microbiologists (Computational Series) (London: Springer Verlag London).

Vial, L., Chapalain, A., Groleau, M.C., and Déziel, E. (2011). The various lifestyles of the *Burkholderia cepacia* complex species: a tribute to adaptation. Environ. Microbiol. *13*, 1–12.

Molecular Epidemiology and Population Structure of Clinically Relevant *Burkholderia* Species

3

Tom Coenye, Eshwar Mahenthiralingam and John J. LiPuma

Abstract

The aim of this chapter is to provide a comparative overview of the various methods used in epidemiological studies of *Burkholderia* species, including ribotyping, pulsed-field gel electrophoresis of genomic macrorestriction fragments, randomly amplified polymorphic DNA fingerprinting, Rep- and BOX-PCR fingerprinting, multilocus-based approaches and variable-number tandem repeat analysis. Subsequently, the epidemiology of the most-commonly encountered *Burkholderia* species is discussed. The focus will be on the *Burkholderia cepacia* complex and infection in people with cystic fibrosis and the epidemiology of *Burkholderia pseudomallei* and *Burkholderia mallei*. Recent advances in our understanding of the population genetics of *B. cepacia* complex and *B. pseudomallei*, based on results obtained with multilocus-based approaches like multilocus sequence typing and multilocus restriction typing is also discussed. In a final section we describe the use of whole genome sequencing and typing as novel approach and illustrate its potential with several examples.

Introduction

Interest in the epidemiology of members of the genus *Burkholderia*, comprising over 70 validly described species occurring in a wide range of ecological niches, has increased tremendously in the last decades. In this chapter we will first provide an overview of genotypic methods used in epidemiological studies of *Burkholderia* species and subsequently discuss the epidemiology of the most commonly encountered *Burkholderia* species. The primary focus will be on the *Burkholderia cepacia* complex and infection in people with cystic fibrosis (CF), but we will also discuss the epidemiology of other important pathogens (including *Burkholderia pseudomallei* and *Burkholderia mallei*). In addition, we will also present recent advances in our understanding of the population genetics of various *Burkholderia* species. In a final section we will discuss the use of whole genome sequences in studies of the molecular epidemiology of *Burkholderia* species.

Methods for epidemiological studies of *Burkholderia* species

Phenotypic methods

Numerous phenotypic typing methods have been used to establish relationships between *Burkholderia* strains, including serotyping, antimicrobial susceptibility typing, phage typing and bacteriocin typing (Rabkin *et al.*, 1987, 1989; Arbeit, 1995). In general, these methods suffer from technical complexity, poor reproducibility and/or relatively low discriminatory power and have been replaced by genotypic methods (Arbeit, 1995; LiPuma, 1998; Olive and Bean, 1999; Zaidi *et al.*, 2003).

Genotypic methods

Ribotyping

In ribotyping, chromosomal restriction fragment length polymorphisms (RFLP) are detected by probing restriction digested genomic DNA with rRNA (Stull *et al.*, 1988; Bingen *et al.*, 1994) and

RFLPs result mainly from nucleotide sequence variation in regions flanking rRNA genes. A modification of ribotyping, referred to as PCR-ribotyping, employs PCR to amplify the 16S–23S intergenic spacer region of the bacterial rRNA operon to detect sequence length polymorphisms therein (Kostman *et al.*, 1995). Both ribotyping and PCR-ribotyping were used in early investigations of *B. cepacia* complex epidemiology (LiPuma *et al.*, 1988, 1990, 1994; Dasen *et al.*, 1994), but have since been replaced by methods with higher discriminatory power.

Pulsed-field gel electrophoresis of genomic macrorestriction fragments

Pulsed-field gel electrophoresis (PFGE) of macrorestriction fragments of chromosomal DNA has long been considered to be the gold standard in typing of bacterial pathogens (Arbeit, 1995; Tenover *et al.*, 1995; Olive and Bean, 1999). It has been applied to numerous studies of the molecular epidemiology of *B. cepacia* complex (see, for example, Segonds *et al.*, 1997; Vandamme *et al.*, 2000; Agodi *et al.*, 2001; Chen *et al.*, 2001; Coenye and LiPuma, 2002; Biddick *et al.*, 2003; De Boeck *et al.*, 2004; Drevinek *et al.*, 2005) and *B. pseudomallei* (see, for example, Vadivelu *et al.*, 1997; Koonpaew *et al.*, 2000; Inglis *et al.*, 2002; Cheng *et al.*, 2005). The most frequently used restriction enzymes for *Burkholderia* spp. are *Xba*I and *Spe*I. A recent study in which 146 clinical *B. pseudomallei* isolates were included demonstrated that at least for this species the use of *Spe*I results in a higher discriminatory power than the use of *Xba*I (Chua *et al.*, 2011). The major drawbacks of macrorestriction of chromosomal DNA followed by PFGE are that it is time-consuming and requires specialized (relatively expensive) equipment. This is due to the inability of standard DNA isolation procedures to recover DNA fragments that are large enough to be suitable for analysis (which makes it necessary to isolate DNA from cells embedded in an agarose matrix, a slow procedure) and the long electrophoresis times required for PFGE. The entire procedure can easily take up to four days and overall PFGE suffers from a lack of portability between different laboratories (Arbeit, 1995; Olive and Bean, 1999). In addition, little is known about the stability of macrorestriction patterns. For example, Drevinek *et al.* (2010a) recently showed that oxidative stress induced genomic rearrangements (through movement of insertion sequences) in *B. cenocepacia*. These rearrangements can result in changes in PFGE patterns and lead to an overestimation of diversity and misinterpretation of relationships between strains.

Randomly amplified polymorphic DNA fingerprinting

PCR-based fingerprinting using short random primers (randomly amplified polymorphic DNA fingerprinting, or RAPD) is a widely used alternative to PFGE (van Belkum, 1994; Arbeit, 1995; Power, 1996; LiPuma, 1998; Olive and Bean, 1999). This method has been used extensively for molecular typing of *B. cepacia* complex (Bingen *et al.*, 1993; Liu *et al.*, 1995; Miyawake *et al.*, 1995; Mahenthiralingam *et al.*, 1996; Segonds *et al.*, 1997; Okazaki *et al.*, 1999; Chen *et al.*, 2001) and *B. pseudomallei* (Haase *et al.*, 1995; Leelayuwat *et al.*, 2000; Radua *et al.*, 2000). While considerably cheaper, easier to perform and faster than PFGE, RAPD fingerprinting has been found to be less reproducible (Coenye *et al.*, 2002).

Rep-PCR fingerprinting

The genomes of most bacteria contain specific conserved repetitive sequences. Versalovic *et al.* (1991) showed that these repeats can be used to amplify the intervening DNA sequences. Three main sets of repeats have been used for typing: (i) repetitive extragenic palindromic (REP) elements (35–40 bp), (ii) enterobacterial repetitive intergenic consensus (ERIC) sequences (124–127 bp), and (iii) BOX elements (154 bp) (Olive and Bean, 1999). Depending on which sequences are targeted, these assays are referred to as REP-, ERIC- or BOX-PCR, respectively. Collectively, these methods are referred to as rep-PCR. Targets most commonly used for the typing of *B. cepacia* complex organisms are the ERIC (Liu *et al.*, 1995; Cimolai and Trombely, 1996; Seo and Tsuchiya, 2005) and BOX elements (Chen *et al.*, 2001; LiPuma *et al.*, 2002; Biddick *et al.*, 2003; Coenye *et al.*, 2004; Seo and Tsuchiya, 2005). BOX-PCR has also been used to type *B. pseudomallei* (Currie *et al.*, 2007a). Overall, rep-PCR shows good reproducibility and is relatively easy to use. Its discriminatory power is somewhat lower than that of PFGE and RAPD fingerprinting (Coenye *et al.*, 2002).

Multilocus-based approaches

Most of the methods described above are well suited to compare relatively small sets of isolates obtained during outbreak investigations; their utility for answering questions regarding long-term epidemiology or population structure, however, is limited (Maiden *et al.*, 1998; Enright and Spratt, 1999). Historically, multilocus enzyme electrophoresis (MLEE) was the tool used to address these questions. In MLEE allelic variation in sets of selected housekeeping genes is indexed through the electrophoretic mobility of the corresponding enzymes (Musser, 1996). Several studies have documented the utility of MLEE to assess the molecular epidemiology of *B. cepacia* complex isolates (McArthur *et al.*, 1988; Carson *et al.*, 1991; Johnson *et al.*, 1994; Wise *et al.*, 1995, 1996). MLEE has also been used in small-scale studies regarding the epidemiology and grouping of *B. pseudomallei* (Norton *et al.*, 1998) and *Burkholderia unamae* (Caballero-Mellado *et al.*, 2004).

In the late 1990s, multilocus sequence typing (MLST) emerged as a powerful and portable replacement to MLEE (Maiden *et al.*, 1998). With MLST, allelic variation at chromosomal loci is examined by direct nucleotide sequencing of approximately 500 bp fragments of several (typically seven) housekeeping genes. The sequence of each gene fragment can be compared to all known alleles to obtain an allelic profile that unambiguously defines a strain (Feil and Enright, 2004). An important aspect of MLST is that, unlike conventional phylogenetic analyses, the number of nucleotide differences between alleles is ignored and sequences are given different allele numbers whether they differ at a single nucleotide site or at many sites (Maiden *et al.*, 1998; Spratt, 1999). This is because a single genetic event resulting in a new allele can occur by a point mutation (altering only a single nucleotide site), or by a recombinational replacement (that will often change multiple sites). Weighting according to the number of nucleotide differences between alleles would imply that an allele altered by recombination is more distantly related to the original allele than an allele altered by point mutation. This would be true if all nucleotide changes occurred by mutation, but not if changes can also be caused by a recombinational replacement. In MLST, different sequences for each locus are assigned an arbitrary integer; and each unique allelic profile (consisting of the allele designation for each locus) is also assigned an arbitrary integer (the sequence type, ST). Alternatively, the sequences of all housekeeping genes can be concatenated into a single sequence that can then be used to assess relationships between isolates.

Currently MLST schemes are available for a large number of bacterial pathogens (Urwin and Maiden, 2003; Feil and Enright, 2004; Maiden, 2006; see also http://www.mlst.net and http://pubmlst.org/). The major advantages of MLST over most other typing methods is its high discriminatory power, the unambiguous nature of the data and the ease with which these sequence data can be stored and transmitted (Feil and Enright, 2004). Within the genus *Burkholderia*, MLST schemes have been developed for *B. pseudomallei* (this scheme also includes *Burkholderia thailandensis* and *B. mallei*) (Godoy *et al.*, 2003) and the *B. cepacia* complex (Baldwin *et al.*, 2005). In order to keep up with the increasing taxonomic complexity of the *B. cepacia* complex, and to be able to include other *Burkholderia* species, an 'expanded' MLST scheme was developed and published in 2009 (Spilker *et al.*, 2009). The use of degenerate primers allowed expansion of the *B. cepacia* complex MLST scheme to include other clinically relevant (e.g. *Burkholderia gladioli*) or important environmental (e.g. *Burkholderia glumae*) species. In addition, the *B. cepacia* complex MLST scheme was shown to allow culture-independent strain typing directly form sputum, a huge advantage compared to several pattern-based methods that require pure culture (e.g. PFGE) (Drevinek *et al.*, 2010b).

An alternative to MLST is multilocus restriction typing (MLRT) in which variation at several loci is indexed by restriction analysis of PCR-amplified genes. A MLRT scheme has been developed for *Burkholderia cenocepacia* (Coenye and LiPuma, 2002). However, the continuously decreasing cost of sequencing, the higher resolution of sequencing-based approaches, and the lack of portability of MLRT data make it unlikely this method will continue to be used.

In the near future, the improvements in genome sequencing and use of next generation sequencing technologies will likely supersede MLST as conventionally performed with 7 loci being amplified and sequenced individually. Genome sequencing methods for genotyping and studying strain evolution are described in more detail below. However, for researchers lacking resources to perform full MLST analysis with 7 genes, partial MLST using two genes can serve as rapid and cost effective means of firstly identifying the species of a *Burkholderia* and secondly, determining whether the strain is likely to be unique in the context of genotypes already deposited in the MLST database (Fig. 3.1). The two MLST genes most suited to this form of partial MLST analysis are (i) *recA*, as an accurate and widely used taxonomic marker for both the Bcc (Mahenthiralingam *et al.*, 2000; Baldwin *et al.*, 2005) and *Burkholderia* (Payne *et al.*, 2006; Spilker *et al.*, 2009), and (ii) *gyrB*, as the MLST gene with the greatest number of alleles and polymorphic sites (Baldwin *et al.*, 2005; Table 3.2), and hence the most discriminatory genes in terms of providing some differentiation below the species level. While this partial analysis is not completely discriminatory to the strain level, it will provide accurate *Burkholderia* species identification and allows the likelihood of the strain being novel or belonging to an epidemic lineage to be established. The sequences of the amplified *recA* and *gyrB* genes can be rapidly searched at the MLST database to accurately identify the *Burkholderia* species and potential strain matches (Fig. 3.1). The outcome of this partial analysis can then be assessed in the context of the resulting matches or novelty of these two MLST genes and a decision to perform full MLST made.

Variable-number tandem repeat analysis

In certain genetically homogeneous species like *Bacillus anthracis* or *Mycobacterium tuberculosis*, the low level of nucleotide variation prevents the use of classical MLST schemes for epidemiological investigations (Cooper and Feil, 2004). To assess the intraspecies diversity of these homogeneous species, alternative genomic methods have been developed. In variable-number tandem repeat (VNTR) analysis, the variation in copy number of short nucleotide sequences that are repeated multiple times (resulting in length polymorphisms) is detected by PCR (van Belkum *et al.*, 1997; Frothingham *et al.*, 1998; Keim *et al.*, 2000). The resolution of this technique can be further improved by simultaneously studying multiple loci (multiple-locus VNTR analysis, MLVA) (Keim *et al.*, 2000; Johansen *et al.* 2004; Schouls *et al.*, 2004). It should be noted that this method has also been used to type strains of more variable species, like *Pseudomonas aeruginosa* (Onteniente *et al.*, 2003).

In 2006, Liu *et al.* published a MLVA procedure for *B. pseudomallei* based on five VNTR loci. The loci used were selected from 32 VNTR loci that were identified in the *B. pseudomallei* K96243 genome. The method was subsequently used to investigate an outbreak in Singapore and showed that 30 different VNTR types could be found among 32 clinical isolates, excluding the possibility that the outbreak was due to a single source. Shortly after this initial publication, U'Ren *et al.* (2007a) identified additional polymorphic loci in the *B. pseudomallei* genome (104 in total), and selected 32 loci to be used in an expanded MLVA typing scheme. Selection of the loci was based on variation within the screening panel (consisting of *B. pseudomallei*, *B. mallei* and *B. thailandensis*), robust PCR amplification (i.e. success rate > 80%) and highly discrete PCR amplicon sizes. In subsequent studies, not all 32 loci were used. For example, Pearson *et al.* (2007) did not use two loci because of inconsistent PCR amplification within their set of isolates, in the study of U'Ren *et al.* (2007b) a subset of 26 loci was used, and Price *et al.* (2010) used 'only' 23 loci. According to U'Ren and co-workers, this MLVA procedure has a higher discriminatory power than MLST, as there were multiple instances where isolates were identical based on MLST although the MLVA profiles were different, e.g. four *B. mallei* isolates that were identical by MLST showed four different MLVA profiles (U'Ren *et al.*, 2007a).

Subsequent studies confirmed this extremely high discriminatory power and highlighted the potential of MLVA to provide insights into the

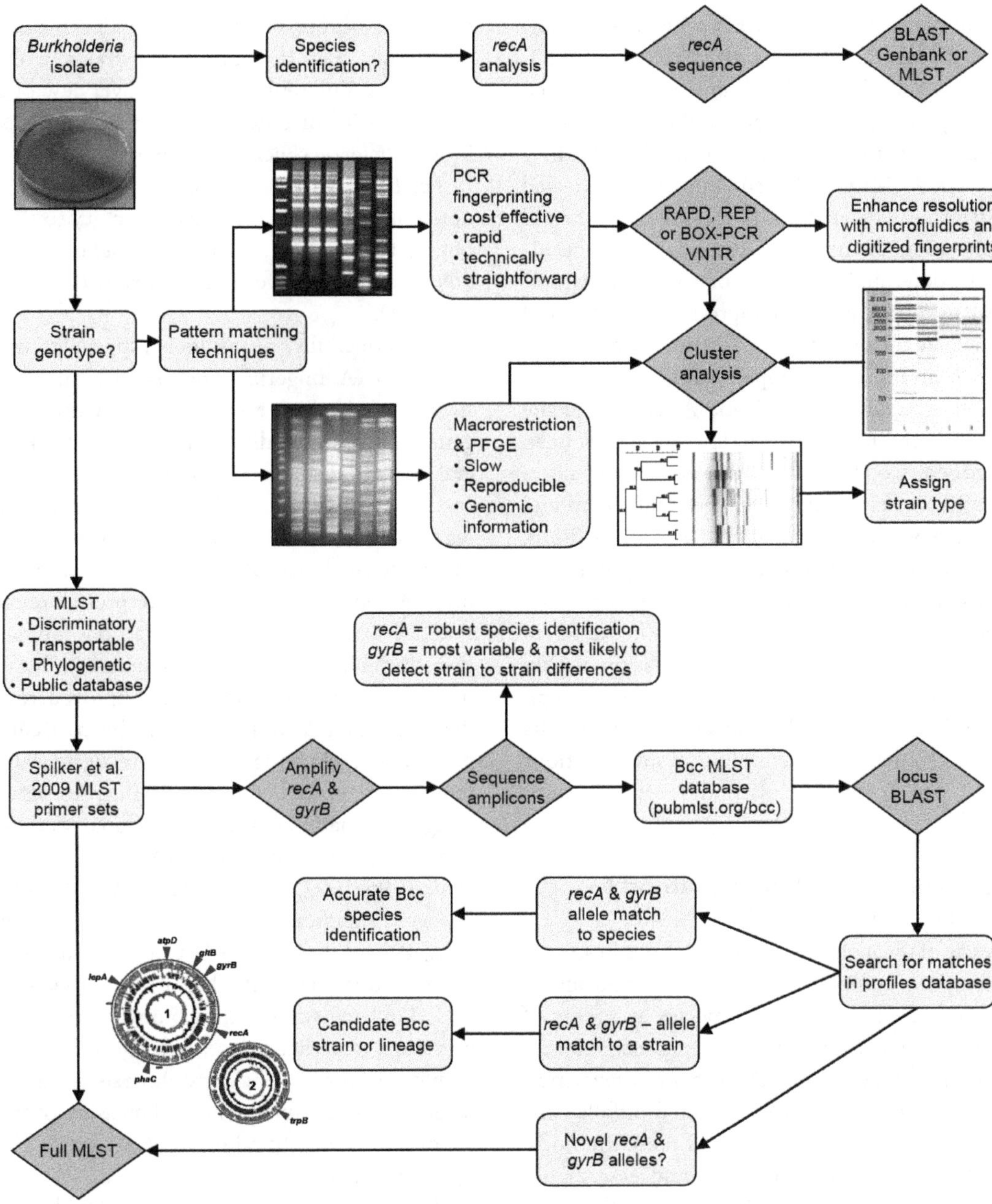

Figure 3.1 Molecular methods for species identification and strain typing of Bcc bacteria. The flow chart illustrates the different types of genetic methods that may be used to identify and genotype Bcc bacteria. Sequence analysis of the *recA* gene is shown at the top as a method of species identification, followed by pattern matching strain typing methods in the middle panel, and then both a partial and full MLST analysis at bottom.

fine-scale genetic heterogeneity of individual strains (U'Ren *et al.*, 2007b; Pearson *et al.*, 2007; Price *et al.*, 2010). A simplified MLVA (based on four of the above mentioned 32 loci and aptly termed MLVA-4) was described by Currie *et al.* (2009). In contrast to the first MLVA developed by Liu *et al.* (2006), it makes use of fluorescently labelled primers and capillary electrophoresis to separate PCR amplicons, which allows multiplexing and reduces the time-to-result to 8 h (within receipt of the strain). In a set of 65 *B. pseudomallei* isolates previously typed with PFGE and MLST, MLVA-4 was able to discriminate between each MLST genotype and in general there was a good agreement between PFGE and MLVA-4 (Currie *et al.*, 2009). MLVA-4 has subsequently been used to determine diversity of *B. pseudomallei* in rural water supplies in northern Australia (Draper *et al.*, 2010), determine the cause of a meliodosis outbreak in captive birds (Hampton *et al.*, 2011) and investigate the molecular epidemiology of *B. pseudomallei* in India (Mukhopadhyay *et al.*, 2011).

A thoroughly evaluated MLVA scheme for the *B. cepacia* complex has to our knowledge not been described. Cooper *et al.* (2004) identified more than 40 candidate regions (larger than 50 bp and with a repeat number higher than six) in the *B. cenocepacia* J2315 genome that could potentially be useful for VNTR-typing. Preliminary work showed that VNTR-typing could be used to demonstrate fine-scale genetic variation within specific *B. cenocepacia* strains (including the epidemic PHDC strain), but further evidence of its potential value in epidemiological investigations of *B. cepacia* complex isolates has not been published.

Discussion: what typing method to select?

The use of the methods described above in studies of *B. cepacia* complex epidemiology has advanced our understanding of these species as agents of human disease as well as providing insights into their environmental distribution and prevalence. Nevertheless, a number of potential pitfalls exist that could limit the use of any of these methods. In addition to considering cost and ease of use (including the ease of performance of the assay but also the ease of interpretation of the data), attention must be paid to how each method performs in terms of typeability, reproducibility, and discriminatory power (Arbeit, 1995; LiPuma, 1998). Typeability describes the ability of a method to provide a readable result for each isolate analysed. Reproducibility measures the ability of a technique to yield the same result when replicate assays are performed on the same isolate. Discriminatory power defines the ability of a typing method to distinguish different strains, and can be expressed as an index that measures probability that two unrelated strains will be placed into different groups. An overview of the characteristics of different typing methods used for *Burkholderia* species is given in Table 3.1.

For *Burkholderia* isolates, typeability is usually not an issue, although strains refractory to PFGE typing (probably due to the presence of high DNase activity) have been reported (Coenye *et al.*, 2002).

Reproducibility of results is especially important if DNA fingerprinting patterns are being stored in a database for comparison with profiles generated in several different experiments. In general, PFGE offers excellent reproducibility, but in certain circumstances where extensive genomic rearrangement within *Burkholderia* isolates is seen, the reproducibility of macrorestriction profiles may be poor and difficult to interpret (Drevinek *et al.*, 2010a). The PCR based methods are much more variable in terms of reproducibility. Because of the low stringency conditions employed, reproducibility of RAPD typing may be particularly problematic. Although some investigators have found good reproducibility of RAPD patterns enabling local, national and international studies of Bcc molecular epidemiology (Mahenthiralingam *et al.*, 1996, 1997, Speert *et al.* 2002), others have observed significant day-to-day variation (Bingen *et al.*, 1993; Liu *et al.*, 1995; Coenye *et al.*, 2002). Overall, this variability has limited the use of RAPD in large-scale studies, but when performed by a single reference centre, with tightly controlled amplification and gel electrophoresis conditions, it can be very useful as an initial measure of strain genotype (Mahenthiralingam *et al.*, 1996, 1997; Speert *et al.*, 2002).

The lack of pattern reproducibility between laboratories has been an obstacle to implementing a uniform RAPD, REP or BOX PCR typing schemes with broad utility. Recent studies have shown that lack of reproducibility with these PCR-fingerprinting methods is frequently due to the variation in separating the DNA polymorphisms using agarose gels as opposed to a lack of reproducibility in the actual PCR amplification (Healy *et al.*, 2005). If high resolution DNA amplicon separation methods such as acrylamide gels or microfluidics using instruments such as an

Table 3.1 Characteristics of currently used typing methods used for *Burkholderia* species (based on references cited in text)

Methodology	Typeability	Reproducibility	Discriminatory power	Ease of use	Set-up cost	Cost per test	Time to result
PFGE	High	High	High	Low	High	Moderate	4 days
RAPD	High	Variable	High	High	Low	Low	1 day
Rep-PCR	High	High	Moderate	High	Low	Low	2 days
MLST*	High	Very high	Very high	Moderate	High	Moderate	2 days
MLVA	Moderate	High	Very high	Moderate	Low	Low	1 day
WGST*	Very high	Very high	Very high	Moderate	High	Moderate	2 days

*Set-up cost are for when sequencing is done in the own laboratory.

Table 3.2 Properties of the loci used in the *B. pseudomallei* and *B. cepacia* complex MLST schemes

Scheme	Locus	Function	No. of alleles
*B. pseudomallei** (2798 isolates, 947 ST)	ace	Acetyl coenzyme A reductase	31
	gltB	Glutamate synthase	50
	gmhD	ADP glycerol-mannoheptose epimerase	55
	lepA	GTP-binding elongation factor	44
	lipA	Lipoic acid synthetase	48
	nark	Nitrate extrusion protein	63
	ndh	NADH dehydrogenase	39
B. cepacia complex† (1172 isolates, 721 ST)	atpD	ATP synthase β-subunit	233
	gltB	Glutamate synthase	296
	gyrB	Gyrase B subunit	407
	recA	Recombination protein A	259
	lepA	GTP-binding elongation factor	284
	phaC	Polyhydroxyalkanoate synthase	221
	trpB	Tryptophan synthase subunit B	277

*Based on data downloaded from http://bpseudomallei.mlst.net/on 20 August 2012.
†Based on data downloaded from http://pubmlst.org/bcc/on 20 August 2012.

Agilent BioAnalyser 2100 (Agilent Technologies Inc. USA) (Healy *et al.*, 2005; White *et al.*, 2011) are used to separate PCR fingerprints then high levels of reproducibility can be obtained (LiPuma, unpublished observations). Commercial PCR-fingerprinting kits such as the DiversiLab system (bioMerieux Inc., Durham, NC, USA) which incorporate microfluidics as a means to improve reproducibility offers one solution to this issue of poor fingerprint reproducibility. Other researchers have modified a conventional RAPD method (Mahenthiralingam *et al.*, 1996) by using rapid single colony-based DNA extraction procedures and microfluidics to greatly improve the reproducibility of RAPD. For example, Fothergill *et al.* (2010) compared RAPD and PFGE analysis for strain typing of *P. aeruginosa* and White *et al.* (2011) genotyped a range of fuel contaminating bacteria including *B. cepacia* complex isolates using such microfluidics enhanced PCR-fingerprinting protocols. However, using microfluidics chips is an expensive solution to the issue of PCR fingerprint reproducibility. Some of this expense is offset by the fact it allows for very rapid analysis of amplicons within approximately 20 min, as well as the direct import of digitized strain fingerprint into cluster analysis software.

As MLST relies on DNA sequencing, it shows excellent reproducibility, and since sequence data are unambiguous, the construction of central databases is possible (Chan *et al.*, 2001). The *Burkholderia* MLST databases (Fig. 3.1) enable rapid searching of full or partial MLST data, allowing comparison to well-characterized panels of strains, including systematically identified reference isolates and species type strains.

Discriminatory power must be carefully considered when choosing a typing method. It is important to appreciate that the degree to which a given method is able to detect differences between isolates is often a function of the species being investigated. In a comparative assessment of typing methods for *B. cenocepacia*, it was observed that PFGE and RAPD had higher discriminatory power than BOX-PCR (Coenye *et al.*, 2002). Provided there is sufficient variation within the population, and provided that suitable genes are picked for analysis, the discriminatory power of MLST is tremendous (Maiden *et al.*, 1998; Enright and Spratt, 1999; Urwin and Maiden, 2003; Cooper and Feil, 2004). In the MLST schemes developed so far, including those for *B. pseudomallei* and the *B. cepacia* complex, it is unlikely that two unrelated isolates will have the same allelic profile by chance. For example, the schemes for *B. pseudomallei* and the *B. cepacia* complex have the potential to distinguish >3.4 million and $>2.1 \times 10^{13}$ different genotypes, respectively (Godoy *et al.*, 2003; Baldwin *et al.*, 2005). It is also important to understand that different degrees of discrimination may be appropriate for different epidemiological questions. In other words, higher discriminatory power does not necessarily always result in a more accurate representation of epidemiologic relatedness. For example, when investigating a hospital-outbreak, a genotyping method with high discriminatory power may be needed to distinguish between outbreak-related and non-related strains. On the other hand, the same genotyping method may be too discriminatory for studies assessing long term or global epidemiology wherein single genetic events that may alter DNA banding profiles (e.g. insertions, deletions and inversions) may obscure similarity among epidemiologically related isolates.

The different techniques also vary in regard to ease of interpretation of the data. Whereas consensus guidelines for interpreting DNA restriction patterns generated by PFGE have been published (Tenover *et al.*, 1995) comparable criteria for interpretation of RAPD and BOX-PCR patterns, and MLST and MLVA data are not available. However, the PFGE interpretative criteria were intended to be used only as an aid in the visual examination of small sets of isolates related to putative outbreaks of disease. These criteria are impractical and quite limited for analysing larger sets of isolates where multiple pairwise comparisons are required. Indeed, for all the genotyping methods under consideration, visual comparison of large number of complex fingerprint patterns is not only time consuming but also highly subjective. The use of equipment to digitize patterns and software to perform numerical analysis of these patterns are necessary for studies involving typing of a significant number of isolates. The same is true for MLST data: while it is possible to manually compare a few selected sequences, comparing thousands of sequences can not be done reliably without computer assistance, and software specifically designed for the analysis of MLST data is available from the MLST website (http://www.mlst.net) (Chan *et al.*, 2001).

Finally, in the absence of epidemiological data, the interpretation of data derived from molecular typing can be far from straightforward. This is illustrated by the finding of unexpected similarity in PFGE and RAPD patterns, and MLST allelic profiles obtained from a diverse collection of *B. stabilis* isolates (Vandamme *et al.*, 2000; Baldwin *et al.*, 2005). Based on the results of genotyping alone it could have been concluded that there was extensive inter-patient spread of a single strain. However, most isolates in this study were epidemiologically unrelated suggesting that the similarities observed in banding patterns or sequence were due to a low level of genomic diversity within this species. Thus, interpreting similarity of DNA profiles for epidemiological investigation must be done cautiously, and, whenever possible, within the context of the available epidemiological data.

Epidemiology of *Burkholderia* infections and infection control

Epidemiology of *B. pseudomallei* and *B. mallei*

B. pseudomallei

B. pseudomallei is the etiologic agent of melioidosis, an infection of animals and humans characterized by a broad spectrum of clinical manifestations, ranging from asymptomatic colonization to fulminant sepsis. Melioidosis is endemic in Southeast Asia and tropical northern Australia, and is being increasingly recognized in the Indian subcontinent and Central and South America (Cheng and Currie, 2005). It is an important cause of morbidity and mortality in Thailand, where the frequency of disease is most likely underestimated due to lack of access to adequate health care (Chaowagul *et al.*, 1989).

B. pseudomallei is endemic in rodents and can be found in moist soil, surface water, and on fruits and vegetables. It is found in particularly high concentrations in rice paddy surface water in the rice-growing regions of northern Thailand and southern and central Vietnam (Parry *et al.*, 1999). Infection occurs by inoculation through cut or abraded skin or by inhalation. In endemic regions, up to 85% of human infection occurs during the monsoon wet season. The association of severe weather with respiratory infection and high mortality rates may be attributed to a shift from percutaneous inoculation to inhalation (Currie and Jacups, 2003). *B. pseudomallei* also infects a variety of animals, including livestock, resulting in significant economic losses (Dance, 1991). Acquisition of infection from animals has been described, but is believed to be very uncommon, as is person to person transmission and laboratory-acquired infection (Dance, 2000; Cheng and Currie, 2005).

Seropositivity rates in excess of 20% occur in some endemic regions (Kanaphun *et al.*, 1993). The majority of infected persons remain asymptomatic; however, latent infection with subsequent reactivation is well recognized (Ngauy *et al.*, 2005). Nevertheless, the vast majority of clinical cases of melioidosis are associated with recent infection. The incubation period ranges from 1 to 21 days, with a mean of 9 days (Currie *et al.*, 2000). Risk factors for disease after acquisition of *B. pseudomallei* include diabetes, alcohol excess, chronic renal disease and chronic lung disease (Currie *et al.*, 2004). Approximately 20% of persons with disease have no apparent risk factors. Children infrequently have clinical disease. Mortality rates from melioidosis vary from 15% in centres where intensive care is available to greater than 50% in regions with poor health care (White, 2003; Cheng and Currie, 2005).

Sporadic cases of human melioidosis occur in regions outside the endemic area, such as China, Korea, the Philippines, Indonesia, India and West Africa. Most cases in Europe and North America occur in immigrants or travellers returning from endemic regions (Dorman *et al.*, 1998; Dance *et al.*, 1999). Several recent reports describe melioidosis in European CF patients, who most likely acquired *B. pseudomallei* during travel to Thailand (Schulin and Steinmetz, 2001; Visca *et al.*, 2001; O'Carroll *et al* 2003). In a review of melioidosis in North America, Dorman *et al.* (1998) reported systemic lupus erythematosus and chronic granulomatous disease as predisposing factors. *B. pseudomallei* is frequently misidentified in cases of sporadic infection due to a lack of familiarity with the disease and/or inappropriate identification methods. Melioidosis should be considered in the differential diagnosis of fever of unknown origin or tuberculosis-like illness in individuals with a history of travel to endemic areas. Referral of the organism to a reference laboratory with experience with this species is advisable. Owing to the potential use of *B. pseudomallei* as an agent of bioterrorism, the isolation of this species from patients who do not give a history of travel to a disease-endemic area should be reported to local or state public health authorities.

B. mallei

B. mallei is the aetiological agent of glanders, a highly communicable disease of livestock, particularly horses, mules, and donkeys. The disease is characterized by fever, inflammation of the nasal mucosa, necrosis and obstruction of the oropharynx. *B. mallei* is a host-adapted pathogen that does not persist in the environment outside its equine host. *B. mallei* can be transmitted to humans via contact with infected animals. Human infection can be acute or chronic, and limited to subcutaneous tissues or disseminated to cause sepsis. The contagious and potentially lethal nature of human infection with *B. mallei* makes this species well suited for use as an agent of bioterrorism. In fact, *B. mallei* was one of the first biologic weapons of the 20th century, being used by Germany during the First World War (Wheelis, 1998). Glanders has been eradicated from most countries due to stringent infection control measures, including the immediate slaughter of affected animals. However, enzootic foci persist in the Middle East, Asia, Africa and South America. The only human case of glanders in the United States in the past 50 years was a laboratory-acquired case in a biodefence scientist (Srinivasan *et al.*, 2001).

Epidemiology of *B. cepacia* complex in CF and infection control

Emergence of *B. cepacia* in CF

The first reports of *B. cepacia* complex infection in CF patients appeared in the late 1970s and early 1980s (Laraya-Cuasay *et al.*, 1977; Rosenstein and Hall, 1980; Nolan *et al.*, 1983; Gold *et al.*, 1983). In 1984 Isles *et al.* documented the increasing prevalence of *B. cepacia* complex infection among patients receiving care in the Toronto CF centre, and described the occurrence in a minority of patients of a rapidly progressive deterioration in respiratory function after infection (Isles *et al.*, 1984). This so-called 'cepacia syndrome' was characterized by necrotizing pneumonia, bacteraemia and sepsis. Similar increases in the incidence of *B. cepacia* complex infection were noted subsequently in other CF treatment centres in North America (Thomassen *et al.*, 1985; Tablan *et al.*, 1985, 1987). Risk factors for acquisition of *B. cepacia* complex included increasing age, underlying severe lung disease, use of aminoglycoside antibiotics, previous hospitalization and the presence of a *B. cepacia* complex colonized sibling.

Evidence for person-to-person transmission

The relatively high prevalence of *B. cepacia* complex infection in some CF care centres, with the sparing of others, and the decrease in incidence of infection after the institution of strict infection control (Thomassen *et al.*, 1986) suggested that *B. cepacia* complex could be transmitted between CF patients. Further evidence for inter-patient spread was provided by ribotyping analyses that documented the existence of strains common to several patients in large CF care centres (LiPuma *et al.*, 1988). Ribotyping analysis subsequently confirmed person-to-person transmission of *B. cepacia* complex among CF patients attending a summer internship programme (LiPuma *et al.*, 1990, 1994). Several additional reports followed that also provided strong epidemiological and genotyping evidence for inter-patient spread of *B. cepacia* complex either within hospitals and clinics or through social contact outside of the hospital (Millar-Jones *et al.*, 1992; Smith *et al.*, 1993; Govan *et al.*, 1993; John *et al.*, 1994; Pegues *et al.*, 1994a; Whiteford *et al.*, 1995).

There are several likely means by which *B. cepacia* complex may be transmitted between patients. Hands contaminated during coughing allow for direct transmission of bacteria between patients (Pegues *et al.*, 1994b). Infected patients can also contaminate hospital rooms and respiratory therapy equipment (Pitchford *et al.*, 1987; Burgde *et al.*, 1993; Hutchinson *et al.*, 1996; Doring *et al.*, 1996; Moore *et al.*, 2002a, b). *B. cepacia* complex has been isolated at low levels from air in rooms occupied by infected patients (Hymphreys *et al.*, 1994; Ensor *et al.*, 1996). Therefore, indirect transmission of *B. cepacia* complex, which can survive for long periods in respiratory droplets on environmental surfaces, is also likely (Drabick *et al.*, 1996; Govan *et al.*, 1996).

The environment as a source of infection

Strict infection control measures have reduced but not entirely eliminated new *B. cepacia* complex infection among CF patients. This suggests that the natural environment also serves as a reservoir for acquisition of *B. cepacia* complex. Small scale studies in the early 1990s failed to recover *B. cepacia* complex strains in significant numbers from select environmental sources (Fisher *et al.*, 1993; Mortensen *et al.*, 1995; Butler *et al.*, 1995). Subsequent studies, however, employing improved isolation and identification methods, have shown recovery of *B. cepacia* complex species from a variety of agricultural soils (Gonzalez *et al.*, 2000; Miller *et al.*, 2000; Balandreau *et al.*, 2001; Fiore *et al.*, 2001; Bevivino *et al.*, 2002; Salles *et al.*, 2002; Vanlaere *et al.*, 2005; Dalmastri *et al.*, 2005). This includes *B. cenocepacia* strain PHDC (see below), which infects a large number of CF patients in the mid-Atlantic region of the eastern US (LiPuma *et al.*, 2002). Interestingly, *B. multivorans*, which is being increasingly recovered from CF patients (Reik *et al.*, 2005), has been only infrequently recovered from the natural environment (Vermis *et al.*, 2003; Ramette *et al.*, 2005). Although typically not animal pathogens, *B. cepacia* complex have been detected in unpasteurized milk (Moore *et al.*, 2001), and implicated as a cause of subclinical mastitis in sheep (Berriatua *et al.*, 2001). The elucidation of the preferred environmental niches of each species of the *B. cepacia* complex is required to better determine the risk posed by these sources for persons with CF.

Epidemic strains

Numerous studies employing isolate genotyping analyses have identified *B. cepacia* complex strains that are common to multiple CF patients, suggesting that certain so-called 'epidemic' strains are more transmissible or better adapted to infection than are others, and/or that such strains are numerically more dominant in the natural environment. In 1994, Johnson *et al.* used MLEE and ribotyping to identify a '*B. cepacia*' strain, designated electrophoretic type (ET) 12, which predominated among CF patients in Ontario, Canada (Johnson *et al.*, 1994). This strain apparently spread, via patient contact at a CF summer camp, to the United Kingdom (Govan *et al.*, 1993), where it was later recognized as accounting for nearly one-half of *B. cepacia* complex-infected CF patients (Pitt *et al.*, 1996). ET12 is a *B. cenocepacia* strain characterized by the expression of distinctive 'cable pili' that mediate adherence to respiratory epithelial cells (Sun *et al.*, 1995). This strain also contains a genomic fragment referred to as the '*B. cepacia* epidemic strain marker' (BCESM; Mahenthiralingam *et al.*, 1997). This 1.4 kb sequence contained an open reading frame with homology to negative transcriptional regulators of other species; however, no specific function of the BCESM in virulence or strain transmissibility was determined at the time. During completion of the *B. cenocepacia* J2315 genome, further analysis of the region encoding the BCESM demonstrated that it was part of a genomic island encoding both virulence and metabolism genes (Baldwin *et al.*, 2004). Analysis of deletion mutants of genes within this island demonstrated that the region did play a role during infection, with the porin and *N*-acylhomoserine lactone (AHL) synthase mutants being altered in virulence, and an amidase mutant being less persistant during infection (Baldwin *et al.*, 2004).

In another study, Kumar *et al.* (1997) used cellular fatty acid methyl ester and PFGE analyses to demonstrate that 29 of 32 '*B. cepacia*' isolates from five CF centres in Michigan were the same strain type. This strain, subsequently designated the *B. cenocepacia* 'Midwest clone' (Coenye and LiPuma, 2002), accounts for considerable infection in the Midwestern region of the US.

B. cenocepacia strain PHDC has been recovered from the majority of *B. cepacia* complex-infected CF patients in the mid-Atlantic region of the US (Chen *et al.*, 2001). PHDC has also been recovered from agricultural soil (LiPuma *et al.*, 2002) and CF patients in three European countries (Coenye *et al.*, 2004; Campana *et al.*, 2005), suggesting that this strain is widely distributed and that acquisition from the environment may contribute to its prevalence in CF.

Other 'epidemic' *B. cenocepacia* strains have also been described. Manno *et al.* (2004) showed that two *B. cenocepacia* strains (type 01 belonging to *B. cenocepacia* IIIA and type 03 belonging to *B. cenocepacia* IIID) were responsible for the majority of infections in a large Italian CF centre. Similarly, Drevinek *et al.* (2005) reported the widespread occurrence of clone CZ1 among CF patients in the Czech Republic. This clone apparently was closely related to the Canadian epidemic strain type RAPD 01, one of the four genetic lineages of *B. cenocepacia* that cluster by geographic region in Canada (Speert *et al.*, 2002). Strains responsible for nosocomial pulmonary tract infections among non-CF patients in a French ICU (designated the B&B clone) were found to be very similar (but not identical) to ET12 strains (Graindorge *et al.*, 2010). These strains were found to belong to *B. cenocepacia* ST32, another globally distributed clone recovered from patients (CF and non-CF) and the environment in Canada, the US, the UK, Czech Republic, Germany and Italy (Drevinek and Mahenthiralingam, 2010).

Non-*B. cenocepacia* strains appear to be less frequently shared among CF patients. A notable exception is *B. dolosa* strain SLC6, which accounts for approximately one-third of *B. cepacia* complex-infected patients in a large US CF care centre (Biddick *et al.*, 2003). Nevertheless, epidemic strains belonging to other species, including *B. multivorans* (Biddick *et al.*, 2003; Whiteford *et al.*, 1995; Segonds *et al.*, 1999), *B. pyrrocinia* (Campana *et al.*, 2005) and *B. contaminans* (Campana *et al.*, 2005; Vanlaere *et al.*, 2009), have been described.

Infection control measures

The contention that some *B. cepacia* complex strains are transmitted between persons with CF is supported by firm epidemiologic data, particularly the observed decrease in new acquisition coincident with the institution of stringent infection control practices (Chen *et al.*, 2001). Bacterial genotyping analyses demonstrating 'epidemic' strains common to multiple patients (see above) further support this contention. These observations have prompted strict infection control measures aimed at decreasing the incidence of *B. cepacia* complex infection in CF. The implementation of infection control as well as patient education led to a considerable reduction in the incidence of *B. cenocepacia* (classified as Bcc *recA* group III-A at the time) epidemic strains in a retrospective study which followed *B. cepacia* complex infection from 1981 to 1998 among CF patients in Vancouver, Canada (Mahenthiralingam *et al.*, 2001). Published infection control guidelines (Saiman *et al.*, 2003) focus on hand hygiene, the appropriate use of specific physical barriers (gloves and gowns), and sterilization/disinfection of patient care equipment and environmental surfaces. Data collected in the CF population attending the Manchester Adult CF Centre (UK) are also a clear illustration that infection control measures do have an impact on *B. cepacia* complex epidemiology (France *et al.*, 2008). After introducing complete cohort segregation, incidence and prevalence of *B. cepacia* complex started to decline, and the relative contribution of transmissible strains to overall prevalence gradually decreased, as new infections are now predominantly caused by unique strains.

Two aspects of infection control included in the published guidelines bear emphasis. First, although *B. cepacia* complex species (and specific strains) appear to vary considerably with respect to their capacity to be transmitted among CF patients, it is not yet possible to reliably determine prospectively the 'transmissibility' of a given strain. Genetic elements such as the BCESM and the cable pilin subunit gene, *cblA*, lack sufficient sensitivity and specificity to identify transmissible strains (Agodi *et al.*, 2002; Chen *et al.*, 2001; Manno *et al.*, 2004). Thus, stratifying infection control measures based on specific strains or the presence or absence of specific markers is not justified. Second, super-infection or the supplanting of an initial infecting *B. cepacia* complex strain by another strain has been documented in 'chronically infected' CF patients. For example, Mahenthiralingam *et al.* (2001) identified several patients in whom *B. cenocepacia* strains apparently supplanted initial infection with *B. multivorans*. A similar observation was made by Ledson *et al.* (1998), who demonstrated superinfection with *B. cenocepacia* ET12 in patients who were previously infected with another *B. cepacia* complex strain. In an analysis of serial isolates recovered from 379 CF patients receiving care in 112 CF treatment centres in the US, Bernhardt *et al* (2003) demonstrated a change in infecting species or strain in approximately 7% of individuals. Thus, cohorting CF patients on the basis of *B. cepacia* complex infection status may place patients at risk of superinfection by a more transmissible (and potentially more virulent) strain; therefore, this practice should be avoided.

Population genetics and molecular epidemiology of *Burkholderia* species based on MLST

Multilocus sequence typing of *B. pseudomallei*, *B. mallei* and *B. thailandensis*

B. mallei and *B. pseudomallei* are phylogenetically very closely related and these two species were shown to represent a single genomic species by DNA–DNA hybridization criteria (Coenye and Vandamme, 2003). Several avirulent, environmental *B. pseudomallei*-like organisms were formally classified as *B. thailandensis* (Brett *et al.*, 1998).

Godoy *et al.* (2003) reported the use of MLST to elucidate the relationships between these closely related species. In this study they determined the sequence of internal fragments (approximately 550 bp) of seven housekeeping genes (Table 3.2) for a large collection of *B. mallei*, *B. pseudomallei* and *B. thailandensis* isolates. Overall, sequence diversity was less than generally seen in housekeeping genes (Table 3.2). These sequences were then concatenated into one large sequence of 3399 bp and this sequence was used to construct a phylogenetic tree (Fig. 3.2). The analysis showed that all *B. pseudomallei* isolates were tightly clustered and well-separated from the *B. thailandensis* isolates. In addition, all *B. mallei* isolates had identical sequences and clustered with the *B. pseudomallei* isolates. These data confirm that (i) *B. pseudomallei* and *B. thailandensis* are different species (even though they have virtually identical 16S rRNA gene sequences; Coenye and Vandamme, 2003) and (ii) *B. mallei* is a clone of *B. pseudomallei*. This means that on taxonomic and population genetic grounds, *B. pseudomallei* and *B. mallei* should not be given separate species names, although the major differences in their biochemical activities and in the clinical symptoms and epidemiology justify this two-species concept.

Besides its value in taxonomic studies, this MLST scheme has also been shown to be extremely valuable for epidemiological studies. Early studies (Godoy *et al.*, 2003; Chen *et al.*, 2004) already showed that there was very good congruence between results obtained with MLST and those obtained with PFGE. MLST has since then been used in various studies addressing local and global *B. pseudomallei* epidemiology (McCombie *et al.*, 2006; Currie *et al.*, 2007b; Cheng *et al.*, 2008; Chantratita *et al.*, 2008; Engelthaler *et al.*, 2011a). For example, MLST revealed that there are no sequence types (STs) shared between Australia and Southeast Asia (McCombie *et al.*, 2006; Currie *et al.*, 2007b). In addition, the investigation of close to 300 isolates recovered in northern Australia (Northern Territory and Queensland) showed that there was extensive diversity within this collection (277 isolates grouped in 159 STs) and significant differentiation between isolates from different regions was observed (Cheng *et al.*, 2008). Detailed analysis of the sequence diversity at the different MLST alleles demonstrated that this sequence diversity was very limited compared to the high number of STs observed. This indicates that diversity in *B. pseudomallei*, at least in this population, primarily arises from high rates of genetic recombination, rather than from point mutations.

Multilocus sequence typing and multilocus restriction typing of *B. cepacia* complex

An MLST scheme for the entire *B. cepacia* complex was developed by Baldwin *et al.* (2005) and used sequence fragments (between 301 and 454 bp in size) of seven loci (Table 3.2). Nucleotide sequence divergence was extensive in all seven loci, resulting in a large number of different alleles for each locus (Table 3.2). Within the current collection of 1172 isolates investigated with MLST, 597 different STs can be identified, that group in 64 clonal complexes (CC). Mathematical analyses of a subset of these isolates indicated that different *B. cepacia* complex species appeared to have different 'index of association' values (I_A) (a measure of clonality of a population) and thus different mechanisms of evolution (Baldwin *et al.*, 2005).

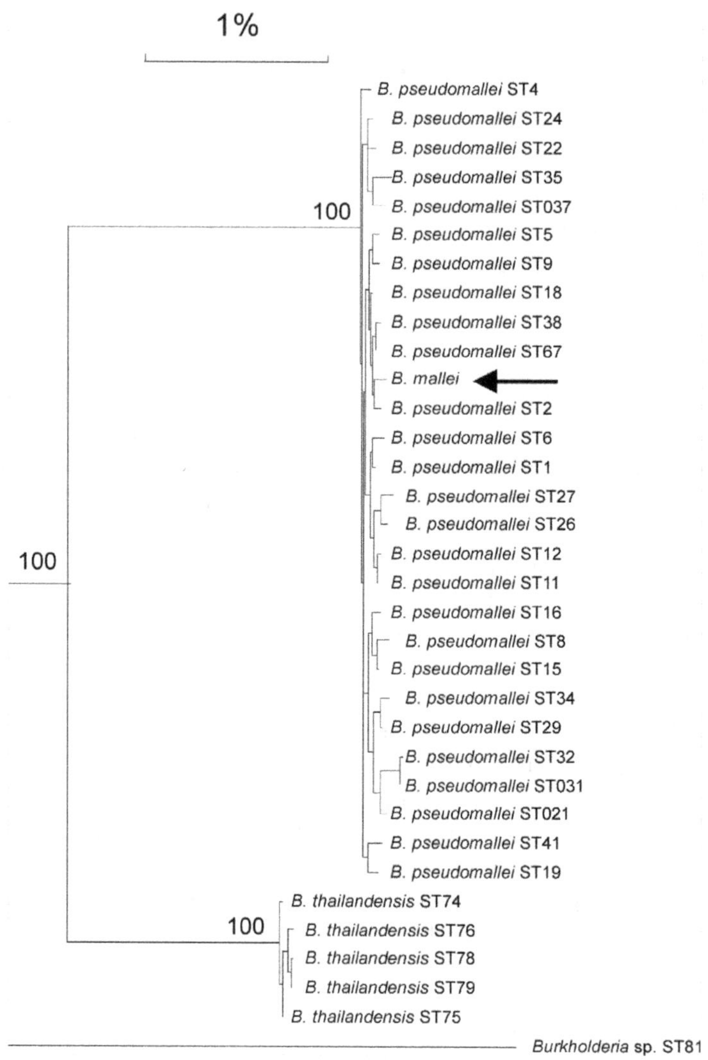

Figure 3.2 Neighbour-joining tree constructed from the concatenated sequences of *ace*, *gltB*, *gmhD*, *lepA*, *lipA*, *narK* and *ndh*. Sequences were obtained from the MLST website, http://www.mlst.net. The position of the *B. mallei* clone within *B. pseudomallei* is indicated with an arrow. The per cent recoveries of the major nodes in 1000 bootstrap replicates are shown.

Although MLST analysis was primarily developed as a means to examine the population biology of the Bcc at the strain level, it also contributed to the taxonomic classification of several new Bcc species including: *B. latens*, *B. diffusa*, *B. arboris*, *B. seminalis* and *B. metallica* (Vanlaere *et al.*, 2008), and *B. lata* and *B. contaminans* (Vanlaere, *et al.*, 2009). The close evolutionary relationships between Bcc species, subdivision of several species such as *B. cenocepacia* into lineages, and presence of novel groups that may comprise new species (Mahenthiralingam *et al.*, 2008) can be depicted in a clonal frame analysis (Didelot and Falush, 2007) of the complex (Fig. 3.3). This analysis illustrates both the utility of MLST data as a means to explore the evolutionary biology of bacterial species, and the fact that we have a long way to go in terms of full exploration of the diversity within the Bcc (Fig. 3.3).

Since its publication in 2005, the *B. cepacia* complex MLST scheme has been used in various studies, addressing several aspects of the molecular epidemiology of this group of organisms. One of the first questions that was tackled was the long

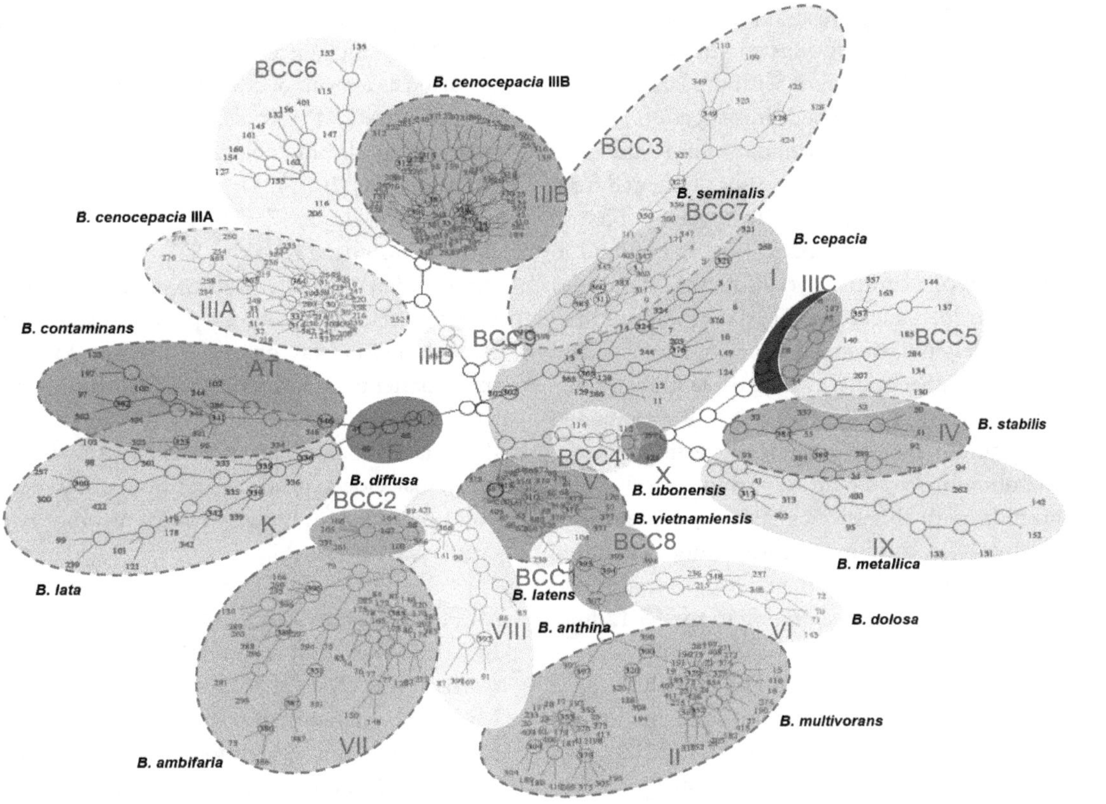

Key:

I - *B. cepacia*	VI - *B. dolosa*	BCC1 - *B. latens*	AT - *B. lata*
II - *B. multivorans*	VII - *B. ambifaria*	BCC2 - *B. diffusa*	K - *B. contaminans*
IIIA-D - *B. cenocepacia*	VIII - *B. anthina*	BCC3 - *B. arboris*	
IV - *B. stabilis*	IX - *B. pyrrocinia*	BCC7 - *B. seminalis*	
V - *B. vietnamiensis*	X - *B. ubonensis*	BCC9 - *B. metallica*	BCC 4, 5, 6 & 8 Novel groups

Figure 3.3 Prediction of evolutionary relationships within the *B. cepacia* complex. Sequence data from the MLST database were examined using clonal frame analysis to predict the evolutionary relationships between each Bcc species and STs within them. The 17 currently named Bcc species as well as novel MLST groups are indicated adjacent to their respective clonal frame analysis clusters and also listed in the key. Data analysis was performed by Dr Adam Baldwin using the methods described by Didelot and Falush (2007). A colour version of this figure is located in the plate section at the back of the book.

outstanding issue whether there are differences between environmental and clinical *B. cepacia* complex isolates. The taxonomic studies of Peter Vandamme and co-workers already had shown that the majority of *B. cepacia* complex species can be found in both clinical and environmental niches and LiPuma *et al.* (2002) showed that the epidemic *B. cenocepacia* strain PHDC is common in both niches. In a large-scale study, Baldwin *et al.* (2007) compared 381 clinical and 233 environmental *B. cepacia* complex strains. They found that 21.5% of the clinical isolates (encompassing 15 STs) were identical to environmental isolates. Remarkable cases include the strains ATCC 25416 (*B. cepacia* type strain recovered from onion in 1948) and J1050 (non-CF clinical *B. cepacia* isolate from the US) that both belong to ST10, and *B. ambifaria* strains AMMD (well-known biocontrol strain) and AU0212 (CF isolate from the US) that both belong to ST77. These data confirm that 'clinical' and 'environmental' strains are not necessarily different. As already discussed above, the implementation of strict infection control measures reduced the spread of 'epidemic' *B. cenocepacia* strains but the continued emergence of *B. multivorans* strains and the proportional rise of infections with the latter species prompted a MLST-based investigation of the global epidemiology of *B. multivorans* (Baldwin *et al.*, 2008). Three main observations emerged from this study (which included 107 isolates). First of all, 12 globally distributed STs (i.e. STs found in more than one country over more than a decade) were identified, including ST16 that was found in France, Belgium, Australia, Canada, New Zealand and the US. Secondly, two of these globally distributed STs (ST21 and ST375) also contained environmental isolates, again suggesting that the environment may be an important reservoir for infection with this species. Thirdly, evidence for interspecies recombination was found, including recombination between *B. multivorans* strains and *B. cenocepacia* strains belonging to the ET12 lineage.

In addition to epidemic strains, which have clearly spread as a result of the transmission of infections between people with CF, MLST has also shown certain strains to be globally distributed without clear links between the locations of each incident of infection or isolate source. The *Burkholderia* SAR-1 metagenome found in Sargasso sea water was matched to a widely distributed strain of sequence type ST102 that had caused incidences of CF infection, sheep mastitis, non-sterile pharmaceutical product contamination and renal infection at multiple globally disparate sites. This ST102 belonged to a novel group of Bcc bacteria designated Bcc group K at the time. Links between each incidence of ST102 infection were not apparent, however, for several of the cases, contamination was a common trait associated with the recovered isolates. Subsequently, this led to the species name for Bcc group K being designated as *B. contaminans* and one of the ST102 sheep mastitis isolates, LMG 23361^T becoming the type strain for this new Bccc species. The ability to contaminate man-made products, pharmaceutical drugs and instruments is a major feature of Bcc bacteria (Mahenthiralingam, *et al.*, 2008). A recent review by Torbeck *et al.* (2011) has highlighted why Bcc bacteria are considered one of the most objectionable contaminants in pharmaceutical products and reviewed U.S. Food and Drug Administration (FDA) reports which record the incidence of problems which have occurred as a result of contamination.

Finally, MLST was also used to study evolution during chronic colonization of CF patients (Waine *et al.*, 2007). Of the 20 patients (mean colonization time: 7.07 years) investigated, 15 showed no changes in the ST of the colonizing strain and in one patient, strain replacement likely occurred. However, in four patients evidence for recombination was observed (at one locus for three patients and at two loci for one patient). Although the sample size was relatively small, these data indicate that the MLST loci have a slow molecular clock speed. In addition, a low rate of point mutation and higher rate of recombination events was observed.

MLRT of *B. cenocepacia*

In MLRT variation at several loci is indexed by restriction analysis of PCR-amplified genes and a MLRT scheme, which used five loci, was developed for studying the epidemiology and population structure of *B. cenocepacia* (Coenye and LiPuma, 2002, 2003). It was shown that results obtained by MLRT were in good agreement with PFGE, and that MLRT can be used to compare *B. cenocepacia* isolates collected in larger-scale (both temporal and spatial) studies. It was subsequently used to investigate the population structure of *B. cenocepacia*, using a collection of 375 *B. cenocepacia* isolates recovered from CF-patients, other clinical sources and the environment between 1966 and 2001 (Coenye and LiPuma, 2003). Eleven CCs were delineated, but over 86% of all *B. cenocepacia* isolates investigated belonged to three major CCs (designated RT-6-complex, RT-46-complex and RT-88-complex), comprising the epidemic ET12, Midwest and PHDC clonal lineages, respectively. The I_A was significantly different from zero, suggesting that the overall *B. cenocepacia* population is clonal and that recombination within this population has been rare (Maynard-Smith *et al.*, 1993). However, when the I_A was calculated for each major CC separately, major differences could be observed. While the RT-88-complex shows evidence of clonality, the opposite is true for the RT-46-complex; the RT-6-complex shows evidence of an epidemic population (providing evidence that this population is recombining in the long term but that a few RTs recently have become more widespread). The underlying reasons for the observed differences in population structure have not been elucidated yet. Most likely the availability and efficiency of genetic exchange mechanisms, and ecological factors, as well as inherent genomic plasticity, play a crucial role.

MLRT was also used by Bevivino *et al.* (2011) to investigate the genetic relationships among Italian and Mexican maize-rhizosphere isolates belonging to *B. cenocepacia* or a provisonal *B. cepacia* complex group designated 'BCC6' (Dalmastri *et al.*, 2006; Fig. 3.3). They observed a clear relationship between the origin of the isolates and their grouping in different CCs and concluded that both the *B. cenocepacia* and the BCC6 population investigated had epidemic population structures, in which particular clones can emerge and spread. They also observed that identical types were found in both locations, leading to the suggestion that ecological factors and niche-specific selective pressure could promote the spread of a particular clone over a broad geographic area.

The use of whole genome sequences in molecular epidemiology of *Burkholderia* species

The increased availability and decreasing costs of next generation sequencing technologies, combined with the availability of dedicated and user-friendly analysis software make it relatively straightforward to sequence, annotate and analyse a bacterial genome. The use of whole-genome sequences (WGS) for typing purposes allows the use of all information present in the genome, instead of relying on the sequence variation in a very small fraction of the genome (as is the case with most other typing techniques). The availability of WGS for a set of isolates allows one to deduce relationships based on the presence of a large number of single nucleotide polymorphisms (SNPs) and this approach ('whole genome sequencing and typing', WGST) has been validated in the bacteria *Staphylococcus aureus* (Harris *et al.*, 2010; Köser *et al.*, 2012a), *Acinetobacter baumanii* (Lewis *et al.*, 2010), *Bacillus anthracis* (Kuroda *et al.*, 2010), *Vibrio cholerae* (the 2010 Haiti outbreak) (Chin *et al.*, 2011) and *Streptococcus pyogenes* (Ben Zakour *et al.*, 2012), as well as in the fungus *Coccidioides immitis* (Engelthaler *et al.*, 2011b) (for a complete overview see Köser *et al.*, 2012b). While mostly restricted to a high-end research setting, the investigation of a neonatal MRSA outbreak by Köser *et al.* (2012a) demonstrated that WGST can provide clinically relevant data within a reasonable time frame (i.e. a time frame that allows to make decisions that can have an impact on patient care).

To our knowledge there are two examples of WGST studies in the genus *Burkholderia*. Engelthaler *et al.* (2011a) used WGST to determine the likely source of a case of melioidosis in southern Arizona (US). To this end they sequenced the genome of the case isolate, mapped it to the sequence of reference isolate K92643 and compared it to the WGS of 32 *B. pseudomallei* and *B. mallei* strains. 54135 orthologous SNP loci were identified, including 31202 parsimony informative SNPs, and a maximum parsimony phylogenetic analysis grouped the isolate recovered in Arizona with isolates from Southeast Asia, although no epidemiological link between the patient and the potential source of the infection was found.

B. dolosa strain SLC6 infected 39 CF patients in Boston (Biddick *et al.*, 2003; Kalish *et al.*, 2006). In a retrospective study the genome of 112 isolates recovered from 14 of these patients over 16 years was sequenced (Lieberman *et al.*, 2011). The main goal was to study adaptive evolution during an infection, but SNP analysis also allowed the investigators to study the mode of transmission. 492 polymorphic loci were identified and the evolutionary relationships between the isolates (i.e. the epidemiology) were investigated by a maximum likelihood analysis. The resulting phylogenetic trees suggested a network of transmission between subjects, while isolates from a particular patient formed genetically related clusters. Lieberman *et al.* went on to reconstruct the last common ancestor (LCA) for the set of isolates from each patient, and by inferring the relationships between these LCA they were able to identify the transmission network between the 14 patients. While inherently limited due to the fact that only 14/39 patients involved in this outbreak are represented, this analysis shows that the epidemic SLC6 strain was transmitted through several people during its spread. For 3/14 patients, multiple blood isolates were available, and this allowed an assessment of the relationships between lung and blood isolates. In two patients, pairs of blood isolates were found that evolved from distinct lung isolates, providing evidence for multiclonal transmission from the lung to the bloodstream. Whether this is due to punctuated transmission of multiple clones or multiple transmissions over time remains to be determined.

Concluding remarks

Interest in the epidemiology of *Burkholderia* species has arisen from the emergence of certain of these species as opportunistic pathogens in persons with CF and the potential of other species as agents of bioterrorism. This interest has stimulated the application and assessment of a variety of genotypic methods of stain typing. These methods vary with respect to key features, such as reproducibility, discriminatory power, ease of use and cost. Nevertheless, their use has proven useful in elucidating a great deal about the epidemiology of *Burkholderia* species. In particular, much has been learned regarding the epidemiology of *B. cepacia* complex infection in persons with CF. This has led to the institution of more effective infection control measures that have become integral to the management of persons with CF. The more recent focus on *B. pseudomallei* and *B. mallei* will almost certainly lead to a better understanding of the ecology, epidemiology and natural history of human infection due to these species in the next several years. The application of newer methods, such as MLST, has revealed a great deal about the phylogeny and population genetic composition of human pathogenic *Burkholderia* species. This represents an important advance in our understanding of the biology of these species. Finally, the use of WGST holds great promise for the field of molecular epidemiology and bacterial population structure.

Acknowledgement

We thank Dr Adam Baldwin for providing and giving us permission to use the Bcc clonal frame analysis figure.

References

Agodi, A., Mahenthiralingam, E., Barchitta, M., Giannino, V., Sciacca, A., and Stefani, S. (2001). *Burkholderia cepacia* complex infection in Italian patients with cystic fibrosis: prevalence, epidemiology, and genomovar status. J. Clin. Microbiol. *39*, 2891–2896.

Arbeit R.D. (1995). Laboratory procedures for the epidemiologic analysis of microorganisms. In Manual of Clinical Microbiology. 6th edition. Murray, P.R., Baron, E.J., Pfalle, M.A., Tenover, F.C., and Yolken, R.H., eds. (Washington DC, USA: ASM Press), pp. 190–208.

Balandreau, J., Viallard, V., Cournoyer, B., Coenye, T., Laevens, S., and Vandamme, P. (2001). *Burkholderia cepacia* genomovar III Is a common plant-associated bacterium. Appl. Environ. Microbiol. *67*, 982–985.

Baldwin, A., Sokol, P.A., Parkhill, J., and Mahenthiralingam, E. (2004). The *Burkholderia cepacia* epidemic strain marker is part of a novel genomic island encoding both virulence and metabolism-associated genes in *Burkholderia cenocepacia*. Infect. Immun. *72*, 1537–1547.

Baldwin, A., Mahenthiralingam, E., Thickett, K.M., Honeybourne, D., Maiden, M.C.J., Govan, J.R.W., Speert, D.P., LiPuma, J.J., Vandamme, P., and Dowson, C.G. (2005). Multilocus sequence typing scheme for the *Burkholderia cepacia* complex. J. Clin. Microbiol. *43*, 4665–4673.

Baldwin, A., Mahenthiralingam, E., Drevinek, P., Vandamme, P., Govan, J.R., Waine, D.J., LiPuma, J.J., Chiarini, L., Dalmastri, C., Henry, D.A., *et al.* (2007). Environmental *Burkholderia cepacia* complex isolates in human infections. Emerg. Infect. Dis. *13*, 458–461.

Baldwin, A., Mahenthiralingam, E., Drevinek, P., Pope, C., Waine, D.J., Henry, D.A., Speert, D.P., Carter, P., Vandamme, P., LiPuma, J.J., and Dowson, C.G. (2008). Elucidating global epidemiology of *Burkholderia multivorans* in cases of cystic fibrosis by multilocus sequence typing. J. Clin. Microbiol. *46*, 290–295.

Ben Zakour, N.L., Venturini, C., Beatson, S.A., and Walker, M.J. (2012). Analysis of a *Streptococcus pyogenes* puerperal sepsis cluster by use of whole-genome sequencing. J. Clin. Microbiol. *50*, 2224–2228.

Bernhardt, S.A., Spilker, T., Coffey, T., and LiPuma, J.J. (2003). *Burkholderia cepacia* complex in cystic fibrosis: frequency of strain replacement during chronic infection. Clin. Infect. Dis. *37*, 780- 785.

Berriatua, E., Ziluaga, I., Miguel-Virto, C., Uribarren, P., Juste, R., Laevens, S., Vandamme, P., and Govan, J.R. (2001). Outbreak of subclinical mastitis in a flock of dairy sheep associated with *Burkholderia cepacia* complex infection. J. Clin. Microbiol. *39*, 990–994.

Bevivino, A., Dalmastri, C., Tabacchioni, S., Chiarini, L., Belli, M.L., Piana, S., Materazzo, A., Vandamme, P., and Manno, G. (2002). *Burkholderia cepacia* complex bacteria from clinical and environmental sources in Italy: genomovar status and distribution of traits related to virulence and transmissibility. J. Clin. Microbiol. *40*, 846–851.

Bevivino, A., Costa, B., Cantale, C., Cesarini, S., Chiarini, L., Tabacchioni, S., Caballero-Mellado, J., and Dalmastri, C. (2011). Genetic relationships among Italian and Mexican maize-rhizosphere *Burkholderia cepacia* complex (BCC) populations belonging to *Burkholderia cenocepacia* IIIB and BCC6 group. BMC Microbiol. *11*, 228.

Biddick, R., Spilker, T., Martin, A., and LiPuma, J.J. (2003). Evidence of transmission of *Burkholderia cepacia, Burkholderia multivorans* and *Burkholderia dolosa* among persons with cystic fibrosis. FEMS Microbiol. Lett. *228*, 57–62.

Bingen, E.H., Weber, M., Derelle, J., Brahimi, N., Lambert-Zechovsky, N.Y., Vidailhet, M., Navarro, J., and Elion, J. (1993). Arbitrarily primed polymerase chain reaction as a rapid method to differentiate crossed from independent *Pseudomonas cepacia* infections in cystic fibrosis patients. J. Clin. Microbiol. *31*, 2589–2593.

Bingen E.H., Denamur, E., and Elion, J. (1994). Use of ribotyping in epidemiological surveillance of nosocomial outbreaks. Clin. Microbiol. Rev. *7*, 311–327.

Brett, P.J., DeShazer, D., and Woods, D.E. (1998). *Burkholderia thailandensis* sp. nov., a *Burkholderia pseudomallei*-like species. Int. J. Syst. Bacteriol. *48*, 317–320.

Burdge, D.R., Nakielna, E.M., and Noble, M.A. (1993). Case–control and vector studies of nosocomial acquisition of *Pseudomonas cepacia* in adult patients with cystic fibrosis. Infect. Control Hosp. Epidemiol. *14*, 127–130.

Butler, S.L., Doherty, C.J., Hughes, J.E., Nelson, J.W., and Govan, J.R. (1995). *Burkholderia cepacia* and cystic fibrosis: do natural environments present a potential hazard? J. Clin. Microbiol. 33, 1001–1004.

Caballero-Mellado, J., Martinez-Aguilar, L., Paredes-Valdez, G., and Santos, P.E. (2004). *Burkholderia unamae* sp. nov., an N_2-fixing rhizospheric and endophytic species. Int. J. Syst. Evol. Microbiol. *54*, 1165–1172.

Campana, S., Taccetti, G., Ravenni, N., Favari, F., Cariani, L., Sciacca, A., Savoia, D., Collura, A., Fiscarelli, E., De Intinis, G., *et al.* (2005). Transmission of *Burkholderia cepacia* complex: evidence for new epidemic clones infecting cystic fibrosis patients in Italy. J. Clin. Microbiol. *43*, 5136–5142.

Carson, L.A., Anderson, R.L., Panlilio, A.L., Beck-Sague, C.M., and Miller, J.M. (1991). Isoenzyme analysis of *Pseudomonas cepacia* as an epidemiologic tool. Am. J. Med. *91*, 252S-255S.

Chan, M.S., Maiden, M.C.J., and Spratt, B.G. (2001). Database-driven multi locus sequence typing (MLST) of bacterial pathogens. Bioinformatics *17*, 1077–1083.

Chantratita, N., Wuthiekanun, V., Limmathurotsakul, D., Vesaratchavest, M., Thanwisai, A., Amornchai, P., Tumapa, S., Feil, E.J., Day, N.P., and Peacock, S.J. (2008). Genetic diversity and microevolution of *Burkholderia pseudomallei* in the environment. PLoS Negl. Trop. Dis. *2*, e182.

Chaowagul, W., White, N.J., Dance, D.A., Wattanagoon, Y., Naigowit, P., Davis, T.M., Looareesuwan, S., and Pitakwatchara, N. (1989). Melioidosis: a major cause of community-acquired septicemia in northeastern Thailand. J. Infect. Dis. *159*, 890–899.

Chen, J.S., Witzmann, K.A., Spilker, T., Fink, R.J., and LiPuma, J.J. (2001). Endemicity and inter-city spread of *Burkholderia cepacia* genomovar III in cystic fibrosis. J. Pediatr. *139*, 643–649.

Cheng, A.C., and Currie, B.J. (2005). Melioidosis: epidemiology, pathophysiology, and management. Clin. Microbiol. Rev. *18*, 383–416.

Cheng, A.C., Godoy, D., Mayo, M., Gal, D., Spratt, B.G., and Currie, B.J. (2004). Isolates of *Burkholderia pseudomallei* from Northern Australia are distinct by multilocus sequence typing, but strain types do not correlate with clinical presentation. J. Clin. Microbiol. *42*, 5477–5783.

Cheng, A.C., Day, N.P., Mayo, M.J., Gal, D., and Currie, B.J. (2005). *Burkholderia pseudomallei* strain type, based on pulsed-field gel electrophoresis, does not determine disease presentation in melioidosis. Microbes Infect. *7*, 104–109.

Cheng, A.C., Ward, L., Godoy, D., Norton, R., Mayo, M., Gal, D., Spratt, B.G., and Currie, B.J. (2008). Genetic diversity of *Burkholderia pseudomallei* isolates in Australia. J. Clin. Microbiol. *46*, 249–254.

Chin, C.S., Sorenson, J., Harris, J.B., Robins, W.P., Charles, R.C., Jean-Charles, R.R., Bullard, J., Webster, D.R., Kasarskis, A., Peluso, P., Paxinos, *et al.* (2011). The origin of the Haitian cholera outbreak strain. N. Engl. J. Med. *364*, 33–42.

Chua, K.H., See, K.H., Thong, K.L., and Puthucheary, S.D. (2011). *Spe*I restriction enzyme displays greater discriminatory power than *Xba*I enzyme does in a pulsed-field gel electrophoresis study on 146 clinical *Burkholderia pseudomallei* isolates. Jpn. J. Infect. Dis. *64*, 228–233.

Cimolai, N., and Trombley, C. (1996). Enterobacterial intergenic consensus sequence polymerase chain reaction as a typing method for *Burkholderia (Pseudomonas) cepacia*. Clin. Microbiol. Infect. *2*, 59–62.

Coenye, T., and LiPuma, J.J. (2002). Multilocus restriction typing, a novel tool for studying global epidemiology of *Burkholderia cepacia* complex infection in cystic fibrosis. J. Infect. Dis. *185*, 1454–1462.

Coenye, T., and LiPuma, J.J. (2003). Population structure analysis of *Burkholderia cepacia* genomovar III: varying degrees of genetic recombination characterize major clonal complexes. Microbiology *149*, 77–88.

Coenye, T., and Vandamme, P. (2003). Diversity and significance of *Burkholderia* species occupying diverse ecological niches. Env. Microbiol. *5*, 719–729.

Coenye T., Spilker, T., Martin, A., and LiPuma, J.J. (2002). Comparative assessment of genotyping methods for epidemiologic study of *Burkholderia cepacia* genomovar III. J. Clin. Microbiol. *40*, 3300–3307.

Coenye, T., Spilker, T., Van Schoor, A., LiPuma, J.J., and Vandamme, P. (2004). Recovery of *Burkholderia cenocepacia* strain PHDC from cystic fibrosis patients in Europe. Thorax *59*, 952–954.

Cooper, J.E., and Feil, E.J. (2004). Multilocus sequence typing – what is resolved.? Trends Microbiol. *12*, 373–377.

Cooper, V.S., Romero, E.O., and LiPuma, J.J. (2004). VNTR-based typing of *Burkholderia cenocepacia*. Abstract presented at the 9th IBCWG Meeting. Published online at http://go.to/cepacia.

Currie, B.J., and Jacups, S.P. (2003). Intensity of rainfall and severity of melioidosis, Australia. Emerg. Infect. Dis. *9*, 1538–1542.

Currie, B.J., Fisher, D.A., Howard, D.M., Burrow, J.N., Selvanayagam, S., Snelling, P.L., Anstey, N.M., and Mayo, M.J. (2000). The epidemiology of melioidosis in Australia and Papua New Guinea. Acta. Trop. *74*, 121–127.

Currie, B.J., Jacups, S.P., Cheng, A.C., Fisher, D.A., Anstey, N.M., Huffam, S.E., and Krause, V.L. (2004). Melioidosis epidemiology and risk factors from a prospective whole-population study in northern Australia. Trop. Med. Int. Health *9*, 1167–1174.

Currie, B.J., Gal, D., Mayo, M., Ward, L., Godoy, D., Spratt, B.G., and LiPuma, J.J. (2007a). Using BOX-PCR to exclude a clonal outbreak of melioidosis. BMC Infect. Dis. *7*, 68.

Currie, B.J., Thomas, A.D., Godoy, D., Dance, D.A., Cheng, A.C., Ward, L., Mayo, M., Pitt, T.L., and Spratt, B.G. (2007b). Australian and Thai isolates of *Burkholderia pseudomallei* are distinct by multilocus sequence typing: revision of a case of mistaken identity. J. Clin. Microbiol. *45*, 3828–3829.

Currie, B.J., Haslem, A., Pearson, T., Hornstra, H., Leadem, B., Mayo, M., Gal, D., Ward, L., Godoy, D., Spratt, B.G., and Keim, P. (2009). Identification of melioidosis outbreak by multilocus variable number tandem repeat analysis. Emerg. Infect. Dis. *15*, 169–174.

Dalmastri, C., Pirone, L., Tabacchioni, S., Bevivino, A., and Chiarini, L. (2005). Efficacy of species-specific *recA* PCR tests in the identification of *Burkholderia cepacia* complex environmental isolates. FEMS Microbiol. Lett. *246*, 39–45.

Dalmastri, C., Baldwin A., Tabacchioni, A., Bevivino, A., Mahenthiralingam, E., Chiarini, L., and Dowson, C. (2007). Investigating *Burkholderia cepacia* complex populations recovered from Italian maize rhizosphere by multilocus sequence typing. Env. Microbiol. *9*, 1632–1639.

Dance, D.A. (1991). Melioidosis: the tip of the iceberg? Clin. Microbiol. Rev. *4*, 52–60.

Dance, D.A. (2000). Ecology of *Burkholderia pseudomallei* and the interactions between environmental *Burkholderia* spp. and human-animal hosts. Acta. Trop. *74*, 159–168.

Dance, D.A., Smith, M.D., Aucken, H.M., and Pitt, T.L. (1999). Imported melioidosis in England and Wales. Lancet *353*, 208.

Dasen, S.E., LiPuma, J.J., Kostman, J.R., and Stull, T.L. (1994). Characterization of PCR-ribotyping for *Burkholderia (Pseudomonas) cepacia*. J. Clin. Microbiol. *32*, 2422–2424.

De Boeck, K., Malfroot, A., Van Schil, L., Lebecque, P., Knoop, C., Govan, J.R., Doherty, C., Laevens, S., Vandamme, P., and the Belgian *Burkholderia cepacia* Study Group. (2004). Epidemiology of *Burkholderia cepacia* complex colonisation in cystic fibrosis patients. Eur. Respir. J. *23*, 851–856.

Didelot, X., and Falush, D. (2007). Inference of bacterial microevolution using multilocus sequence data. Genetics. *175*, 1251–1266.

Doring, G., Jansen, S., Noll, H., Grupp, H., Frank, F., Botzenhart, K., Magdorf, K., and Wahn, U. (1996). Distribution and transmission of *Pseudomonas aeruginosa* and *Burkholderia cepacia* in a hospital ward. Pediatr. Pulmonol. *21*, 90–100.

Dorman, S.E., Gill, J.J., Gallin, J.I., and Holland, S.M. (1998). *Burkholderia pseudomallei* infection in a Puerto Rican patient with chronic granulomatouos disease: case report and review of occurrences in the Americas. Clin. Infect. Dis. *26*, 889–894.

Drabick, J.A., Gracdly, E.J., Heidecker, G.J., and LiPuma, J.J. (1996). Survival of *Burkholderia cepacia* on environmental surfaces. J. Hosp. Infect. *32*, 267–276.

Draper, A.D., Mayo, M., Harrington, G., Karp, D., Yinfoo, D., Ward, L., Haslem, A., Currie, B.J., and Kaestli, M. (2010). Association of the melioidosis agent *Burkholderia pseudomallei* with water parameters in rural water supplies in Northern Australia. Appl. Environ. Microbiol. *76*, 5305–5307.

Drevinek, P., and Mahenthiralingam, E. (2010). *Burkholderia cenocepacia* in cystic fibrosis: epidemiology and molecular mechanisms of virulence. Clin. Microbiol. Infect. *16*, 821–830.

Drevinek, P., Vosahlikova, S., Cinek, O., Vavrova, V., Bartosova, J., Pohunek, P., and Mahenthiralingam, E. (2005). Widespread clone of *Burkholderia cenocepacia* in cystic fibrosis patients in the Czech Republic. J. Med. Microbiol. *54*, 655–659.

Drevinek, P., Baldwin, A., Lindenburg, L., Joshi, L.T., Marchbank, A., Vosahlikova, S., Dowson, C.G., and Mahenthiralingam, E. (2010a). Oxidative stress of *Burkholderia cenocepacia* induces insertion sequence-mediated genomic rearrangements that interfere with macrorestriction-based genotyping. J. Clin. Microbiol. *48*, 34–40.

Drevinek, P., Vosahlikova, S., Dedeckova, K., Cinek, O., and Mahenthiralingam, E. (2010b). Direct culture-independent Strain typing of *Burkholderia cepacia* complex in sputum samples from patients with cystic fibrosis. J. Clin. Microbiol. *48*, 1888–1891.

Engelthaler, D.M., Bowers, J., Schupp, J.A., Pearson, T., Ginther, J., Hornstra, H.M., Dale, J., Stewart, T., Sunenshine, R., Waddell, V., *et al.* (2011a). Molecular investigations of a locally acquired case of melioidosis in Southern AZ, USA. PLoS Negl. Trop. Dis. *5*, e1347.

Engelthaler, D.M., Chiller, T., Schupp, J.A., Colvin, J., Beckstrom-Sternberg, S.M., Driebe, E.M., Moses, T., Tembe, W., Sinari, S., Beckstrom-Sternberg, J.S., *et al.* (2011b). Next-generation sequencing of *Coccidioides immitis* isolated during cluster investigation. Emerg. Infect. Dis. *17*, 227–232.

Enright, M.C., and Spratt, B.G. (1999). Multilocus sequence typing. Trends Microbiol. 7, 482–487.

Ensor, E., Humphreys, H., Peckham, D., Webster, C., and Knox, A.J. (1996). Is *Burkholderia (Pseudomonas) cepacia* disseminated from cystic fibrosis patients during physiotherapy? J. Hosp. Infect. 32, 9–15.

Feil, E.J., and Enright, M.C. (2004). Analyses of clonality and the evolution of bacterial pathogens. Curr. Opin. Microbiol. 7, 308–313.

Feil, E.J., Li, B.C., Aanensen, D.M., Hanage, W.P., and Spratt, B.G. (2004). eBURST: inferring patterns of evolutionary descent among clusters of related bacterial genotypes from multilocus sequence typing data. J. Bacteriol. *186*, 1518–1530.

Fiore, A., Laevens, S., Bevivino, A., Dalmastri, C., Tabacchioni, S., Vandamme, P., and Chiarini, L. (2001). *Burkholderia cepacia* complex: distribution of genomovars among isolates from the maize rhizosphere in Italy. Environ. Microbiol. 3, 137–143.

Fisher, M.C., LiPuma, J.J., Dasen, S.E., Caputo, G.C., Mortensen, J.E., McGowan, K.L., and Stull, T.L. (1993). Source of *Pseudomonas cepacia*: ribotyping of isolates from patients and from the environment. J. Pediatr. *123*, 745–747.

Fothergill, J.L., White, J., Foweraker, J.E., Walshaw, M.J., Ledson, M.J., Mahenthiralingam, E., and Winstanley, C. (2010). Impact of *Pseudomonas aeruginosa* genomic instability on the application of typing methods in chronic cystic fibrosis infections. J. Clin. Microbiol. *48*, 2053–2059.

France, M.W., Dodd, M.E., Govan, J.R., Doherty, C.J., Webb, A.K., and Jones, A.M. (2008). The changing epidemiology of Burkholderia species infection at an adult cystic fibrosis centre. J. Cyst. Fibros. 7, 368–3672.

Frothingham, R., and Meeker-O'Connell, W.A. (1998). Genetic diversity in the *Mycobacterium tuberculosis* complex based on variable numbers of tandem DNA repeats. Microbiol. *144*, 1189–1196.

Gilligan, P.H. (1995). *Pseudomonas* and *Burkholderia*. In Manual of Clinical Microbiology, Murray, P.R., Baron, E.J., and Pfaller, M.A., eds. (Washington, D.C.: ASM Press), pp. 509–519.

Godoy, D., Randle, G., Simpson, A.J., Aanensen, D.M., Pitt, T.L., Kinoshita, R., and Spratt, B.G. (2003). Multilocus sequence typing and evolutionary relationships among the causative agents of melioidosis and glanders, *Burkholderia pseudomallei* and *Burkholderia mallei*. J. Clin. Microbiol. *41*, 2068–2079.

Gold, R., Jin, E., Levison, H., Isles, A., and Fleming, P.C. (1983). Ceftazidime alone and in combination in patients with cystic fibrosis: lack of efficacy in treatment of severe respiratory infections caused by *Pseudomonas cepacia*. J. Antimicrob. Chemother. *12 (Suppl A)*, 331–336.

Gonzalez, C.F., Mark, G.L., Mahenthiralingam, E., and LiPuma, J.J. (2000). Isolation of soil-borne genomovar III *Burkholderia cepacia* and lytic phages with inter-species host range. (Abstract). Pediatr. Pulmonol. *30*, 288–289.

Govan, J.R., Brown, P.H., Maddison, J., Doherty, C.J., Nelson, J.W., Dodd, M., Greening, A.P., and Webb, A.K. (1993). Evidence for transmission of *Pseudomonas cepacia* by social contact in cystic fibrosis. Lancet *342*, 15–19.

Govan, J.R., Hughes, J.E., and Vandamme, P. (1996). *Burkholderia cepacia*: medical, taxonomic and ecological issues. J. Med. Microbiol. *45*, 395–407.

Graindorge, A., Menard, A., Neto, M., Bouvet, C., Miollan, R., Gaillard, S., de Montclos, H., Laurent, F., and Cournoyer, B. (2010). Epidemiology and molecular characterization of a clone of *Burkholderia cenocepacia* responsible for nosocomial pulmonary tract infections in a French intensive care unit. Diagn. Microbiol. Infect. Dis. *66*, 29–40.

Haase, A., Smith-Vaughan, H., Melder, A., Wood, Y., Janmaat, A., Gilfedder, J., Kemp, D., and Currie, B. (1995). Subdivision of *Burkholderia pseudomallei* ribotypes into multiple types by random amplified polymorphic DNA analysis provides new insights into epidemiology. J. Clin. Microbiol. *33*, 1687–1690.

Hampton, V., Kaestli, M., Mayo, M., Choy, J.L., Harrington, G., Richardson, L., Benedict, S., Noske, R., Garnett, S.T., Godoy, D., Spratt, B.G., and Currie, B.J. (2011). Melioidosis in birds and *Burkholderia pseudomallei* dispersal, Australia. Emerg. Infect. Dis. *17*, 1310–1312.

Harris, S.R., Feil, E.J., Holden, M.T., Quail, M.A., Nickerson, E.K., Chantratita, N., Gardete, S., Tavares, A., Day, N., Lindsay, J.A., *et al.* (2010). Evolution of MRSA during hospital transmission and intercontinental spread. Science. *327*, 469–474.

Healy, M., Huong, J., Bittner, T., Lising, M., Frye, S., Raza, S., Schrock, R., Manry, J., Renwick, A., Nieto, R., Woods, C., Versalovic, J., and Lupski, J.R. (2005). Microbial DNA typing by automated repetitive-sequence-based PCR. J. Clin. Microbiol. *43*,199–207.

Hutchinson, G.R., Parker, S., Pryor, J.A., Duncan-Skingle, F., Hoffman, P.N., Hodson, M.E., Kaufmann, M.E., and Pitt, T.L. (1996). Home-use nebulizers: a potential primary source of *Burkholderia cepacia* and other colistin-resistant, Gram-negative bacteria in patients with cystic fibrosis [published erratum appears in J. Clin. Microbiol. 1996 Jun; 34(6):1601]. J. Clin. Microbiol. *34*, 584–587.

Hymphreys, H., Peckham, D., Patel, P., and Knox, A. (1994). Airborne dissemination of *Burkholderia (Pseudomonas) cepacia* from adult patients with cystic fibrosis. Thorax *59*, 1157–1159.

Inglis, T.J., O'Reilly, L., Foster, N., Clair, A., and Sampson, J. (2002). Comparison of rapid, automated ribotyping and DNA macrorestriction analysis of *Burkholderia pseudomallei*. J. Clin. Microbiol. *40*, 3198–3203.

Isles, A., Maclusky, I., Corey, M., Gold, R., Prober, C., Fleming, P., and Levison, H. (1984). *Pseudomonas cepacia* infection in cystic fibrosis: an emerging problem. J. Pediatr. *104*, 206–210.

John, M., Ecclestone, E., Hunter, E., Couroux, P., and Hussain, Z. (1994). Epidemiology of *Pseudomonas cepacia* colonization among patients with cystic fibrosis. Pediatr. Pulmonol. *18*, 108–113.

Johansson, A., Farlow, J., Larsson, P., Dukerich, M., Chambers, E., Bystrom, M., Fox, J., Chu, M., Forsman, M., Sjostedt, A., and Keim, P. (2004). Worldwide genetic relationships among *Francisella tularensis* isolates determined by multiple-locus variable-number tandem repeat analysis. J. Bacteriol. *186*, 5808–5818.

Johnson, W.M., Tyler, S.D., and Rozee, K.R. (1994). Linkage analysis of geographic and clinical clusters in *Pseudomonas cepacia* infections by multilocus enzyme electrophoresis and ribotyping. J. Clin. Microbiol. 32, 924–930.

Kalish, L.A., Waltz, D.A., Dovey, M., Potter-Bynoe, G., McAdam, A.J., LiPuma, J.J., Gerard, C., and Goldmann, D. (2006). Impact of *Burkholderia dolosa* on lung function and survival in cystic fibrosis. Am. J. Respir. Crit. Care Med. *173*, 421–425.

Kanaphun, P., Thirawattanasuk, N., Suputtamongkol, Y., Naigowit, P., Dance, D.A., Smith, M.D., and White, N.J. (1993). Serology and carriage of *Pseudomonas pseudomallei*: a prospective study in 1000 hospitalized children in northeast Thailand. J. Infect. Dis. *167*, 230–233.

Keim, P., Price, L.B., Klevytska, A.M., Smith, K.L., Schupp, J.M., Okinaka, R., Jackson, P.J., and Hugh-Jones, M.E. (2000). Multiple-locus variable-number tandem repeat analysis reveals genetic relationships within *Bacillus anthracis*. J. Bacteriol. *182*, 2928–2936.

Koonpaew, S., Ubol, M.N., Sirisinha, S., White, N.J., and Chaiyaroj, S.C. (2000). Genome fingerprinting by pulsed-field gel electrophoresis of isolates of *Burkholderia pseudomallei* from patients with melioidosis in Thailand. Acta. Trop. *74*, 187–191.

Köser, C.U., Holden, M.T., Ellington, M.J., Cartwright, E.J., Brown, N.M., Ogilvy-Stuart, A.L., Hsu, L.Y., Chewapreecha, C., Croucher, N.J., Harris, S.R., *et al.* (2012a). Rapid whole-genome sequencing for investigation of a neonatal MRSA outbreak. N. Engl. J. Med. *366*, 2267–2275.

Köser, C.U., Ellington, M.J., Cartwright, E.J., Gillespie, S.H., Brown, N.M., Farrington, M., Holden, M.T., Dougan, G., Bentley, S.D., Parkhill, J., and Peacock, S.J. (2012b). Routine use of microbial whole genome sequencing in diagnostic and public health microbiology. PloS Pathog. *8*, e1002824.

Kostman J.R., Alden, M.B., Mair, M., Edlind, T.D., LiPuma, J.J., and Stull, T.L. (1995). A universal approach to bacterial molecular epidemiology by polymerase chain reaction ribotyping. J. Infect. Dis. *171*, 204–208.

Kumar, A., Dietrich, S., Schneider, W., Jacobson, R., Downes, F.P., Robinson-Dunn, B.E., Honicky, R., Smith, J., and Martin, R. (1997). Genetic relatedness of *Burkholderia (Pseudomonas) cepacia* isolates from five cystic fibrosis centers in Michigan. Respir. Med. *91*, 485–492.

Kuroda, M., Serizawa, M., Okutani, A., Sekizuka, T., Banno, S., and Inoue, S. (2010). Genome-wide single nucleotide polymorphism typing method for identification of *Bacillus anthracis* species and strains among *B. cereus* group species. J. Clin. Microbiol. *48*, 2821–2829.

Laraya-Cuasay, L.R., Lipstein, M., and Huang, N.N. (1977). *Pseudomonas cepacia* in the respiratory flora of patients with cystic fibrosis. Pediatr. Pulmonol. *11*, 502.

Ledson, M.J., Gallagher, M.J., Corkill, J.E., Hart, C.A., and Walshaw, M.J. (1998). Cross infection between cystic fibrosis patients colonised with *Burkholderia cepacia*. Thorax *53*, 432–436.

Leelayuwat, C., Romphruk, A., Lulitanond, A., Trakulsomboon, S., and Thamlikitkul, V. (2000). Genotype analysis of *Burkholderia pseudomallei* using randomly amplified polymorphic DNA (RAPD): indicative of genetic differences amongst environmental and clinical isolates. Acta. Trop. *77*, 229–237.

Lewis, T., Loman, N.J., Bingle, L., Jumaa, P., Weinstock, G.M., Mortiboy, D., and Pallen, M.J. (2010). High-throughput whole-genome sequencing to dissect the epidemiology of Acinetobacter baumannii isolates from a hospital outbreak. J. Hosp. Infect. *75*, 37–41.

Lieberman, T.D., Michel, J.B., Aingaran, M., Potter-Bynoe, G., Roux, D., Davis, M.R. Jr., Skurnik, D., Leiby, N., LiPuma, J.J., Goldberg, J.B., *et al.* (2011). Parallel bacterial evolution within multiple patients identifies candidate pathogenicity genes. Nat. Genet. *43*, 1275–1280.

LiPuma, J.J. (1998). Molecular tools for epidemiological study of infectious diseases. Pediatr. Infect. Dis. J. *17*, 667–675.

LiPuma J.J., Mortensen, J.E., Dasen, S.E., Edlind, T.D., Schidlow, D.V., Burns, J.L., and Stull, T.L. (1988). Ribotype analysis of *Pseudomonas cepacia* from cystic fibrosis treatment centers. J. Pediatr. *113*, 859–862.

LiPuma J.J., Dasen, S.E., Nielson, D.W., Stern, R.C., and Stull, T.L. (1990). Person-to-person transmission of *Pseudomonas cepacia* between patients with cystic fibrosis. Lancet *336*, 1094–1096.

LiPuma, J.J., Marks-Austin, K.A., Holsclaw, D.S., Winnie, G.B., Gilligan, P.H., and Stull, T.L. (1994). Inappararent transmission of *Pseudomonas (Burkholderia) cepacia* among patients with cystic fibrosis. Pediatr. Infect. Dis. J. *13*, 716–719.

LiPuma, J.J., Spilker, T., Coenye, T., and Gonzalez, C.F. (2002). An epidemic *Burkholderia cepacia* complex strain identified in soil. Lancet *359*, 2002–2003.

Liu, P.Y.F., Dhi, Z.Y., Lau, Y.J., Hu, B.S., Shyr, J.M., Tsai, W.S., Lin, Y.H., and Tseng, C.Y. (1995). Comparison of different PCR approaches for characterization of *Burkholderia (Pseudomonas) cepacia* isolates. J. Clin. Microbiol. *33*, 3304–3307.

Liu, Y., Loh, J.P., Aw, L.T., Yap, E.P.H., Lee, M.A., and Ooi, E.E. (2006). Rapid molecular typing of *Burkholderia pseudomallei*, isolated in an outbreak of melioidosis in Singapore in 2004, based on variable-number tandem repeats. Trans. Royal Soc. Trop. Med. Hyg. *100*, 687–692.

McCombie, R.L., Finkelstein, R.A., and Woods, D.E. (2006). Multilocus sequence typing of historical *Burkholderia pseudomallei* isolates collected in Southeast Asia from 1964 to 1967 provides insight into the epidemiology of melioidosis. J. Clin. Microbiol. *44*, 2951–2962.

Maiden, M.C. (2006). Multilocus sequence typing of bacteria. Annu. Rev. Microbiol. *60*, 561–588.

Maiden, M.C.J., Bygraves, J.A., Feil, E., Morelli, G., Russell, J.E., Urwin, R., Zhang, Q., Zhou, J., Zurth, K., Caugant, D.A., Feavers, I.M., Achtman, M., and Spratt, B.G. (1998). Multilocus sequence typing: a portable approach to the identification of clones within populations of pathogenic microorganisms. Proc. Natl. Acad. Sci. U.S.A. *95*, 3140–3145.

Mahenthiralingham, E., Campbell, M.E., Henry, D.A., and Speert, D.P. (1996). Epidemiology of *Burkholderia cepacia* infection in patients with cystic fibrosis: analysis by randomly amplified polymorphic DNA fingerprinting. J. Clin. Microbiol. *34*, 2914–2920.

Mahenthiralingam, E., Simpson, D.A., and Speert, D.P. (1997). Identification and characterization of a novel DNA marker associated with epidemic *Burkholderia cepacia* strains recovered from patients with cystic fibrosis. J. Clin. Microbiol. *35*, 808–816.

Mahenthiralingam, E., Bischof, J., Byrne, S.K., Radomski, C., Davies, J.E., Av-Gay, Y., and Vandamme, P. (2000). DNA-Based diagnostic approaches for identification of *Burkholderia cepacia* complex, *Burkholderia vietnamiensis, Burkholderia multivorans, Burkholderia stabilis*, and *Burkholderia cepacia* genomovars I and III. J. Clin. Microbiol *38*, 3165–3173.

Mahenthiralingam E., Vandamme, P., Campbell, M.E., Henry, D.A., Gravelle, A.M., Wong, L.T.K., Davidson, A.G.F., Wilcox, P.G., Nakielna, B., and Speert, D.P. (2001). Infection with *Burkholderia cepacia* complex genomovars in patients with cystic fibrosis: virulent transmissible strains of genomovar III can replace *Burkholderia multivorans*. Clin. Infect. Dis. *33*, 1469–1475.

Mahenthiralingam, E., Baldwin, A., Drevinek, P., Vanlaere, E., Vandamme, P., LiPuma J.J., and Dowson, C.G. (2006). Multilocus sequence typing breathes life into a microbial metagenome. PLoS One 1:e17.

Mahenthiralingam, E., Baldwin, A., and Dowson, C.G. (2008). *Burkholderia cepacia* complex bacteria: opportunistic pathogens with important natural biology. J. Appl. Microbiol. *104*,1539–1551.

Manno, G., Dalmastri, C., Tabacchioni, S., Vandamme, P., Lorini, R., Minicucci, L., Romano, L., Giannattasio, A., Chiarini, L., and Bevivino, A. (2004). Epidemiology and clinical course of *Burkholderia cepacia* complex infections, particularly those caused by different *Burkholderia cenocepacia* strains, among patients attending an Italian cystic fibrosis center. J. Clin. Microbiol. *42*, 1491–1497.

Maynard-Smith, J.M., Smith, N.H., O'Rourke, M., and Spratt, B.G. (1993). How clonal are bacteria? Proc. Natl. Acad. Sci. U.S.A. *90*, 4384–4388.

McArthur J.V., Kovacic, D.A., and Smith, M.H. (1988). Genetic diversity in natural populations of a soil bacterium across a landscape gradient. Proc. Natl. Acad. Sci. U.S.A. *85*, 9621–9624.

Millar-Jones, L., Paull, A., Saunders, Z., and Goodchild, M.C. (1992). Transmission of *Pseudomonas cepacia* among cystic fibrosis patients. [letter; comment]. Lancet *340*, 491.

Miller, S.M., Parke, J.L., Bies, S., and LiPuma, J.J. (2000). Detection, recovery and identification of *Burkholderia cepacia* from the natural environment [abstract]. Pediatr. Pulmonol. Suppl. *20*, 288.

Miyawaki, H., Fujita, J., Takigawa, K., Negayama, K., Yamagishi, Y., Yamaji, Y., Ouchi, K., Nakazawa, T., Kawanishi, K., and Takahara, J. (1995). Investigation of nosocomial respiratory infection due to *Pseudomonas cepacia* by arbitrarily primed polymerase chain reaction. Diagn. Microbiol. Infect. Dis. *23*, 77–83.

Moore, J.E., Thompson, I., Crowe, M., Xu, J., Shaw, A., Millar, B.C., Redmond, A.O.B., Reid, A.J.M., Clarke, C., and Elborn, J.S. (2002a). *Burkholderia cepacia* from a sink drain. J. Hosp. Infect. *50*, 235–237.

Moore, J.E., McIlhatton, B., Buchanan, J., Gilpin, D., Shaw, A., Hall, V., Murphy, P.G., and Elborn, J.S. (2002b). Occurrence of *Burkholderia cepacia* in the hospital environment. Ir. J. Med. Sci. *171*, 131–133.

Mortensen, J.E., Fisher, M.C., and LiPuma, J.J. (1995). Recovery of *Pseudomonas cepaci*a and other *Pseudomonas* species from the environment. Infect. Control Hosp. Epidemiol. *16*, 30–32.

Mukhopadhyay, C., Kaestli, M., Vandana, K.E., Sushma, K., Mayo, M., Richardson, L., Tuanyok, A., Keim, P., Godoy, D., Spratt, B.G., and Currie, B.J. (2011). Molecular characterization of clinical *Burkholderia pseudomallei* isolates from India. Am. J. Trop. Med. Hyg. *85*, 121–123.

Musser J.M. (1996). Molecular population genetic analysis of emerged bacterial pathogens: selected insights. Emerg. Infect. Dis. *2*, 1–17.

Ngauy, V., Lemeshev, Y., Sadkowski, L., and Crawford, G. (2005). Cutaneous melioidosis in a man who was taken as a prisoner of war by the Japanese during World War II. J. Clin. Microbiol. *43*, 970–972.

Nolan, G., Moivor, P., Levison, H., Fleming, P.C., Corey, M., and Gold, R. (1982). Antibiotic prophylaxis in cystic fibrosis: inhaled cephaloridine as an adjunct to oral cloxacillin. J. Pediatr. *101*, 626–630.

Norton, R., Roberts, B., Freeman, M., Wilson, M., Ashhurst-Smith, C., Lock, W., Brookes, D., and La Brooy, J. (1998). Characterisation and molecular typing of *Burkholderia pseudomallei*: are disease presentations of melioidosis clonally related? FEMS Immunol. Med. Microbiol. *20*, 37–44.

O'Carroll, M.R., Kidd, T.J., Coulter, C., Smith, H.V., Rose, B.R., Harbour, C., and Bell, S.C. (2003). *Burkholderia pseudomallei*: another emerging pathogen in cystic fibrosis. Thorax *58*, 1087–1091.

Okazaki, M., Watanabe, T., Morita, K., Higurashi, Y., Araki, K., Shukuya, N., Baba, S., Watanabe, N., Egami, T., Furuya, N., *et al.* (1999). Molecular epidemiological investigation using a randomly amplified polymorphic DNA assay of *Burkholderia cepacia* isolates from nosocomial outbreaks. J. Clin. Microbiol. *37*, 3809–3814.

Olive, D.M., and Bean, P. (1999). Principles and applications of methods for DNA-based typing of microbial organisms. J. Clin. Microbiol. *37*, 1661–1669.

Onteniente, L., Brisse, S., Tassios, P.T., and Vergnaud, G. (2003). Evaluation of the polymorphisms associated with tandem repeats for *Pseudomonas aeruginosa* strain typing. J. Clin. Microbiol. *41*, 4991–4997.

Parry, C.M., Wuthiekanun, V., Hoa, N.T., Diep, T.S., Thao, L.T., Loc, P.V., Wills, B.A., Wain, J., Hien, T.T., White, N.J., and Farrar, J.J. (1999). Melioidosis in southern Vietnam: clinical surveillance and environmental sampling. Clin. Infect. Dis. *29*, 1323–1326.

Payne, G.W., Vandamme, P., Morgan, S.H., Lipuma, J.J., Coenye, T., Weightman, A.J., Jones, T.H., and Mahenthiralingam, E. (2005). Development of a *recA* gene-based identification approach for the entire *Burkholderia* genus. Appl. Environ. Microbiol. *71*, 3917–3927.

Pearson, T., U'Ren, J.M., Schupp, J.M., Allan, G.J., Foster, P.G., Mayo, M.J., Gal, D., Choy, J.L., Daugherty, R.L., Kachur, S., *et al.* (2007). VNTR analysis of selected outbreaks of *Burkholderia pseudomallei* in Australia. Infect. Genet. Evol. *7*, 416–423.

Pegues, D.A., Carson, L.A., Tablan, O.C., FitzSimmons, S.C., Roman, S.B., Miller, J.M., and Jarvis, W.R. (1994a). Acquisition of *Pseudomonas cepacia* at summer camps for patients with cystic fibrosis. Summer Camp Study Group. J. Pediatr. *124*, 694–702.

Pegues, D.A., Schidlow, D.V., Tablan, O.C., Carson, L.A., Clark, N.C., and Jarvis, W.R. (1994b). Possible nosocomial transmission of *Pseudomonas cepacia* in patients with cystic fibrosis. Arch. Ped. Adolesc. Med. *148*, 805–812.

Pitchford, K.C., Corey, M., Highsmith, A.K., Perlman, R., Bannatyne, R., Gold, R., Levison, H., and Ford-Jones, E.L. (1987). *Pseudomonas* species contamination of cystic fibrosis patients' home inhalation equipment. J. Pediatr. *111*, 212–216.

Pitt, T.L., Kaufmann, M.E., Patel, P.S., Benge, L.C., Gaskin, S., and Livermore, D.M. (1996). Type characterization and antibiotic susceptibility of *Burkholderia (Pseudomonas) cepacia* isolates from patients with cystic fibrosis in the United Kingdom and the Republic of Ireland. J. Med. Microbiol. *44*, 203–210.

Power, E.G. (1996). RAPD typing in microbiology – a technical review. J. Hosp. Infect. *34*, 247–265.

Price, E.P., Hornstra, H.M., Limmathurotsakul, D., Max, T.L., Sarovich, D.S., Vogler, A.J., Dale, J.L., Ginther, J.L., Leadem, B., Colman, R.E., *et al.* (2010). Within-host evolution of *Burkholderia pseudomallei* in four cases of acute melioidosis. PLoS Pathog. *6*, e1000725.

Rabkin, C.S., Jarvis, W.R., and Martone, W.J. (1987). Current status of *Pseudomonas cepacia* typing systems. Eur. J. Epidemiol. 3, 343–346.

Rabkin C.S., Jarvis, W.R., Anderson, R.L., Govan, J., Klinger, J., LiPuma, J.J., Martone, W.J., Monteil, H., Richard, C., Shigeta, S., Sosa, A., Stull, T., Swenson, J., and Woods, D. (1989). *Pseudomonas cepacia* typing systems: collaborative study to assess their potential in epidemiologic investigations. Rev. Infect. Dis. *11*, 600–607.

Radua, S., Ling, O.W., Srimontree, S., Lulitanond, A., Hin, W.F., Yuherman, Lihan, S., Rusul, G., and Mutalib, A.R. (2000). Characterization of *Burkholderia pseudomallei* isolated in Thailand and Malaysia. Diagn. Microbiol. Infect. Dis. *38*, 141–145.

Ramette, A., LiPuma, J.J., and Tiedje, J.M. (2005). Species abundance and diversity of *Burkholderia cepacia* complex in the environment. Appl. Environ. Microbiol. *71*, 1193–1201.

Reik, R., Spilker, T., and LiPuma, J.J. (2005). Distribution of *Burkholderia cepacia* complex species recovered from persons with or without cystic fibrosis. J. Clin. Microbiol. *43*, 2926–2928.

Rosenstein, B.J., and Hall, D.E. (1980). Pneumonia and septicemia due to *Pseudomonas cepacia* in a patient with cystic fibrosis. Johns Hopkins Med. J. *147*, 188–189.

Saiman, L., and Siegel, J. (2003). Infection control recommendations for patients with cystic fibrosis: microbiology, important pathogens, and infection control practices to prevent patient-to-patient transmission. Am. J. Infect. Control. *31* (Suppl.), S1–62.

Salles, J.F., De Souza, F.A., and van Elsas, J.D. (2002). Molecular method to assess the diversity of Burkholderia species in environmental samples. Appl. Environ. Microbiol. *68*, 1595–1603.

Schouls, L.M., van der Heide, H.G., Vauterin, L., Vauterin, P., and Mooi, F.R. (2004). Multiple-locus variable-number tandem repeat analysis of Dutch *Bordetella pertussis* strains reveals rapid genetic changes with clonal expansion during the late 1990s. J. Bacteriol. *186*, 5496–5505.

Schulin, T., and Steinmetz, I. (2001). Chronic melioidosis in a patient with cystic fibrosis. J. Clin. Microbiol. *39*, 1676–1677.

Segonds, C., Bingen, E., Couetdic, G., Mathy, S., Brahimy, N., Marty, N., Plesiat, P., Michel-Briand, Y., and Chabanon, G. (1997). Genotypic analysis of *Burkholderia cepacia* isolates from 13 French cystic fibrosis centers. J. Clin. Microbiol. 35, 2055–2060.

Segonds, C., Heulin, T., Marty, N., and Chabanon, G. (1999). Differentiation of *Burkholderia* species by PCR-restriction fragment length polymorphism analysis of the 16S rRNA gene and application to cystic fibrosis isolates. J. Clin. Microbiol. 37, 2201–2208.

Seo, S.T., and Tsuchiya, K. (2005). Genotypic characterization of *Burkholderia cenocepacia* strains by rep-PCR and PCR-RFLP of the fliC gene. FEMS Microbiol. Lett. *245*, 19–24.

Smith, D.L., Gumery, L.B., Smith, E.G., Stableforth, D.E., Kaufmann, M.E., and Pitt, T.L. (1993). Epidemic of *Pseudomonas cepacia* in an adult cystic fibrosis unit: evidence of person-to-person transmission. J. Clin. Microbiol. *31*, 3017–3022.

Speert, D.P., Henry, D., Vandamme, P., Corey, M., and Mahenthiralingam, E. (2002). Epidemiology of *Burkholderia cepacia* complex in patients with cystic fibrosis, Canada. Emerg. Infect. Dis *8*, 181–187.

Spilker, T., Baldwin, A., Bumford, A., Dowson, C.G., Mahenthiralingam, E., and LiPuma, J.J. (2009).

Expanded multilocus sequence typing for *Burkholderia* species. J. Clin. Microbiol. *47*, 2607–2610.

Spratt, B.G. (1999). Multilocus sequence typing: molecular typing of bacterial pathogens in an era of rapid DNA sequencing and the Internet. Curr. Opin. Microbiol. *2*, 312–316.

Srinivasan, A., Kraus, C.N., DeShazer, D., Becker, P.M., Dick, J.D., Spacek, L., Bartlett, J.G., Byrne, W.R., and Thomas, D.L. (2001). Glanders in a military research microbiologist. N. Engl. J. Med. *345*, 256–258.

Stull, T.L., LiPuma, J.J., and Edlind, T.D. (1988). A broad-spectrum probe for molecular epidemiology of bacteria: ribosomal RNA. J. Infect. Dis. *157*, 280–286.

Sun, L., Jiang, R.Z., Steinbach, S., Holmes, A., Campanelli, C., Forstner, J., Sajjan, U., Tan, Y., Riley, M., and Goldstein, R. (1995). The emergence of a highly transmissible lineage of cbl+ *Pseudomonas (Burkholderia) cepacia* causing CF centre epidemics in North America and Britain. Nat. Med. *1*, 661–666.

Tablan, O.C., Chorba, T.L., Schidlow, D.V., White, J.W., Hardy, K.A., Gilligan, P.H., Morgan, W.M., Carson, L.A., Martone, W.J., and Jason, J.M. (1985). *Pseudomonas cepacia* colonization in patients with cystic fibrosis: risk factors and clinical outcome. J. Pediatr. *107*, 382–387.

Tablan, O.C., Martone, W.J., Doershuk, C.F., Stern, R.C., Thomassen, M.J., Klinger, J.D., White, J.W., Carson, L.A., and Jarvis, W.R. (1987). Colonization of the respiratory tract with *Pseudomonas cepacia* in cystic fibrosis. Risk factors and outcomes. Chest *91*, 527–532.

Tenover F.C., Arbeit, R.D., Goering, R.V., Mickelsen, P.A., Murray, B.E., Persing, D.H., and Swaminathan, B. (1995). Interpreting chromosomal DNA restriction patterns produced by pulsed-field gel electrophoresis: criteria for bacterial strain typing. J. Clin. Microbiol. *33*, 2233–2239.

Thomassen, M.J., Demko, C.A., Klinger, J.D., and Stern, R.C. (1985). *Pseudomonas cepacia* colonization among patients with cystic fibrosis. A new opportunist. Am. Rev. Respir. Dis. *131*, 791–796.

Thomassen, M.J., Demko, C.A., Doershuk, C.F., Stern, R.C., and Klinger, J.D. (1986). *Pseudomonas cepacia*: decrease in colonization in patients with cystic fibrosis. Am. Rev. Respir. Dis. *134*, 669–671.

Torbeck, L., Raccasi, L., Guilfoyle, D.E., Friedman, R.L., and Hussong, D. (2011). *Burkholderia cepacia*: This decision is overdue. PDA J. Pharm. Sci. Technol. *65*, 535–543.

U'Ren, J.M., Hornstra, H., Pearson, T., Schupp, J.M., Leadem, B., Georgia, S., Sermswan, R.W., and Keim, P. (2007a). Fine-scale genetic diversity among *Burkholderia pseudomallei* soil isolates in northeast Thailand. Appl. Environ. Microbiol. *73*, 6678–6681.

U'Ren, J.M., Schupp, J.M., Pearson, T., Hornstra, H., Friedman, C.L., Smith, K.L., Daugherty, R.R., Rhoton, S.D., Leadem, B., Georgia, S., *et al.* (2007b). Tandem repeat regions within the *Burkholderia pseudomallei* genome and their application for high resolution genotyping. BMC Microbiol. *7*, 23.

Urwin, R., and Maiden, M.C. (2003). Multi-locus sequence typing: a tool for global epidemiology. Trends Microbiol. *11*, 479–487.

Vadivelu, J., Puthucheary, S.D., Mifsud, A., Drasar, B.S., Dance, D.A., and Pitt, T.L. (1997). Ribotyping and DNA macrorestriction analysis of isolates of *Burkholderia pseudomallei* from cases of melioidosis in Malaysia. Trans. R. Soc. Trop. Med. Hyg. *91*, 358–360.

van Belkum, A. (1994). DNA fingerprinting of medically important microorganisms by use of PCR. Clin. Microbiol. Rev. *7*, 174–184.

van Belkum, A., Melchers, W.J., Ijsseldijk, C., Nohlmans, L.,Verbrugh, H., and Meis, J.F. (1997). Outbreak of amoxicillin-resistant *Haemophilus influenzae* type b: variable number of tandem repeats as novel molecular markers. J. Clin. Microbiol. *35*, 1517–1520.

Vandamme, P., Mahenthiralingam, E., Holmes, B., Coenye, T., Hoste, B., De Vos, P., Henry, D., and Speert, D.P. (2000). Identification and population structure of *Burkholderia stabilis* sp. nov. (formerly *Burkholderia cepacia* genomovar IV). J. Clin. Microbiol. *38*, 1042–1047.

Vanlaere, E., Coenye, T., Samyn, E., Van den Plas, C., Govan, J., De Baets, F., De Boeck, K., Knoop, C., and Vandamme, P. (2005). A novel strategy for the isolation and identification of environmental *Burkholderia cepacia* complex bacteria. FEMS Microbiol. Lett. *249*, 303–307.

Vanlaere, E., Lipuma, J.J., Baldwin, A., Henry, D., De Brandt, E., Mahenthiralingam, E., Speert, D., Dowson, C., and Vandamme, P. (2008). *Burkholderia latens* sp. nov., *Burkholderia diffusa* sp. nov., *Burkholderia arboris* sp. nov., *Burkholderia seminalis* sp. nov. and Burkholderia metallica sp. nov., novel species within the *Burkholderia cepacia* complex. Int. J. Syst. Evol. Microbiol. *58*, 1580–1590.

Vanlaere, E., Baldwin, A., Gevers, D., Henry, D., De Brandt, E., LiPuma, J.J., Mahenthiralingam, E., Speert, D.P., Dowson, C., and Vandamme, P. (2009). Taxon K, a complex within the *Burkholderia cepacia* complex, comprises at least two novel species, *Burkholderia contaminans* sp. nov. and *Burkholderia lata* sp. nov. Int. J. Syst. Evol. Microbiol. *59*, 102–111.

Visca, P., Cazzola, A., Petrucca, A., and Vraggion, V. (2001). Travel-associated *Burkholderia pseudomallei* infection (melioidosis) in a patient with cystic fibrosis: a case report. Clin. Infect. Dis. *32*, 315-e16.

Vermis, K., Brachkova, M., Vandamme, P., and Nelis, H. (2003). Isolation of *Burkholderia cepacia* complex genomovars from waters. Syst. Appl. Microbiol. *26*, 595–600.

Versalovic, J., Koeuth, T., and Lupski, J.R. (1991). Distribution of repetitive DNA sequences in eubacteria and application to fingerprinting of bacterial genomes. Nucleic Acids Res. *19*, 6823–6831.

Waine, D.J., Henry, D.A., Baldwin, A., Speert, D.P., Honeybourne, D., Mahenthiralingam, E., and Dowson, C.G. (2007). Reliability of multilocus sequence typing of the *Burkholderia cepacia* complex in cystic fibrosis. J. Cystic Fibros. *6*, 215–219.

Wheelis, M. (1998). First shots fired in biological warfare. Nature *395*, 213.

White, N.J. (2003). Melioidosis. Lancet *361*, 1715–1722.

White, J., Gilbert, J., Hill, G., Hill, E., Huse, S.M., Weightman, A.J., and Mahenthiralingam, E. (2011). Culture-independent analysis of bacterial fuel contamination provides insight into the level of concordance with the standard industry practice of aerobic cultivation. Appl. Env. Microbiol. *77*, 4527–4538.

Whiteford, M.L., Wilkinson, J.D., McColl, J.H., Conlon, F.M., Michie, J.R., Evans, T.J., and Paton, J.Y. (1995). Outcome of *Burkholderia (Pseudomonas) cepacia* colonisation in children with cystic fibrosis following a hospital outbreak. Thorax *50*, 1194–1198.

Wise M.G., Shimkets, L.J., and McArthur, J.V. (1995). Genetic structure of a lotic population of *Burkholderia* (*Pseudomonas*) *cepacia*. Appl. Env. Microbiol. *61*, 1791–1798.

Wise M.G., McArthur, J.V., Wheat, C., and Shimkets, L.V. (1996). Temporal variation in genetic diversity and structure of a lotic population of *Burkholderia* (*Pseudomonas*) *cepacia*. Appl. Env. Microbiol. *62*, 1558–1562.

Zaidi, N., Konstantinou, K., and Zervos, M. (2003). The role of molecular biology and nucleic acid technology in the study of human infection and epidemiology. Arch. Pathol. Lab. Med. *127*, 1098–1105.

Comparative Transcriptomics in *Burkholderia*

4

Andrea Sass, Pavel Drevinek and Pamela A. Sokol

Abstract

The advance in sequencing technologies and the availability of whole genome sequences enabled studying the entire transcriptome of organisms, either by microarray analysis or RNA sequencing. *Burkholderia* species for which such gene expression studies have been performed include *B. cenocepacia*, *B. pseudomallei*, *B. mallei*, *B. xenovorans*, *B. ambifaria*, and *B. multivorans*. This review focuses on findings in the areas of resistance to antibiotics, to biocides and to oxidative stress, iron transport, quorum sensing, mucoidy and host adaptation in these *Burkholderia* sp., obtained by means of global differential transcriptome analysis.

Introduction

New high-throughput sequencing technologies have greatly facilitated the deciphering of whole genome sequences. To date, more than 80 annotated genome sequences of the genus *Burkholderia* are deposited at public databases, with many more to follow. In the wake of this development, transcriptome profiling, analysis of gene expression on a whole genome scale, became possible. While genome sequence profiling can identify evolutionary changes, core and unique genes, transcriptome profiling generates knowledge about expression and regulation of genes.

Transcriptome profiling may be performed by two major methods (Sorek and Cossart, 2010), hybridization-based techniques using high-density oligonucleotide microarrays or sequencing-based techniques using high-throughput next-generation sequencing of RNA (RNA-seq). Both microarray hybridization and RNA-seq can be used as comparative methods, providing a measure for the relative abundance of transcripts in test samples compared with a control sample. Microarray technology obligately requires a reference genome sequence from which the microarray design is made. In general, a particular microarray design can only be used for the strain the microarray probes have been designed for. However, closely related strains and even species may be analysed but this may lead to problems with the reliability and interpretation of the expression data. Based on whole-genome DNA–DNA hybridization to oligonucleotide microarrays, it has been recommended that only strains/species with greater than 90% average nucleotide identity are analysed by a conventional microarray approach (Oh *et al.*, 2010).

RNA-seq can in principle be performed without knowledge of the genome sequence of the investigated bacterial strain, but an available reference sequence will facilitate clearer designation of sequencing reads. While microarray based approaches are somewhat constricted and can detect only transcripts for which probes are present on the chip, RNA-sequencing can reveal additional information about gene acquisitions, mutations, un-annotated genes and non-coding RNA. The majority of studies on transcriptome profiling discussed in this review have been performed using high-density microarrays designed against the available *Burkholderia* genomes. Microarray-based transcriptomics have been more cost effective to date, with a range of analytical methods and computer software already developed to process the expression data. However, reduced sequencing costs and the availability of free software to analyse RNA-seq data is likely to result in the majority of future expression studies being conducted using this method.

For meaningful transcriptome comparisons, experimental conditions need to be chosen carefully to ensure results can be attributed to a known and controlled experimental variable. Transcriptome comparisons are largely performed as variations of two major experimental approaches (Conway and Schoolnik, 2003) (Table 4.1) in which (1) the same strain or isolate are compared after growth in different (stress) conditions or (2) different isolates or mutant strains are compared after growth under the same conditions.

A variation of the first approach is exposure of lab cultures to a chemical agent or a stressor with comparison to untreated controls; such an approach can be used to investigate which genes are involved in resistance and adaptation to the stressor (Table 4.1). Furthermore, bacterial lab cultures grown under different conditions can be compared to each other or bacterial cells retrieved from infected hosts can be compared with *in vitro*-grown cells, to investigate physiological adaptation to different environments or to a host.

A variation of the second microarray comparison approach is the study of gene expression profiles of mutant strains with targeted gene inactivation compared to their parent strains, grown under the same conditions (Table 4.1). This approach is particularly powerful for the investigation of regulatory networks, for example if transcriptional regulator mutants are being studied. Spontaneous mutants isolated from selective media, or strain derivatives isolated from the same source, but with variations in their phenotype can also be investigated using this approach, as well as closely related strains from different sources. For spontaneous mutants and strain derivatives, transcriptome profiling can indicate which genes are potentially involved in phenotype variation. However, such expression analysis does not usually determine the underlying genetic cause for any observed differential regulation, and other methods such as genome resequencing need to be employed to reveal these traits.

The results of expression profiling are lists of genes that change expression. This is often described as a fold change value relative to the control condition. In addition to describing differential regulation, transcriptome comparisons can define regulons via analysis of co-regulation for linked and genomically disparate genes, as well as identify putative operon structures through analysis of co-transcription of adjacent genes. Transcriptome profiling usually generates a large amount of data, but the differentially expressed genes it identifies are not necessarily responsible for the given phenotype that is possessed by the mutants or elicited by the growth conditions being compared. For confidence in biological significance of observed gene expression changes, transcriptomics should be accompanied by more traditional microbiological approaches, like targeted gene inactivation combined with phenotypic assays.

Table 4.1 Experimental approaches for transcriptome comparisons

Approach no.	Experimental agent and design
1. Comparison of different exogenous conditions	
1.a	Culture exposed to chemicals compared to unexposed control culture
1.b	Growth in test conditions compared to control condition
1.c	*In vivo* growth compared to *in vitro* growth
2. Comparison of identical growth conditions	
2.a	Mutant with targeted or random gene inactivation compared to parent strain
2.b	*In vitro* selected spontaneous mutant compared to parent strain
2.c	Closely related strains with phenotype variations, isolated from the same source or from different sources

For statistical significance, both microarray and RNA-seq technologies require that experiments are performed with biological replicates, with a minimum of three replicates necessary for the assembly of robust expression datasets (Yang and Speed, 2002). However, statistical significance is not equivalent to biological significance, and no level of statistical stringency can ensure all false positive results are filtered out. In contrast, results with low statistical significance can still be biologically significant. Induction of otherwise silent genes tends to result in large values for fold change, whereas changes in transcript abundance of constitutively expressed genes usually result in lower fold values. Therefore, the biological significance of differential gene expression cannot be inferred from values for fold change alone. Comparability between studies conducted by different laboratories is also generally limited. Even small differences in cultivation and processing, prior to RNA extraction can impact on gene expression profiles. In particular, the total number of differentially expressed genes and their fold change can only be compared between RNA samples processed and analysed in exactly the same way, as statistical methods and thresholds for reporting differential expression have a great impact on the results.

Generally, transcriptomic approaches operate under the assumption that increased transcript levels correspond to an increased abundance and activity of the translated protein. However, post-transcriptional regulation may further influence protein stability, abundance and functional phenotype. Integrated approaches, combining proteomics and transcriptomics, can be used to assess the impact of transcriptome changes on the proteome (Zhang *et al.*, 2010). When applied to *B. cenocepacia*, such integrated approaches produced lists of genes with limited overlap between differentially expressed transcripts and proteins; if overlap between the proteome and transcriptome did occur, it was generally only observed for genes with large changes in transcript abundance (Inhülsen *et al.*, 2012; Madeira *et al.*, 2011; Zlosnik and Speert, 2010). In studies on pathways of aromatic compound degradation with *B. xenovorans*, proteomic approaches could largely confirm gene expression changes observed by transcriptomics (Denef *et al.*, 2004, 2005, 2006; Patrauchan *et al.*, 2011). The effect of post-transcriptional regulation on differential gene expression therefore varies depending on strain and growth condition.

Transcriptomics is also generally more sensitive in detecting expression changes than proteomics, giving useful insight into gene regulation and genetic base of physiological adaptation and allowing analysis of untranslated RNA. Proteomics investigates the actual abundance of translated gene products in the cell and can reveal additional information about the location of proteins by separating the proteome in cytoplasmic, membrane associated or extracellular fractions. Proteomics and transcriptomics should therefore be considered complementary to each other, and not necessarily suitable as a means to validate each other.

RNA extraction and enrichment for bacterial mRNA

The first step in any gene expression analysis is the generation of a high-quality RNA sample, which, owing to the unstable and fragile nature of mRNA, requires certain precautions during extraction to be accounted for. RNA extraction methods retrieve total cellular nucleic acid and can generally not distinguish between RNA and DNA. Therefore, raw extracts need to be treated with DNase to remove DNA contamination. The fraction of RNA carrying gene expression information is coding RNA or mRNA. mRNA is relatively unstable and in bacteria it has a short half life of only a few minutes before it is degraded by bacterial RNases (La *et al.*, 2008). However, genes can also be induced within just a few minutes, and harvesting procedures need to be robust and quick to capture a specific snapshot of gene expression. The transcription processes and bacterial RNases need to be rapidly inactivated by either cooling or treating cells with RNA-stabilizing agents. Harvested cells should be kept frozen, extracted RNA should be kept cold at all times. Contamination with RNase from environmental sources like human skin needs to be prevented.

Bacterial mRNA constitutes less than 5% of total RNA as the majority is non-coding functional RNA such as ribosomal RNAs and tRNAs

(La *et al.*, 2008; Sorek and Cossart, 2010). Several transcriptomic approaches, particularly RNA-seq, require enrichment of mRNA. Since bacterial mRNA lacks the poly-A tails found in eukaryotic mRNA, this prohibits poly-A-based methods to selectively transcribe only bacterial mRNA into cDNA. Available methods for mRNA enrichment are selective capture of large bacterial RNAs, selective degradation of abundant RNAs and selective polyadenylation of prokaryotic mRNAs (Sorek and Cossart, 2010). Selective capture is the most commonly used approach and it works by subtractive hybridization, using probes complementary to conserved regions of the bacterial 16S and 23S rRNA to pull out these abundant RNAs from the total nucleic acid extraction.

The depletion of ribosomal RNA in samples can be assessed by electrophoresis; however, preservation of original mRNA composition cannot be determined by this method. Current methods for mRNA enrichment have been optimized for bacteria other than *Burkholderia* sp. Protocols for commercially available mRNA enrichment kits specify compatibility with *B. mallei* and *B. pseudomallei* (Chen and Duan, 2011), and mRNA enrichment kits have been used for *B. cenocepacia* (O'Grady and Sokol, 2011). However, the suitability of these kits has not been widely assessed across the whole *Burkholderia* genus and optimization might be required for specific applications.

In vivo transcriptomic profiling poses extra challenges due to the scarcity of bacterial transcripts among eukaryotic RNA. A bacterial cell contains between 0.05 and 0.10 pg of RNA, while a mammalian cell contains around 20 to 30 pg RNA (La *et al.*, 2008; Leroy and Raoult, 2010). Therefore, bacterial RNA represents only a small fraction of total RNA extracted from infected eukaryotic cells/tissues, and enrichment of bacterial RNA prior to further processing is essential when examining bacterial transcriptomics in these experimental models. Approaches that have been used for *in vivo* grown bacteria include selective capture of eukaryotic large RNAs using host specific probes (La *et al.*, 2008) and selective capture of transcribed sequences (SCOTS) using probes specific for the infecting bacteria to enrich for target mRNA (Tolman and Valvano, 2012).

Microarray method

Microarray analysis is based on hybridization of processed RNA to complementary oligonucleotide probes immobilized on a microarray chip. Low density microarrays consisting of presynthesized probes spotted on membranes or glass slides have widely been outdated by the advance of high-density arrays with *in situ* synthesized probes, as these allow for a sufficient number of probes to cover all features of an entire genome. High-density microarrays usually consist of glass slides carrying thousands or even millions of probe spots per array, and probe locations, or features, are typically spaced less than 12 µm apart. Major platforms for high-density oligonucleotide microarrays used for prokaryotes are provided by Affymetrix and Agilent. Affymetrix provides the highest density arrays, produced by photolithography, with 25mer oligonucleotides built up on a silica support. Agilent arrays have a lower feature density, but longer 60mer probes are immobilized on a glass support. Agilent arrays are produced by *in situ* printing and allow custom microarray designs.

Microarray probes are designed to be representative of all genes within a genome. Non-coding RNA and other intergenic regions can also be included in the probe set. Probes consist of sense-strand sequences, and the directionality of transcription needs to be considered when designing probes for non-coding RNA. Each sequence has to be unique and specific to avoid cross-hybridization. Longer oligonucleotides increase specificity (Leiske *et al.*, 2006), but also increase probability of cross-hybridization. Probe design should therefore take into account the formation of secondary structures in the target sequence and also ensure the probe GC-content does not exceed certain limits (Kreil *et al.*, 2006;Leparc *et al.*, 2009). The relatively high GC-content of *Burkholderia* genomes, 62–68.5%, can make probe design more challenging (Leiske *et al.*, 2006). Genomic tiling microarrays overcome the need to target annotated genome features with specific probes, as instead they represent the entire sequence of both strands of the genome, by placing a high coverage of staggered probes, ideally starting at every single base on the microarray

(Sorek and Cossart, 2010). Tiling array designs therefore can enable the identification of un-annotated genome features. However, genomic tiling arrays have not been used for transcriptome studies on *Burkholderia* spp., probably due to cost and practicability issues resulting from the high number of probes necessary for this type of array.

Prior to microarray hybridization, RNA is transcribed into fluorescently labelled cDNA or cRNA, using random priming with short oligomers to initiate transcription of bacterial RNAs. cDNA can be labelled directly by adding fluorescently labelled dNTPs to the reverse transcription reaction, or via a post-labelling of modified dNTPs (Ehrenreich, 2006). Direct labelling can lead to bias caused by different incorporation efficiencies for the respective fluorescent dyes (Ehrenreich, 2006). In a direct-comparison experimental design (see below) this effect needs to be quantified by dye-swapping the RNA samples on different arrays. Post-labelling reduces or even prevents dye-bias, and a reference experimental design (see below) circumvents any dye bias issues altogether.

Typically, a two-colour approach is used for microarray analyses, with two RNA samples each labelled with a different fluorescent dye, hybridized simultaneously on one array. The two-colour approach can be conducted in various experimental designs, two of which are direct comparison design and reference design (Ehrenreich, 2006; Kim *et al.*, 2002). For the direct comparison design, control and test samples are labelled with different dyes and hybridized on the same array. Differential expression is inferred from differences in fluorescence intensities between two colours on the same feature. For the reference design, control and test samples are labelled with the same dye and hybridized against a common reference labelled with a different dye. Here, differential expression is inferred from differences in fluorescence intensities between features of the same probe on different arrays. The reference design is advantageous in that it allows the determination of variation between biological replicates and cross-comparison of all samples within one experiment. For experiments consisting of more than two conditions, a reference design also enables a maximum number of comparisons to be made with a minimum number of arrays.

The amount of bacterial RNA obtained from infected hosts is often too low for microarray applications, and therefore RNA amplification is required before labelling (La *et al.*, 2008; Leroy and Raoult, 2010). Although the amplification step introduces a bias, this can be generally tolerated as long as all steps are performed in a thoroughly reproducible manner (Francois *et al.*, 2007). The amplification and subsequent labelling procedure have to be adjusted to ensure the end product consists of labelled anti-sense DNA or RNA to complement the sense-strand probe sequence (Goff *et al.*, 2004).

Microarray hybridization requires strict adherence to stringent conditions for maximum annealing specificity of the complex sequence mix (Ehrenreich, 2006). After hybridization, the microarray chips are scanned with a laser, using wavelengths appropriate for the fluorescent dyes used and a resolution in the range of several micrometres, appropriate for the size of the features. The resulting image is analysed with specific software that assigns each feature on the array values for fluorescent intensities. Further analysis steps must include a normalization procedure, which scales the fluorescent intensities of all features on the chip to an average intensity to minimize the effect of variations in target sequence quantities loaded onto the chip (Quackenbush, 2002). Gene expression changes between samples are then inferred from signal intensity variations. Statistical analysis is applied to reduce the incidence of false positive differential expression, taking signal intensity and variation between biological replicates into account (Ehrenreich, 2006). Microarray analysis cannot provide a reliable absolute quantification of transcript abundance, as annealing efficiency varies between transcripts and cross-hybridization events can interfere with total signal intensity.

To rule out false positive differential expression caused by non-specific hybridization (Chuaqui *et al.*, 2002; Morey *et al.*, 2006), microarray data should be validated using other molecular methods like qPCR or Northern blot. Fold changes in expression measured by qPCR are often higher than measured by microarrays, indicating that microarray analysis might underestimate expression changes (Conway and Schoolnik, 2003; Morey *et al.*, 2006).

In addition, to the methods provided in the multiple studies discussed above and within the chapter, detailed technical protocols for microarray analysis of *B. cepacia* complex bacteria using a series of custom microarrays designed on the Agilent platform, are available at the International *B. cepacia* Working Group webpage (http://go.to/cepacia).

RNA sequencing

Transcriptome profiling via RNA sequencing is based on producing a very large number of short DNA sequences from an RNA sample by high throughput next generation sequencing (NGS) (Febrer *et al.*, 2011; Metzker, 2010). NGS platforms currently dominating this form of genomic analysis are Illumina, 454 pyrosequencing (Roche) and SOLiD sequencing (Applied Biosystems). All three platforms require production of a cDNA library which is sequenced after clonal amplification of single fragments. The technology produces millions of sequence reads per RNA sample, typically 30–400 bp long, depending on the platform used. The high number of reads enables gene expression profiling with high coverage for the whole transcriptome. Before the sequencing procedure, extracted and mRNA-enriched RNA is generally fragmented and then converted into double-stranded cDNA. cDNA fragments are ligated to platform specific adapter sequences and single adaptor-ligated DNA fragments are separated and immobilized onto a flow cell where they are clonally amplified. Amplified fragments are then sequenced using real time signal detection approaches. The exact method, workflow order and sequencing chemistry vary depending on the platform (Febrer *et al.*, 2011; Metzker, 2010).

cDNA libraries are often PCR-amplified before separation for clonal amplification and sequencing. The sequence read is then essentially derived from an amplified single nucleic acid molecule and the usually rather high error rate for PCR-amplification is then carried over into a rather high error rate for individual sequence reads. The conversion to double stranded cDNA prior to PCR amplification can result in loss of information about strand specificity of the original transcript, which is disadvantageous for the study of non-coding RNA. PCR-amplification before sequencing can also introduce bias regarding transcript abundance and discriminate against high GC content fragments (Aird *et al.*, 2011), a particularly problematic characteristic of GC-rich *Burkholderia* transcriptomes. To overcome all these potential shortcomings, protocols for PCR amplification-free (Hoeijmakers *et al.*, 2011; Linnarsson, 2010) and strand-specific library preparation (Mamanova *et al.*, 2010; Mamanova and Turner, 2011) are available. Even transcriptome analysis of single cells is possible (Kang *et al.*, 2011), enabling study of bacterial subpopulations. However, such methods can be more costly and labour intensive, and are not bias free themselves (Sorek and Cossart, 2010).

Evaluation of transcript abundance is essentially performed by counting reads, the number of reads mapping to a specific sequence gives a measure of expression. Before counting, sequencing reads have to be assembled into transcription units to assign each read to a specific gene or untranscribed RNA (Garber *et al.*, 2011). This requires mapping of reads to a reference genome or, in absence of a reference genome, ab initio assembly of reads. Ab initio assembly needs more computational power and is more sensitive to sequencing errors (Martin and Wang, 2011), whereas mapping to a reference genome can miss genome rearrangements. A combination of mapping to reference and ab initio assembly can give the best results, as abundance of known sequences can be reliably evaluated and novel sequences can be identified.

The method of library preparation and sequencing platform should be chosen according to the aim of the study and the availability of a reference genome. Where a reference sequence is available, short single reads can be sufficient. When reads have to be assembled before read counting, paired reads and longer reads can facilitate read assembly (Martin and Wang, 2011). A high coverage of sequencing reads is important where the aim is detection of rare transcripts and distinction between sequencing errors and true transcript variation (Garber *et al.*, 2011).

After read counting, a normalization procedure is necessary to compensate for differences in read numbers between transcripts of different lengths and for differences in library size (Garber *et al.*, 2011; Oshlack *et al.*, 2010). This is commonly achieved using the 'kilobase of transcript per million mapped reads' (RPKM, for single reads) or 'fragments per kilobase of transcript per million mapped reads' (FPKM, for paired-end reads) metric. Evaluation of differential expression is performed after normalization, using statistical methods appropriate for the discontinuous distribution of RNA-seq data (Oshlack *et al.*, 2010). RNA-seq is better at estimating total transcript abundance than microarray analysis (Fu *et al.*, 2009), but biological replicates are still essential to account for biological variation and minimize false positives. Compared with microarray analysis, fold changes in gene expression seen with RNA-seq are greater (Lipscomb and Schell, 2011) and RNA-seq offers a higher dynamic range (Wang *et al.*, 2011). Owing to the potential biases caused by the multiple steps of RNA processing and mRNA enrichment needed for RNA-seq analysis, just like microarray-based results, the data should be validated by other molecular methods.

Transcriptomic analysis of *Burkholderia*

Since the first *Burkholderia* genomes were completed, a large number of global, whole-genome gene expression studies of these bacteria have been carried out, using a range of experimental designs (Table 4.2). The following sections summarize findings gained in the clinically relevant areas of resistance to antibiotics and oxidative stress, iron transport, quorum sensing, mucoidy and host adaptation.

Resistance mechanisms to antibiotics, the biocide chlorhexidine and exogenous oxidative stress were studied in *B. cenocepacia* (Table 4.2). Gene expression changes induced by exposure to the respective agent and stable gene expression changes occurring in spontaneous strain derivatives with increased resistance to the agent were analysed to identify factors involved in resistance. Gene expression during growth in media with different iron concentrations was investigated in *B. pseudomallei*, *B. mallei* and *B. cenocepacia* (Table 4.2), to evaluate which genes are associated with iron transport in these species. Quorum sensing was studied in *B. cenocepacia*. Gene expression in targeted mutants with inactivated quorum sensing regulators, signalling molecule synthases or putative sensor kinases (Table 4.2) was compared to that of their parent strains. Differences in transcript abundance then define which genes are under control of the respective quorum sensing system component. The phenomenon of mucoidy was investigated by comparing isolates with mucoid phenotype with non-mucoid isolates from CF lungs, in *B. cenocepacia* and *B. multivorans* (Table 4.2). Stable differences in gene expression between these spontaneous mutants give an impression of which factors are involved in causing a mucoid phenotype in *Burkholderia*. Gene expression changes occurring during growth in a host environment were investigated with *B. pseudomallei*, *B. mallei* and *B. cenocepacia* (Table 4.2). Bacterial RNA was extracted from infected organs such as the rat lung, and the transcript abundance compared to an *in vitro* control to determine factors necessary for establishing infection. Gene expression changes occurring during internalization into macrophages were investigated in *B. pseudomallei* and *B. cenocepacia* (Table 4.2), to identify genes necessary for intracellular survival.

Altogether, these studies substantially increased our understanding of *Burkholderia* gene regulation, helped assigning putative new functions to genes and pathways, and initiated further investigations of these new functions.

Resistance to antibiotics, biocides and oxidative stress

Many *Burkholderia* spp., Bcc bacteria in particular, are highly resistant to multiple antibiotics and biocides, with major implications for treatment and prevention of infections. Understanding the genetic base for this resistance is of importance for finding new anti-infective target genes, improving treatments strategies, and preventing increase of resistance in bacterial populations. Known resistance mechanisms are membrane impermeability,

Table 4.2 Comparative transcriptome studies with *Burkholderia* spp.

Species	Strains	Approach[1]	Experimental condition	Reference
B. ambifaria	AMMD	1b	Production of enacyloxin	Mahenthiralingam *et al.* (2011)
B. cenocepacia	H111	2a, 2b	Virulence attenuated strains, loss of entire small replicon	Agnoli *et al.* (2012)
B. cenocepacia	J2315	2a	Deletion of efflux pumps (BCAL2820–2822 and BCAM1945–1948), as single and double mutants	Bazzini *et al.* (2011)
B. cenocepacia	J2315	1a	Exposure to chlorhexidine	Coenye *et al.* (2011)
B. cenocepacia	J2315	1b	Growth in CF sputum	Drevinek *et al.* (2008)
B. cenocepacia	H111	2a	Deletion of quorum sensing regulator *cepR* (BCAM1868)	Inhülsen *et al.* (2012)
B. cenocepacia	K56–2	1a, 2b	Exposure to polymyxin	Loutet *et al.* (2011)
B. cenocepacia	K56–2	2a	Deletion of orphan quorum sensing regulator (BCAM0188)	Malott *et al.* (2009)
B. cenocepacia	J2315	2a	Deletion of quorum sensing signal synthase (BCAM0581) and putative sensor kinase (BCAM0227)	McCarthy *et al.* (2010)
B. cenocepacia	IST439, IST4113	2c	Resistance phenotype variation	Mira *et al.* (2011)
B. cenocepacia	K56–2	1c	Infection of rat lungs	O'Grady and Sokol (2011)
B. cenocepacia	K56–2	2a	Deletion of quorum sensing genes *cepR* (BCAM1868), *cciR* (BCAM0240), and *cciI* (BCAM0239A), as single and double mutants	O'Grady *et al.* (2009)
B. cenocepacia	K56–2	2a	Deletion of shiny colony variant transcriptional regulator (BCAS0225)	O'Grady *et al.* (2011)
B. cenocepacia	J2315	1a	Exposure to oxidative stress	Peeters *et al.* (2010)
B. cenocepacia	J2315, BCC1616, BCC1617	1a, 2b, 2c	Exposure to antibiotics	Sass *et al.* (2011)
B. cenocepacia	K56–2	1c	Internalization into macrophages	Tolman and Valvano (2012)
B. cenocepacia	HI2424, AU1054	1b, 2c	Growth in synthetic CF medium and in soil extract	Yoder-Himes *et al.* (2009)
B. cenocepacia	HI2424, AU1054	1b, 2c	Growth in synthetic CF medium and in soil extract	Yoder-Himes *et al.* (2010)
B. cenocepacia	C9343, C8963	2c	Mucoid phenotype variation	Zlosnik and Speert (2010)
B. mallei	ATCC23344	1c	Infection of mouse spleen and liver	Kim *et al.* (2005)
B. mallei *B. pseudomallei*	ATCC23344 1026b	1c	Infection of mouse spleen and lung	Song *et al.* (2010)
B. mallei *B. pseudomallei*	ATCC23344 K96243	1b	Growth under iron limitation	Tuanyok *et al.* (2005)
B. multivorans	D2095, D2214	2c	Mucoid phenotype variation	Silva *et al.* (2011)
B. pseudomallei	K96243	1b	Growth under iron limitation	Alice *et al.* (2006)
B. pseudomallei	D286	1c	Internalization into macrophages	Chieng *et al.* (2012
B. pseudomallei	K96243	1b	Growth under oxygen limitation	Hamad *et al.* (2011

Species	Strains	Approach[1]	Experimental condition	Reference
B. pseudomallei	1026b	2a	Deletion of type III secretion system regulator (BPSS1611)	Lipscomb and Schell (2011)
B. pseudomallei	406e	1b	Growth on arabinose	Moore *et al.* (2004)
B. pseudomallei	K96243	1b	Growth under salt stress	Pumirat *et al.* (2010)
B. pseudomallei	K96243	1b	Logarithmic growth and stationary phase	Rodrigues *et al.* (2006)
B. pseudomallei	1026b	1c	Infection of hamster liver, spleen and lung	Tuanyok *et al.* (2006)
B. pseudomallei	1026b	1c	Infection of rat lung	van Schaik *et al.* (2008)
B. thailandensis	E264	1a	Exposure to glyphosate	Kang *et al.* (2011)
B. xenovorans	LB400	1b	Growth on biphenyl and benzoate	Denef *et al.* (2004)
B. xenovorans	LB400	1b	Growth on biphenyl and benzoate	Denef *et al.* (2005)
B. xenovorans	LB400	1b, 2a	Growth on benzoate	Denef *et al.* (2006)
B. xenovorans	LB400	1b	Growth on phenylacetic acid, benzoate and biphenyl	Patrauchan *et al.* (2011)
B. xenovorans	LB400	1b	Growth on abietane	Smith *et al.* (2007)

[1]For number designation, refer to Table 4.1.

efflux, enzymatic drug degradation, target modification and biofilm formation. *Burkholderia* spp. intrinsically harbour multiple resistant determinants, collectively responsible for their high innate resistance. *Burkholderia* species are also very resistant to oxidative stress, making *Burkholderia* contaminants difficult to eliminate from surfaces that are potential sources of infection with reactive oxygen species (ROS)-containing disinfectants. ROS production is a major defence mechanism of the host in bacterial infection, and the ability to scavenge and detoxify ROS therefore constitutes a virulence factor of *Burkholderia* spp.

Experimental approaches for transcriptomic profiling of resistance involve exposing bacteria *in vitro* to the respective effector, usually in sub-inhibitory concentrations of antimicrobials to ensure total mRNA content and RNA quality are not compromised by reduced viability, and to minimize the impact any differences in growth rate may have. Other approaches compare baseline gene expression of wild type parental strains to derivatives with altered resistance phenotypes. For example, the strain derivatives can be either genetically modified, spontaneous mutants or naturally evolved clones that have an altered antimicrobial resistance phenotype.

In transcriptomic studies where *B. cenocepacia* has been exposed to an antimicrobial agent in sub-inhibitory concentrations, known resistance determinants are often not prominent among differentially expressed genes, with the exception of beta-lactamases (Sass *et al.*, 2011) and efflux pumps (Coenye *et al.*, 2011; Drevinek *et al.*, 2008; Loutet *et al.*, 2011; Sass *et al.*, 2011). This suggests that most known resistance determinants must be constitutively expressed to account for the high intrinsic resistance of *B. cenocepacia*. The frequent differential regulation of efflux pumps confirms their importance in the adaptive or lifestyle specific resistance elicited by *Burkholderia*. Drug efflux is mediated by pumps of several major classes, such as tripartite efflux pumps of the resistance–nodulation–cell division-type (RND), and ATP-binding cassette (ABC) transport systems and major facilitator superfamily (MFS) proteins (Piddock, 2006); all these classes are represented among genes differentially regulated under exposure to antimicrobials in *Burkholderia*.

Efflux may also play a role in self-resistance for antibiotic-producing bacteria. In the enacyloxin-producing *B. ambifaria* AMMD (Mahenthiralingam *et al.*, 2011) an MFS transport protein (Bamb_5933) is co-transcribed with genes involved in enacyloxin synthesis, that were collectively encoded within a large genomic island; these data suggest that enacyloxin may be transported outside the cell via this route. Expression of efflux pumps is not necessarily controlled by their substrates. An example of strong induction of a tripartite efflux pump by its putative substrate is up-regulation of RND-8 (genes BCAM0925–0927; Guglierame *et al.*, 2006) in *B. cenocepacia* J2315 by the biocide chlorhexidine (Coenye *et al.*, 2011) and by chlorpromazine (Sass *et al.*, 2011). Both compounds are cationic and known to interact with membranes, suggesting a common induction mechanism linked to membrane integrity might exist. An example for an efflux pump known to transport a particular substance but not induced by presence of that substance is the *B. cenocepacia* ceo efflux system or RND-10 (genes BCAM2549–2551). ceo is known to transport trimethoprim (Nair *et al.*, 2004) but it is not induced in the presence of trimethoprim in *B. cenocepacia* J2315 (Sass *et al.*, 2011). However, absence of induction does not rule out that the efflux pumps in question are expressed constitutively and thus contribute to resistance.

Efflux pumps are not only important for resistance to antimicrobials, their presence also influences lifestyle specific traits such as biofilm formation, quorum sensing, and motility/expression of flagella (Bazzini *et al.*, 2011). Presumably, this is mediated by influence of efflux pumps on the intracellular concentration of certain effectors such as quorum sensing signal molecules or toxic molecules.

In addition to encoding factors that actively contribute to resistance, genes induced by exposure to antimicrobials may also be involved in damage control and compensation for impaired functions (Kang *et al.*, 2011; Sass *et al.*, 2011). Several constitute potential new resistance determinants, warranting further investigation of their function. These include previously uncharacterized efflux pumps, drug transport and drug modification proteins, and genes involved in altering membrane composition, such as a squalene hopene cyclase (BCAS0167) and outer membrane lipoproteins (Loutet *et al.*, 2011; Sass *et al.*, 2011).

Baseline expression profiling of *B. cenocepacia* strain derivatives with increased antibiotic resistance has demonstrated that this phenotype is generally mediated by a larger number of differentially expressed genes compared to when the parent strain is exposed to sub-inhibitory antibiotics (Loutet *et al.*, 2011; Sass *et al.*, 2011). Observed gene expression changes are constitutive and stable even in the absence of the antibiotic and are generally not inducible by exposure of the parent strain to the respective antibiotics. Various efflux pumps are among constitutive de-repressed genes (Loutet *et al.*, 2011; Sass *et al.*, 2011), as are genes involved in synthesis of membrane components and in degradation of aromatic compounds (Loutet *et al.*, 2011; Sass *et al.*, 2011) and degradation of branched chain amino acids (Loutet *et al.*, 2011). Stable deregulation of resistance determinants such as efflux pumps and beta-lactamases can also arise *in vivo* during adaptation to infection and administered antibiotics, as shown with *B. cenocepacia* and *B. multivorans* sequential isolates from CF sputum (Mira *et al.*, 2011; Sass *et al.*, 2011; Silva *et al.*, 2011; Zlosnik and Speert, 2010).

Aromatic compound degradation pathways are up-regulated in multiple strain derivatives with elevated antibiotic resistance (Loutet *et al.*, 2011; Sass *et al.*, 2011). *Burkholderia* spp. have diverse metabolic capacities and can degrade a multitude of aromatic compounds; this trait could facilitate modification or degradation of antibiotics and biocides via non-specific enzymatic activity, regardless of whether they contain aromatic moieties or not. A putative link between metabolic diversity and the ability to degrade antibiotics was established in a study in which soil bacteria were isolated with antibiotics as the sole carbon source. It was observed that the majority of isolates belonged to the order *Burkholderiales* (Dantas *et al.*, 2008). An example for a metabolic pathway potentially contributing to antibiotic resistance is the pathway for phenylacetic acid (paa) degradation, which is widespread among *Burkholderiales* (Pérez-Pantoja *et al.*, 2012) and has been associated with a central role in adaptation to antimicrobial or *in vivo* related stress. The paa degradation pathway is required for full pathogenicity of *B. cenocepacia* in *Caenorhabditis elegans* (Law *et al.*, 2008), induced in synthetic cystic fibrosis (CF) sputum medium in *B. cenocepacia* (Yoder-Himes *et al.*, 2009), implicated in resistance to meropenem in *B. cenocepacia* (Sass *et al.*, 2011) and may contribute to physiological adaptation to the rhizosphere in *B. xenovorans* (Patrauchan *et al.*, 2011).

Exposure of *B. cenocepacia* J2315 to exogenous ROS simultaneously induces a multitude of ROS detoxifying enzymes (Peeters *et al.*, 2010), with both inorganic and organic peroxide ROS inducing largely the same spectrum of genes (Sass, unpublished data). The large number of activated genes and the indiscriminate up-regulation by different oxidative compounds can explain the high resistance of *B. cenocepacia* to oxidative stress. Other functions induced in the presence of exogenous ROS are involved in damage repair: thioredoxin, glutathione synthase, DNA polymerases, SOS stress response, proteases and peptidases. Mobile genetic elements such as phage genes and transposases are also activated, suggesting oxidative stress leads to higher rates of genetic rearrangement in *B. cenocepacia*. *B. multivorans* ATCC 17616, more susceptible to oxidative stress than *B. cenocepacia* J2315, activates the same homologous major detoxifying enzymes and damage repair genes in the presence of exogenous ROS (Sass, unpublished data). The cause of the increased susceptibility of *B. multivorans* compared to *B. cenocepacia* is therefore not apparent from transcriptome profiling.

ROS-detoxifying enzymes are also up-regulated in the presence of sputum (Drevinek *et al.*, 2008), and can be de-regulated in isolates from CF sputum (Zlosnik and Speert, 2010), indicating that bacteria are subjected to oxidative stress in the CF lung. The genes up-regulated in a *B. cenocepacia* CF isolate (Zlosnik and Speert, 2010) and in *B. cenocepacia* J2315 exposed to sputum (Drevinek *et al.*, 2008) represented subsets of those activated under exogenous oxidative stress, suggesting that more than one regulatory mechanism is responsible for induction of the full oxidative stress response.

A multitude of genes are induced in response to both antimicrobials and ROS. However, the exposure to ROS appears to specifically activate most of the known resistance determinants, whereas exposure to antimicrobials does not always induce known resistance genes or pathways. Long-term adaptation of *Burkholderia* strains to exposure to antimicrobials and oxidative stress appears to be mediated by stable constitutive modulation of gene expression, increasing resistance without the need for gene acquisition.

Iron transport

Iron is needed for many cellular processes; however, the availability of free iron is generally low in oxygenated environments, and bacteria possess specific uptake mechanisms such as siderophore production to ensure sufficient iron intake. Concentration of free iron is particularly low in host environments, as iron sequestration is one of the major defence mechanisms employed against infection. Bacterial iron uptake mechanisms are therefore regarded as important virulence factors and have been the subject of numerous studies. The transcriptomic response to reduced iron concentrations in growth medium has been studied in *B. pseudomallei* (Alice *et al.*, 2006; Tuanyok *et al.*, 2005), *B. mallei* (Tuanyok *et al.*, 2005; Table 4.2). Studies comparing growth of *B. cenocepacia* strains in an artificial sputum to growth in a soil extract medium (Yoder-Himes *et al.*, 2009, 2010) reported induction of genes involved in iron uptake, these studies were therefore included in this section.

When grown in medium with reduced iron content, all investigated *Burkholderia* strains induce production of hydroxamate-type siderophores ornibactin (Bcc strains) and malleobactin (*B. pseudomallei* and *B. mallei*) and TonB-dependent receptors (Alice *et al.*, 2006; Tuanyok *et al.*, 2005; Yoder-Himes *et al.*, 2009, 2010). Genes for production of salicylate-type siderophore pyochelin, present in *B. cenocepacia* and *B. pseudomallei*, are also activated under iron-limiting conditions, with a lower fold change than hydroxamate-type siderophores. All investigated strains activated genes for haem transport (BCAM2626–2628, BMAA1826–1828, BPSS0242–0244), making haemin utilization a common feature of the

Burkholderia response to iron-limiting conditions. Ornibactin and malleobactin are the major siderophores in *Burkholderia* species; they have a higher affinity for iron than pyochelin, and only ornibactin is required *in vivo* for *B. cenocepacia* to survive in lung infection (Visser *et al.*, 2004). Despite the apparent redundance in siderophore synthesis genes, both types of siderophores are induced under the conditions used in these studies. The co-regulation of haemin uptake with siderophore production suggests both are under the same iron-mediated regulatory control.

Siderophore production is under control of the *fur* gene (ferric uptake-regulator), however, this regulator does not change expression under iron-limited conditions in the investigated strains. An extracytoplasmatic function subfamily sigma factor, involved in transcriptional control of siderophore production in *B. pseudomallei* (Alice *et al.*, 2006), is activated under low iron in all investigated strains (BCAL1688, BMA1191, BPSL1787), suggesting it is commonly involved in induction of iron sequestration and transport in *Burkholderia.* In *Escherichia coli*, the corresponding iron regulation system is the FecIR-system.

The simultaneous activation of multiple iron uptake mechanism under iron-limiting conditions shows that *Burkholderia* spp. are well adapted to cope with reduced iron concentrations that might occur during infection. Induction of pyochelin production in *B. cenocepacia* J2315 grown in CF sputum (Drevinek *et al.*, 2008), and de-repression of iron transport genes in a *B. cenocepacia* strain isolated from CF sputum (Mira *et al.*, 2011) indicate that iron availability in CF sputum is low and that *B. cenocepacia* can easily adapt to grow at these levels. In several pathogenic bacteria, such as *Pseudomonas aeruginosa* and *Neisseria meningitidis*, reduced iron concentration can serve as environmental cue for activation of virulence factors, namely exotoxins (Ratledge and Dover, 2000). Activation of virulence factors other than siderophores does not appear to occur in *Burkholderia*, not even for the obligate pathogen *B. mallei*. In contrast, other environmental cues in addition to iron concentration can influence expression of genes for siderophore and haem-mediated iron transport. Regulatory systems acting on other cues and studied in most detail for their role in iron regulation in *B. cenocepacia* are the quorum sensing systems described below (Lewenza and Sokol, 2001; Malott *et al.*, 2009; O'Grady *et al.*, 2009).

Quorum sensing

Quorum sensing (QS) in bacteria signifies inter- or intraspecies cell-to-cell communication mediated by production of and response to a diffusible signalling molecule in a cell density dependent manner (see chapter 10 for details). In Gram-negative bacteria, most QS systems consist of a LuxI homologue, an *N*-acylhomoserine lactone (AHL) synthase, and its cognate receptor, a LuxR homologue. Increasing cell density leads to accumulation of the diffusible AHL. When AHL levels reach a certain threshold the cognate receptor is activated which then induces or represses genes under its regulatory control.

Studies of QS in *Burkholderia* species by transcriptomic methods have been performed for *B. cenocepacia* (Table 4.2), where QS controls expression of numerous virulence factors and is necessary for full virulence (Sokol *et al.*, 2007). *B. cenocepacia* possesses two LuxIR QS systems, CepIR (genes BCAM1870 and BCAM1868), that is widely distributed among all members of the Bcc, and CciIR (genes BCAM0239A and BCAM0240), that is present only in the highly transmissible *B. cenocepacia* ET12 lineage and other *B. cenocepacia* strains with the cenocepacia genomic island (cci) (O'Grady *et al.*, 2009). *B. cenocepacia* strains also possess an orphan LuxR homologue without an associated AHL synthase, designated CepR2 (BCAM0188; Malott *et al.*, 2009). Recently a non-AHL cell-to-cell signalling molecule was discovered in *B. cenocepacia*, the *Burkholderia* diffusible signalling factor (BDSF). BDSF is a cis-2-dodecenoic acid, synthesized by an enoyl CoA hydratase family protein encoded by *rpfF*, (BCAM0581; McCarthy *et al.*, 2010), and universally present in Bcc species.

Expression profiles of mutant strains with inactivated QS regulatory genes have been assembled for *cepR*, *cepR2*, *cciR*, a double-mutant *cepRcciIR*, BDSF synthase and a potential BDSF sensor

(BCAM0227, Table 4.2). All these mutants were compared to their parent strains after growth to stationary phase and high cell density, with the overall aim to define genes that are under regulatory control of these QS systems. Transcriptome analyses of *cepR* mutants compared to their parent strains have been performed in representatives of two *B. cenocepacia* III-A lineages, K56–2, an ET12 strain, and H111, which lacks the cci and hence *cciIR* genes.

CepR and CciR are global regulators, with regulated genes proportionally distributed on all major replicons. CepR is primarily a positive regulator in *B. cenocepacia* K56-2 and H111 (Inhülsen *et al.*, 2012; O'Grady *et al.*, 2009). CepR positively regulates *cepI* and the co-transcribed adjacent gene BCAM1871 which also influences QS gene expression and virulence (O'Grady *et al.*, 2012) as well as *cciRI*. Regulation of virulence factors by CepR can be strain specific. Fimbrial genes, lectins (BCAM0184-0186), extracellular proteases *zmpA* (BCAS0409) and *zmpB* (BCAM2307), and the nematocidal protein *aidA* (BCAS0293) are positively regulated by CepR in both K56–2 and H111. A type I secretion system (BCAM2140–2142) with an adjacently encoded large surface protein (BCAM2143, *bapA* or *adhA*; Inhülsen *et al.*, 2012), lipases (BCAM0949-0950), flagella and a type VI secretion system (BCAL0337–351; O'Grady *et al.*, 2009) are positively regulated in either H111 or K56-2, respectively.

CciR, present in *B. cenocepacia* K56–2, but not in *B. cenocepacia* H111, is primarily a negative regulator (O'Grady *et al.*, 2009). One third of CciR-regulated genes overlap with CepR-regulated genes. The majority of these overlapping genes are regulated reciprocally and in most cases induced by CepR and repressed by CciR. In the case of genes reciprocally regulated by both regulators, CepR is dominant over CciR (O'Grady *et al.*, 2009). Virulence factors negatively regulated by CciR that are positively regulated by CepR include fimbrial genes, lectins, ZmpA, ZmpB, type VI secretion genes, and flagella (O'Grady *et al.*, 2009). Other virulence factors, however, are positively regulated by CciR and negatively regulated by CepR, including genes involved in synthesis and uptake of the siderophore ornibactin and genes involved in haem transport (O'Grady *et al.*, 2009).

CepR2 is both a positive and negative regulator in *B. cenocepacia* K56-2. While CepR and CciR are global regulators, the majority of CepR2-regulated genes are located on the second replicon. The most strongly CepR2 negatively regulated genes are in a cluster of 14 genes directly adjacent to *cepR2* (BCAM0189-0203), which includes an *araC*-type transcriptional regulator. The strong regulation of this gene cluster may be caused indirectly by CepR2, via repression of this potential *araC* positive regulator. CepR2 executes negative autoregulation and shows some overlap in virulence factor regulation with CciR; namely lectin, *aidA* and *zmpA* are negatively regulated by both CepR2 and CciR. CepR2 does not influence the expression of either *cepR* or *cciR*, but CepR can positively regulate *cepR2* (Inhülsen *et al.*, 2012), and CciR can negatively regulate *cepR2* (Malott *et al.*, 2009).

CepR positively regulates *shvR* (BCAS0225), a transcriptional regulator located on the small replicon, ShvR on the other hand represses the expression of *cepIR* and *cciIR* (O'Grady *et al.*, 2011). Mutations in *shvR* result in the formation of shiny colony variants (shv) with reduced virulence in *B. cenocepacia* K56-2 (O'Grady *et al.*, 2011). ShvR itself is a global regulator (O'Grady *et al.*, 2011), with equal proportions of positively and negatively regulated genes. A genomic region with 23 genes (BCAS0202–0224) on the small replicon, directly adjacent to *shvR*, is strongly positively regulated by ShvR. The genomic region includes the *afcA* and *afcC* operons necessary for antifungal compound production and gene BCAS0208 required for rough colony morphology and biofilm production (Subramoni *et al.*, 2011). One third of ShvR-regulated genes are co-regulated by CepR and CciR; of these the vast majority is commonly positively regulated by CepR and commonly negatively regulated by CciR. This common regulation can be attributed to the influence of ShvR on *cepIR* and *cciIR* expression and the resulting modulation of AHL accumulation (O'Grady *et al.*, 2011).

BDSF signalling has been investigated by inactivation of the BDSF synthase gene, *rpfF* in *B.*

cenocepacia J2315 (McCarthy *et al.*, 2010). Mutation of this gene globally influences expression of genes on all three major replicons of J2315 with almost equal proportions of genes with increased and decreased expression. A putative BDSF sensor kinase, is required for expression of a subset of the genes regulated by *rpfF* suggesting that it is involved in BDSF perception. BDSF negatively regulates flagella and chemotaxis genes and positively regulates type III secretion system genes, cable pili and several efflux and drug transport proteins. Efflux pumps are implicated in homoserine lactone efflux (Bazzini *et al.*, 2011) and could also be responsible for transport of BDSF out of the cell. *rpfF* does not influence expression of other QS regulators, but it positively regulates expression of *shvR* and its adjacent genes.

The study of transcriptome profiles of *cepR* and *cciR* mutants in *B. cenocepacia* K56-2 confirmed many results achieved by other approaches like promoter:lux fusions, random promoter libraries and a bioinformatics search for a *cepR* box motif (Chambers *et al.*, 2006). In other cases, transcriptomic data could not confirm results from other experimental approaches. A complicating factor in studying QS by microarrays is the growth phase dependent nature of the investigated gene regulation. Costs usually prohibit an exhaustive analysis of all growth stages and in all studies discussed above only a single time point was used for microarray analysis, which can account for discrepancies between observations. Establishing regulons for QS components is further complicated by the occurrence of multiple interacting QS systems, each with overlapping regulons. Transcriptome profiles of single QS mutants represent a combination of direct effects of gene inactivation, as well as indirect effects through expression changes of other QS systems and regulatory proteins. Overlaps between regulons can therefore be a result of direct regulation by multiple regulators, or of indirect regulation.

In *B. cenocepacia* J2315, *cepIR* can be induced not only at a higher cell density, but also at reduced oxygen concentrations and increased temperature, resulting in induction of CepIR-regulated genes such as the *zmpA* protease and *aidA* nematocidal protein (Sass, unpublished data). *B. pseudomallei* also induces autoinducer synthase *pmlI* (BPSS0885) at reduced oxygen concentration (Hamad *et al.*, 2011). Expression of *rpfF* in *B. cenocepacia* J2315 decreases at reduced oxygen concentration and does not increase at higher cell densities. In *B. multivorans*, *cepIR* and *rpfF* are induced at higher cell density but neither changes expression at reduced oxygen concentration (Sass, unpublished data). These results indicate that factors other than cell density and growth phase can influence expression of QS in *Burkholderia*, and expression and regulation of QS varies between *Burkholderia* species.

Mucoidy

In CF, long-term infections with *Pseudomonas aeruginosa* are often accompanied by switching from non-mucoid to mucoid phenotype, characterized by increased production of exopolysaccharide (EPS) and linked to increased mortality of the patients. Bcc bacteria also produce various types of EPS and a mucoid phenotype, characterized by increased EPS production. However in CF lung infection, *B. cenocepacia* and *B. multivorans* switch their phenotype typically from mucoid to non-mucoid. The mucoid phenotype of Bcc bacteria was associated with increased virulence and persistence in infection models (Conway *et al.*, 2004), whereas the non-mucoid phenotype was associated with more severe outcome in CF infections (Zlosnik *et al.*, 2011).

To further investigate the role of mucoidy in Bcc infections, transcriptome comparisons of non-mucoid and mucoid isogenic isolates were performed for *B. cenocepacia* (Zlosnik and Speert, 2010) and *B. multivorans* (Silva *et al.*, 2011) (Table 4.2). In both cases isolate pairs were sequential isolates retrieved from the same respective patient with CF.

In comparison to the non-mucoid isolates, the mucoid isolates of both *Burkholderia* species over-expressed genes for exopolysaccharide synthesis as well as carbohydrate transport and processing (Silva *et al.*, 2011; Zlosnik and Speert, 2010). Up-regulated genes for exopolysaccharide synthesis in both species belonged to two clusters involved in cepacian synthesis (BCAM0854–0864, BCAM1003–1011, Bmul_4910–4920, Bmul_4604–4613; Ferreira *et al.*, 2010). Of the induced carbohydrate transport genes and processing, representatives of one homologous cluster were induced in both species (BCAM2014–2018, Bmul_3783–3787), whereas other induced genes were specific for each species.

All other transcriptomic differences between mucoid and non-mucoid isolates were strain specific (Silva *et al.*, 2011; Zlosnik and Speert, 2010). In the mucoid *B. cenocepacia* strain, CepR-induced genes including *cepI*, *bapA*, *adhA*, *aidA*, *zmpA* and *shvR*, and genes involved in fimbria synthesis and type I secretion, (Inhülsen *et al.*, 2012), were down-regulated. Subsequent PCR analysis revealed that the mucoid strain had mutations that resulted in loss of function in *cepR*, *cciI* and *cciR* (Sokol, personal communication), which apparently caused the observed down-regulation of these genes. Beta-lactamase encoding genes, which are not QS regulated, were also down-regulated in the mucoid *B. cenocepacia* isolate (Zlosnik and Speert, 2010). In contrast, in the mucoid *B. multivorans* isolate the above mentioned genes did not change expression (Silva *et al.*, 2011).

A switch from mucoid to non-mucoid phenotype in *Burkholderia* spp. may also be associated with increased or decreased expression of virulence factors such as surface proteins. In the latter studies (Silva *et al.*, 2011; Zlosnik and Speert, 2010) transcriptomic changes other than decreased cepacian production and carbohydrate turnover varied and did not show common trends towards specific adaptation to the CF lung environment. Others have noted a lack of consistent correlations between EPS production and virulence or biofilm formation in Bcc bacteria (Cunha *et al.*, 2004), and the transcriptomic studies provide further support for the variable nature of these phenotypes.

Several of the expression profiling results suggest a link between QS and mucoid phenotype in *Burkholderia cenocepacia*, which could be mediated by differential expression of the CepR-induced *shvR*. Decreased expression of *shvR* often coincides with increased expression of genes involved in production of cepacian and other EPS; this was observed in a *B. cenocepacia* BDSF mutant (McCarthy *et al.*, 2010), a mucoid *B. cenocepacia* isolate with inactivated *cepR* (Zlosnik and Speert, 2010) and a *B. cenocepacia* K56-2 *shvR* mutant (O'Grady *et al.*, 2011; Sokol, unpublished data). On the other hand, increased expression of *shvR* can coincide with decreased EPS production (Coutinho *et al.*, 2011; Mira *et al.*, 2011).

Changes in growth conditions also influence transcription of the various gene clusters involved in EPS production in *B. cenocepacia* J2315 and *B. multivorans* ATCC 17616 (Sass, unpublished data). In *B. multivorans*, an EPS cluster on the medium replicon was induced under conditions associated with higher cell density and stationary phase while reduced oxygen and iron concentrations induced an EPS cluster on the large replicon. In *B. cenocepacia*, only cepacian synthesis genes consistently change expression with considerable fold change and only under heat stress conditions. These data suggest that *B. multivorans* may more readily induce EPS production than *B. cenocepacia* and that in *B. multivorans* EPS other than cepacian plays a role in adaption to varying growth conditions.

Host adaptation

The genus *Burkholderia* consists of metabolically very versatile bacteria that can be found in diverse environments; including in soil, particularly the rhizosphere, and in freshwater bodies. Members of the Bcc and *B. pseudomallei* can also act as opportunistic pathogens in humans, whereas *B. mallei* is an obligate pathogen of animals (Coenye and Vandamme, 2003). The infection process involves entry into host environment, survival of host defence mechanisms and growth in the host environment. Transcriptome profiling of bacteria during infection can provide understanding of the metabolic adaptation to the change in growth conditions, and identify virulence factors important for survival in host cells and for establishing infection.

While some studies have attempted to identify stable changes in gene expression that occur during long term chronic infections by comparing transcriptional profiles of isogenic strains with different phenotypes (Sass *et al.*, 2011; Silva *et al.*, 2011; Zlosnik and Speert, 2010), gene expression changes of bacterial populations have also been explored using a variety of infection models. *In vivo* gene expression during infection of lungs, liver and spleen in different rodent infection models has been studied with *B. pseudomallei* (Tuanyok *et al.*, 2006; van Schaik *et al.*, 2008), and with *B. mallei* (Kim *et al.*, 2005) (Table 4.2). *B. cenocepacia* CF lung infections have been modelled using a chronic rat lung model and studied using transcriptomics (O'Grady and Sokol, 2011). The experimental design employed has varied considerably between these studies. In addition, enriching bacterial RNA from tissues can introduce a bias that is not seen when working directly with a pure culture *in vitro*. Despite this, several commonalities in differential gene regulation have been observed in these *in vivo* studies.

Gene expression profiles for the closely related species *B. pseudomallei* and *B. mallei* during infection were very similar when comparing patterns for conserved orthologues (Song *et al.*, 2010). Genes linked to a type III secretion system (T3SS) important for establishing infection were up-regulated in all investigated infection models for both species (BPSS1523–52, BCAA1520–1552; Kim *et al.*, 2005; Tuanyok *et al.*, 2006; van Schaik *et al.*, 2008). T3SSs in *B. pseudomallei* can be co-regulated with genes for invasion proteins and haemagglutinin or haemolysin-like proteins (Lipscomb and Schell, 2011; Moore *et al.*, 2004). Accordingly, it was also observed that the latter genes were also induced in infected tissues (Kim *et al.*, 2005; van Schaik *et al.*, 2008). Transcription of T3SS (BCAM2040–2057) was activated by *B. cenocepacia* during growth in sputum (Drevinek *et al.*, 2008). The homologous *B. pseudomallei* T3SS (BPSS1523–52) is induced by growth under salt stress (Pumirat *et al.*, 2010), suggesting that changes in sodium chloride concentration can play a role in regulation of T3SS expression.

In *B. cenocepacia*, a type IV secretion system (T4SS, pBCA020-059) located on the plasmid was up-regulated in several infection models (O'Grady and Sokol, 2011). This secretion system has been shown to be involved in causing disease in plants and for survival in macrophages. Several other *B. cenocepacia* secretion systems were down-regulated in bacteria recovered from lung infection models compared to the *in vitro* control (O'Grady and Sokol, 2011).

Other putative virulence factors induced during growth in infected tissues include phospholipases (Tuanyok *et al.*, 2006) and iron transport genes (Kim *et al.*, 2005; O'Grady and Sokol, 2011; Tuanyok *et al.*, 2006). In a study comparing transcriptional profiles of *B. cenocepacia cepRcciR* and *cepRcciRcepR2* mutants to the parent strain during rat lung infections, iron transport genes including those involved in synthesis and transport of ornibactin and pyochelin were induced in the parent strain suggesting that iron is limited and QS regulation is functional (McKeon, O'Grady, and Sokol, unpublished data). In contrast to what was found *in vitro*, CciR induction rather than CepR repression of ornibactin genes appeared to occur *in vivo*. Mira *et al.* (2011) also reported an increase in *cciR* expression in a CF patient's isolates over time, suggesting that the CciIR QS system may play an important role *in vivo*.

Genes for flagella, capsule and lipopolysaccharide production are regulated during infection, but the results appear less consistent across the different transcriptomic studies. In *B. pseudomallei* and *B. mallei* isolated from infected tissue, more flagella biosynthesis genes were up-regulated than down-regulated (Kim *et al.*, 2005; Tuanyok *et al.*, 2006; van Schaik *et al.*, 2008). In *B. cenocepacia* genes for flagella synthesis were generally down-regulated in various infection models (O'Grady and Sokol, 2011). In contrast, growth of *B. cenocepacia* in CF sputum induced flagella gene expression (Drevinek *et al.*, 2008). Genes for capsule and lipopolysaccharide synthesis have variable expression during *B. pseudomallei* infection (Tuanyok *et al.*, 2006; van Schaik *et al.*, 2008), while the capsule synthesis genes in *B. mallei* were down-regulated (Kim *et al.*, 2005). In *B. cenocepacia* QS and QS-regulated virulence factors are expressed to a higher level *in vitro*, possibly due to a higher cell density in the control cultures compared to infected lung tissue (O'Grady and Sokol, 2011).

Functions such as energy production, replication and transcription tend to be down-regulated during infection for all *Burkholderia* species studied (Kim *et al.*, 2005; O'Grady and Sokol, 2011; Tuanyok *et al.*, 2006; van Schaik *et al.*, 2008). This observation indicates that cells perhaps do not replicate as much in infected tissues compared to control cultures. Up-regulated metabolic functions also indicate that carbon and nitrogen sources used during infection are different from those used in the *in vitro* grown controls. Genes relating to methylcitrate cycle and lipid metabolism were induced during *B. cenocepacia* infection (O'Grady and Sokol, 2011), whereas genes relating to putative anaerobic metabolism were induced in *B. mallei* (Kim *et al.*, 2005) and *B. pseudomallei* (Tuanyok *et al.*, 2006).

In vivo gene expression of bacteria internalized into macrophages has been studied in *B. cenocepacia* and *B. pseudomallei* (Chieng *et al.*, 2012; Tolman and Valvano, 2012). Both species activate a type VI secretion system (T6SS, BPSS1496–1508 and BCAL0337–0351) intracellularly within macrophages, confirming that T6SS plays a role in intracellular survival and growth. Other functions commonly induced in intracellular bacteria relate to carbohydrate transport and metabolism. This indicates that carbohydrates are less available inside macrophages than under control conditions, where bacteria are cultivated in the respective macrophage growth medium. Commonly down-regulated functions relate to iron uptake, suggesting that iron is not a growth limiting factor inside macrophages, and to QS, which again may be a reflection of cell density. In intracellular *B. cenocepacia*, the *N*-acylhomoserine lactone dependent regulator *cciR* is repressed (Tolman and Valvano, 2012), and in intracellular *B. pseudomallei*, *N*-acylhomoserine lactone synthase *pmlI* is down-regulated (Chieng *et al.*, 2012).

Intracellular *B. pseudomallei* grow slower than in extracellular control cultures, and correspondingly down-regulation of genes involved in energy production and replication was observed (Chieng *et al.*, 2012). Also down-regulated during intracellular growth were virulence factors involved in motility such as flagella and chemotaxis, as well as capsule and lipopolysaccharide production. Up-regulated genes include those for aromatic compound degradation, anaerobic metabolism and virulence factors such as haemolysin and haemagglutinin precursors (Chieng *et al.*, 2012). Intracellular *B. cenocepacia* appear to regulate genes involved in metabolism as an adaptation to different nutrient availability inside macrophages (Tolman and Valvano, 2012). In addition to carbohydrate turnover, amino acid transport is activated. Energy production is apparently not down-regulated, and flagella are up-regulated. Genes involved in membrane modification are up-regulated, possibly as a response to stress affecting the bacterial cytoplasmic membrane. For example, squalene hopene cyclase, implicated in resistance to polymyxin, is possibly involved in altering membrane fluidity (Schmerk *et al.*, 2011).

In general, *in vivo* transcription profiles do not show extensive up-regulation of genes known to be involved in virulence and host adaptation, with the exception of up-regulation of certain secretion systems. For example, none of the investigated *Burkholderia* spp. appeared to activate an oxidative stress response, although production of reactive oxygen and nitrogen species is generally regarded as a major host defence mechanism. The high intrinsic resistance of *Burkholderia* to ROS may be a reason behind this observation, although growth in CF sputum did elicit an oxidative stress response (Drevinek *et al.*, 2008).

Summary and outlook

Comparative transcriptomics on *Burkholderia* spp. has greatly increased knowledge about gene expression, expression regulation, adaptation to environmental changes and stress response. Transcriptomics has led to discovery of multiple new virulence factors and resistance determinants, which could result in new therapeutic targets and treatment strategies for *Burkholderia* infections. It has also provided us with a much clearer picture of gene organization and regulation, defining regulons such as those involved in QS. However, the function of many genes that are highly regulated is not yet known, but gene expression data might initiate further investigation of these proteins with unknown function.

It is likely that for future *Burkholderia* transcriptome investigations, RNA-seq will become the favoured approach over microarray analysis, once disadvantages such as higher cost and the lack of well established analysis pipelines have been overcome. RNA-seq ultimately offers more information and more flexibility, as it can be performed without a reference genome and enables analysis of gene mutations, deletions, insertions and acquisitions. In addition, RNA-seq can discover new previously not annotated genes, non-coding RNAs (ncRNAs) and antisense activity (Filiatrault, 2011), and can define transcription start sites, operons and untranslated regions (Sorek and Cossart, 2010; Toledo-Arana and Solano, 2010).

In particular, areas where RNA-seq will be providing more possibilities lie in the detection of small RNAs or non-coding RNAs, which are emerging as key regulators in bacterial gene expression on a single cell level. Expression of ncRNAs in Bcc species has to date only been investigated in *B. cenocepacia*, detected by differential expression of intergenic regions compared to neighbouring genes, using microarrays (Coenye *et al.*, 2011; Peeters *et al.*, 2010) or by RNA-seq (Yoder-Himes *et al.*, 2009). ncRNA detection with microarrays has been somewhat random, as these ncRNAs were un-annotated and therefore not included in the systematic probe design. The probes for intergenic regions included on many microarrays have revealed several ncRNAs; however, because they do not cover these regions in their entirety in both strand orientations, they will miss ncRNAs which are generally small (50–500 base pairs) and can be encoded on both strands (Gottesman and Storz, 2010). RNA-seq is a therefore a more appropriate method for analysis of ncRNA expression and it can identify all RNA, particularly if strand information is preserved in the data. Future investigations using RNA-seq will undoubtedly improve knowledge about the entire transcriptome of *Burkholderia* spp.

References

Agnoli, K., Schwager, S., Uehlinger, S., Vergunst, A., Viteri, D.F., Nguyen, D.T., Sokol, P.A., Carlier, A., and Eberl, L. (2012). Exposing the third chromosome of *Burkholderia cepacia* complex strains as a virulence plasmid. Mol. Microbiol. *83*, 362–378.

Aird, D., Ross, M.G., Chen, W., Danielsson, M., Fennell, T., Russ, C., Jaffe, D.B., Nusbaum, C., and Gnirke, A. (2011). Analyzing and minimizing PCR amplification bias in Illumina sequencing libraries. Genome Biol. *12*, R18.

Alice, A.F., López, C.S., Lowe, C.A., Ledesma, M.A., and Crosa, J.H. (2006). Genetic and transcriptional analysis of the siderophore malleobactin biosynthesis and transport genes in the human pathogen *Burkholderia pseudomallei* K96243. J. Bacteriol. *188*, 1551–1566.

Bazzini, S., Udine, C., Sass, A., Pasca, M.R., Longo, F., Emiliani, G., Fondi, M., Perrin, E., Decorosi, F., Viti, C., *et al.* (2011). Deciphering the role of rnd efflux transporters in *Burkholderia cenocepacia*. PLoS ONE *6*, e18902.

Chambers, C.E., Lutter, E.I., Visser, M.B., Law, P.P.Y., and Sokol, P.A. (2006). Identification of potential CepR regulated genes using a *cep* box motif-based search of the *Burkholderia cenocepacia* genome. BMC Microbiol. *6*, 104.

Chen, Z., and Duan, X. (2011). Ribosomal RNA depletion for massively parallel bacterial RNA-sequencing applications. Methods Mol. Biol. (Clifton, NJ) *733*, 93–103.

Chieng, S., Carreto, L., and Nathan, S. (2012). *Burkholderia pseudomallei* transcriptional adaptation in macrophages. BMC Genom. *13*, 328.

Chuaqui, R.F., Bonner, R.F., Best, C.J.M., Gillespie, J.W., Flaig, M.J., Hewitt, S.M., Phillips, J.L., Krizman, D.B., Tangrea, M.A., Ahram, M., *et al.* (2002). Post-analysis follow-up and validation of microarray experiments. Nature Genet. *32*, 509–514.

Coenye, T., and Vandamme, P. (2003). Diversity and significance of *Burkholderia* species occupying diverse ecological niches. Environ. Microbiol. *5*, 719–729.

Coenye, T., Van Acker, H., Peeters, E., Sass, A., Buroni, S., Riccardi, G., and Mahenthiralingam, E. (2011). Molecular mechanisms of chlorhexidine tolerance in *Burkholderia cenocepacia* biofilms. Antimicrob. Agents Chemother. *55*, 1912–1919.

Conway, T., and Schoolnik, G.K. (2003). Microarray expression profiling: capturing a genome-wide portrait of the transcriptome. Mol. Microbiol. *47*, 879–889.

Conway, B.A.D., Chu, K.K., Bylund, J., Altman, E., and Speert, D.P. (2004). Production of exopolysaccharide by *Burkholderia cenocepacia* results in altered cell-surface interactions and altered bacterial clearance in mice. J. Infect. Dis. *190*, 957–966.

Coutinho, C.P., de Carvalho, C.C.C.R., Madeira, A., Pinto-de-Oliveira, A., and Sá-Correia, I. (2011). *Burkholderia cenocepacia* phenotypic clonal variation during a 3.5-year colonization in the lungs of a cystic fibrosis patient. Infect. Immun. *79*, 2950–2960.

Cunha, M.V., Sousa, S.A., Leitão, J.H., Moreira, L.M., Videira, P.A., and Sá-Correia, I. (2004). Studies on the involvement of the exopolysaccharide produced by cystic fibrosis-associated isolates of the *Burkholderia cepacia* complex in biofilm formation and in persistence of respiratory infections. J. Clin. Microbiol. *42*, 3052–3058.

Dantas, G., Sommer, M.O.A., Oluwasegun, R.D., and Church, G.M. (2008). Bacteria subsisting on antibiotics. Science *320*, 100–103.

Denef, V.J., Park, J., Tsoi, T.V., Rouillard, J.M., Zhang, H., Wibbenmeyer, J.A., Verstraete, W., Gulari, E., Hashsham, S.A., and Tiedje, J.M. (2004). Biphenyl and benzoate metabolism in a genomic context: Outlining genome-wide metabolic networks in *Burkholderia xenovorans* LB400. Appl. Environ. Microbiol. *70*, 4961–4970.

Denef, V.J., Patrauchan, M.A., Florizone, C., Park, J., Tsoi, T.V., Verstraete, W., Tiedje, J.M., and Eltis, L.D. (2005). Growth substrate- and phase-specific expression of biphenyl, benzoate, and C1 metabolic pathways in

Burkholderia xenovorans LB400. J. Bacteriol. *187*, 7996–8005.

Denef, V.J., Klappenbach, J.A., Patrauchan, M.A., Florizone, C., Rodrigues, J.L.M., Tsoi, T.V., Verstraete, W., Eltis, L.D., and Tiedje, J.M. (2006). Genetic and genomic insights into the role of benzoate-catabolic pathway redundancy in *Burkholderia xenovorans* LB400. Appl. Environ. Microbiol. 72, 585–595.

Drevinek, P., Holden, M.T.G., Ge, Z., Jones, A.M., Ketchell, I., Gill, R.T., and Mahenthiralingam, E. (2008). Gene expression changes linked to antimicrobial resistance, oxidative stress, iron depletion and retained motility are observed when *Burkholderia cenocepacia* grows in cystic fibrosis sputum. BMC Infect. Dis. *8*, 121.

Ehrenreich, A. (2006). DNA microarray technology for the microbiologist: An overview. Appl. Microbiol. Biotechnol. 73, 255–273.

Febrer, M., McLay, K., Caccamo, M., Twomey, K.B., and Ryan, R.P. (2011). Advances in bacterial transcriptome and transposon insertion-site profiling using second-generation sequencing. Trends Biotechnol. *29*, 586–594.

Ferreira, A.S., Leitão, J.H., Silva, I.N., Pinheiros, P.F., Sousa, S.A., Ramos, C.G., and Moreira, L.M. (2010). Distribution of cepacian biosynthesis genes among environmental and clinical *Burkholderia* strains and role of cepacian exopolysaccharide in resistance to stress conditions. Appl. Environ. Microbiol. *76*, 441–450.

Filiatrault, M.J. (2011). Progress in prokaryotic transcriptomics. Curr. Opin. Microbiol. *14*, 579–586.

Francois, P., Garzoni, C., Bento, M., and Schrenzel, J. (2007). Comparison of amplification methods for transcriptomic analyses of low abundance prokaryotic RNA sources. J. Microbiol. Methods *68*, 385–391.

Fu, X., Fu, N., Guo, S., Yan, Z., Xu, Y., Hu, H., Menzel, C., Chen, W., Li, Y., Zeng, R., *et al.* (2009). Estimating accuracy of RNA-Seq and microarrays with proteomics. BMC Genom. *10*, 161.

Garber, M., Grabherr, M.G., Guttman, M., and Trapnell, C. (2011). Computational methods for transcriptome annotation and quantification using RNA-seq. Nature Methods *8*, 469–477.

Goff, L.A., Bowers, J., Schwalm, J., Howerton, K., Getts, R.C., and Hart, R.P. (2004). Evaluation of sense-strand mRNA amplification by comparative quantitative PCR. BMC Genom. *5*, 76.

Gottesman, S., and Storz, G. (2010). Bacterial small RNA regulators: Versatile roles and rapidly evolving variations. Cold Spring Harb. Perspect. Biol. *3*, a003798.

Guglierame, P., Pasca, M.R., De Rossi, E., Buroni, S., Arrigo, P., Manina, G., and Riccardi, G. (2006). Efflux pump genes of the resistance-nodulation-division family in *Burkholderia cenocepacia* genome. BMC Microbiol. *6*, 66.

Hamad, M.A., Austin, C.R., Stewart, A.L., Higgins, M., Vázquez-Torres, A., and Voskuil, M.I. (2011). Adaptation and antibiotic tolerance of anaerobic *Burkholderia pseudomallei*. Antimicrob. Agents Chemother. *55*, 3313–3323.

Hoeijmakers, W.A.M., Bártfai, R., Françoijs, K.J., and Stunnenberg, H.G. (2011). Linear amplification for deep sequencing. Nature Prot. *6*, 1026–1036.

Inhülsen, S., Aguilar, C., Schmid, N., Suppiger, A., Riedel, K., and Eberl, L. (2012). Identification of functions linking quorum sensing with biofilm formation in *Burkholderia cenocepacia* H111. Microbiol. Open *1*, 225–242.

Kang, Y., Norris, M.H., Zarzycki-Siek, J., Nierman, W.C., Donachie, S.P., and Hoang, T.T. (2011). Transcript amplification from single bacterium for transcriptome analysis. Genome Res. *21*, 925–935.

Kim, H., Zhao, B., Snesrud, E.C., Haas, B.J., Town, C.D., and Quackenbush, J. (2002). Use of RNA and genomic DNA references for inferred comparisons in DNA microarray analyses. BioTechniques *33*, 924–930.

Kim, H.S., Schell, M.A., Yu, Y., Ulrich, R.L., Sarria, S.H., Nierman, W.C., and DeShazer, D. (2005). Bacterial genome adaptation to niches: Divergence of the potential virulence genes in three *Burkholderia* species of different survival strategies. BMC Genom. *6*, 174.

Kreil, D.P., Russell, R.R., and Russell, S. (2006). Microarray oligonucleotide probes. Methods Enzymol. *410*, 73–98.

La, M.V., Raoult, D., and Renesto, P. (2008). Regulation of whole bacterial pathogen transcription within infected hosts. FEMS Microbiol. Rev. *32*, 440–460.

Law, R.J., Hamlin, J.N.R., Sivro, A., McCorrister, S.J., Cardama, G.A., and Cardona, S.T. (2008). A functional phenylacetic acid catabolic pathway is required for full pathogenicity of *Burkholderia cenocepacia* in the *Caenorhabditis elegans* host model. J. Bacteriol. *190*, 7209–7218.

Leiske, D.L., Karimpour-Fard, A., Hume, P.S., Fairbanks, B.D., and Gill, R.T. (2006). A comparison of alternative 60-mer probe designs in an in-situ synthesized oligonucleotide microarray. BMC Genom. *7*, 72.

Leparc, G.G., Tuchler, T., Striedner, G., Bayer, K., Sykacek, P., Hofacker, I.L., and Kreil, D.P. (2009). Model-based probe set optimization for high-performance microarrays. Nucleic Acids Res. *37*, e18.

Leroy, Q., and Raoult, D. (2010). Review of microarray studies for host-intracellular pathogen interactions. J. Microbiol. Methods *81*, 81–95.

Lewenza, S., and Sokol, P.A. (2001). Regulation of ornibactin biosynthesis and *N*-acyl-L-homoserine lactone production by cepR in *Burkholderia cepacia*. J. Bacteriol. *183*, 2212–2218.

Linnarsson, S. (2010). Recent advances in DNA sequencing methods – general principles of sample preparation. Experimental Cell Res. *316*, 1339–1343.

Lipscomb, L., and Schell, M.A. (2011). Elucidation of the regulon and cis-acting regulatory element of HrpB, the AraC-type regulator of a plant pathogen-like type III secretion system in *Burkholderia pseudomallei*. J. Bacteriol. *193*, 1991–2001.

Loutet, S.A., Di Lorenzo, F., Clarke, C., Molinaro, A., and Valvano, M.A. (2011). Transcriptional responses of *Burkholderia cenocepacia* to polymyxin B in isogenic strains with diverse polymyxin B resistance phenotypes. BMC Genom. *12*, 472.

McCarthy, Y., Yang, L., Twomey, K.B., Sass, A., Tolker-Nielsen, T., Mahenthiralingam, E., Dow, J.M., and Ryan, R.P. (2010). A sensor kinase recognizing the cell–cell signal BDSF (cis-2-dodecenoic acid) regulates virulence in *Burkholderia cenocepacia*. Mol. Microbiol. *77*, 1220–1236.

Madeira, A., Santos, P.M., Coutinho, C.P., Pinto-de-Oliveira, A., and Sá-Correia, I. (2011). Quantitative proteomics (2-D DIGE) reveals molecular strategies employed by *Burkholderia cenocepacia* to adapt to the airways of cystic fibrosis patients under antimicrobial therapy. Proteomics *11*, 1313–1328.

Mahenthiralingam, E., Song, L., Sass, A., White, J., Wilmot, C., Marchbank, A., Boaisha, O., Paine, J., Knight, D., and Challis, G.L. (2011). Enacyloxins are products of an unusual hybrid modular polyketide synthase encoded by a cryptic *Burkholderia ambifaria* genomic island. Chem. Biol. *18*, 665–677.

Malott, R.J., O'Grady, E.P., Toller, J., Inhülsen, S., Eberl, L., and Sokol, P.A. (2009). A *Burkholderia cenocepacia* orphan LuxR homolog is involved in quorum-sensing regulation. J. Bacteriol. *191*, 2447–2460.

Mamanova, L., Andrews, R.M., James, K.D., Sheridan, E.M., Ellis, P.D., Langford, C.F., Ost, T.W.B., Collins, J.E., and Turner, D.J. (2010). FRT-seq: Amplification-free, strand-specific transcriptome sequencing. Nature Methods *7*, 130–132.

Mamanova, L., and Turner, D.J. (2011). Low-bias, strand-specific transcriptome Illumina sequencing by on-flowcell reverse transcription (FRT-seq). Nature Prot. *6*, 1736–1747.

Martin, J.A., and Wang, Z. (2011). Next-generation transcriptome assembly. Nature Rev. Genet. *12*, 671–682.

Metzker, M.L. (2010). Sequencing technologies the next generation. Nature Rev. Genet. *11*, 31–46.

Mira, N.P., Madeira, A., Moreira, A.S., Coutinho, C.P., and Sá-Correia, I. (2011). Genomic expression analysis reveals strategies of *Burkholderia cenocepacia* to adapt to cystic fibrosis patients' airways and antimicrobial therapy. PLoS ONE *6*, e28831.

Moore, R.A., Reckseidler-Zenteno, S., Kim, H., Nierman, W., Yu, Y., Tuanyok, A., Warawa, J., DeShazer, D., and Woods, D.E. (2004). Contribution of gene loss to the pathogenic evolution of *Burkholderia pseudomallei* and *Burkholderia mallei*. Infect. Immun. 72, 4172–4187.

Morey, J.S., Ryan, J.C., and Van Dolah, F.M. (2006). Microarray validation: Factors influencing correlation between oligonucleotide microarrays and real-time PCR. Biological Proc. Online *8*, 175–193.

Nair, B.M., Cheung Jr, K.J., Griffith, A., and Burns, J.L. (2004). Salicylate induces an antibiotic efflux pump in *Burkholderia cepacia* complex genomovar III (*B. cenocepacia*). J. Clin. Invest. *113*, 464–473.

O'Grady, E.P., and Sokol, P.A. (2011). *Burkholderia cenocepacia* differential gene expression during host–pathogen interactions and adaptation to the host environment. Front. Cell. Infect. Microbiol. *1*, 15.

O'Grady, E.P., Viteri, D.F., Malott, R.J., and Sokol, P.A. (2009). Reciprocal regulation by the CepIR and CciIR quorum sensing systems in *Burkholderia cenocepacia*. BMC Genom. *10*, 441.

O'Grady, E.P., Nguyen, D.T., Weisskopf, L., Eberl, L., and Sokol, P.A. (2011). The *Burkholderia cenocepacia* LysR-Type transcriptional regulator ShvR influences expression of quorum-sensing, protease, type II secretion, and afc genes. J. Bacteriol. *193*, 163–176.

O'Grady, E.P., Viteri, D.F., and Sokol, P.A. (2012). A unique regulator contributes to quorum sensing and virulence in *Burkholderia cenocepacia*. PLoS ONE *7*, e37611.

Oh, S., Yoder-Himes, D.R., Tiedje, J., Park, J., and Konstantinidis, K.T. (2010). Evaluating the performance of oligonucleotide microarrays for bacterial strains with increasing genetic divergence from the reference strain. Appl. Environ. Microbiol. *76*, 2980–2988.

Oshlack, A., Robinson, M.D., and Young, M.D. (2010). From RNA-seq reads to differential expression results. Genome Biol. *11*, 220.

Patrauchan, M.A., Jacob Parnell, J., McLeod, M.P., Florizone, C., Tiedje, J.M., and Eltis, L.D. (2011). Genomic analysis of the phenylacetyl-CoA pathway in *Burkholderia xenovorans* LB400. Arch. Microbiol. *193*, 641–650.

Peeters, E., Sass, A., Mahenthiralingam, E., Nelis, H., and Coenye, T. (2010). Transcriptional response of *Burkholderia cenocepacia* J2315 sessile cells to treatments with high doses of hydrogen peroxide and sodium hypochlorite. BMC Genom. *11*, 90.

Pérez-Pantoja, D., Donoso, R., Agulló, L., Córdova, M., Seeger, M., Pieper, D.H., and González, B. (2012). Genomic analysis of the potential for aromatic compounds biodegradation in *Burkholderiales*. Environ. Microbiol. *14*, 1091–1117.

Piddock, L.J.V. (2006). Clinically relevant chromosomally encoded multidrug resistance efflux pumps in bacteria. Clin. Microbiol. Rev. *19*, 382–402.

Pumirat, P., Cuccui, J., Stabler, R.A., Stevens, J.M., Muangsombut, V., Singsuksawat, E., Stevens, M.P., Wren, B.W., and Korbsrisate, S. (2010). Global transcriptional profiling of *Burkholderia pseudomallei* under salt stress reveals differential effects on the Bsa type III secretion system. BMC Microbiol. *10*, 171.

Quackenbush, J. (2002). Microarray data normalization and transformation. Nature Genet. Suppl. *32*, 496–501.

Ratledge, C., and Dover, L.G. (2000). Iron metabolism in pathogenic bacteria. Ann. Rev. Microbiol. *54*, 881–941.

Rodrigues, F., Sarkar-Tyson, M., Harding, S.V., Siew, H.S., Hui, H.C., Chi, H.L., Han, X., Karuturi, R.K.M., Sung, K., Yu, K., *et al.* (2006). Global map of growth-regulated gene expression in *Burkholderia pseudomallei*, the causative agent of melioidosis. J. Bacteriol. *188*, 8178–8188.

Sass, A., Marchbank, A., Tullis, E., LiPuma, J.J., and Mahenthiralingam, E. (2011). Spontaneous and evolutionary changes in the antibiotic resistance of *Burkholderia cenocepacia* observed by global gene expression analysis. BMC Genom. *12*, 373.

Schmerk, C.L., Bernards, M.A., and Valvano, M.A. (2011). Hopanoid production is required for low-pH tolerance, antimicrobial resistance, and motility in *Burkholderia cenocepacia*. J. Bacteriol. *193*, 6712–6723.

Silva, I.N., Ferreira, A.S., Becker, J.D., Zlosnik, J.E.A., Speert, D.P., He, J., Mil-Homens, D., and Moreira, L.M. (2011). Mucoid morphotype variation of *Burkholderia multivorans* during chronic cystic fibrosis lung infection is correlated with changes in metabolism, motility, biofilm formation and virulence. Microbiology *157*, 3124–3137.

Smith, D.J., Park, J., Tiedje, J.M., and Mohn, W.W. (2007). A large gene cluster in *Burkholderia xenovorans* encoding abietane diterpenoid catabolism. J. Bacteriol. *189*, 6195–6204.

Sokol, P.A., Malott, R.J., Riedel, K., and Eberl, L. (2007). Communication systems in the genus *Burkholderia*: global regulators and targets for novel antipathogenic drugs. Future Microbiol. *2*, 555–563.

Song, H., Hwang, J., Yi, H., Ulrich, R.L., Yu, Y., Nierman, W.C., and Kim, H.S. (2010). The early stage of bacterial genome-reductive evolution in the host. PLoS Pathogens *6*, e1000922.

Sorek, R., and Cossart, P. (2010). Prokaryotic transcriptomics: a new view on regulation, physiology and pathogenicity. Nature Rev. Genet. *11*, 9–16.

Subramoni, S., Nguyen, D.T., and Sokol, P.A. (2011). *Burkholderia cenocepacia* ShvR-regulated genes that influence colony morphology, biofilm formation, and virulence. Infect. Immun. *79*, 2984–2997.

Toledo-Arana, A., and Solano, C. (2010). Deciphering the physiological blueprint of a bacterial cell: Revelations of unanticipated complexity in transcriptome and proteome. BioEssays *32*, 461–467.

Tolman, J.S., and Valvano, M.A. (2012). Global changes in gene expression by the opportunistic pathogen *Burkholderia cenocepacia* in response to internalization by murine macrophages. BMC Genom. *13*, 63.

Tuanyok, A., Kim, H.S., Nierman, W.C., Yu, Y., Dunbar, J., Moore, R.A., Baker, P., Tom, M., Ling, J.M.L., and Woods, D.E. (2005). Genome-wide expression analysis of iron regulation in *Burkholderia pseudomallei* and *Burkholderia mallei* using DNA microarrays. FEMS Microbiol. Lett. *252*, 327–335.

Tuanyok, A., Tom, M., Dunbar, J., and Woods, D.E. (2006). Genome-wide expression analysis of *Burkholderia pseudomallei* infection in a hamster model of acute melioidosis. Infect. Immun. *74*, 5465–5476.

van Schaik, E., Tom, M., DeVinney, R., and Woods, D.E. (2008). Development of novel animal infection models for the study of acute and chronic *Burkholderia pseudomallei* pulmonary infections. Microbes Infect. *10*, 1291–1299.

Visser, M.B., Majumdar, S., Hani, E., and Sokol, P.A. (2004). Importance of the ornibactin and pyochelin siderophore transport systems in *Burkholderia cenocepacia* lung infections. Infect. Immun. *72*, 2850–2857.

Wang, Y., Mehta, G., Mayani, R., Lu, J., Souaiaia, T., Chen, Y., Clark, A., Yoon, H.J., Wan, L., Evgrafov, O.V., *et al.* (2011). RseqFlow: Workflows for RNA-Seq data analysis. Bioinformatics *27*, 2598–2600.

Yang, Y.H., and Speed, T. (2002). Design issues for cDNA microarray experiments. Nature Rev. Genet. *3*, 579–588.

Yoder-Himes, D.R., Chain, P.S.G., Zhu, Y., Wurtzel, O., Rubin, E.M., Tiedje, J.M., and Sorek, R. (2009). Mapping the *Burkholderia cenocepacia* niche response via high-throughput sequencing. Proc. Natl. Acad. Sci. U.S.A. *106*, 3976–3981.

Yoder-Himes, D.R., Konstantinidis, K.T., and Tiedje, J.M. (2010). Identification of potential therapeutic targets for *Burkholderia cenocepacia* by comparative transcriptomics. PLoS ONE *5*, e8724.

Zhang, W., Li, F., and Nie, L. (2010). Integrating multiple 'omics' analysis for microbial biology: Application and methodologies. Microbiology *156*, 287–301.

Zlosnik, J.E.A., and Speert, D.P. (2010). The role of mucoidy in virulence of bacteria from the *Burkholderia cepacia* complex: A systematic proteomic and transcriptomic analysis. J. Infect. Dis. *202*, 770–781.

Zlosnik, J.E.A., Costa, P.S., Brant, R., Mori, P.Y.B., Hird, T.J., Fraenkel, M.C., Wilcox, P.G., Davidson, A.G.F., and Speert, D.P. (2011). Mucoid and nonmucoid *Burkholderia cepacia* complex bacteria in cystic fibrosis infections. Am. J. Respirat. Crit. Care Med. *183*, 67–72.

Burkholderia Proteomics: Methodologies, Challenges and Applications

5

Isabel Sá-Correia, Sandra C. dos Santos and Andreia Madeira

Abstract

Complete genome sequencing for a large number of microorganisms has paved the way for large-scale proteomic studies to investigate global protein expression levels and regulation and to assist in functional analysis. Although microarray technologies can provide invaluable information on genomic expression, it is also known that mRNA levels often do not correlate directly with protein levels. Proteomics is thus necessary to provide complementary quantitative and functional data to genomic information. In this review we will provide a comparative overview of the various proteomic approaches currently available and give a brief tutorial on the methodology and application of two-dimensional electrophoresis-based expression proteomics. Subsequently, the exploitation of proteomic analyses in studies dedicated to different *Burkholderia* species with a broad range of objectives will be discussed, based on the relevant scientific literature. These include biomedical research and applications (e.g. elucidation of pathogenic mechanisms and interaction with the host, improvement of diagnostic tools, identification of immunoreactive proteins and candidate vaccine targets) and environmental studies, namely on biodegradation of organic pollutants by *Burkholderia* strains.

Proteomic analyses in microbiological research

The implementation of genome-wide approaches together with the possibility of performing relatively low-cost and rapid genome sequencing led to a major revolution in biological research, creating new opportunities to address biological questions at a systems level. This has been particularly relevant in the field of microbiological research, given that most of the genome sequences currently available belong to bacterial species/strains; according to the Genomes Online Database (www.genomesonline.org) in September 2012, 3366 out of 3708 complete genome projects were from bacteria. It is now possible to look at a bacterial phenotype as the result of variations at different cellular levels (DNA, RNA, protein, and metabolite) and integrate the resulting *omic* information to obtain a systems biology understanding of a cell's response to environmental or genetic alterations relevant to medical and environmental research (Ge *et al.*, 2003).

Proteomics, the large-scale study of proteins, is a key technology in this post-genomic era, where improved methods of protein separation, quantification, and identification can provide a bridge between the genome and protein expression. Proteomics also provides insights into protein modification, subcellular localization and regulatory mechanisms. Indeed, one important goal of proteomics can be defined as the ability to obtain a quantitative description of protein expression and its changes under the influence of genetic or environmental perturbations (Cox and Mann, 2011; Elliott *et al.*, 2009). Quantitative or expression proteomics is similar in concept to the analysis of the transcriptome by microarrays; however, it possesses the advantage of measuring the level of the end product of the gene expression cascade, allowing a more accurate understanding of the biological functions involved. Furthermore, expression proteomics can determine the distribution of specific proteins within the cell as well as the occurrence of post-translational modifications. In the next section we will provide an overview of the currently available methodologies for proteomic analysis that can be applied to the study of *Burkholderia* species/strains, their implementation, limitations and advantages. A particular emphasis is given to proteomics based on two-dimensional electrophoresis (2-DE) coupled with mass spectrometry (MS), which remains one of the most widely used quantitative proteomics approaches in microbiology laboratories. Several recent reviews on the topic are indicated for further reading. The third section will describe specific applications of proteomics to the study of *Burkholderia* species in a systems biology perspective.

Currently available methodologies for proteome analyses

Unlike other post-genomic approaches, the history of proteomics precedes the release of the first genome sequence, when a revolutionary technique, 2-DE, was developed in 1975 (O'Farrell, 1975). However, the methodology was obviously hindered by the lack of genome sequence knowledge and of adequate protein identification methods. Today, the development of new methodologies is one of the most challenging and rapidly changing fields of proteomics, with a range of either gel-based or gel-free approaches available (Fig. 5.1) (Cox and Mann, 2011; Elliott *et al.*, 2009; Ong and Mann, 2005; Otto *et al.*, 2012). The selection of a method for a particular microbiology project depends on multiple factors, including the source and number of samples, number of conditions being compared, the type of equipment available, and the corresponding expertise, cost and time required.

Gel-based 2-D electrophoresis analyses

2-DE has come a long way since 1975, but is still the most commonly used gel-based technique, coupled with identification by mass spectrometry (Lopez, 2007; Rabilloud and Lelong, 2011). This combined approach has been used extensively in the field of microbiology, as reviewed elsewhere (Cash, 2000; Hecker *et al.*, 2008; Phillips and Bogyo, 2005; Sá-Correia and Teixeira, 2010). In fact, the majority of the work reviewed here in the context of *Burkholderia* studies is an application of 2-DE-based expression proteomics. The technique allows the separation of a complex mixture of proteins with a high discriminatory power, first by charge, using isoelectric focusing (IEF), and second by molecular weight, using sodium dodecyl sulphate-polyacrylamide gel electrophoresis (SDS-PAGE).

Compared with alternative protein profiling methods, 2-DE presents numerous advantages that are responsible for its sustained popularity. It is the most cost-effective and simple method, and can be implemented in virtually any microbiology laboratory. MS analysis of protein spots of interest can be performed in an external MS platform, deeply facilitating the routine application of 2D-based expression proteomics in microbiology laboratories. Furthermore, 2-DE separation allows the generation of protein reference maps for different organisms, facilitating the identification of protein spots produced under the same conditions and reducing costs (Sá-Correia and Teixeira, 2010); reference maps have been completed for a number of important human pathogens (Cash, 2000; Phillips and Bogyo, 2005). Moreover, since gel images can be stored in a computer, a 2-DE gel can be reanalysed later if needed, and new samples can be compared with previously analysed samples that may no longer be available (Lopez, 2007; Sá-Correia and Teixeira, 2010). Another significant advantage of 2-DE is the ability to detect proteomic variability resulting from post-translational modifications, as discussed below. Post-translational modifications are often overlooked in bacteria, but recent reports (including studies on *Burkholderia* species) highlight their possible involvement in pathogenicity (Inhülsen *et al.*, 2012; Ou *et al.*, 2005; Wongtrakoongate *et al.*, 2007).

Many of the advantages and limitations of 2-DE are relevant for all general applications of proteomics. However, some aspects are specific to proteomic applications in microbiology research with impact in environmental and biomedical fields. For instance, protein heterogeneity between bacterial isolates can make the cross-matching of 2-D gels difficult and unreliable. Difficulties can

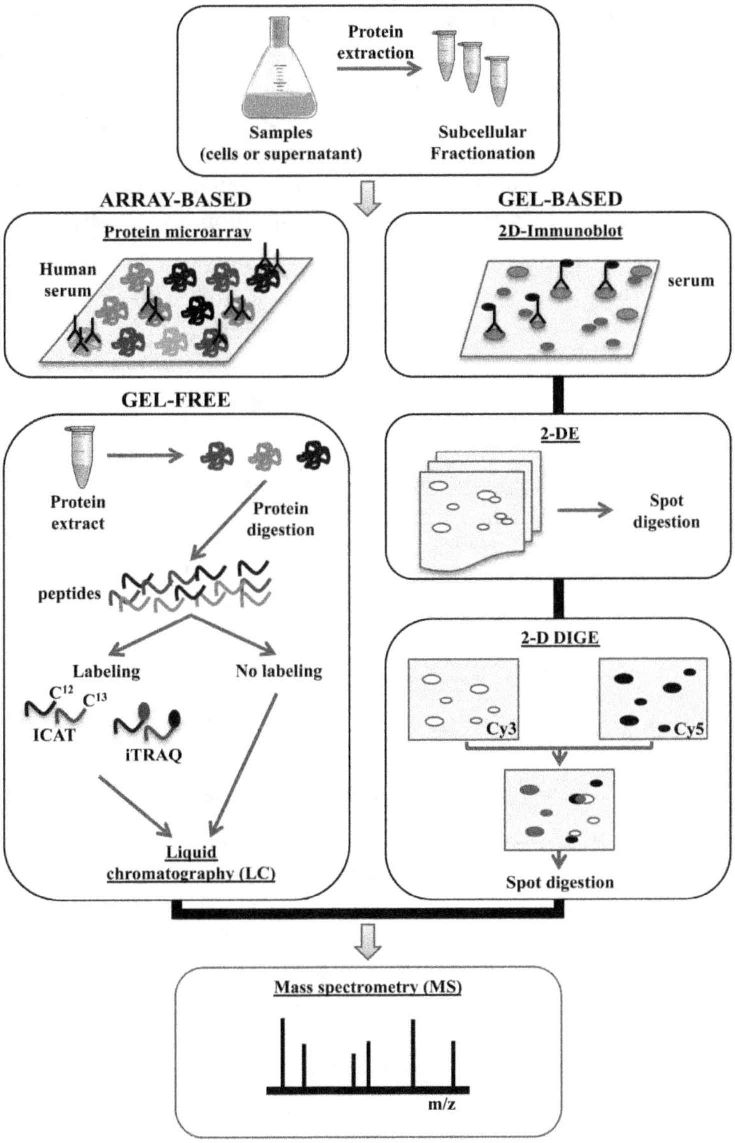

Figure 5.1 Schematic representation of currently available methodologies for expression proteomic analysis. After sample preparation, proteins are extracted and can be submitted to fractionation. Several expression proteomic strategies can be adopted: array-based, gel-based or gel-free. The final objective in all cases is the quantification, identification and characterization of proteins of interest.

also arise when comparing well-characterized and sequenced laboratory strains with clinical or environmental isolates of interest, since their genomic sequence might differ (Cash, 2000). However, modern 2-DE-based approaches can even be applied to species whose genomes are not available. For example, mapping the proteome response to phenol in *Pseudomonas* spp. M1 was possible by matching a large number of proteins identified by peptide mass fingerprinting with *P. aeruginosa* proteins present in databases (Santos *et al.*, 2007). In some cases where close homologues cannot be found, alternative MS approaches or *de novo* sequencing may be required.

Basic methodology and variations

The design of a 2-DE-based quantitative bacterial proteomics experiment basically involves the following steps (Fig. 5.2): (i) cultivating an adequate amount of cells, from the same organism exposed to different environmental conditions or different bacterial isolates grown under the same conditions; (ii) proteome extraction and/or concentration followed by fractionation; (iii) solubilization of proteins in the IEF buffer and isoelectric focusing (IEF – 1st dimension); (iv) separation by molecular weight (SDS-PAGE – 2nd dimension); (v) staining of proteins, image acquisition and comparative gel analysis; (vi) identification of proteins with a significantly different content between samples; (vii) protein identification, and finally, (viii) data analysis (Fig.

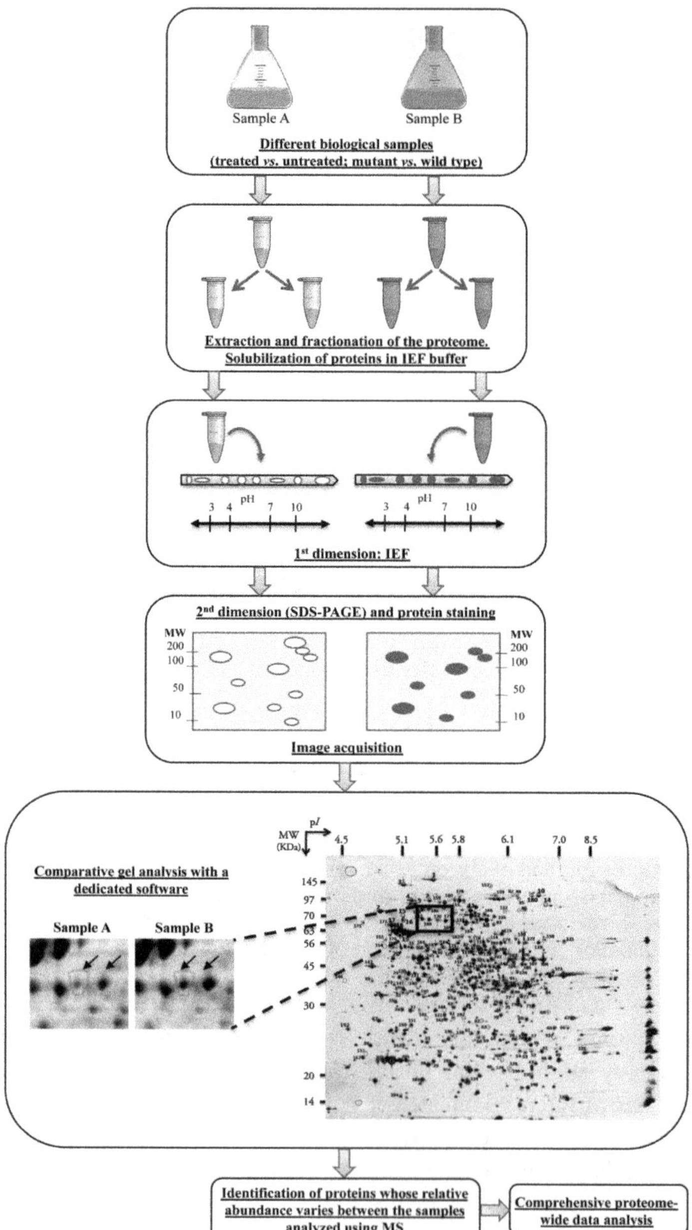

Figure 5.2 Schematic representation of the typical workflow required for expression proteomic analysis using 2-DE.

5.2). Some proteins occur in 2-DE gels as single spots and others as multiple spots, corresponding to different levels and types of post-translational modifications, and the relative intensity of each of these species can be quantified individually (Lopez, 2007; Sá-Correia and Teixeira, 2010). Theoretically, up to 10,000 proteins can be resolved in a single gel, although only around 2000 can be detected using fluorescent protein staining (Lopez, 2007). After analysis of the gels, each protein spot can be excised as plugs and digested overnight with trypsin. Identification of the resulting protein fragments is fairly easy and inexpensive when the genome sequence is known, for example using peptide mass fingerprinting or tandem mass spectrometry (MS/MS) (Elliott *et al.*, 2009).

Modifications of several of the steps described above have been developed and implemented in recent years, successfully overcoming some of the shortcomings of 2-DE. The commercialization of immobilized pH (IPG) gradient gel strips allows greater reproducibility and separation of complex protein samples in larger amounts. Sample application for the 1st dimension can be carried out by in-gel rehydration or using a manifold system (GE Healthcare) with cup or paper loading, facilitating the separation of very basic proteins. Protein staining has also been dramatically improved in recent years with the development of protein staining with fluorescent dyes such as SYPRO® Ruby (Molecular Probes, USA) and Flamingo™ (BioRad, USA), which are more sensitive and capable of detecting around 1–1000 ng of protein. Certain dyes also exhibit specificity towards different protein classes and modification states (e.g. ProQ® Diamond, Invitrogen, USA). These dyes represent a major improvement over classical protein staining methods based on Coomassie Blue or silver nitrate which have much narrower dynamic ranges. They also help overcome the reduced ability of 2-DE to detect low abundance proteins, one of its major weaknesses. The improvement of image acquisition systems such as fluorescence scanners (e.g. Typhoon series) and computer-aided gel analysis tools has further increased the sensitivity, reliability, reproducibility and range of applicability of 2-DE-based proteomics (Lopez, 2007; Rabilloud and Lelong, 2011; Sá-Correia and Teixeira, 2010). Software packages such as those marketed by Progenesis or GE Healthcare include automated spot alignment and refined statistical analyses (e.g. one-way ANOVA, PCA) between groups of gels, increasing the degree of confidence in the assessment of protein content variation.

An important drawback of 2-DE-based proteomics is the relatively narrow range of proteins that can be analysed; proteins that are very alkaline or acidic, hydrophobic, and/or have low or high molecular weight are often not captured by the analysis (Lopez, 2007; Sá-Correia and Teixeira, 2010). Several efforts have been made to overcome these limitations, including sample pre-fractionation, optimization of the IEF pH range, manipulation of gel porosity, and use of new detergents, surfactants and chaotropes that allow the solubilization of a higher number of proteins (Braun *et al.*, 2007; Lopez, 2007; Sá-Correia and Teixeira, 2010). Fractionation approaches have been applied successfully to isolate proteins associated with specific cellular structures, allowing the detection of proteins present at lower relative abundance in whole-cell extracts (Braun *et al.*, 2007; Cordwell *et al.*, 2000; Roma-Rodrigues *et al.*, 2010). Pre-fractionation in Gram-negative bacteria is of particular interest since, using 2-DE-compatible detergents, it is possible to fractionate the total proteome into the soluble proteome, and inner-membrane and outer-membrane subproteomes. Around two-thirds of all drugs exert their effects through membrane proteins (Hopkins and Groom, 2002). Analysis of these subproteomes might shed light into detoxification mechanisms involved for example in drug resistance in a clinical setting (Braun *et al.*, 2007). However, although pre-fractionation can dramatically increase the total number of proteins that can be identified and quantified, it still remains that 2-DE has a strong bias against hydrophobic membrane proteins. Several approaches have been developed to reduce this bias, but despite the efforts and the fact that 2-DE can be used to separate membrane-enriched fractions, it is still inadequate to detect integral membrane proteins such as multiple spanner transporters (Braun *et al.*, 2007; Roma-Rodrigues *et al.*, 2010).

A recent evolution of conventional 2-DE is represented by 2D-Difference in Gel Electrophoresis (2D-DIGE), which allows sample multiplexing and achieves more accurate and sensitive quantitative proteomic determinations, detecting proteins over a 20,000-fold concentration range (Minden, 2007, 2012). The technique relies on pre-labelling of different protein samples with different fluorophores, such as cyanine (Cy) dyes CyDye DIGE Fluor (GE Healthcare, USA), followed by mixing and 2-DE separation, and scanning of the gel under wavelengths specific for each dye. This enables a simultaneous multiplexing analysis of individual alterations in abundance changes for thousands of intact proteins in low amounts of replicate samples (Friedman and Lilley, 2008; Lilley and Friedman, 2004; Minden, 2007). The use of an internal protein standard, and the fact that up to three samples can be resolved in a single gel, contributes to improve gel-to-gel spot matching and in-gel alignment of protein spots.

Identification of post-translational modifications

Several proteins are under tight regulation by a series of post-transcriptional modifications, and therefore their activity may not correlate with protein abundance alone. 2-DE is unique not only for its ability to simultaneously separate thousands of proteins but also for the straightforward separation of protein species that differ only at the level of post-translational modifications, which cannot be predicted from genome sequences (Lopez, 2007; Rabilloud and Lelong, 2011). Many 2-DE-based methods for studying post-translation modifications rely on selective and specific probes for the detection of modified proteins in the gel.

Protein phosphorylation, one of the most common regulatory mechanisms in nature, although still underestimated in bacterial species, can be studied by phosphoproteomics. In a 2-DE gel, the occurrence of phosphoproteins is usually detected by multiple p*I* variants of a single protein. The development of ProQ® Diamond (Invitrogen, USA), a fluorescent dye specific for phosphoproteins, allowed the easy and selective detection and quantification of phosphoproteins following 2-DE separation. A multiplex approach in combination with sequential SYPRO Ruby or Flamingo post-staining of all proteins in the same gels enables the dichromatic fluorescent detection of phosphorylated and non-phosphorylated proteins separated in 2-DE gels (Agrawal and Thelen, 2005; Steinberg *et al.*, 2003). Using a similar multiplex approach, the glycoproteome can be studied using ProQ® Emerald (Invitrogen, USA), a fluorescent dye that is conjugated to glycoproteins prior to 2-DE separation and later detected in the gel using a fluorescence scanner (Steinberg *et al.*, 2001). Other post-translational modifications that can also be quantified by 2-DE include carbonylation or S-thiolation, two irreversible modifications that occur under oxidative stress (Sheehan, 2006; Sheehan *et al.*, 2010). The use of affinity selection methods to trap sub-proteomes can improve current 2-DE-based redox proteomics approaches and allow the analysis of reversible and low-frequency modifications (Sheehan *et al.*, 2010).

These experimental approaches may shed light on the mechanisms mediating post-translation modifications as the result of a given stress or as an adaptive regulatory mechanism, highlighting the fact that the assessment of protein concentration per se is often not enough, and that the study of post-translational modifications at the proteome-wide scale is crucial for understanding biological phenomena.

Gel-free/MS-based analyses

Gel-free/MS-based approaches have been developed in recent years as an alternative to the classical method of 2-DE (Fig. 5.1). These include liquid chromatography/mass spectrometry (LC/MS), which has evolved as a major analytical platform for proteomic studies due to its sensitivity, selectivity, accuracy, speed and output (Chen and Pramanik, 2009). Proteins are digested to peptides, preferably by a sequence-specific enzyme such as trypsin, and the resulting peptide mixture is separated with a gradient of aqueous/organic solvent in LC and analysed by MS. Chemical protein derivatization using isobaric labelling reagents can be implemented to improve the analysis, namely tandem mass tags and isobaric tags for relative and absolute quantitation (iTRAQ), as well as isotope-coded affinity tags (ICAT) (Chen and Pramanik, 2009; Cox and Mann, 2011; Rodriguez-Suarez and Whetton, 2013). LC/MS methodologies are of particular interest in the analysis of samples that are difficult to process with 2-DE, for example containing very acidic, very basic, or hydrophobic integral membrane proteins. However, the LC/MS platform is a less accessible alternative for standard microbiology laboratories. Moreover, and unlike 2-DE, LC/MS analysis of new samples is difficult to compare with others analysed before (Lopez, 2007).

In addition, LC/MS-based approaches still lack the high resolving power of 2-DE, especially for highly complex mixtures in particular those containing different forms of the same protein, such as those resulting from post-translational modifications (Chen and Pramanik, 2009; Lopez, 2007; Otto *et al.*, 2012). Whole protein characteristics such as isoelectric point and molecular weight are typically not preserved when using gel-free approaches, and findings concerning protein alterations are often missed. Nevertheless, advances are being made towards identification of post-translational modifications by LC/MS approaches, although prior knowledge of the modifications is required for this, and thus its applicability for the identification of novel or unexpected protein alterations is limited (Chen and Pramanik, 2009). Protein microarrays constructed from the predicted proteome of an organism are another highly useful gel-free proteomic approach currently available. These arrays can be used to elucidate the interaction between a pathogen and the host immune system, and identify proteins that can be potentially used as diagnostic markers or be included in vaccines (Davies *et al.*, 2005; Felgner *et al.*, 2009). Multidimensional protein identification technology (MudPIT) is an alternative gel-free technique in which two-dimensional liquid chromatography is fed directly into a mass spectrometer, maximizing sensitivity and eliminating sample manipulation (Washburn, 2004). Pathogens are suitable candidates for MudPIT analysis, and reports of its use with *Burkholderia* species are now emerging (Wickramasekara *et al.*, 2011).

Applications of quantitative proteomics in *Burkholderia* studies

Proteomic analyses have provided new insights that complement other genomic-wide investigations of *Burkholderia* species in fields ranging from environmental to medical microbiology. This literature review identified 31 original research papers where proteomic approaches were applied to obtain a systems level understanding of *Burkholderia* bacteria, making use of varied methodological strategies depending on the application, although 2-DE-based quantitative proteomics was found to predominate (Table 5.1). The main results and key findings emerging from these studies are critically discussed below.

Biomedical applications

Proteomics has been shown to play a role in the elucidation of mechanisms of bacterial pathogenicity and antibiotic resistance. In particular, proteomics has contributed to these studies in the important opportunistic human pathogens of the *Burkholderia cepacia* complex (Bcc). This heterogeneous group of bacteria comprises at least 17 closely related species that are highly antibiotic resistant and can cause serious respiratory tract infections in immunocompromised individuals (Mahenthiralingam *et al.*, 2005). People who have AIDS or cancer patients under chemotherapy are at a high risk of Bcc infection, as well as patients with cystic fibrosis (CF), the most common life-threatening autosomal recessive disease among Caucasians (Drevinek and Mahenthiralingam, 2010; Mahenthiralingam *et al.*, 2002). There is therefore a marked medical interest in the characterization of the pathogenic mechanisms involved in the action of Bcc bacteria during infection of a human host, and the information provided by proteomic studies can be crucial.

Another relevant pathogen of the *Burkholderia* species is *B. pseudomallei*, an environmental saprophyte that is currently included in the category B list of potential biological warfare agents of the United States Center for Disease Control and Prevention (Rotz *et al.*, 2002). *B. pseudomallei* is the causative agent of melioidosis, a serious infectious disease characterized by severe pulmonary

Table 5.1 Proteomic analyses applied to *Burkholderia* species relevant in the context of human and plant pathogenesis and in environmental applications

Species/strain	Objective	Quantitative proteomics approach	Fraction[a]	Main conclusions	Reference
Biomedical applications					
Identification of diagnostic markers and vaccine targets					
Burkholderia cepacia complex (Bcc) bacteria	To use a MS approach for rapid identification of Bcc strains and biochemically similar bacteria	MALDI-TOF MS	WC (intact)	The mass spectra obtained allowed the discrimination of 65 of the Bcc and *Burkholderia* strains analysed	Vanlaere *et al.* (2008)
B. pseudomallei K96243	To identify immunogenic surface proteins for exploitation as vaccine candidates and diagnostic targets	2-DE/MS	OM	Identification of nine proteins as potential vaccine and diagnostic targets	Harding *et al.* (2007)
	To identify serodiagnostic and cross-reactive antigens to distinguish melioidosis cases	Protein microarray	WC	A total of 49 antigens were identified as significantly more reactive in melioidosis patients compared to healthy individuals or with other types of bacterial infections	Felgner *et al.* (2009)
B. pseudomallei CMS	To identify immunogenic proteins in bacterium secretome that might be potential diagnosis molecules	2-DE/2D-immunoblot/MS	EC	Identification of 12 immunoreactive secretory proteins	Vellasamy *et al.* (2011)
B. pseudomallei 844 and *B. thailandensis* E257	To identify potential protein markers and virulence factors by comparison of *B. pseudomallei* and *B. thailandensis*	2-DE/MS	WC	Two hypothetical proteins found only in *B. pseudomallei* were proposed as potential markers to distinguish the two species. Six proteins were suggested to be involved in *B. pseudomallei* virulence	Wongtrakoongate *et al.* (2007)
B. cenocepacia J2315 and *B. multivorans* ATCC 17616	To identify species-specific secretory proteins as biomarkers for diagnostic assays and targets for therapeutic agents or vaccines	2-DE/MS	EC	Nine discriminatory proteins between the two species analysed were defined	Mott *et al.* (2010)
B. cepacia CQK	To identify immunogenic proteins in the secretome that might be potential diagnostic molecules	2-DE/2D-immunoblot/MS	EC	Identification of 18 immunoreactive secretory proteins	Mariappan *et al.* (2010)

Burkholderia–host interactions and pathogenic mechanisms					
B. cenocepacia H111	To identify the quorum sensing system CepIR-controlled gene products	2-DE/MS	IC, SF, EC	CepIR was identified as a global regulatory system in this strain. Identification of 55 cep-regulated proteins	Riedel *et al*. (2003)
	To provide a detailed 2D reference map	2-DE/MS	IC, SF, EC	In total 642 proteins spots were identified corresponding to 390 different protein species	Riedel *et al*. (2006)
	To identify functions linking the CepIR quorum sensing system with biofilm formation using a combined transcriptomic and proteomic approach	iTRAQ/MS	E√C, EC, WC	A post-transcriptional level regulation of quorum sensing-regulated proteins was suggested. The role of 3 loci (BclACB, BapA and a type I pilus) in biofilm development was studied in detail	Inhülsen *et* al. (2012)
B. cenocepacia C1394	To identify candidate proteins involved in B. cenocepacia survival in a susceptible murine host	2-DE/MS, LC-MS/MS, immunoblot	IM, OM, WC	The proteins alkyl hydroperoxide reductase subunit C (AhpC) and flagellin were identified	Chung and Speert (2007)
B. cenocepacia C8963 and C9343	To gain insights into the differences between mucoid and non-mucoid isolates from a CF patient using proteomic and transcriptomic analyses	iTRAQ/MS		The proteome of a non-mucoid isolate was marked by a strong overexpression of several virulence factors, namely AidA, AhpC and the metalloprotease ZmpA, also detected in microarray data	Zlosnik and Speert (2010)
B. cenocepacia IST439 and IST4113	To reveal molecular strategies employed in adaptation to cystic fibrosis airways under antimicrobial therapy by comparison of the expression programmes of sequential clonal isolates	2-D DIGE/MS	IC, M	Protein expression differences were related with bacterial resistance to several antimicrobials, iron uptake, motility, biofilm formation and colony morphology. Potential targets for inactivation to enhance microbial activity and/or to avoid bacterial persistence were suggested	Madeira *et al*. (2011)

Table 5.1 continued

Species/strain	Objective	Quantitative proteomics approach	Fraction[a]	Main conclusions	Reference
B. cepacia CQK	To compare *B. cepacia* secretomes at mid-logarithmic and early stationary phases of growth	2-DE/MS	EC	Forty proteins only expressed at the early stationary phase were correlated with the higher cytotoxic activity shown by this supernatant. Most of these proteins were known virulence factors	Mariappan *et al*. (2011)
B. vietnamiensis G4	To identify *B. vietnamiensis* low abundance virulence factors expressed under growth in minimal medium	LC-MS/MS, MudPIT analysis	EC, SF, M, IC	Low abundance virulence factors were mainly identified in cell surface, cell wall or extracellular fractions, such as flagellin, porins, peroxiredoxins and zinc proteases	Wickramasekara *et al*., 2011)
B. pseudomallei K96243 and ATCC 15682	To investigate the molecular mechanisms contributing to *B. pseudomallei* virulence phenotypes using a combination of genomic, transcriptomic and proteomic approaches	2-DE/MS	WC	Around 20% of the proteome was different between ATCC 15682 and K96243. Up to 38% of the proteomic variation between strains is associated with strain-specific protein isoforms	(Ou *et al*., 2005)
B. pseudomallei DD503 and B. mallei ATCC 23344	To assess the outer membrane proteome of *B. pseudomallei* and *B. mallei* under different growth conditions	LC-MS/MS	OM	Identification of 90 highly probable outer membrane proteins in both pathogens. The 20 most abundant outer membrane proteins detected were linked with potential roles in growth and survival within the host	Schell *et al*. (2011)
B. pseudomallei K96243	To explore the mechanisms by which the alternative sigma factor σ^E (RpoE) operon controls stress tolerance and survival	2-DE/MS	WC	Several components of stress response proteins and known virulence factors were down regulated in the rpoE mutant, providing new indications for the role of rpoE operon in *B. pseudomallei* virulence	Thongboonkerd *et al*. (2007)

B. pseudomallei PP844	To identify RpoS regulon and RpoS-dependent promoter sequences using both proteomic and computational analyses	2-DE/MS	WC	Proteomic analysis showed that 29% of RpoS-dependent genes were related to stress response. The number of RpoS-controlled genes were expanded based on proteomic data, promoter and operon prediction, and gene organization	Osiriphun *et al*. (2009)
B. pseudomallei 10276	To evaluate changes in bacterial secretome under exposure to a high-salt environment	2-DE/MS	EC	Forty-two proteins were altered in response to salt stress. The results provided the first evidences that salt stress enhances the secretion of virulence factors	Pumirat *et al*. (2009)
B. pseudomallei 844	To construct a proteome reference map from early stationary phase bacteria to facilitate comparative proteomics with sibling species	2-DE/MS	WC	A total of 220 different proteins were identified and many of them were previously shown to be implicated in bacterial pathogenesis	Wongtrakoongate *et al*. (2011)
B. pseudomallei 153	To expand the number of proteins known to undergo differential expression during colony morphology switching	2-DE/MS	WC	Morphotype switching involved at least 19 proteins. The most prominent proteins found (ArcA, ArcC) were studied in relation to bacterial survival under adverse conditions	Chantratita *et al*., 2012)
B. glumae BGR1	To identify the tofIR quorum sensing system-dependent proteome	2-DE/MS	WC, EC	A total of 79 proteins were found to be differently expressed in the wild type compared to tofI mutant, including proteins secreted through the type II secretion system (T2SS) with a putative role in virulence	Goo *et al*. (2010)
	To identify proteins regulated by the hypersensitive response and pathogenicity type III protein secretion system (T3SS) transcriptional activator (HrpB) *in vitro*	2-DE/MS	WC, EC	Under hrpB constitutive expression, 46 cellular (n=12) and extracellular (n=34) proteins with altered content were identified. Further studies showed that 21 extracellular proteins were secreted by T2SS	Kang *et al*. (2008)

Table 5.1 continued

Species/strain	Objective	Quantitative proteomics approach	Fraction[a]	Main conclusions	Reference
Environmental applications					
B. xenovorans LB400	To provide insights into the coordinated use of multiple biphenyl and benzoate catabolic pathways	2-DE/MS	WC	Characterization of the conditional regulation of the three major benzoate-catabolic pathways (ben-cat, box_M and box_C) depending on substrate and growth phase	Denef *et al*. (2005)
	To investigate the functional redundancy of benzoate-catabolic pathways (ben-cat, box_M and box_C) using a combined approach of genetics, transcriptomics and proteomics	2-DE/MS	WC	All three benzoate pathways are able to degrade benzoate, while the two box pathways are preferentially active at reduced oxygen tension	Denef *et al*. (2006)
	To study the effects of 4-chlorobiphenyl and biphenyl on physiology, morphology and proteome of LB400	2-DE/MS	WC	Catabolic enzymes, molecular chaperones, alkyl hydroperoxide reductase and other proteins involved in energy production, general metabolism and transport were induced in response to 4-chlorobiphenyl or biphenyl	Agulló *et al*. (2007)
	To study the effects of 4-chlorobenzoate on physiology, morphology and proteome of LB400	2-DE/MS	WC	The 11 proteins induced by 4-CBA were mainly enzymes of benzoate and catechol catabolic pathways, molecular chaperones and Krebs' cycle enzymes	Martínez *et al*. (2007)

B. vietnamiensis PR1301	To compare protein expression in response to the toxic metal nickel at pH 5 and 7, to identify potential tolerance mechanisms	2-DE/MS	IC, M	The higher tolerance of *B. vietnamiensis* to nickel at pH 5 was related to increased expression of proteins involved in cell shape and membrane composition	Van Nostrand *et al.* (2008)
B. cepacia MBA4	To investigate the MBA4 protein expression profile in the presence of monochloroacetate (MCA) and glycolate, a product of MCA metabolism	2-DE/MS	WC	Degradation enzymes were induced in the presence of MCA but not of glycolate. MCA was not found to be a stressing substrate active on strain MB4A	Kwok *et al.* (2007)
B. cepacia WZ1	To study the molecular response of WZ1 during biodegradation of the herbicide quinclorac	2-DE/MS	WC, EC	Results provided insights into stress response mechanisms and into key enzymes for herbicide degradation. A model for quinclorac degradation was proposed	Li *et al.* (2009)

[a]Cell compartments: I, intracellular; EC, extracellular; S, surface; WC, whole-cell; IM, inner membrane; OM, outer membrane; M, membrane.
Abbreviations: SDS-PAGE, sodium dodecyl sulphate polyacrylamide gel electrophoresis; 2-DE, two-dimensional electrophoresis; MS, mass spectrometry; iTRAQ, isobaric Tag for relative and absolute quantitation; 2-D DIGE, two-dimensional difference gel electrophoresis; LC-MS/MS, reverse phase liquid chromatography tandem mass spectrometry; MudPIT, multidimensional protein identification technique.

distress with frequent progression to septicaemia and death (Cheng and Currie, 2005; Dance, 2002). Melioidosis is an important cause of death across Southeast Asia, tropical regions of India and Northern Australia, and is of worldwide concern (Dance, 2002). No vaccine is currently available for *B. pseudomallei*, and its high mortality rate is probably related to the multiple clinical presentations in melioidosis patients, often resulting in misdiagnosis and inadequate treatment (Stevens *et al.*, 2004).

Proteomics has helped elucidate factors pertinent to virulence, adaptation and *in vivo* survival of *Burkholderia* pathogens, and has also contributed to identify potential diagnostic markers or vaccine candidates, as discussed below. An effort has been made towards working with clinical isolates, and advances in genome sequencing technology are facilitating the access of microbiology laboratories to genomic information that can be used to complement proteomic results.

Identification of diagnostic markers and vaccine candidates

Understanding the interaction of the immune system with bacteria is crucial for elucidating the pathogenesis of an infectious disease. The recent progress in proteomic technologies combined with genomics has made possible the comprehensive profiling of bacterial immunogenic proteins on a genome-wide scale. The identification of bacterial antigen biomarkers is expected to provide potential candidates for diagnosis, prognosis indicators and even vaccine development. Our current literature survey found seven research works where proteomic approaches were exploited to this end in the context of *Burkholderia*-related infections (Table 5.1).

Different Bcc species differ in their persistence, epidemiology, and pathogenic potential in CF, and the same strain can be associated with very different clinical outcomes (Drevinek and Mahenthiralingam, 2010; Mahenthiralingam *et al.*, 2002). Correct identification and classification of different Bcc species in the clinical laboratory is therefore of paramount importance for the implementation of infection control measures and adequate surgical and clinical management of CF patients. However, the genetic diversity of Bcc bacteria makes correct identification problematic at the level of sensitivity and specificity, and misidentification occurs frequently in routine diagnostic laboratories (Vandamme and Dawyndt, 2011). Currently, accurate identification and differentiation of Bcc species require molecular typing techniques, such as *recA* gene restriction fragment length polymorphism (RFLP) or sequencing, and multilocus sequence typing (MLST). Because these are expensive and technically challenging assays, implementation of molecular-based assays in the clinical laboratory is usually not considered, and proper patient management is often delayed as a consequence. These limitations have resulted in the need for alternative methods that lead to an accurate identification of Bcc and biochemically similar bacteria.

Two recent studies have applied proteomic approaches to tackle this problem. The first one examined the potential of an intact cell assay using matrix-assisted laser desorption/ionization time-of-flight mass spectrometry (MALDI-TOF MS) for rapid differentiation of Bcc bacteria (Vanlaere *et al.*, 2008). MALDI-TOF MS is reported as a rapid, simple, reproducible and inexpensive method for identification, characterization, and typing of bacteria (Keys *et al.*, 2004; Ruelle *et al.*, 2004). The fact that intact cells are used simplifies the method, and the resulting spectra are easy to interpret and can be used to identify genus, species or strain-specific peaks. Vanlaere *et al.* (2008) examined 75 strains representing established Bcc species and other closely related or biochemically similar species that are commonly misidentified in the context of CF. Overall, 65 strains were clustered correctly; the diversity observed among bacterial spectra was found to be species dependent, and all non-Bcc species were distinguished from Bcc species, with the exception of *B. gladioli*. The mass spectra of Bcc and the other *Burkholderia* strains were demonstrated to be sufficiently divergent to allow the discrimination of species, and only one Bcc strain did not cluster in its species–specific cluster (*B. pyrrocinia* R-22918). MALDI-TOF MS is being increasingly used in clinical laboratories and hospitals throughout the world, and was here shown to be a powerful alternative for the rapid and accurate identification of Bcc bacteria and species level.

The second study took on a different approach to address the problem of Bcc identification. A 2-DE-based expression proteomics approach was adopted for the comparative analysis of the secretome of *B. cepacia* and *B. multivorans* (Mott *et al.*, 2010). Secretory proteins control and regulate host biological and physiological processes, playing an important role in the pathogenesis of many bacterial infections (Hathout, 2007; Pallen *et al.*, 2003). Bcc species are known to produce secretory proteins such as proteases, lipases, phosphatases, and haemolysins, which have been identified as virulence factors (Kooi *et al.*, 2006; Mullen *et al.*, 2007). But more importantly, proteomic analyses of the Bcc species *B. cenocepacia, B. vietnamiensis, B. multivorans,* and *B. ambifaria* have revealed a species-dependent differential expression of secretory proteins (Carvalho *et al.*, 2007). Therefore, species-specific secretory proteins might represent biomarkers for diagnostic assays, and even targets for vaccines and therapeutic agents. Mott *et al.* (2010) compared the culture supernatants of *B. cepacia* and *B. multivorans* by 2-DE, and spots that were unique or differently expressed were selected for MALDI-TOF MS identification. The 2-DE analysis identified 23 spots that were discriminatory for either *B. multivorans* or *B. cenocepacia,* which after analysis gave a total of nine distinct proteins. These include the chaperonin GroEL, a transcription elongation factor, metabolic enzymes, an extracellular ligand-binding receptor, and a flagellin-like protein. Although these proteins represent potential diagnostic markers, further studies need to be carried out in conditions that better simulate the CF lung environment in order to ascertain their potential role in disease or survival during infection.

Secretory proteins are also at the forefront of the host–pathogen interaction, and are thus the primary antigen targets of the host immune system (Kaufmann and Hess, 1999). These proteins can induce protective immunity or elicit an immune response, and represent not only potential diagnostic markers, but also putative candidate vaccines against bacterial infections. In order to identify the different immunogenic proteins present in the secretome of a non-CF clinical isolate of *B. cepacia,* a recent study employed a combination of 2-DE with Western blot analysis using polyclonal antisera raised in mice, followed by MALDI-TOF MS identification of proteins of interest (Mariappan *et al.*, 2010). *B. cepacia* extracellular proteins were extracted from the supernatants of stationary-phase cultures and resolved by 2-DE. Of the 265 spots obtained initially on the 2-DE gels, 18 immunogen protein spots were detected in common after Western blot analysis with three different antisera. The proteins were identified and found to be mainly involved in metabolic functions. Remarkably, only one protein was predicted as secretory and belonging to the classical Sec pathway. Although metabolic proteins are known to play major roles in energy production and survival, an involvement of metabolic proteins in virulence has also been suggested (Vellasamy *et al.*, 2011; Wongtrakoongate *et al.*, 2007), as described later in this section. Immunogenic proteins that are conserved in different *Burkholderia* species, as is the case of a metalloprotease identified in the 2-DE – Western blot study of *B. pseudomallei* K96243 (Mariappan *et al.*, 2010), might be used to confer cross-protection against Bcc bacteria and *B. pseudomallei*. However, further *in vivo* studies are required to validate their immunogenicity.

In the context of melioidosis, the secretome of *B. pseudomallei* was also probed by a proteomic analysis (Vellasamy *et al.*, 2011). The experimental design and objectives were similar to those described above by Mariappan *et al.* (2010), combining 2-DE analysis of stationary phase supernatants of a *B. pseudomallei* melioidosis clinical isolate followed by Western blot and MALDI-TOF identification of spots of interest. Out of 113 distinct spots observed in the 2-DE gel, 54 were identified using MS to construct a secretome reference map of *B. pseudomallei*. Remarkably, *in silico* analysis predicted that only two proteins were extracellular, with the remaining having cytoplasmic, periplasmic, plasma membrane-associated, or unknown location. Moreover, only 26% of the identified proteins had a predicted signal peptide. In total, 12 proteins were found to be reactive to specific mice antisera raised to *B. pseudomallei* secreted proteins, including flagellin, a cell invasion protein, the chaperonin GroEL, a cell division protein, the hypothetical protein BPSL1622, and several proteins with metabolic functions. One particularly interesting finding was that BPSL1622 does not have any orthologue in *B. thailandensis*, a closely related non-pathogenic species, suggesting that this protein may be involved in the virulence of *B. pseudomallei*. GroEL was singled out as a potential vaccine candidate, since this protein has been previously described as being highly immunoreactive and produces a strong antigen–antibody response with melioidosis patient's sera (Woo *et al.*, 2001). Glyceraldehyde 3-phosphate dehydrogenase (GAPDH), a glycolytic enzyme, might also play a role in bacterial pathogenesis, as described for other Gram-negative bacteria (Egea *et al.*, 2007).

It is interesting to note that the studies discussed above describe the presence of non-secretory, cytoplasmic and periplasmic proteins in the secretomes analysed, even though the authors applied tests to assess cell lysis, which indicated that the proteins identified were indeed from the secretome (Mariappan *et al.*, 2010; Mott *et al.*, 2010; Vellasamy *et al.*, 2011). In fact, Riedel *et al.* (2006) had previously reported that nearly half of the exoproteome of a *B. cenocepacia* clinical strain was composed of intracellular and/or surface associated proteins. This phenomenon has been previously reported in similar studies carried out in other pathogenic bacterial species (Malen *et al.*, 2007; Sibbald *et al.*, 2006; Trost *et al.*, 2005), and suggests that some cytoplasmic proteins, such as the commonly identified chaperonin GroEL (Mott *et al.*, 2010; Vellasamy *et al.*, 2011), also have extracellular roles. Non-classical secretion pathways or signal peptide-independent pathways may be involved in the presence of cytoplasmic proteins in the bacterial secretome.

An alternative to the study of the secretome is the analysis of immunogenic surface proteins, which are often good vaccine candidates. One study explored two proteomic-based approaches in parallel, involving a biotinylation technique for the detection of surface located proteins and a subsequent screen with human sera following 2-DE separation (Harding *et al.*, 2007). The objective of this study was to identify antigens that could function as components of a subunit vaccine against *B. pseudomallei* infection (Harding *et al.*, 2007). Of the 35 biotinylated proteins identified, only three were predicted to be located in the outer membrane. This might be related to the protective effect of the bacterial capsule, uptake of biotin into the cell, and post-secretion labelling. Nine proteins were found to be simultaneously biotinylated and immunoreactive, including the chaperones GroEL and DnaK. In fact, a high number of proteins identified were associated with the stress response and defence against stress, possibly because the cells were cultured in minimal medium to mimic the nutrient limitation encountered within the host. The authors plan to evaluate a subset of nine of these proteins as vaccine candidates and diagnostic targets.

The differences between *B. pseudomallei* and its non-virulent counterpart *B. thailandensis* have also been addressed via a proteomics strategy (Wongtrakoongate *et al.*, 2007). The authors implemented a 2-DE-based comparative proteomics approach to analyse intracellular 2-DE profiles corresponding to clinical isolates of *B. pseudomallei* and *B. thailandensis* E257 in an effort to identify distinctively expressed proteins, which theoretically could play a role as potential virulence factors or protein markers. A reference map for *B. pseudomallei* grown in stationary phase was constructed with the aid of MALDI-TOF MS, and the existence of post-translational modifications was confirmed by the presence of several protein isoforms, including three isoforms of the chaperonin GroEL. Although post-translational modifications are known to play a major role in eukaryotes, not much is known about their role in bacterial physiology and possible involvement in pathogenicity. The comparative analysis yielded 14 spots with significantly different levels of expression in *B. pseudomallei* compared with *B. thailandensis*, of which 12 were absent or at very low levels in *B. thailandensis*. These proteins were found to be involved in energy metabolism, cellular processes, and the stress response; surface proteins and hypothetical proteins were also identified. Six of the 12 proteins were indicated to be involved in bacterial virulence, including the chaperonin GroEL.

A different approach was applied by Felgner *et al.* (2009) for the same objective of identifying serodiagnostic and cross-reactive antigens in *B. pseudomallei*. The authors constructed a protein microarray containing over 1,200 *B. pseudomallei* proteins and probed it with sera from 88 melioidosis patients. The authors initially identified 170 reactive antigens that were subsequently probed with a collection of sera derived from well-defined melioidosis positive and negative patients. This experimental design led to the identification of 49 antigens that were significantly more reactive in melioidosis patients than in healthy individuals or patients with other types of bacterial infections. To test the applicability of these antigens as serodiagnostic agents, the 10 melioidosis immunoreactive proteins from the microarray assay (including the chaperonin GroEL) were selected and printed on nitrocellulose membranes. Using these so-called 'immunostrips', melioidosis patients reacted strongly while the remainder of the test subjects had low reactivity. The authors concluded that their experimental design allowed a comprehensive analysis of the immune response to *B. pseudomallei* and was capable of identifying serodiagnostic agents that accurately distinguish melioidosis cases from controls. Their serodiagnostic test was shown to be a good alternative to the indirect haemagglutination assay (IHA). It classified melioidosis positive and negative cases with a sensitivity of 95% and 83%, respectively, which was a substantial improvement over currently available clinical assays. It would be interesting to verify the feasibility and discriminatory power of this methodology when applied to other *Burkholderia* species including Bcc bacteria.

In conclusion, although several potential diagnostic markers have been identified, the use of these immunogenic proteins as serodiagnostic candidates is dependent on their uniqueness in the species of interest; if close homologues are present in other Bcc or related species, it is very difficult to avoid cross-reactivity. The chaperonin GroEL has emerged as a common denominator in several of these proteomic studies. It is universally highly immunoreactive and has been proposed as a potential vaccine candidate or serodiagnostic marker for both Bcc (Mott *et al.*, 2010) and *B. pseudomallei* (Felgner *et al.*, 2009; Vellasamy *et al.*, 2011; Wongtrakoongate *et al.*, 2007). A role in virulence for GroEL has been proposed previously (Woo *et al.*, 2001), but the identification of isoform structures, where one is highly expressed in a virulent species (Wongtrakoongate *et al.*, 2007), raises the hypothesis of post-translational modifications contributing additively to pathogenicity. Metabolic enzymes were also frequently identified as immunogenic and potential diagnostic markers in both Bcc (Mariappan *et al.*, 2010) and *B. pseudomallei* (Felgner *et al.*, 2009; Vellasamy *et al.*, 2011; Wongtrakoongate *et al.*, 2007) species, and roles in virulence have been suggested for these proteins.

Elucidation of host–pathogen interactions

One of the main goals of medical microbiology is to understand how pathogens interact with their hosts and persist following infection. Proteomics has played an important role in unravelling *Burkholderia* pathogenic mechanisms of disease in humans or plants, at the molecular level (Table 5.1). Members of the genus *Burkholderia* can occupy a wide range of ecological niches and also opportunistically infect different hosts. One such example is *B. pseudomallei*, which is also a saprophytic microorganism that can survive in soil and water for many years (Cheng and Currie, 2005). Understanding the adaptive mechanisms involved in the survival of *B. pseudomallei* in these environments is crucial for the overall goal of preventing and treating melioidosis. Several studies have employed 2-DE-based quantitative proteomics approaches to elucidate these adaptive mechanisms in the presence of relevant stresses (Chantratita *et al.*, 2012; Osiriphun *et al.*, 2009; Pumirat *et al.*, 2009; Thongboonkerd *et al.*, 2007; Wongtrakoongate *et al.*, 2011). For example, Pumirat *et al.* (2009) identified a total of 37 secreted proteins whose content was altered in the supernatant of *B. pseudomallei* under salt stress. Most of the altered proteins identified were involved in metabolism, transcription/translation regulators, potential virulence factors, namely exoproteases, adhesins and components of type I, II, III and IV secretion systems, and drug resistance proteins (Pumirat *et al.*, 2009).

The alternative sigma-factor RpoS is a major regulator of the response to several stress conditions in Gram-negative bacteria (Hengge-Aronis, 2002). Osiriphun *et al.* (2009) performed the characterization of the RpoS regulon and of RpoS-dependent promoters in *B. pseudomallei*, using a proteomic approach combined with computational analysis. In total, 70 proteins forms were found to have an altered content between *rpoS*$^+$ and *rpoS*$^-$ strains. Most of the proteins that were suggested to be under positive regulation by RpoS belonged to the 'stress response' category and were mainly involved in the response to oxidative and osmotic stress. Furthermore, RpoS was also found to regulate several proteins involved in cell envelope biogenesis, virulence and energy metabolism. The proteomic data coupled with the computational analysis performed allowed the expansion and definition of RpoS-controlled genes and helped to determine whether they were directly or indirectly controlled by RpoS. Stress tolerance in *B. pseudomallei* has also been addressed by comparing the proteomes of a *rpoE* operon knockout mutant and *B. pseudomallei* K96243 wild type, in order to elucidate how the alternative sigma factor σ^E (RpoE), controls stress tolerance and biofilm formation in this species (Thongboonkerd *et al.*, 2007). Several stress response proteins were down-regulated in the *rpoE* mutant, in particular oxidative stress response proteins and chaperones. Potential virulence factors were also reduced in expression, such as type I-IV secretion system, lipopolysaccharide synthesis, exoproteins and adhesins, suggesting that the *rpoE* operon might also be involved in *B. pseudomallei* virulence. These proteomic results were confirmed by *in vivo* experiments using BALB/c mice, where animals infected with the *rpoE* mutant strain had a marked delay in time to death compared to wild type. Although most of the altered proteins were cytosolic, several periplasmic and secreted proteins were found to have an altered content in the *rpoE* mutant, suggesting that other secreted proteins might also be regulated by the *rpoE* operon (Thongboonkerd *et al.*, 2007). Proteomic analysis of the secretome is of wide interest in terms of virulence factor production, however, only a few secreted proteins were noted in latter study. This may have been due to the number of spots visualized in each gel being limited ($n = 450$), which was probably related to the 2D gel-size (8×9.5 cm) and the Coomassie Brilliant Blue R-250 staining method used (Thongboonkerd *et al.*, 2007). A larger format of 2D gels and more sensitive staining could have increased the number of spots detected (Sá-Correia and Teixeira, 2010).

Examples are two more recent studies, also performed in *B. pseudomallei*, where the whole-cell proteome was separated by 2-DE and protein spots were stained using the more sensitive Colloidal Coomassie Blue G-250 method (Chantratita *et al.*, 2012; Wongtrakoongate *et al.*, 2011). In both studies, a total of 700 protein spots were detected, representing a significant improvement over that of Thongboonkerd *et al.* (2007). Wongtrakoongate and co-workers reported the construction of a reference map containing 282 proteins identified by MALDI-TOF MS, using *B. pseudomallei* cells recovered from the early stationary phase of growth (Wongtrakoongate *et al.*, 2011), when bacteria undergo carbon starvation and oxidative stress (Subsin *et al.*, 2003). Several of the proteins identified had been previously shown to be implicated in bacterial virulence and are regulated by the stationary phase sigma factor RpoS (Osiriphun *et al.*, 2009) and the sigma factor RpoE (Thongboonkerd *et al.*, 2007), which plays a pivotal role in stress tolerance and biofilm formation in *B. pseudomallei* (Korbsrisate *et al.*, 2005). Nine protein spots were found to correspond to phasin isoforms, in agreement with a previous study by the same authors who described phasin as the major protein isoform expressed in the proteome of late stationary phase *B. pseudomallei* (Wongtrakoongate *et al.*, 2007). Chantratita *et al.* (2012) compared the protein expression profiles of two isogenic colony morphology types, II and III, obtained in laboratory from a *B. pseudomallei* strain 153 type I clinical isolate. The rationale behind this study was based on the hypothesis that *B. pseudomallei* colony morphology switching led to phenotypic changes that enhanced bacterial survival under unfavourable environmental or host conditions. Several proteins were found to be differentially expressed, including flagellin and components of the arginine deiminase system (ArcA, ArcC), which showed a higher content in isogenic types II and III, compared with parental type I. The results of the proteomic analysis were further explored to provide proof of concept that one or more of the differently expressed proteins, namely ArcA and ArcC, were involved in survival under adverse conditions. However, although the comparison of *B. pseudomallei* strain 153 type I with *arcA* and *arcC* defective mutants showed that the arginine deiminase system could be associated with bacterial survival under acid stress, the role in intracellular survival or replication within the mouse macrophage cell line J774A.1 was not established.

A systems biology perspective of *B. pseudomallei* pathogenicity and phenotypic variability between two *B. pseudomallei* isolates, Bp15682 and K96243, which exhibit greater virulence towards *C. elegans*, was explored by combining different *omic* strategies: genomic, transcriptomic, and proteomic (Ou *et al.*, 2005). A pathogen's phenotypic diversity has important clinical implications for antibiotic resistance, vaccine design, and the prevention of emerging diseases (Grenfell *et al.*, 2004). Although the focus of this review is the application of proteomics, it should be highlighted that, ultimately, a pathogenic phenotype is the integrated result of variations at different cellular levels. A remarkable finding of Ou *et al.* (2005) was that while almost 50% of the gene expression differences were associated with genes that were differentially present/absent in the two isolates, more than one-third of the global proteomic variability observed could be attributed to proteins that were expressed in both strains, but present as strain-specific isoforms. The most striking example of these isoform differences was PaaZ, a putative phenylacetic acid degradation oxidoreductase, which was identified in 14 and five non-overlapping protein spots in K96243 and Bp15682, respectively. Such a high number of protein isoforms could be the result of strain-specific mechanisms of post-transcriptional or post-translational modifications, including protein truncation or differential translational termination.

During the course of infection, gene expression in *Burkholderia* species is regulated in response to environmental signals. *In vivo* studies of pathogen evolution within a host can provide important insights into the dynamic adaptability of infectious agents. A better understanding of these adaptive mechanisms may improve the management of chronic infections and guide the design of new therapeutic strategies. In this context, proteomic analyses have proven to be extremely useful and have contributed greatly to the elucidation of the mechanisms involved in pathoadaptation during the course of chronic *Burkholderia* infections such as those seen in the CF lung. During chronic infection, the airways represent an evolving ecosystem, with multiple phenotypic variants emerging from a clonal population and becoming established in patients' airways as a result of genetic adaptation (Lieberman *et al.*, 2011). In relation to CF the most widely studied species has been *B. cenocepacia*, one of the most prevalent Bcc species that is commonly associated with increased morbidity and mortality (Drevinek and Mahenthiralingam, 2010). The proteomic profiles of two *B. cenocepacia* clonal variants that demonstrated differential persistence in a mouse model of pulmonary infection were compared using a quantitative 2-DE-based approach followed by MALDI-TOF/MS (Chung and Speert, 2007). Samples that did not yield confident results were further processed by LC-MS/MS. The parental strain C1394, recovered from a CF patient, was rapidly cleared from the murine lung while the variant C1394mp2, which was recovered after sequential passage of C1394 through a mouse model of pulmonary infection (Chung *et al.*, 2003), persisted. The results showed that expression of flagellin proteins was more prominent in the 2D proteome of the persistent variant C1394mp2, whereas the expression of the protein AhpC, an antioxidant enzyme expressed under oxidative stress, was completely abolished, indicating a possible role for these proteins in *B. cenocepacia* adaptation and persistence in the murine lung (Chung and Speert, 2007). However, the authors did not take into account quantitative differences and only registered the presence or absence of 2D protein spots, which represented a small percentage of the total proteome.

Using a combined approach based on transcriptomics and proteomics, the non-mucoid *B. cenocepacia* clinical isolate C8963 was compared with the mucoid C9343 clonal variant, both obtained from a chronically infected CF patient in the year prior to his death (Zlosnik and Speert, 2010). Applying an iTRAQ labelling methodology, a total of 49 differently expressed proteins were identified in the non-mucoid isolate C8963, compared with the mucoid C9343. In particular, several virulence factors were overexpressed, including the nematocidal protein AidA, AhpC and the zinc metalloprotease ZmpA. The authors showed that this overexpression translates into increased resistance to oxidative stress in the non-mucoid isolate, a microbial determinant to host immune system resistance. In general, the results obtained from the transcriptional analysis were consistent with the proteomics data. However, the proteomic analysis also identified proteins that did not display a detectable difference in the transcriptome, including proteins involved in energy metabolism (AceA, AcoD, Fpr, BCAL2197), post-translational modifications, protein turnover and chaperoning (GroEL1, HlsU, ClpB, BCAL2829), lipid metabolism (BktB, BCAS0610), and amino acid metabolism and transport (GcvT, IscC, CarA, ZmpA, BCAL2791, BCAL0600, BCAL1796, BCAM2149). These results reinforce the importance of proteomic studies as a highly complementary *omics* approach for deciphering *Burkholderia* adaptation to the host.

Another proteomic study of *B. cenocepacia* clonal variants isolated from the lungs of a CF patient during long-term infection compared the proteomes of 2 out of 11 *B. cenocepacia* sequential isolates using 2D-DIGE (Madeira *et al.*, 2011). One of the isolates, IST439, was the first isolate recovered from the patient and believed to have initiated the infection, while the other, IST4113, was obtained after 3 years of chronic infection and a period of intravenous antibiotic therapy with ceftazidime and gentamicin. The two isolates were clonal variants and, among other phenotypic differences, IST4113 was much more resistant than IST439 to all antimicrobials tested (Coutinho *et al.*, 2011; Madeira *et al.*, 2011). A total of 79 proteins were found to have a different content when comparing both the cytoplasmic and the membrane-associated protein fractions of the two isolates (Madeira *et al.*, 2011). These results allowed the proposal of a number of molecular strategies employed by *B. cenocepacia* to adapt and persist in the CF airways under antimicrobial therapy. Owing to limitations of 2-DE in analysis of hydrophobic proteins, only 17 membrane proteins with different content in the two isolates under study were identified. IST4113 showed increased levels of proteins mainly involved in energy metabolism, translation and purine synthesis. Several proteins involved in iron uptake, including the TonB-dependent siderophore receptor, the TonB-dependent copper receptor and the FAD-binding 9, siderophore interacting protein (MxcB), and in protein folding were also found to be more abundant in IST4113. In this last category, the most prominent proteins were the heat shock protein ClpB followed by the chaperones DnaK and GroEL. In addition, the level of proteins related to peptidoglycan synthesis, membrane lipids and lipopolysaccharide synthesis were also found to have their content altered in the most resistant isolate, IST4113. In light of the limitations of the 2-DE approach as previously described, a genome-wide transcriptomic analysis was subsequently carried out and in general found that the functional classes identified within the datasets of differentially abundant proteins or transcripts were complementary (Mira *et al.*, 2011).

Extracellular proteins may also be involved in virulence of Bcc bacteria during CF infection and the role of the *B. cepacia* secretome in pathogenesis has been examined via a proteomic approach (Mariappan *et al.*, 2011). Early stationary-phase *B. cepacia* was shown to be more invasive and its culture supernatant more cytotoxic when compared with mid-logarithmic phase cells and supernatant, respectively, in an A549 human lung epithelial cell *in vitro* model. In order to profile the protein composition of the supernatants at mid-logarithmic and early stationary phases of growth, a 2-DE-based quantitative proteomics approach coupled with MALDI-TOF/MS was used. In total, 40 proteins were found to be only expressed during the early stationary phase of growth, namely metabolic proteins and proteins involved in cellular processes. Several putative virulence factors (FliD, FlgK, FusA, AslA, Fiu, Pgk) were identified in the supernatant of *B. cepacia* in the early stationary phase of growth, which may also explain its higher cytotoxicity (Mariappan *et al.*, 2011).

In the context of *Burkholderia* proteomics applied to host–pathogen interactions, only two studies involving a phytopathogenic species have been performed to date, with both using the same experimental design combining 2-DE to separate proteins from intra- and extracellular protein fractions followed by protein identification (Goo *et al.*, 2010; Kang *et al.*, 2008). These studies focused on *Burkholderia glumae*, the causal agent of rice grain rot and the most important bacterial phytopathogen of rice in Japan, Korea, Taiwan and the southern United States (Sayler *et al.*, 2006). Given the economical consequences of this disease it is crucial to elucidate the molecular mechanisms underlying *B. glumae* virulence. The first study was performed using a *B. glumae* derivative with constitutive expression of the hypersensitive response and pathogenicity (Hrp) type III protein secretion system (T3SS) regulator, HrpB (Kang *et al.*, 2008); the involvement of this secretion system in pathogenicity had been previously demonstrated (Cornelis and Van Gijsegem, 2000). By comparative analysis of the intracellular and extracellular protein fraction of a type II secretion system (T2SS) and a Hrp T3SS mutants, 46 proteins were found to exhibit altered expression (Kang *et al.*, 2008). This comparative analysis enabled the authors to verify that, unexpectedly, 16 proteins accumulated in the supernatant were secreted via T2SS and not T3SS. Interestingly, most of the extracellular proteins identified exhibited a high similarity with those from *B. pseudomallei* K96243, suggesting that even though the hosts and ecological niches of the two bacteria are different, their secreted proteins might have important common functions in establishing pathogenic interactions with hosts.

In the second study of *B. glumae* phytopathogenicity, a total of 79 proteins were differentially expressed between the wild type strain BGR1 and a *tofI* mutant (Goo *et al.*, 2010). TofI synthetizes N-octanoyl homoserine lactone, a molecule that is crucial for the quorum sensing (QS) system of *B. glumae*, a signalling system known to be important for successful *Burkholderia* colonization and pathogenesis in several host organisms (Eberl, 2006). Around 75% of the QS-dependent proteins were found in the extracellular fraction, including lipases and proteases secreted through the T2SS. Similarly to what was seen in the *B. cepacia* and *B. pseudomallei* secretome studies (Mariappan *et al.*, 2010; Mott *et al.*, 2010; Vellasamy *et al.*, 2011), both studies involving *B. glumae* found proteins in the culture supernatant that are characterized as cytoplasmic, such as the chaperone DnaK and translation elongation factors (Goo *et al.*, 2010; Kang *et al.*, 2008). Since it was unlikely that the proteins were released due to cell lysis, it is possible that unknown secretory pathways or unusual transporters are involved in the pathogenicity of *B. glumae*. Together, these two studies emphasize the importance of bacterial secretome in bacterial pathogenesis with *B. glumae* extracellular proteins being implicated as virulence factors against plant hosts.

Although transcriptomic analysis can be used to obtain genome-wide information on QS-dependent genes, proteomic analysis was very useful for the identification of proteins under the control of QS at post-transcriptional levels (Goo *et al.*, 2010; Inhülsen *et al.*, 2012; Riedel *et al.*, 2003). In order to characterize the QS regulon in *B. cenocepacia* strain H111, Riedel *et al.* (2003) compared the protein expression pattern of the wild type with an N-acyl homoserine lactone deficient *cepI* mutant by 2-DE. To obtain the maximal amount of proteomic information, protein samples were pre-fractionated into intracellular, extracellular and surface protein fractions. This strategy allowed a better spot resolution and also the detection of several low-abundance proteins. However, owing to the lack of genomic data on strain H111 when the study was performed, comparison of N-terminal sequences to the available *Burkholderia* genomic data only identified 11 out of 55 differently expressed proteins in the *cepI* mutant. An extended analysis of the intra-, surface-associated and extracellular proteome of *B. cenocepacia* H111 was subsequently performed and provided a detailed 2-DE reference map for this organism, based on MALDI-TOF/TOF identification (Riedel *et al.*, 2006). In total 642 proteins spots were identified, corresponding to 390 different protein species.

With the same objective of mapping the QS regulon of *B. cenocepacia* H111, Inhülsen *et al.* employed an integrated *omics* approach by combining transcriptomics based on DNA microarrays and proteomics with iTRAQ labelling coupled with MS/MS (Inhülsen *et al.*, 2012). The majority of the QS-regulated proteins were not found among the QS-regulated genes identified by the microarray analysis, suggesting that the ultimate expression of the identified proteins was being controlled by QS at the post-transcriptional level.

Three loci that had been previously described as being involved in biofilm formation exhibited lower expression levels in the *cepR* mutant H111-R, compared with the wild type strain, at both the transcriptome and proteome levels. These loci (BclACB lectins, the large surface biofilm-associated protein BapA, and the type I pilus FimA) were selected for more detailed analysis with the objective of linking the CepIR QS system with the formation of biofilms by *B. cenocepacia* H111 species. Using deletion mutants, the three lectins were shown to play an important role in the development of mature biofilm's structure, while BapA had an even stronger influence on biofilm development, with the *bapA* mutant forming smaller and less abundant microcolonies compared to the wild type. Considering the type I pilus FimA locus, no relationship was established with the formation of biofilms or biofilm structure. The nematocidal protein AidA was also found to be *cepIR*-regulated in *B. cenocepacia* H111, similarly to previous studies (Inhülsen *et al.*, 2012; Riedel *et al.*, 2003).

Despite the wide application of 2-DE-based approaches, other sophisticated approaches have been adopted to dissect virulence determinants in *Burkholderia* species. Gel-free LC-MS/MS and MudPIT – LC/LC-MS/MS analyses were explored to identify proteins from *B. vietnamiensis* (Wickramasekara *et al.*, 2011). Using this combined experimental strategy, the authors aimed to identify low-abundance virulence factors, using a MS-based approach without the need to over express pathogenic factors that can bias results such as the constitutive secretion system screen used in *B. glumae* (as described in Kang *et al.*, 2008). The *B. vietnamiensis* proteome was fractionated into extracellular, cell surface, cell wall and intracellular protein fractions, allowing in-depth analyses. This combined proteomic approach led to the identification of more than 1200 unique proteins which to date represents the highest number of proteins reported for the *Burkholderia* genus in a proteomic study (Wickramasekara *et al.*, 2011). Multiple virulence factors including flagellin, porin, peroxiredoxin and zinc proteases were identified (Wickramasekara *et al.*, 2011).

To overcome the limitation of 2-DE in the analysis of membrane proteins, Schell *et al.* developed a simple, safe and rapid method to prepare highly purified outer membrane fragments from *B. pseudomallei* and *B. mallei* using trypsin shaving and LC-MS/MS (Schell *et al.*, 2011). Briefly, cells were first treated with lysozyme and then freeze–thawed in the presence of the detergent Triton X-100 and $MgCl_2$. The insoluble outer membrane fragments were recovered by differential centrifugation, treated with RNase and EDTA and finally digested with trypsin. The peptides obtained were analysed by LC-MS/MS, which allowed the identification of 225 proteins. Of these, more than 155 were outer membrane proteins, expressed under several growth conditions, with the majority being porins and efflux pumps. In addition, the authors were able to select a total of 20 proteins from all the outer membrane proteins identified for each *Burkholderia* species analysed, that were always abundant under a wide variety of growth conditions, including growth in serum and in the presence of macrophages, suggesting that these proteins play a role in bacterial survival. Since these outer membrane proteins are exposed to the host's immune system, they may represent vaccine targets. Indeed, 20% of all outer membrane proteins identified in *B. pseudomallei* had been previously found to react with sera of recovered melioidosis patients (Felgner *et al.*, 2009), an indication that they are expressed and targeted *in vivo* during infection. Comparison of the proteomes of *B. pseudomallei* and *B. mallei* revealed that there were a few differences under the growth conditions tested, perhaps reflecting the evolution of *B. mallei* away from environmental survival towards host adaptation (Schell *et al.*, 2011). The number of outer membrane proteins identified in this study was very impressive when compared with the previously described results of Harding *et al.*, who identified only 35 surface-located proteins in *B. pseudomallei* using a biotinylation approach followed by 2-DE/MALDI-TOF (Harding *et al.*, 2007).

Environmental applications

Although much of the attention given to the *Burkholderia* genus is directed towards those species that are human pathogens, the reality is that several species, including those from the Bcc, are often found living in association with the plant rhizosphere (Suarez-Moreno *et al.*, 2012). A number of *Burkholderia* strains have been isolated based on their ability to biodegrade xenobiotic pollutants (O'Sullivan and Mahenthiralingam, 2005). These include *B. vietnamiensis* G4 and *B. kururiensis* KP23 that are both able to degrade trichloroethylene; *B. xenovorans* LB400, one of the most effective aerobic degraders of polychlorinated biphenyls (PCBs); *B. phenoliruptrix* AC1100, which is able to degrade the potent herbicide 2,4,5-trichlorophenoxyacetate; and several as yet unclassified *Burkholderia* sp. strains that can degrade polycyclic aromatic hydrocarbons and 2,4-dichlorophenoxyacetate (O'Sullivan and Mahenthiralingam, 2005; Suarez-Moreno *et al.*, 2012). Several isolates classified as *B. cepacia* can degrade a variety of organic compounds including PCBs, organohalides, trichloroethylene, cyclic aromatic compounds and herbicides; further information on these isolates can be obtained from the Biodegradative Strain Database (http://bsd.cme.msu.edu/). *Burkholderia* species are also suitable for rhizoremediation strategies, that is, microbial degradation of hazardous compounds in the rhizosphere (for further reading, see O'Sullivan and Mahenthiralingam, 2005), however these have not yet been the subject of proteomic applications. The majority of the studies described below do not represent the exploitation of proteomics as a stand-alone approach, but rather as part of an integrated system biology approach encompassing other experimental platforms, such as genomics, transcriptomics, metabolomics, and phenotypic characterization. This is common for studies in this area, since the research groups involved often have a strong component in genomics and are also very interested in the physiological and biochemical alterations induced by environmental changes.

One very important application of *Burkholderia* species is in the aerobic biodegradation of PCBs and related compounds. *B. xenovorans* LB400 has been long known for its exceptional ability to transform these widespread aromatic pollutants (Bopp, 1986; Goris *et al.*, 2004) and has become a model organism for the research of PCB degradation. Biphenyl is transformed via the 'upper' catabolic pathway (*bph*) into benzoate and pentadiene or their chlorinated derivatives (Seeger *et al.*, 1999). PCB-degrading bacteria are generally unable to transform the dead-end chlorobenzoate (CBA) metabolites, and their accumulation is in part responsible for the deleterious effects of PCB degradation on *B. xenovorans* LB400. This strain possesses one of the largest known bacterial genomes (9.7 Mbp), which encodes three putative benzoate-catabolic pathways: the *ben-cat* pathway, and two copies of the *box* pathway (Gescher *et al.*, 2005). However, critical aspects of the mechanism of CBA degradation, including the coordinated regulation of these pathways, remain unknown. Proteomics has provided an invaluable tool for the elucidation of the complex catabolic processes involved in PCB degradation.

A combined approach of genetics, transcriptomics, and proteomics has been explored by one research group to obtain insights into the regulation of the benzoate-catabolic pathways and the biological significance of its apparent redundancy (Denef *et al.*, 2004, 2005, 2006). In agreement with the first transcriptomic study (Denef *et al.*, 2004), the authors reported that expression of the three pathways in *B. xenovorans* LB400 was strongly dependent on the growth substrate and growth phase (Denef *et al.*, 2005). The authors applied a quantitative 2-DE-based approach to analyse the proteomes of cells grown in either biphenyl or benzoate and clarify the role of each pathway as a function of substrate and growth phase. It was found that only *ben-cat* pathway proteins were expressed during exponential growth on benzoate, while the box_M pathway (located on the megaplasmid) was also expressed during the transition to stationary phase. Both *ben-cat* and box_C (the copy located in the large chromosome) were expressed during growth on biphenyl. Six out of seven proteins known to be involved in the transformation of biphenyl via the 'upper' catabolic pathway (*bph*) were identified on the 2-DE gels, but were more abundant in biphenyl-grown cells.

The occurrence of oxidative stress in response to aromatic compounds is well known (Chavez *et al.*, 2004), and this was also seen in *B. xenovorans* by the increased content of the chaperonin GroEL across all conditions tested, as well as other factors presumed to be involved in the oxidative stress response (Denef *et al.*, 2005). Although the simultaneous utilization of more than one benzoate catabolic pathway under conditional regulation was established, the benefit of the functional redundancy of the two *box* pathways remained unclear. To further elucidate this matter and the physiological role of each pathway, a series of mutants with deletions of the *ben-cat* pathway, the box_C pathway, and both pathways were constructed (Denef *et al.*, 2006). The authors performed a physiological characterization of the deletion mutants, transcriptomic analyses, and 2-DE-based proteomics. Overall, the combined microarray and proteomic results provided evidence that all three pathways were able to degrade benzoate. Conditional regulation of the different pathways may ensure optimal metabolic integration depending on the substrate or oxygen tension. It is known that *box* pathways use only half of the amount of molecular oxygen utilized by the *ben-cat* pathway (Gescher *et al.*, 2005), leading to the hypothesis that *box* pathways are preferentially expressed at lower oxygen levels in *B. xenovorans* LB400.

In the same series of studies, a later report described the effects of long-term exposure to biphenyl and 4-chlorobiphenyl on the physiology, morphology, and proteome of *B. xenovorans* LB400 (Agulló *et al.*, 2007). Both compounds induced similar changes in proteins patterns, although the lower solubility of 4-chlorobiphenyl resulted in less pronounced effects. In addition to the expected *bph* biphenyl catabolic enzymes, the 2-DE-based approach also revealed an activation of the stress-responsive molecular chaperones GroEL and DnaK, suggesting that growth on these compounds constitutes a stressful condition for *B. xenovorans* LB400. This was consistent with the observation of irregular outer membranes and cell lysis during incubation with 4-chlorobiphenyl (Agulló *et al.*, 2007). The evidence suggested that this compound, like other membrane-permeant weak acids, can perturb the membrane and lead to an increase of unfolded or precursor proteins. In agreement with the results described previously (Denef *et al.*, 2005, 2006), the authors also observed the induction of an oxidative stress response to biphenyl (Agulló *et al.*, 2007). A continuation of this study was carried out by Martínez *et al.* (2007), who applied the same experimental approaches to study the effects of CBAs on *B. xenovorans* LB400. Exposure to 4-CBA was also found to represent a stressful condition for these bacteria, as evidenced by the increased content of molecular chaperones (e.g. DnaK). However, 4-CBA did not affect the structure of the membrane of LB400 cells, in contrast with the marked effect that had been previously described for 4-chlorobiphenyl (Agulló *et al.*, 2007). 4-CBA was also found to have a more pronounced effect both on growth and on the proteome of LB400 than 2-CBA. It was also found to partially induce the machinery required to degrade CBAs; however, 4-CBA was not catabolized under the study's conditions (Martínez *et al.*, 2007). One insight that can be taken from the integrated studies described above is that future bioremediation strategies for PCBs need to take into account the accumulation of toxic metabolites or by-products such as ROS, since they decrease bacterial viability and degradation efficacy.

B. cepacia MBA4 was isolated from soil and shown to possess 2-haloacid degradation activity via dehalogenation (Tsang *et al.*, 1988). The global protein expression pattern of this strain in response to the haloacid monochloroacetate (MCA) and glycolate, the dehalogenation product of MCA, was examined using a 2-DE-based approach (Kwok *et al.*, 2007). The results show that the 2-haloacid dehalogenase DehIVa is indeed induced by haloacids. Moreover, the dramatic induction of this enzyme in the presence of MCA and absence in glycolate-cultivated cells provided evidence of the importance of this mechanism in the utilization of MCA as an energy source by *B. cepacia* MBA4. Proteins involved in energy production, such as glyceraldehyde 3-phosphate dehydrogenase, were also identified in the presence of MCA, suggesting an important role in the adaptive response to MCA. Remarkably, no stress shock proteins such as GroEL and DnaK were induced during growth in MCA, suggesting that it is likely not a stress substrate for *B. cepacia* MBA4.

Another proteomic application exploring the environmental microbiology of *Burkholderia,* was the study of the response of *B. vietnamiensis* $PR1_{301}$ to the soil contaminant nickel (Van Nostrand *et al.*, 2008). The effect of nickel in protein expression in *B. vietnamiensis* $PR1_{301}$ was assessed by quantitative 2-DE coupled with MS at pH 5 and pH 7, to identify potential mechanisms of tolerance to nickel. Metal contaminated soils are often acidic, and nickel toxicity in strain $PR1_{301}$ at pH 5 is lower than at pH 7. The proteomic results showed an increased expression of proteins involved in cell shape and membrane composition at pH 5 when compared to pH 7. These observations were corroborated by scanning electron microscopy, which indicated an elongation of cells at pH 5, and by an analysis of the percentage of membrane long- and short-chain fatty acids at pH 5 and 7. Taken together, the results suggest that alterations in membrane structure and composition may be a novel mechanism involved in the increased fitness of this strain at pH 5. One possible consequence of these modifications may be the inhibition of transporter function, which might limit the entry of nickel into the cell, allowing *B. vietnamiensis* to tolerate high nickel concentrations.

Herbicide-related soil pollution is another environmentally relevant issue. *B. cepacia* WZ1 is known to degrade the herbicide quinclorac, but the mechanisms and key enzymes involved were unclear. The stress response of WZ1 to quinclorac was examined using a proteomic approach based on the combination of 2-DE and gas chromatography (GC)-MS (Li *et al.*, 2009). The results indicated that the virulence and pathogenicity of these bacteria was attenuated in the presence of the herbicide, which is a positive aspect in terms of safety risk for *in situ* application of the strain in bioremediation. However, protective proteins such as Mn-SOD and heat shock proteins (DnaK- and GroEL-related) were expressed presumably to counteract quinclorac-induced stress, and these stress pathways have been shown to play an important role in *Burkholderia* pathogenicity. The induction of metabolic enzymes such as phthalate dioxygenase reductase and chlorocatechol 1,2-dioxygenase coupled with the identification of metabolites by GC-MS led to the proposition of a pathway for quinclorac degradation by *B. cepacia* WZ1. Overall, the collective findings of the proteomic studies on *Burkholderia* biodegradation of organic pollutants have great practical interest in the design of optimized bioremediation strategies.

Concluding remarks and future perspectives

The increased technical proficiency of proteomics has given rise to a remarkable growth of biologically, biomedically or pharmacologically motivated application fields. In this review, the relevance of proteomics applied to the study of *Burkholderia* biological processes, either in the context of biomedical or environmental applications, was discussed. Compared to other *omics* approaches, proteomics provides not only information at a mechanistic level but can also capture changes in protein activity measured as post-transcriptional/translational modifications, including their interacting partners, temporal expression and subcellular localization. Regarding the identification of protein biomarkers and vaccine targets, proteomics was found to be a powerful tool, notwithstanding the gap between protein discovery and the translation into clinical application. Considerable knowledge has also been gained by the application of proteomics to the study of *Burkholderia*–host interactions, contributing to our understanding of their pathophysiology.

One handicap of the application of proteomics in medical microbiology is that many of the studies use laboratory strains, which can limit the extrapolation of the proteomic information to clinical isolates. This is particularly relevant in the case of Bcc bacteria where horizontal gene transfer between genetically diverse bacteria has greatly contributed to the genomic plasticity of natural isolates (Holden *et al.*, 2009). Moreover, proteomic studies are carried out under defined laboratory conditions that might not be an adequate representation of the environment found by these bacteria during infection of the host, thereby creating a bias in the analysis and extrapolation of the results. In fact, the analysis of the bacterial proteome recovered directly from the host has never been described, presumably due to technical difficulties in isolating sufficient numbers of pure and intact bacteria from infection.

Although proteomics has been shown to provide a global view of *Burkholderia* biological processes and can be implemented as a stand-alone approach, it hugely benefits from the integration with genomics, transcriptomics, and metabolomics. This systems perspective is one of the major directions of future microbiology research. It is expected that the knowledge gathered from these multiple approaches can be integrated into a global view that will contribute to a better understanding of critical issues in *Burkholderia* biology such as their ecological and environmental applications, and biomedical interest as pathogens. A molecular systems biology approach will be of paramount importance to achieve these objectives.

Acknowledgements

Financial support from Fundacão para a Ciência e a Tecnologia (FCT, Portugal) is gratefully acknowledged (contracts PTDC/SAU-MII/69591/2006, PEst-OE/EQB/LA0023/2011, and ERA-PTG/SAU/0001/2008 – this under the EraNet Pathogenomics ADHRES Signature Project, PhD grant to AM (SFRH/BD/37012/2007), and postdoctoral fellowship to SCS (SFRH/BPD/75483/2010)). Research in this context is also performed under the COST Action BM1003 'Microbial cell surface determinants of virulence as targets for new therapeutics in Cystic Fibrosis.'

References

Agrawal, G.K., and Thelen, J.J. (2005). Development of a simplified, economical polyacrylamide gel staining protocol for phosphoproteins. Proteomics *5*, 4684–4688.

Aguló, L., Cámara, B., Martínez, P., Latorre, V., and Seeger, M. (2007). Response to (chloro)biphenyls of the polychlorobiphenyl-degrader *Burkholderia xenovorans* LB400 involves stress proteins also induced by heat shock and oxidative stress. FEMS Microbiol. Lett. *267*, 167–175.

Bopp, L.H. (1986). Degradation of highly chlorinated PCBs by *Pseudomonas* strain LB400. J. Ind. Microbiol. Biotechnol. *1*, 23–29.

Braun, R.J., Kinkl, N., Beer, M., and Ueffing, M. (2007). Two-dimensional electrophoresis of membrane proteins. Anal. Bioanal. Chem. *389*, 1033–1045.

Carvalho, A.P., Ventura, G.M., Pereira, C.B., Leao, R.S., Folescu, T.W., Higa, L., Teixeira, L.M., Plotkowski, M.C., Merquior, V.L., Albano, R.M., *et al.* (2007). *Burkholderia cenocepacia, B. multivorans, B. ambifaria* and *B. vietnamiensis* isolates from cystic fibrosis patients have different profiles of exoenzyme production. APMIS *115*, 311–318.

Cash, P. (2000). Proteomics in medical microbiology. Electrophoresis *21*, 1187–1201.

Chantratita, N., Tandhavanant, S., Wikraiphat, C., Trunck, L.A., Rholl, D.A., Thanwisai, A., Saiprom, N., Limmathurotsakul, D., Korbsrisate, S., Day, N.P., *et al.* (2012). Proteomic analysis of colony morphology variants of *Burkholderia pseudomallei* defines a role for the arginine deiminase system in bacterial survival. J. Proteomics *75*, 1031–1042.

Chavez, F.P., Lunsdorf, H., and Jerez, C.A. (2004). Growth of polychlorinated-biphenyl-degrading bacteria in the presence of biphenyl and chlorobiphenyls generates oxidative stress and massive accumulation of inorganic polyphosphate. Appl. Environ. Microbiol. *70*, 3064–3072.

Chen, G., and Pramanik, B.N. (2009). Application of LC/MS to proteomics studies: current status and future prospects. Drug Discov. Today *14*, 465–471.

Cheng, A.C., and Currie, B.J. (2005). Melioidosis: epidemiology, pathophysiology, and management. Clin. Microbiol. Rev. *18*, 383–416.

Chung, J.W., and Speert, D.P. (2007). Proteomic identification and characterization of bacterial factors associated with *Burkholderia cenocepacia* survival in a murine host. Microbiology *153*, 206–214.

Chung, J.W., Altman, E., Beveridge, T.J., and Speert, D.P. (2003). Colonial morphology of *Burkholderia cepacia* complex genomovar III: implications in exopolysaccharide production, pilus expression, and persistence in the mouse. Infect. Immun. *71*, 904–909.

Cordwell, S.J., Nouwens, A.S., Verrills, N.M., Basseal, D.J., and Walsh, B.J. (2000). Subproteomics based upon protein cellular location and relative solubilities in conjunction with composite two-dimensional electrophoresis gels. Electrophoresis *21*, 1094–1103.

Cornelis, G.R., and Van Gijsegem, F. (2000). Assembly and function of type III secretory systems. Annu. Rev. Microbiol. *54*, 735–774.

Coutinho, C.P., de Carvalho, C.C.C.R., Madeira, A., Pinto-de-Oliveira, A., and Sá-Correia, I. (2011). *Burkholderia cenocepacia* phenotypic clonal variation during a 3.5-year colonization in the lungs of a cystic fibrosis patient. Infect. Immun. *79*, 2950–2960.

Cox, J., and Mann, M. (2011). Quantitative, high-resolution proteomics for data-driven systems biology. Annu Rev Biochem *80*, 273–299.

Dance, D.A. (2002). Melioidosis. Curr. Opin. Infect. Dis. *15*, 127–132.

Davies, D.H., Liang, X., Hernandez, J.E., Randall, A., Hirst, S., Mu, Y., Romero, K.M., Nguyen, T.T., Kalantari-Dehaghi, M., Crotty, S., *et al.* (2005). Profiling the humoral immune response to infection by using proteome microarrays: high-throughput vaccine and diagnostic antigen discovery. Proc. Natl. Acad. Sci. U.S.A. *102*, 547–552.

Denef, V.J., Klappenbach, J.A., Patrauchan, M.A., Florizone, C., Rodrigues, J.L., Tsoi, T.V., Verstraete, W., Eltis, L.D., and Tiedje, J.M. (2006). Genetic and genomic insights into the role of benzoate-catabolic pathway redundancy in *Burkholderia xenovorans* LB400. Appl. Environ. Microbiol. 72, 585–595.

Denef, V.J., Park, J., Tsoi, T.V., Rouillard, J.M., Zhang, H., Wibbenmeyer, J.A., Verstraete, W., Gulari, E., Hashsham, S.A., and Tiedje, J.M. (2004). Biphenyl and benzoate metabolism in a genomic context: outlining genome-wide metabolic networks in *Burkholderia xenovorans* LB400. Appl. Environ. Microbiol. *70*, 4961–4970.

Denef, V.J., Patrauchan, M.A., Florizone, C., Park, J., Tsoi, T.V., Verstraete, W., Tiedje, J.M., and Eltis, L.D. (2005). Growth substrate- and phase-specific expression of biphenyl, benzoate, and C1 metabolic pathways in *Burkholderia xenovorans* LB400. J. Bacteriol. *187*, 7996–8005.

Drevinek, P., and Mahenthiralingam, E. (2010). *Burkholderia cenocepacia* in cystic fibrosis: epidemiology and molecular mechanisms of virulence. Clin. Microbiol. Infect. *16*, 821–830.

Eberl, L. (2006). Quorum sensing in the genus *Burkholderia*. Int. J. Med. Microbiol. *296*, 103–110.

Egea, L., Aguilera, L., Gimenez, R., Sorolla, M.A., Aguilar, J., Badia, J., and Baldoma, L. (2007). Role of secreted glyceraldehyde-3-phosphate dehydrogenase in the infection mechanism of enterohemorrhagic and enteropathogenic *Escherichia coli*: interaction of the extracellular enzyme with human plasminogen and fibrinogen. Int. J. Biochem. Cell Biol. *39*, 1190–1203.

Elliott, M.H., Smith, D.S., Parker, C.E., and Borchers, C. (2009). Current trends in quantitative proteomics. J. Mass Spectrom. *44*, 1637–1660.

Felgner, P.L., Kayala, M.A., Vigil, A., Burk, C., Nakajima-Sasaki, R., Pablo, J., Molina, D.M., Hirst, S., Chew, J.S., Wang, D., *et al.* (2009). A *Burkholderia pseudomallei* protein microarray reveals serodiagnostic and cross-reactive antigens. Proc. Natl. Acad. Sci. U.S.A. *106*, 13499–13504.

Friedman, D.B., and Lilley, K.S. (2008). Optimizing the difference gel electrophoresis (DIGE) technology. Methods Mol. Biol. *428*, 93–124.

Ge, H., Walhout, A.J., and Vidal, M. (2003). Integrating 'omic' information: a bridge between genomics and systems biology. Trends Genet. *19*, 551–560.

Gescher, J., Eisenreich, W., Worth, J., Bacher, A., and Fuchs, G. (2005). Aerobic benzoyl-CoA catabolic pathway in *Azoarcus evansii*: studies on the non-oxygenolytic ring cleavage enzyme. Mol. Microbiol. *56*, 1586–1600.

Goo, E., Kang, Y., Kim, H., and Hwang, I. (2010). Proteomic analysis of quorum sensing-dependent proteins in *Burkholderia glumae*. J. Proteome Res. *9*, 3184–3199.

Goris, J., De Vos, P., Caballero-Mellado, J., Park, J., Falsen, E., Quensen, J.F., 3rd, Tiedje, J.M., and Vandamme, P. (2004). Classification of the biphenyl- and polychlorinated biphenyl-degrading strain LB400T and relatives as *Burkholderia xenovorans* sp. nov. Int. J. Syst. Evol. Microbiol. *54*, 1677–1681.

Grenfell, B.T., Pybus, O.G., Gog, J.R., Wood, J.L., Daly, J.M., Mumford, J.A., and Holmes, E.C. (2004). Unifying the epidemiological and evolutionary dynamics of pathogens. Science *303*, 327–332.

Harding, S.V., Sarkar-Tyson, M., Smither, S.J., Atkins, T.P., Oyston, P.C., Brown, K.A., Liu, Y., Wait, R., and Titball, R.W. (2007). The identification of surface proteins of *Burkholderia pseudomallei*. Vaccine *25*, 2664–2672.

Hathout, Y. (2007). Approaches to the study of the cell secretome. Expert Rev. Proteomics *4*, 239–248.

Hecker, M., Antelmann, H., Buttner, K., and Bernhardt, J. (2008). Gel-based proteomics of Gram-positive bacteria: a powerful tool to address physiological questions. Proteomics *8*, 4958–4975.

Hengge-Aronis, R. (2002). Signal transduction and regulatory mechanisms involved in control of the sigma(S) (RpoS) subunit of RNA polymerase. Microbiol. Mol. Biol. Rev. *66*, 373–395.

Holden, M.T.G., Seth-Smith, H.M.B., Crossman, L.C., Sebaihia, M., Bentley, S.D., Cerdeño-Tárraga, A.M., Thomson, N.R., Bason, N., Quail, M.A., Sharp, S., *et al.* (2009). The genome of *Burkholderia cenocepacia* J2315, an epidemic pathogen of cystic fibrosis patients. J. Bacteriol. *191*, 261–277.

Hopkins, A.L., and Groom, C.R. (2002). The druggable genome. Nat. Rev. Drug Discov. *1*, 727–730.

Inhülsen, S., Aguilar, C., Schmid, N., Suppiger, A., Riedel, K., and Eberl, L. (2012). Identification of functions linking quorum sensing with biofilm formation in *Burkholderia cenocepacia* H111. Microbiol. Open *1*, 225–242.

Kang, Y., Kim, J., Kim, S., Kim, H., Lim, J.Y., Kim, M., Kwak, J., Moon, J.S., and Hwang, I. (2008). Proteomic analysis of the proteins regulated by HrpB from the plant pathogenic bacterium *Burkholderia glumae*. Proteomics *8*, 106–121.

Kaufmann, S.H., and Hess, J. (1999). Impact of intracellular location of and antigen display by intracellular bacteria: implications for vaccine development. Immunol. Lett. *65*, 81–84.

Keys, C.J., Dare, D.J., Sutton, H., Wells, G., Lunt, M., McKenna, T., McDowall, M., and Shah, H.N. (2004). Compilation of a MALDI-TOF mass spectral database for the rapid screening and characterisation of bacteria implicated in human infectious diseases. Infect. Genet. Evol. *4*, 221–242.

Kooi, C., Subsin, B., Chen, R., Pohorelic, B., and Sokol, P.A. (2006). *Burkholderia cenocepacia* ZmpB is a broad-specificity zinc metalloprotease involved in virulence. Infect. Immun. *74*, 4083–4093.

Korbsrisate, S., Vanaporn, M., Kerdsuk, P., Kespichayawattana, W., Vattanaviboon, P., Kiatpapan, P., and Lertmemongkolchai, G. (2005). The *Burkholderia pseudomallei* RpoE (AlgU) operon is involved in environmental stress tolerance and biofilm formation. FEMS Microbiol. Lett. *252*, 243–249.

Kwok, S.-Y., Siu, A.F.-M., Ngai, S.-M., Che, C.-M., and Tsang, J.S.H. (2007). Proteomic analysis of *Burkholderia cepacia* MBA4 in the degradation of monochloroacetate. Proteomics *7*, 1107–1116.

Li, Z., Shao, T., Min, H., Lu, Z., and Xu, X. (2009). Stress response of *Burkholderia cepacia* WZ1 exposed to quinclorac and the biodegradation of quinclorac. Soil Biol. Biochem. *41*, 984–990.

Lieberman, T.D., Michel, J.B., Aingaran, M., Potter-Bynoe, G., Roux, D., Davis, M.R., Jr., Skurnik, D., Leiby, N., LiPuma, J.J., Goldberg, J.B., *et al.* (2011). Parallel bacterial evolution within multiple patients identifies candidate pathogenicity genes. Nat. Genet. *43*, 1275–1280.

Lilley, K.S., and Friedman, D.B. (2004). All about DIGE: quantification technology for differential-display 2D-gel proteomics. Expert Rev. Proteomics *1*, 401–409.

Lopez, J.L. (2007). Two-dimensional electrophoresis in proteome expression analysis. J. Chromatogr. B Analyt. Technol. Biomed. Life Sci. *849*, 190–202.

Madeira, A., Santos, P.M., Coutinho, C.P., Pinto-de-Oliveira, A., and Sá-Correia, I. (2011). Quantitative proteomics (2-D DIGE) reveals molecular strategies employed by *Burkholderia cenocepacia* to adapt to the airways of cystic fibrosis patients under antimicrobial therapy. Proteomics *11*, 1313–1328.

Mahenthiralingam, E., Baldwin, A., and Vandamme, P. (2002). *Burkholderia cepacia* complex infection in patients with cystic fibrosis. J. Med. Microbiol. *51*, 533–538.

Mahenthiralingam, E., Urban, T.A., and Goldberg, J.B. (2005). The multifarious, multireplicon *Burkholderia cepacia* complex. Nat. Rev. Microbiol. *3*, 144–156.

Malen, H., Berven, F.S., Fladmark, K.E., and Wiker, H.G. (2007). Comprehensive analysis of exported proteins from *Mycobacterium tuberculosis* H37Rv. Proteomics *7*, 1702–1718.

Mariappan, V., Vellasamy, K.M., Thimma, J.S., Hashim, O.H., and Vadivelu, J. (2010). Identification of immunogenic proteins from *Burkholderia cepacia* secretome using proteomic analysis. Vaccine *28*, 1318–1324.

Mariappan, V., Vellasamy, K.M., Hashim, O.H., and Vadivelu, J. (2011). Profiling of *Burkholderia cepacia* secretome at mid-logarithmic and early-stationary phases of growth. PLoS ONE *6*, e26518.

Martínez, P., Agulló, L., Hernández, M., and Seeger, M. (2007). Chlorobenzoate inhibits growth and induces stress proteins in the PCB-degrading bacterium *Burkholderia xenovorans* LB400. Arch. Microbiol. *188*, 289–297.

Minden, J. (2007). Comparative proteomics and difference gel electrophoresis. Biotechniques *43*, 739, 741, 743 passim.

Minden, J.S. (2012). DIGE: past and future. Methods Mol. Biol. *854*, 3–8.

Mira, N.P., Madeira, A., Moreira, A.S., Coutinho, C.P., and Sá-Correia, I. (2011). Genomic expression analysis reveals strategies of *Burkholderia cenocepacia* to adapt to cystic fibrosis patients' airways and antimicrobial therapy. PLoS ONE *6*, e28831.

Mott, T., Soler, M., Grigsby, S., Medley, R., and Whitlock, G.C. (2010). Identification of potential diagnostic markers among *Burkholderia cenocepacia* and *B. multivorans* supernatants. J. Clin. Microbiol. *48*, 4186–4192.

Mullen, T., Markey, K., Murphy, P., McClean, S., and Callaghan, M. (2007). Role of lipase in *Burkholderia cepacia* complex (Bcc) invasion of lung epithelial cells. Eur. J. Clin. Microbiol. Infect. Dis. *26*, 869–877.

O'Farrell, P.H. (1975). High resolution two-dimensional electrophoresis of proteins. J. Biol. Chem. *250*, 4007–4021.

O'Sullivan, L.A., and Mahenthiralingam, E. (2005). Biotechnological potential within the genus *Burkholderia*. Lett. Appl. Microbiol. *41*, 8–11.

Ong, S.E., and Mann, M. (2005). Mass spectrometry-based proteomics turns quantitative. Nat. Chem. Biol. *1*, 252–262.

Osiriphun, Y., Wongtrakoongate, P., Sanongkiet, S., Suriyaphol, P., Thongboonkerd, V., and Tungpradabkul, S. (2009). Identification and characterization of RpoS regulon and RpoS-dependent promoters in *Burkholderia pseudomallei*. J. Proteome Res. *8*, 3118–3131.

Otto, A., Bernhardt, J., Hecker, M., and Becher, D. (2012). Global relative and absolute quantitation in microbial proteomics. Curr. Opin. Microbiol. *15*, 364–372.

Ou, K., Ong, C., Koh, S.Y., Rodrigues, F., Sim, S.H., Wong, D., Ooi, C.H., Ng, K.C., Jikuya, H., Yau, C.C., *et al.* (2005). Integrative genomic, transcriptional, and proteomic diversity in natural isolates of the human pathogen *Burkholderia pseudomallei*. J. Bacteriol. *187*, 4276–4285.

Pallen, M.J., Chaudhuri, R.R., and Henderson, I.R. (2003). Genomic analysis of secretion systems. Curr. Opin. Microbiol. *6*, 519–527.

Phillips, C.I., and Bogyo, M. (2005). Proteomics meets microbiology: technical advances in the global mapping of protein expression and function. Cell Microbiol. *7*, 1061–1076.

Pumirat, P., Saetun, P., Sinchaikul, S., Chen, S.T., Korbsrisate, S., and Thongboonkerd, V. (2009). Altered secretome of *Burkholderia pseudomallei* induced by salt stress. Biochim. Biophys. Acta *1794*, 898–904.

Rabilloud, T., and Lelong, C. (2011). Two-dimensional gel electrophoresis in proteomics: a tutorial. J. Proteomics *74*, 1829–1841.

Riedel, K., Arevalo-Ferro, C., Reil, G., Görg, A., Lottspeich, F., and Eberl, L. (2003). Analysis of the quorum-sensing regulon of the opportunistic pathogen *Burkholderia cepacia* H111 by proteomics. Electrophoresis *24*, 740–750.

Riedel, K., Carranza, P., Gehrig, P., Potthast, F., and Eberl, L. (2006). Towards the proteome of *Burkholderia cenocepacia* H111: setting up a 2-DE reference map. Proteomics *6*, 207–216.

Rodriguez-Suarez, E., and Whetton, A.D. (2013). The application of quantification techniques in proteomics for biomedical research. Mass Spectrom. Rev. *32*, 1–26.

Roma-Rodrigues, C., Santos, P.M., Benndorf, D., Rapp, E., and Sá-Correia, I. (2010). Response of *Pseudomonas putida* KT2440 to phenol at the level of membrane proteome. J. Proteomics *73*, 1461–1478.

Rotz, L.D., Khan, A.S., Lillibridge, S.R., Ostroff, S.M., and Hughes, J.M. (2002). Public health assessment of potential biological terrorism agents. Emerg. Infect. Dis. *8*, 225–230.

Ruelle, V., El Moualij, B., Zorzi, W., Ledent, P., and Pauw, E.D. (2004). Rapid identification of environmental bacterial strains by matrix-assisted laser desorption/ionization time-of-flight mass spectrometry. Rapid Commun. Mass Spectrom. *18*, 2013–2019.

Sá-Correia, I., and Teixeira, M.C. (2010). 2D electrophoresis-based expression proteomics: a microbiologist's perspective. Expert Rev. Proteomics *7*, 943–953.

Santos, P.M., Roma, V., Benndorf, D., von Bergen, M., Harms, H., and Sá-Correia, I. (2007). Mechanistic insights into the global response to phenol in the phenol-biodegrading strain *Pseudomonas* sp. M1 revealed by quantitative proteomics. OMICS *11*, 233–251.

Sayler, R.J., Cartwright, R.D., and Yang, Y. (2006). Genetic characterization and real-time PCR detection of *Burkholderia glumae*, a newly emerging bacterial pathogen of rice in the United States. Plant Dis. *90*, 603–610.

Schell, M.A., Zhao, P., and Wells, L. (2011). Outer membrane proteome of *Burkholderia pseudomallei* and *Burkholderia mallei* from diverse growth conditions. J. Proteome Res. *10*, 2417–2424.

Seeger, M., Zielinski, M., Timmis, K.N., and Hofer, B. (1999). Regiospecificity of dioxygenation of di- to pentachlorobiphenyls and their degradation to chlorobenzoates by the bph-encoded catabolic pathway of *Burkholderia* sp. strain LB400. Appl. Environ. Microbiol. *65*, 3614–3621.

Sheehan, D. (2006). Detection of redox-based modification in two-dimensional electrophoresis proteomic separations. Biochem. Biophys. Res. Commun. *349*, 455–462.

Sheehan, D., McDonagh, B., and Barcena, J.A. (2010). Redox proteomics. Expert Rev. Proteomics *7*, 1–4.

Sibbald, M.J., Ziebandt, A.K., Engelmann, S., Hecker, M., de Jong, A., Harmsen, H.J., Raangs, G.C., Stokroos, I., Arends, J.P., Dubois, J.Y., *et al.* (2006). Mapping the pathways to staphylococcal pathogenesis by comparative secretomics. Microbiol. Mol. Biol. Rev. *70*, 755–788.

Steinberg, T.H., Pretty On Top, K., Berggren, K.N., Kemper, C., Jones, L., Diwu, Z., Haugland, R.P., and Patton, W.F. (2001). Rapid and simple single nanogram detection of glycoproteins in polyacrylamide gels and on electroblots. Proteomics *1*, 841–855.

Steinberg, T.H., Agnew, B.J., Gee, K.R., Leung, W.Y., Goodman, T., Schulenberg, B., Hendrickson, J., Beechem, J.M., Haugland, R.P., and Patton, W.F. (2003). Global quantitative phosphoprotein analysis using Multiplexed Proteomics technology. Proteomics *3*, 1128–1144.

Stevens, M.P., Haque, A., Atkins, T., Hill, J., Wood, M.W., Easton, A., Nelson, M., Underwood-Fowler, C., Titball, R.W., Bancroft, G.J., *et al.* (2004). Attenuated virulence and protective efficacy of a *Burkholderia pseudomallei* bsa type III secretion mutant in murine models of melioidosis. Microbiology *150*, 2669–2676.

Suarez-Moreno, Z.R., Caballero-Mellado, J., Coutinho, B.G., Mendonca-Previato, L., James, E.K., and Venturi, V. (2012). Common features of environmental and potentially beneficial plant-associated *Burkholderia*. Microb. Ecol. *63*, 249–266.

Subsin, B., Thomas, M.S., Katzenmeier, G., Shaw, J.G., Tungpradabkul, S., and Kunakorn, M. (2003). Role of the stationary growth phase sigma factor RpoS of *Burkholderia pseudomallei* in response to physiological stress conditions. J. Bacteriol. *185*, 7008–7014.

Thongboonkerd, V., Vanaporn, M., Songtawee, N., Kanlaya, R., Sinchaikul, S., Chen, S.T., Easton, A., Chu, K., Bancroft, G.J., and Korbsrisate, S. (2007). Altered proteome in *Burkholderia pseudomallei* rpoE operon knockout mutant: insights into mechanisms of rpoE operon in stress tolerance, survival, and virulence. J. Proteome Res. *6*, 1334–1341.

Trost, M., Wehmhoner, D., Karst, U., Dieterich, G., Wehland, J., and Jansch, L. (2005). Comparative proteome analysis of secretory proteins from pathogenic and nonpathogenic *Listeria* species. Proteomics *5*, 1544–1557.

Tsang, J.S.H., Sallis, P.J., Bull, A.T., and Hardman, D.J. (1988). A monobromoacetate dehalogenase from *Pseudomonas cepacia* MBA4. Arch. Microbiol. *150*, 441–446.

Van Nostrand, J.D., Arthur, J.M., Kilpatrick, L.E., Neely, B.A., Bertsch, P.M., and Morris, P.J. (2008). Changes in protein expression in *Burkholderia vietnamiensis* PR1 301 at pH 5 and 7 with and without nickel. Microbiology (Reading, England) *154*, 3813–3824.

Vandamme, P., and Dawyndt, P. (2011). Classification and identification of the *Burkholderia cepacia* complex: Past, present and future. Syst. Appl. Microbiol. *34*, 87–95.

Vanlaere, E., Sergeant, K., Dawyndt, P., Kallow, W., Erhard, M., Sutton, H., Dare, D., Devreese, B., Samyn, B., and Vandamme, P. (2008). Matrix-assisted laser desorption ionisation-time-of of-flight mass spectrometry of intact cells allows rapid identification of *Burkholderia cepacia* complex. J. Microbiol. Methods *75*, 279–286.

Vellasamy, K.M., Mariappan, V., Hashim, O.H., and Vadivelu, J. (2011). Identification of immunoreactive secretory proteins from the stationary phase culture of *Burkholderia pseudomallei*. Electrophoresis *32*, 310–320.

Washburn, M.P. (2004). Utilisation of proteomics datasets generated via multidimensional protein identification technology (MudPIT). Brief Funct. Genomic. Proteomic. *3*, 280–286.

Wickramasekara, S., Neilson, J., Patel, N., Breci, L., Hilderbrand, A., Maier, R.M., and Wysocki, V. (2011). Proteomics analyses of the opportunistic pathogen *Burkholderia vietnamiensis* using protein fractionations and mass spectrometry. J. Biomed. Biotechnol. *2011*, 701928.

Wongtrakoongate, P., Mongkoldhumrongkul, N., Chaijan, S., Kamchonwongpaisan, S., and Tungpradabkul, S. (2007). Comparative proteomic profiles and the potential markers between *Burkholderia pseudomallei* and *Burkholderia thailandensis*. Mol. Cell Probes *21*, 81–91.

Wongtrakoongate, P., Roytrakul, S., Yasothornsrikul, S., and Tungpradabkul, S. (2011). A proteome reference map of the causative agent of melioidosis *Burkholderia pseudomallei*. J. Biomed. Biotechnol. *2011*, 530926.

Woo, P.C., Leung, P.K., Wong, S.S., Ho, P.L., and Yuen, K.Y. (2001). groEL encodes a highly antigenic protein in *Burkholderia pseudomallei*. Clin. Diagn. Lab Immunol. *8*, 832–836.

Zlosnik, J.E., and Speert, D.P. (2010). The role of mucoidy in virulence of bacteria from the *Burkholderia cepacia* complex: a systematic proteomic and transcriptomic analysis. J. Infect. Dis. *202*, 770–781.

Molecular Mechanisms of Virulence in *Burkholderia pseudomallei* and *Burkholderia mallei*

6

Catherine Ong and Patrick Tan

Abstract

Burkholderia pseudomallei and *Burkholderia mallei*, causative agents of the diseases melioidosis and glanders, respectively, are two of the deadliest members of the *Burkholderia* genus. Contained within their large genomes is a sizable number of genes that encode numerous redundant mechanisms involved in virulence. The availability of complete genome sequences of these two species has resulted in targeted characterization of multiple novel virulence factors. Some of these virulence factors are arguably niche factors, more likely to aid the bacteria to adapt and survive in its environment. The distinct difference in niches of *B. pseudomallei* and *B. mallei* allows us to relook at some of these established virulence factors and mechanisms in context of their roles in environmental adaptation and survival. In addition, comparisons with the avirulent environmental saprophyte *Burkholderia thailandensis* helps to better understand the evolution of mammalian pathogenesis in these two pathogens. This chapter serves to summarize previously established virulence determinants, as well as highlight new and novel factors that have since been discovered post-genome sequencing.

Introduction

The genus *Burkholderia* currently contains more than six validly described species. These are Gram-negative β-Proteobacteria that occupy very different and diverse niches. Members of the *Burkholderiaceae* are examples of bacteria with a large multireplicon genome and are adapted to highly diverse lifestyles and environments (Ussery *et al.*, 2009). A large number are found naturally in the environment, in the soil and water. Many are plant symbionts, e.g. *B. graminis, B. phymatum, B. phytofirmans, B. vietnamensis* etc., promoting plant growth via carbon and/or nitrogen fixation and even protecting plants against other pathogens. In contrast, some are plant pathogens e.g. *B. glumae* is touted as the next major pathogen of rice plants (Ham *et al.*, 2010). Some live harmlessly in the environment e.g. *B. ubonensis,* yet others are zoonotic pathogens (*B. mallei*) and opportunistic human pathogens (*B. cenocepacia*). Some species are industrially important biodegradation tools, e.g. *B. xenovorans,* and yet others are highly pathogenic and potentially dangerous Category B biowarfare agents (*B. pseudomallei* and *B. mallei*) (Ussery *et al.*, 2009). Arguably, the last two species are the best-known members amongst the *Burkholderia* genus and their potential use as agents for bioterrorism has attracted a lot of interest in recent years.

Burkholderia pseudomallei is an opportunistic pathogen that thrives in diverse environments including soil, contaminated waters and air. It is the causative agent of melioidosis, a tropical infectious disease. Melioidosis is endemic in Southeast Asia and Northern Australia (Cheng and Currie, 2005). In recent years, there have been reports of melioidosis in countries including Laos, Papua New Guinea and Brazil (Rolim *et al.*, 2009; Warner *et al.*, 2007, 2008; Wuthiekanun *et al.*, 2005). The disease is occasionally imported into non-endemic areas like Europe and the US, mainly by travellers who have visited the tropics. The pathogen is also known to infect a broad host range, including birds, reptiles and mammals like dolphins and humans. Human infection can occur through cutaneous cuts and exposure to contaminated environments, especially contaminated waters during the rainy season, or by inhalation of contaminated dust. Human-to-human transmission is rare. Human melioidosis has a high mortality – 20% in northern Australia and up to 50% in Northeast Thailand. Melioidosis accounts for 20% of all community-acquired septicaemias in northern Australia, and 40% of sepsis-related mortality in north-east Thailand (Wiersinga *et al.*, 2006). In highly urbanized Singapore, the mortality rates are similar to that observed in Northern Australia (~20%) (Communicable Diseases Surveillance in Singapore, 2010).

Burkholderia pseudomallei has been called 'the great mimicker' as clinical manifestations of its infection are wide and variable, ranging from asymptomatic infections to severe septic shock. This variability often leads to late diagnosis resulting in delayed treatment. The longest recorded incubation period documented is 62 years, in a Vietnam war veteran who returned to the US (Ngauy *et al.*, 2005). In addition to difficulties in diagnosis due to the varied clinical manifestations, treatment of infections is equally challenging as the pathogen is intrinsically resistant to many antibiotics, including penicillin, first and second-generation cephalosporins, macrolides, rifamycins, colistin and aminoglycosides. *B. pseudomallei* is susceptible to some third-generation cephalosporins, carbapenems and some fluoroquinolones, as well as amoxicillin-β-lactamase inhibitor combinations (e.g. amoxicillin–clavulanic acid), chloramphenicol, tetracyclines and trimethoprim–sulphonamides (e.g. trimethoprim–sulphamethoxazole) *in vitro*. The clinical efficacy of chloramphenicol, trimethoprim–sulphamethoxazole, doxycycline, imipenem, ciprofloxacin and ofloxacin has also been evaluated for the treatment of melioidosis. The current treatment for melioidosis includes a biphasic therapy, consisting of a short-term parenteral acute phase involving intravenous ceftazidime and a long-term oral eradication phase comprising chloramphenicol, trimethoprim–sulphamethoxazole and doxycycline, or amoxicillin–clavulanic acid alone (Cheng, 2010; Estes *et al.*, 2010).

Unlike *B. pseudomallei*, *Burkholderia mallei* is a host-restricted obligate parasite which is unable to live outside its host environment. It causes the zoonotic disease glanders, infecting equines and solipeds. Horses and mules are known natural reservoirs. Disease symptoms range from acute to chronic forms, characterized by ulcerating nodular lesions of the skin and mucous membranes. Infected animals present with general symptoms like fever, malaise, depression and weight loss, such that accurate diagnosis of the disease remains a challenge (Khan *et al.*, 2012). The disease is highly infectious, and easily transmitted between horses, especially under crowded conditions. Infection primarily results from the contamination of wounds, abrasions or breaks in mucous membranes. Recently, a glanders outbreak at Tehran Zoo, Iran, resulted in the deaths of four lions and one tiger. It appears that the use of contaminated solipeds as feed to these valuable carnivores was the cause of the infections (Khaki *et al.*, 2012). Animal-to-human transmission generally occurs due to occupational requirement for close contact with infected animals, for example, in veterinarians (Kim *et al.*, 2005; Nierman *et al.*, 2004).

B. mallei has similar antibiotic susceptibility pattern as *B. pseudomallei*, including susceptibility to doxycycline. In addition, *B. mallei* is susceptible to aminoglycosides due to the absence of the AmrAB–OprA efflux pump, which confers high-level aminoglycoside resistance in *B. pseudomallei*. There are scarce data available on the efficacy of antibiotics for the treatment of human glanders but current treatment options include a mixed antibiotic regimen that is often only partially effective. Single-drug antibiotic therapy is only partially effective and similarly to treating melioidosis, combinations of antibiotics must be used for extended periods. In a laboratory-acquired glanders infection, the patient was treated with imipenem and doxycycline for 2 weeks, followed by treatment with azithromycin and doxycycline for an additional 6 months (Estes *et al.*, 2010; Srinivasan *et al.*, 2001).

Currently there are no licensed vaccines available for both melioidosis and glanders despite considerable research funding. To date, the vaccine candidates identified included live attenuated mutants, inactivated whole cell vaccines, subunit vaccines, naked DNA and dendritic cell vaccines but sterile immunity was rarely reported in animal models (Peacock *et al.*, 2012).

With the completion of full genome sequences and annotation of *B. pseudomallei* K96243 and *B. mallei* ATCC23343 in 2004, the reference genomes have been a useful resource for many researchers to identify new virulence genes and pathways (Holden *et al.*, 2004; Nierman *et al.*, 2004). The 7.2 Mbp complete genome of *B. pseudomallei* is composed of two chromosomal replicons with significant functional partitioning of genes between them. *B. mallei* has a smaller genome (5.8 Mbp) and comparisons with *B. pseudomallei* have revealed extensive deletions and rearrangements of the *B. mallei* genome. Coupled with advances in sequencing technologies and greater access to sequencing platforms, the genomes of many more *B. pseudomallei* and *B. mallei* strains, as well as that of the avirulent relative *B. thailandensis* have since been sequenced. Comparisons between the pathogenic strains with their non-pathogenic counterpart allowed researchers to identify genes that are required for pathogen virulence and provide an insight to the evolution of pathogenesis and niche specialization amongst *Burkholderia* species (Lin *et al.*, 2008; Yu *et al.*, 2006).

The genotypic and phenotypic differences between *B. pseudomallei* and *B. mallei* have been conscientiously tabulated in a recent review by (Galyov *et al.*, 2010). We strongly recommend readers with an interest in the comparisons between these two species to refer to this paper. In addition, the paper focused on unique and shared virulence features of *B. pseudomallei* and *B. mallei* that contribute to bacterial pathogenesis in the context of the intracellular lifestyle of the pathogens. A detailed list of virulence factors of *B. pseudomallei* and *B. mallei* and their relative importance in animal models of glanders and melioidosis is also beautifully drawn up (Galyov *et al.*, 2010). In their review, the capsule (CPSI), type 3 secretion system and type 6 secretion systems are identified as the key virulence factors of *B. pseudomallei* and *B. mallei* during intracellular infection. The intracellular lifestyle of the pathogens, which involves T3SS-dependent escape of bacteria into the host cell cytosol, BimA-dependent actin-based motility, and formation of multinucleated giant cells, enables *B. pseudomallei* and *B. mallei* to evade host immune responses, replicate, and persist within a host.

In this chapter, we summarize previously established molecular mechanisms of virulence in *B. pseudomallei* and *B. mallei*, and highlight recent progress in the elucidation of pathogenic mechanisms in these pathogens.

Adaptation to environmental factors: niche factors that became accidental virulence mechanisms

Traditionally, virulence factors and mechanisms are described as structures or strategies that are employed by infectious microbes to invade, colonize and survive in a host (Madigan *et al.*, 2009). Classic examples include flagella, pili, and Type 3 secretion systems. Increasingly, research on commensals, pathogens and avirulent counterparts has revealed that many of these are also present in non-pathogenic bacteria, especially those that are found in similar niches or environments. For example, the proteinaceous pili implicated in pathogenesis of many Gram-positive pathogens, including *Enterococcus faecalis* are also found in the commercialized probiotic *Lactobacillus rhamnosus* GG (Hill, 2012). Rather than principally evolved for causing infection, such shared factors and mechanisms serve to aid in bacterial adaptation and survival in the environment itself. For example, bile salt hydrolase has been identified as a virulence factor of *Listeria monocytogenes* during intestinal and hepatic phases of listeriosis (Dussurget *et al.*, 2002). In fact, bile salt hydrolase is also present in many commensals which are probiotics to cope with the presence of bile in the gut (Begley *et al.*, 2006). There is increasing interest in the microbial research community now to relook at these factors as 'niche factors', rather than virulence factors (Hill, 2012; Nandi *et al.*, 2010).

Given that *B. pseudomallei* is an environmental saprophyte, many mechanisms that enable *B. pseudomallei* to survive and adapt to the ever-changing environmental factors translate to efficient virulence mechanisms when it enters a host, resulting in persistence and survival in the host cells. Evidence supporting these mechanisms as 'accidental' virulence came from genome comparison between *B. pseudomallei* and the avirulent *B. thailandensis* – many of these mechanisms are present in both of these environmental species, for example, functional mechanisms for chemotaxis and motility (Yu *et al.*, 2006). In contrast, such mechanisms are absent in *B. mallei*, which is restricted to a specialized niche and thus likely to have lost such adaptive mechanisms to the environment. In fact, it is likely that humans are not an integral part of their life cycle and just accidental 'dead-end' hosts for both pathogens as there has been no convincing evidence of strains being passed back into the environment after human infection. We proceed to describe some of these mechanisms below.

Morphological switching

In the environment, *B. pseudomallei* is constantly challenged by adverse and rapidly changing conditions. Phenotypic adaptability and plasticity is critical for bacterial survival. It was found that the colony morphology of *B. pseudomallei* could alternate between rough and smooth colonies when grown under different *in vitro* conditions. A colony morphotyping scheme and typing algorithm developed in 2006 identified up to seven different *B. pseudomallei* morphotypes (Type I to VII) (Chantratita *et al.*, 2007). The authors showed that under environmental stress (starvation, iron limitation and heat stress), Type I (common 'cornflower head' morphology) switches to other morphotypes, most commonly Type II (small, rough colony) or Type III (large, smooth colony) (Chantratita *et al.*, 2007). Removal of the stress results in a reversible switch back to parental Type I. Morphological switching is also associated with differential expression of virulence determinants like flagella and leads to biofilm formation. Though morphological switching is a process of adaptation to environmental factors, the authors hypothesize that such an adaptation helps the bacteria to persist and survive during infection of host cells. Indeed, when challenged using a mouse model, switching between isogenic morphotypes was observed. Type II morphotype was associated with persistence in a low-virulence state *in vivo*. Upon uptake by human respiratory epithelial cells A549 *in vitro*, Type II morphotype exhibited a significant increase in intracellular replication fitness compared with parental Type I. During interactions with the mouse macrophage cell line J774A.1, isogenic Type III morphotype exhibited higher replication fitness and was associated with a switch to Type II (Chantratita *et al.*, 2007). In a follow-up study using human macrophage cell line U937 and exposure to *in vitro* conditions simulating the macrophage milieu, the authors observed that all morphotypes were highly resistant to lysozyme and lactoferrin. In addition, all three isogenic *B. pseudomallei* morphotypes were resistant to antimicrobial peptides including α-defensin HNP-1 and β-defensin HBD-2. Instead, they were susceptible to LL-37. The ability of all three isogenic morphotypes to survive low oxygen concentration and anaerobic conditions for at least 2 weeks demonstrated that these morphotypes allow the bacteria to survive and persist under hypoxia in the host microenvironment. It also appears that Type I and II are more stable morphotypes, whereas Type III switches to the other two morphotypes under various conditions periodically to survive (Chantratita *et al.*, 2007; Tandhavanant *et al.*, 2010).

Whereas there have been no studies to investigate whether *B. mallei* undergoes morphological switching, *B. thailandensis* isolates exhibit changes in colony morphology when the quorum sensing (QS) system is disrupted (Chandler *et al.*, 2009; Ulrich *et al.*, 2004c). Similar changes in colony morphology were also observed in a *B. pseudomallei* QS mutant which has been found to have an attenuated level of virulence in murine model of infection (Valade *et al.*, 2004), Ong *et al.*, unpublished data).

Motility

The *B. mallei* genome encodes numerous flagellar and chemotaxis genes, though the organism is non-motile and lacks a flagellum (Nierman *et al.*, 2004). In-depth analyses revealed the presence of frame shift mutations in a number of these motility genes, including a methyl-accepting chemotaxis gene and the flagellum motor *motB* gene. In addition, the *fliP* gene, which is essential for flagellum biogenesis, is disrupted by a 65 kb insert that is flanked by IS elements. These mutations are likely to be a result of evolutionary changes as the pathogen adapts to a restricted host environment (Nierman *et al.*, 2004).

In contrast, there are five gene clusters encoding for a fully functional flagellum in the *B. pseudomallei* K96243 reference genome (Holden *et al.*, 2004). The role of flagella in *B. pseudomallei* virulence has been demonstrated in previous studies whereby *B. pseudomallei* mutants lacking flagellin, a 43 kDa protein, are attenuated in an intranasal challenge in mice (Chua *et al.*, 2003; Chuaygud *et al.*, 2008). However, the virulence appeared to be dependent on the route of inoculation as intraperitoneal challenge of these mutants did not produce any attenuation in virulence in hamsters or infant diabetic rats (DeShazer *et al.*, 1997).

Like *B. pseudomallei*, the avirulent *B. thailandensis* is motile and has flagellin. However, comparative genomic analyses between the *B. pseudomallei* genome and that of *B. thailandensis* revealed that the flagellar biosynthesis gene cluster in *B. thailandensis* (BTFC) has been replaced in *B. pseudomallei* by a group of fimbriae-related genes whose closest relatives are found in *Yersinia* (YLF). *B. pseudomallei* strains containing the YLF gene cluster were more commonly observed in clinical cases where BTFC gene clusters are found in the environmental isolates, supporting a potential role for the YLF in virulence. Additionally, recent gel-based glycoproteomics studies revealed that the flagellin proteins from *B. pseudomallei* and *B. thailandensis* were postranslationally modified by novel and different glycans (Scott *et al.*, 2011), possibly contributing to the difference in virulence levels between *B. pseudomallei* and *B. thailandensis* (Yu *et al.*, 2006). Given this evidence, it is likely that the flagella is a relatively minor virulence component whose basic function is more likely for bacterial movement in the environment than infection.

Genomic islands

Genomic islands (GIs) are mobile genetic elements that are acquired by bacterial species via horizontal gene transfer (Dobrindt *et al.*, 2004). Whereas genomic islands may carry traits that confer antibiotic resistance, symbiosis and increased fitness and adaptation for non-pathogenic and environmental microbes, pathogenicity islands carry one or more virulence-associated genes and are usually found only in the virulent strains. Examples of pathogenicity islands include *Salmonella* pathogenicity islands (SPI 1 and 2) which encodes the Type 3 secretion systems (Valdez *et al.*, 2009) and the high-pathogenicity island (HPI) of *Yersinia pestis* which is indispensable during murine infection (Schubert *et al.*, 2004).

There are 16 GIs found in the reference *B. pseudomallei* K96243 genome (Holden *et al.*, 2004) and a total of 71 GIs from an additional four sequenced strains of *B. pseudomallei* (1710b, 1106a, MSHR668 and MSHR305) (Tuanyok *et al.*, 2008). These regions are identified based on characteristic definitions of GIs such as atypical GC content, anomalies in dinucleotide frequency signature of the DNA in these regions and flanking by insertion elements. Some of these GIs are found in the same location in *B. pseudomallei* genomes, indicating the presence of recombination hotspots for horizontal gene transfer (HGT). This is supported by the atypical sequence features of the GIs – evidence of acquisition via HGT. Further evidence of HGT has been observed for GI8 in *B. pseudomallei* K96243 – it is found in place of the corresponding xylose metabolism cluster in the highly similar *B. thailandensis* genome (Yu *et al.*, 2006).

The GIs in *B. pseudomallei* can be largely grouped into four categories: (i) prophages, (ii) metabolism, (iii) pathogenicity and (iv) unknown functions (Tuanyok *et al.*, 2008). Many of the GIs appear to be prophages (e.g. GIs 2, 3 and 15 in *B. pseudomallei* K96243). GI2 is known to be a lysogenic phage ϕK96243, that can infect *B. mallei*. Of the 71 GIs identified across the five *B. pseudomallei* strains, 19 of them (26.7%) encode metabolic genes, which may be related to adaptation and survival in specific environmental niches. More than 20 of these GIs (32%) encode mainly hypothetical proteins and have unknown functions. A handful of GIs appear to be pathogenicity islands (PAIs) and could have functional roles especially in bacterial adherence. A group of these GIs carry genes that encode for filamentous haemagglutinin proteins (FHA). Homologues of *fha* genes have been reported in other Gram-negative bacteria (e.g. *Bordetella pertusis*), where they have been implicated in pathogenesis. In *B. pseudomallei*, *fha* genes are located on different GIs and varied among the strains. Experimental mutation of one of these genes, *fhaB* (*BPSS2053*) in *B. pseudomallei* DD503 resulted in the reduction of microbial adherence to human epithelial cells (Sim *et al.*, 2008).

In some rare *B. pseudomallei* clinical isolates from Australia, GIs are found to carry the genes *bpaA* and *bpaB*, encoding two-partner secretion systems (Brown *et al.*, 2004; Tuanyok *et al.*, 2008). These genes are similar to adhesin-encoding genes and are thought to play a role in virulence. However, no evidence for such a role was found in cell cultures and animal models of *B. pseudomallei* infection (Brown *et al.*, 2004).

Two of the GIs (GI7 and 14 from *B. pseudomallei* K96243) are 'stable' and consistently found across all other *B. pseudomallei* genomes studied (Nandi *et al.*, 2010; Sim *et al.*, 2008). Fragments of GI7 can also be found in *B. mallei* (Holden *et al.*, 2004). In fact, these fragments are the closest to a GI in *B. mallei*. In stark contrast to *B. pseudomallei*, there is no further evidence of other GIs in *B. mallei*.

It was previously hypothesized that *B. mallei* evolved from a single strain of *B. pseudomallei*, that upon infection of a host, the ancestral strain lost the genes that are not required for host survival and thus evolved into an obligate parasite (Godoy *et al.*, 2003; Nierman *et al.*, 2004). The highly similar genomes of *B. pseudomallei* and *B. mallei* support this hypothesis – both pathogens have two circular chromosomal replicons and almost all *B. mallei* genes have orthologues in *B. pseudomallei*. Given that *B. mallei* is adapted to a specialized niche, it was not surprising that the *B. mallei* genome was found to be smaller (5.8 Mb) than that of *B. pseudomallei* (7.2 Mb) and *B. thailandensis* (6.7 Mb) (Nierman *et al.*, 2004). The *B. mallei* genome contains numerous insertion sequence elements that have mediated extensive deletions and rearrangements of the genome relative to *B. pseudomallei*. A recent study looking into the mechanism of genome reduction reveals the loss of all *B. pseudomallei* genomic islands in parasitic *B. mallei* (Losada *et al.*, 2010). Once again, this is unlike the environmental counterpart *B. thailandensis*, which has 15 GIs (similar to *B. pseudomallei* K96243) (Yu *et al.*, 2006).

In animal models of infection, different *B. pseudomallei* strains are known to exhibit varied levels of virulence. *B. pseudomallei* strains are also considered to be more virulent than *B. mallei*, since they generally have a lower lethal does (LD50) in these infection models (Galyov *et al.*, 2010). Such variation may be directly related to the number of *B. pseudomallei* GIs and their contents, which are absent in *B. mallei*.

Bacterial cell communication and quorum-sensing systems

QS is a complex process utilized by many Gram-negative bacteria to regulate gene expression in a cell density-dependent manner. In this process, *N*-acyl-homoserine lactone (HSL) signal molecules are produced, which play a role in cell–cell communication, diffusing out of and into cells and binding to transcriptional regulators upon reaching a threshold concentration, controlling the expression of specific sets of genes (Duerkop *et al.*, 2007; Eberl, 2006). In fact, all members of the *Burkholderia* genus investigated thus far are known to have at least one QS system which utilizes HSL signalling.

B. pseudomallei and *B. mallei* are known to harbour multiple LuxI/LuxR homologues in their genomes, and produce numerous HSL signal molecules. The presence of multiple QS systems has also been observed in other *Burkholderia* species (see chapter 10). *B. mallei* has two QS pairs BmaR1-BmaI1 and BmaR3-BmaI3 and two orphan LuxR receptors BmaR4 and BmaR5 (Ulrich *et al.*, 2004a). In addition to having these homologues in *B. pseudomallei*, there is an additional LuxI/LuxR homologue in the *B. pseudomallei* genome (BpsI1-BpsR1, BpsI2-BpsR2, BpsII3-BpsR3, BpsR4 and BpsR5) (Kiratisin and Sanmee, 2008). It appears that BpsI2-BpsR2 has been deleted in *B. mallei* during evolution (Ong *et al.*, 2004; Ulrich *et al.*, 2004b). In both pathogens, QS is required for full virulence in various animal models (Ulrich *et al.*, 2004a; Valade *et al.*, 2004). *B. pseudomallei* produces at least seven quorum sensing molecules for signalling

– C8-HSL, 3-hydroxy-C8-HSL, 3-oxo-C8-HSL, C10-HSL, 3-hydroxy-C10-HSL, 3-oxo-C10-HSL, 3-hydroxy-C10-HSL and 3-oxo-C14-HSL (Chan *et al.*, 2007; Kiratisin and Sanmee, 2008; Lumjiaktase *et al.*, 2006; Song *et al.*, 2005; Ulrich *et al.*, 2004a,b; Valade *et al.*, 2004).

The complexity of the QS systems in *B. pseudomallei* is compounded by the different nomenclatures used in naming the system that has been characterized in different *B. pseudomallei* strains – PmlI-PmlR in *B. pseudomallei* 008 (Valade *et al.*, 2004) and *B. pseudomallei* DD503 (Ulrich *et al.*, 2004a), BpmIR in *B. pseudomallei* DD503 (Ulrich *et al.*, 2004a), BpsIR in *B. pseudomallei* KHW (Song *et al.*, 2005) and *B. pseudomallei* PP844 (Lumjiaktase *et al.*, 2006). We checked the nucleotide sequences of the respective QS systems in these *B. pseudomallei* strains and found that all of them to correspond to *BPSS0885* and *BPSS0887* in the reference *B. pseudomallei* K96243 genome. As such, we will use BpsI-BpsR to describe the systems in accordance to a recent publication on QS in *B. pseudomallei* K96243 (Kiratisin and Sanmee, 2008).

BpsI1-BpsR1 is currently the best characterized QS system and a mutant defective in this system exhibits attenuated virulence in a murine model of infection. However, the role of QS in *B. pseudomallei* virulence is more complex as the system also regulates many virulence-associated genes and mechanisms. Notably, the QS system negatively regulate MprA metalloprotease, a virulence determinant in a rat model of lung infection at stationary phase (Valade *et al.*, 2004). In a recent study using tiling microarrays to explore the global transcriptional profile of the QS mutant, we observed further regulation of genes involved in motility, biofilm formation, efflux pumps, antimicrobial compounds, secretion systems including Type 3 and Type 6 secretion systems, virulence factors (phospholipase C, haemolysin), sigma factors (RpoS, RpoH and RpoN) and other QS homologues as well (Ooi *et al.*, 2013). The microarray data support various phenotypic observations of the QS mutant, which is non-motile, exhibits atypical capsule and is less virulent than the parental wild type. The positive regulation of BpsI by RpoS is recently published (Wongtrakoongate *et al.*, 2012).

In *B. mallei*, the production of signalling molecules in BmaR1-BmaI1 and BmaR3-BmaI3 have been characterized (Duerkop *et al.*, 2008; Duerkop *et al.*, 2007). The systems produce C8-HSL, C10-HSL, 3-hydroxy-C8-HSL, and 3-hydroxy-C10-HSL for signalling. Mutants defective in these systems are unable to produce the signalling molecules and exhibit attenuated virulence in hamsters and mice (Duerkop *et al.*, 2008; Duerkop *et al.*, 2007; Ulrich *et al.*, 2004a; Ulrich *et al.*, 2004b). However, the regulation of other virulence determinants and mechanisms by the QS systems in *B. mallei* remains largely unknown and are still under investigation (Galyov *et al.*, 2010).

It is apparent that these complex quorum systems in both *B. pseudomallei* and *B. mallei* are required for bacterial cell-to-cell communication during daily growth, as well as for virulence. Perturbing these systems affects global gene expression and results in considerable pleiotrophy.

Host–pathogen interactions

The interplay and regulation of many of the virulence factors and mechanisms is critical when the pathogen encounters a host. Both *B. pseudomallei* and *B. mallei* are facultative intracellular pathogens during mammalian infections, and the adhesion to and invasion of host cells, evasion of host immune systems, and intracellular survival and bacterial replication are mediated by a series of virulence factors and mechanisms. Their roles in pathogenesis are even more obvious considering their absence in the avirulent *B. thailandensis*. A detailed review on their infection biology has been published by Galyov *et al.* (2010) and key findings from this work are summarized below.

Adhesion, invasion and replication

Capsule (CPSI)

Both *B. pseudomallei* and *B. mallei* are encapsulated, rod-shaped bacteria. Both pathogens produce an extracellular polysaccharide capsule comprising a homopolymer of –3)–2-*O*-acetyl-6-deoxy-β-D-*manno*-heptopyranose-(1- (Isshiki *et al.*, 2001). This capsule CPSI has been shown to be a major virulence determinant for both pathogens. In various animal models of infection, including mice, hamsters and horses, *B. mallei* capsule mutants are avirulent (DeShazer *et al.*, 2001).

Similarly, *B. pseudomallei* capsule mutants exhibit high levels of attenuation in mice and hamsters (Atkins *et al.*, 2002; Cuccui *et al.*, 2007; Wikraiphat *et al.*, 2009). The *B. pseudomallei* capsule has been shown to be a critical virulence determinant in respiratory tract infections and an important antigen for generating Th1 immune responses commonly observed in systemic melioidosis (Warawa *et al.*, 2009). Additionally, the capsule is required for optimal survival and replication in macrophages, resisting phagocytosis and contributes to survival in serum by reducing complement factor C3b deposition on the surface of the bacterium (Reckseidler-Zenteno *et al.*, 2005; Wikraiphat *et al.*, 2009). Interestingly, one of the major genes belonging to CPSI gene cluster is down-regulated in a *B. pseudomallei* quorum sensing mutant (Ooi *et al.*, 2013). However, it remains unknown whether such a down-regulation can also be observed in *B. mallei* quorum sensing mutants.

The capsule was generally considered to be absent in *B. thailandensis* until recent genome analyses revealed that variant *B. thailandensis* isolate E555 had acquired a *B. pseudomallei*-like capsule. Like *B. pseudomallei*, the capsular *B. thailandensis* E555 exhibited similar colony morphology, resistance to human complement binding, and intracellular survival in macrophages. However, it did not exhibit enhanced virulence in murine infection assays. This implied that the capsule is important for virulence but other factors are required for successful mammalian infection (Sim *et al.*, 2010). Additionally, the presence of the capsule in avirulent *B. thailandensis* E555 is promising for vaccine development.

LPS O-antigen and other polysaccharide gene clusters

Several surface lipopolysaccharide gene clusters (e.g. capsular polysaccharide, type-II OPS) are also found in *B. pseudomallei* and *B. mallei* and the resulting LPS can function to protect the bacterium against host immune defences.

The LPS O-antigen has been well studied in both *B. pseudomallei* and *B. mallei*. Structurally, it is similar in both pathogens – an unbranched heteropolymer –3)-β-D-glucopyranose-(1–3)–6-deoxy-α-L-talopyranose-(1- (Perry *et al.*, 1995). In both pathogens, the presence of O-acetyl or O-methyl substitutions at the 2 position of the talose residue can be found in the O-antigen. However, the *B. mallei* O-antigen lacks an O-acetyl group at the 4 position of the talose residue. The O-antigen is critical for the pathogens to survive in host cells, conferring resistance against macrophage killing and preventing bacteriocidal action of serum by the alternative pathway of complement (Arjcharoen *et al.*, 2007; DeShazer *et al.*, 1998; Wikraiphat *et al.*, 2009).

In addition to the capsule and LPS O-antigen, there are two clusters of genes with putative functions related to polysaccharide biosynthesis and transport in *B. pseudomallei* (*BPSS0427* – *BPSS0429* and *BPSS1825* to *BPSS1832*) (Holden *et al.*, 2004). These putative polysaccharides type III O-PS and type IV O-PS, respectively, are important for virulence as mutants defective in these polysaccharides exhibit attenuation in mice challenged via the intraperitoneal route. Interestingly, immunization of mice with these killed polysaccharide mutant strains conferred different degrees of protection, demonstrating the immunological importance of the *B. pseudomallei* polysaccharide clusters, and their potential use as vaccine candidates (Sarkar-Tyson *et al.*, 2007). In contrast, these gene clusters are absent in *B. mallei* (Nierman *et al.*, 2004).

Type IV pili

B. pseudomallei has several type IV pili (TFP) gene and fimbrial clusters which might be involved in adherence and recognition of host cells. Experimental disruption of *pilA*, a TFP gene, has been shown to cause reduced *B. pseudomallei* adherence to human epithelial cells and decreased virulence in murine models of melioidosis (Essex-Lopresti *et al.*, 2005). Based on sequence similarity, a putative TFP gene was found in the *B. mallei* genome. Whilst the *B. pseudomallei pilA* mutant exhibits slight attenuation when administered through the intranasal challenge route (but not the intraperitoneal challenge route), the purified PilA of *B. mallei* was highly immunogenic in mice but failed to provide protection against a lethal *B. mallei* aerosol challenge (Essex-Lopresti *et al.*, 2005; Fernandes *et al.*, 2007).

Adhesins

Both *B. pseudomallei* and *B. mallei* also contain several genes with conserved domains or regions associated with haemagglutinins and surface proteins in their genomes. One such group is the Hep-Hag repeat family proteins (Pfam domain PF05658) that are associated with haemagglutinins and invasins (Bateman *et al.*, 2002). Gene homologues in *Streptococcus* spp. appear to be involved in both adhesion and virulence. Factors involved in adhesion are also likely to play a role in persistence in the environment. For example, in a study looking at the expression of different adhesins in Shiga toxin-producing *E. coli* O157 strains and non-O157 environmental strains, biofilm formation and expression of curli were observed in both pathogenic and non-pathogenic strains. In the latter, the presence of curli and ability to form biofilm is further associated with other adhesins like type 1 fimbriae and autotransporter proteins that contribute to the persistence of these organisms in the environment (Biscola *et al.*, 2011). The trimeric autotransporter adhesin gene, *bimA* (*BMAA0749*; *BPSS1492*), is one the best characterized genes involved in adhesion in both *B. pseudomallei* and *B. mallei*. This gene is involved in intracellular actin-based motility during host cell infection (Stevens *et al.*, 2005). Notably, the role of BimA in the pathogenesis and virulence of *B. pseudomallei* and *B. mallei* appears to differ; BimA in *B. pseudomallei* contributes to pathogenesis in murine models of melioidosis and confers protection against homologous challenge (Lazar Adler *et al.*, 2012) but BimA of *B. mallei* is dispensable for virulence in a Syrian hamster model of glanders (Schell *et al.*, 2007).

Secretion systems in pathogens

Many bacterial pathogens employ multi-component protein complexes to deliver macromolecules directly into the eukaryotic host cell to promote infection. Many of these secretion systems have been discovered and characterized in *B. pseudomallei* and *B. mallei*. We proceed to describe notable examples of these implicated in virulence in *B. pseudomallei* and *B. mallei* below.

Type three secretion systems (T3SS)

Arguably the best characterized secretion system in both *B. pseudomallei* and *B. mallei*, the T3SS is a needle complex assembled by many Gram-negative pathogens to deliver bacterial proteins directly into host cell cytosol to facilitate infection. *B. pseudomallei* carries three distinct T3SSs which can operate to facilitate pathogen invasion, intracellular survival, and intercellular spread. In a hamster infection model, it has been shown that T3SS3, which is homologous to the *Salmonella* SPI-1 mammalian pathogenicity island, is required for full *B. pseudomallei* virulence (Warawa and Woods, 2005). Many of the genes of the *B. pseudomallei* T3SS3 have been well characterized, with the recent addition of BimA, BopC (Muangman *et al.*, 2012), and BipD, of which the crystal structure has been recently elucidated (Pal *et al.*, 2012). *B. mallei* has a T3SS that is similar to T3SS3 of *B. pseudomallei* and mutants defective in this T3SS are avirulent in mice and hamsters (Ulrich and DeShazer, 2004). The trimeric autotransporter adhesion BimA is associated with this T3SS in both pathogens.

The other two T3SS in *B. pseudomallei* (T3SS1 and T3SS2) are homologous to the T3SS of the plant pathogen *Ralstonia solanacearum*. *B. pseudomallei* T3SS1 and T3SS2 have been demonstrated to be required for plant infection in a tomato plant model (Lee *et al.*, 2010b). Recently, T3SS2 has been shown to be regulated by HrpB, which also regulates the production of other virulence factors like TFP (Lipscomb and Schell, 2011).

Despite the association with plant infection, recent reports seem to suggest that *B. pseudomallei* T3SS1 is involved in mammalian infection. A mutant lacking the *bpscN* gene from T3SS1 was significantly attenuated for virulence in a respiratory melioidosis mouse model of infection. The mutant also exhibited diminished survival and replication in RAW264.7 cells and an increased level of co-localization with the autophagy marker protein LC3 but an unhindered ability to escape from phagosomes (D'Cruze *et al.*, 2011). Yet another recent report revealed a protein with papain-like hydrolytic activity, capable of blocking host cell cycle progression, in close proximity and likely to be associated with *B. pseudomallei* T3SS1 (Yao *et al.*, 2009). This protein, named CHBP (*BPSS1385*), is a homologue of the type 3 effector Cif which is also found in pathogens like *Yersinia* and enteropathogenic *Escherichia coli*. CHBP has been further characterized and shown to be a potent inhibitor of the eukaryotic ubiquitination pathway. CHBP acted as a deamidase that specifically and efficiently deamidated Gln40 in ubiquitin and ubiquitin-like protein NEDD8 both *in vitro* and during *Burkholderia* infection (Cui *et al.*, 2010). Given the universal role of ubiquitination in many important cellular processes, this discovery suggests a possible pleiotropic function of the CHBP family of effectors in bacterial pathogenesis.

Type six secretion systems (T6SS)

B. pseudomallei harbours six T6SS. The *B. mallei* T6SS gene cluster 1 (*BMAA074 –BMAA0730*) has been characterized and shown to be important for actin-based motility, multinucleated giant cell (MNGC) formation, intracellular growth in murine macrophages, and virulence in hamsters (Burtnick *et al.*, 2010; Schell *et al.*, 2007). One of the deletion mutants defective in *hcp* (Δ*hcp1*) was attenuated in a hamster model of infection and exhibited a growth defect in murine macrophages (Burtnick *et al.*, 2010). The homologous *B. pseudomallei* T6SS, T6SS-1 (*BPSS1496 – BPSS1511*) has also been shown to be induced intracellularly in murine macrophages (Shalom *et al.*, 2007).

The regulation of T3SS3 and T6SS-1 in *B. pseudomallei* has since been elucidated (Fig. 6.1). A TetR family regulator, BspR acts through BprP, a novel transmembrane regulator which is located adjacent to the T3SS3 gene cluster. BprP in turn regulates the expression of structural and secretion components of T3SS3 and the AraC family regulator *bsaN*. BsaN and BicA are likely to form a complex to regulate the expression of T3SS3 effectors and other regulators which in turn affect the expression of T6SS (Sun *et al.*, 2010; Sun and Gan, 2010). The elucidation of this regulatory cascade provides a framework for further understanding of pathogenesis in *B. pseudomallei*.

Type two secretion systems (T2SS)

The T2SS is conserved in Gram-negative bacteria and usually involves a set of 12 to 16 different proteins. It is unique in its ability to promote secretion of large and sometimes multimeric proteins that are folded in the periplasm (Douzi *et al.*, 2012; Korotkov *et al.*, 2012). The system is present in *B. pseudomallei*, and is required for the secretion of several exoproducts, including protease, lipase and phospholipase C, a virulence determinant (DeShazer *et al.*, 1999). These proteins are poorly secreted in *B. mallei*, a result likely due to mutations in two T2SS genes (*gspJ* and *gspL*) in *B. mallei* (Nierman et al., 2004).

Formation of multinucleated giant cells (MNGCs) and autophagy

It has been observed that *B. pseudomallei* induces host cell fusion during *in vitro* cell culture infection, leading to formation of MNGCs (Kespichayawattana *et al.*, 2000). This novel phenotype appears to be unique to intracellular bacterial pathogens and may be relevant to *B. pseudomallei* pathogenesis, as granuloma formation and generation of MNGCs are observed in clinical cases of melioidosis (Wong *et al.*, 1995). Further research has attributed this phenotype to be mediated by a T3SS3 protein, BipD. Genetic inactivation of *bipB* results in reduced MNGC formation, cell-to-cell spreading of bacteria, and induction of apoptosis of mouse macrophages. The *bipB* mutant was also significantly attenuated in a murine model via the intranasal route of infection (Suparak *et al.*, 2005).

New kids on the block

With the availability of many complete *B. pseudomallei*, *B. mallei* and *B. thailandensis* genome sequences, researchers have been able to conduct more targeted research to characterize the structures and functions of many more genes in these pathogens. In the last few years, coupled with increased funding of research on biowarfare agents, there have been several breakthrough findings on some of these newly identified genes and

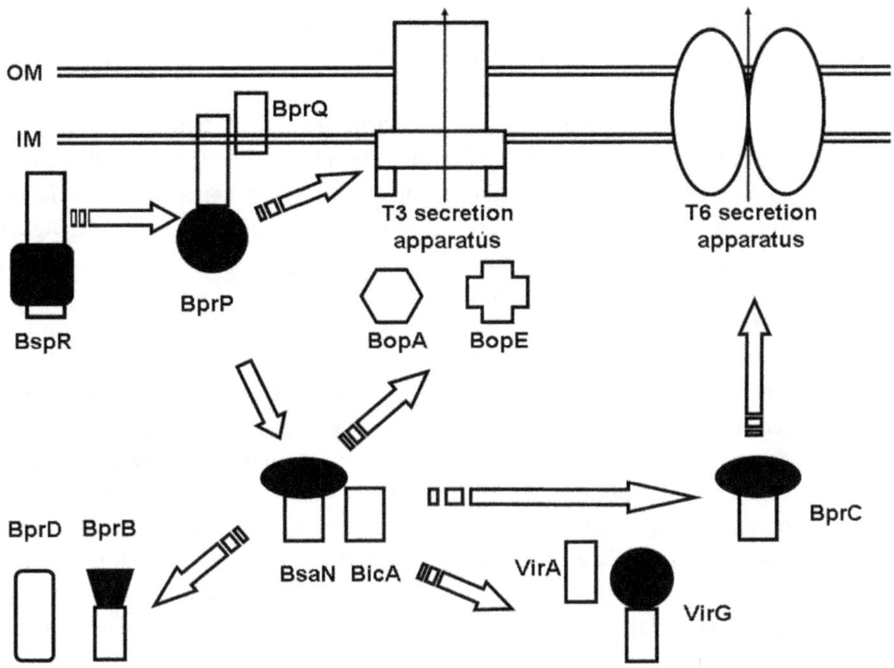

Figure 6.1 Model of BspR-dependent regulation of T3SS3 and T6SS-1 in *Burkholderia pseudomallei*. BspR activates the expression of BprP which in turn activates the expression of T3SS3 genes. BsaN/BicA activate the expression of effector genes, virAG and bprC. BprC is required for expression of T6SS5 genes such as tssA. Broad arrows indicate the direction of transcriptional activation. Thin arrows indicate the direction of secretion. Inner membrane (IM) and outer membrane (OM) of bacteria are indicated. The DNA-binding domain of a transcription regulator is denoted by a filled shape. Broad arrows with broken sections denote uncertainty on whether regulation is direct. (Reproduced from Figure 7, Sun *et al.*, 2010, with permission from John Wiley and Sons.)

their roles in virulence mechanisms. We describe a few of them here.

Toxin

For many years, despite the characterization of various secretion systems in *B. pseudomallei* and *B. mallei*, the hunt for a specific toxin remained elusive. The first *B. pseudomallei* toxin, *Burkholderia* Lethal Factor (BLF1; *BPSL1549*) was recently identified and characterized from *B. pseudomallei* (Cruz-Migoni *et al.*, 2011; Hautbergue and Wilson, 2012). Structurally, *BPSL1549*, previously an unknown protein, was found to be similar to *E. coli* cytotoxic necrotizing factor 1 (CNF1). The protein also shares conserved key catalytic residues with CNF1, suggesting that it is a glutamine deaminase. *BPSL1549* was found to promote deamidation of glutamine-339 of the translation initiation factor eIF4A in the eukaryotic translation machinery, abolishing its helicase activity and inhibiting translation. Furthermore, *BPSL1549* was toxic to J774 macrophage cells and was lethal upon administration to BALB/c mice via intraperitoneal injection. Using a customized *B. pseudomallei* tiling microarray interrogating more than 80 conditions, the expression levels of *BPSL1549* was found to correlate with conditions mimicking the conditions encountered during mammalian infection (e.g. high expression during exposure to 30% human serum, the animal-specific amino acid taurine and physiological insulin concentrations). However, it remains unknown how the protein is secreted into the cytoplasm of host cells. Notably, similar nucleotide sequences are also found in *B. mallei* isolates but not in *B. thailandensis*.

Novel metabolites

A good example of the reference genome being put to good use for the elucidation of new virulence mechanisms was the discovery of cryptic metabolites in *B. pseudomallei*. From the reference *B. pseudomallei* K96243 genome, 13 potential biosynthetic operons were identified based on the putative functions of their genes, including enzymes involved in secondary metabolism like polyketide synthases, non-ribosomal peptide synthases, acyl-CoA ligases, functional-group transferases and redox-active proteins (Biggins *et al.*, 2011). Using *Pseudomonas aeruginosa* as a host for heterologous expression studies, the authors identified seven new metabolites, including a compound produced from the non-ribosomal peptide synthase operon *BPSS0130 – BPSS0133*. Promisingly, this compound inhibited eukaryotic type-4 phosphodiesterases, suggesting that it is utilized to disrupt host signalling pathways during infection, leading to impaired host defence against the pathogenic bacteria. However, the homologous operon is also found in *B. thailandensis* (*BTH-II0204–BTH-II0207*) and even in fungi (40% match), implying that the compound is probably required for survival in the environment in the first place (Biggins *et al.*, 2011).

In their latest publication, the same authors identified yet another metabolite that may play a role in *Burkholderia* virulence. This metabolite, malleilactone, is a polyketide synthase-derived cytotoxic siderophore encoded by the MAL gene cluster. Mutants defective in malleilactone production were unable to produce infection symptoms caused by the wild-type in both the nematode *Caenorhabditis elegans* and the amoeba *Dictyostelium discoideum*. When assayed against a panel of cell types, malleilactone shows low micromolar cytotoxicity to human cell lines. Interestingly, malleilactone efficiently inhibited the growth of Gram-positive bacteria in disk diffusion assays (Biggins *et al.*, 2012). This is in contrast to the dispensable iron-scavenging systems reported by Kvitko *et al.* (2010). Given that this gene is conserved across *B. mallei* (*BMAA1446*), *B. pseudomallei* (*BPSS0311*) and *B. thailandensis* (*BTH-II2088*), the results seem to suggest that malleilactone is a niche factor that provides the environmental *B. pseudomallei* and *B. thailandensis* a competitive edge to survive in the environment. As this competitive advantage is also useful during host infection, it is likely to be retained and thus evolutionarily conserved even in the host-restricted *B. mallei*.

Signalling molecules

In addition to the HSL signalling molecules produced by QS systems in *B. pseudomallei*, recent research has discovered other signalling molecules that play critical roles in the pathogenesis of *B. pseudomallei*.

Cyclic nucleotides are second messenger molecules and can be found in all kingdoms of life. Bacterial genome sequencing has revealed the abundance of proteins with GGDEF/EAL domains. Such domains are involved in the turnover of cyclic diguanylic acid (c-di-GMP), which is an intracellular signalling molecule involved in regulation of cellular functions such as motility, biofilm formation and virulence in various bacteria like *Escherichia coli*, *Pseudomonas aeruginosa* and pathogens like *Salmonella enteric* serovar *Typhimurium* (Kulasakara *et al.*, 2006; Simm *et al.*, 2004). The reference genome of *B. pseudomallei* K96243 reveals several putative proteins with GGDEF-EAL domains (Holden *et al.*, 2004). In their paper, Lee at al. (2010) identified and characterized CdpA (*BPSL1263*), a protein with an EAL domain. The authors found that CdpA is a major c-di-GMP-specific phosphodiesterase in *B. pseudomallei*. Their *cdpA*-null mutant was found to exhibit an 8-fold higher c-di-GMP level, as compared to the wild type, which was further associated with increased production of exopolysaccharides, increased cell–cell aggregation, a non-motile phenotype, and increase biofilm production. The mutant also exhibited attenuated virulence during *in vitro* infections, showing a 3-fold reduction in invasion of human lung epithelial cells and a 6-fold reduction in cytotoxicity on human macrophage cells. Furthermore, the authors also found that CdpA positively regulates components of the *B. pseudomallei* T3SS (*bsaN* and *bipB*) and the expression of *fliC*, another virulence determinant in *B. pseudomallei*. The results suggest that the protein plays an important role as a regulator of virulence in response to environmental signals (Lee *et al.*, 2010a).

Additionally, small signalling molecules involved in global stress responses appears to be implicated in the regulation of *B. pseudomallei* pathogenesis mechanisms as well. These are small modified nucleotides known as hyperphosphorylated guanosine molecules. Guanosine tetra- and pentaphosphates (ppGpp and pppGpp) are produced in bacteria in response to various environmental stresses, including amino acid starvation. It has been well documented that the absence of these molecules in various pathogens like *Franciscella tularensis, Campylobacter jejuni, Helicobacter pylori* and *Brucella* spp. results in attenuated virulence in animal models of disease (Muller *et al.*, 2012). Genes identified to play key roles in the production of ppGpp in bacterial systems included RelA and SpoT. In *B. pseudomallei*, homologues of these two genes have been identified in the genome of *B. pseudomallei* K96243 (*BPSL1946* and *BPSL2561*, respectively) (Muller *et al.*, 2012). Both proteins are also conserved in *B. mallei* and *B. thailandensis*, though their roles in these bacteria remain unclear. Characterization of a double mutant (Δ*relA*Δ*spoT*) in *B. pseudomallei* revealed a defect in stationary-phase survival and intracellular replication in murine macrophages. In addition, the mutant exhibits attenuated virulence in an insect model of melioidosis (*Galleria mellonella*), as well as in both acute and chronic murine models of melioidosis. Notably, these targets appear to be potential vaccine candidates as vaccination of mice with the double mutant resulted in partial protection against infection with wild-type *B. pseudomallei*.

Iron uptake systems

Interestingly, with the availability of the *B. pseudomallei* K96243 reference genome, it has also been found that virulence factors previously assumed to be critical appear to be dispensable, as the highly redundant genome carries backup copies that are able to complement any loss of function. One such mechanism is the iron-uptake system in *B. pseudomallei*.

The ability to acquire iron efficiently from the environment appears to be a critical factor for survival for bacteria since iron is essential to life yet is only available in limited quantities. In *B. pseudomallei*, it has been established that the pathogen possesses a number of efficient iron uptake systems to allow for survival under iron-limiting conditions, e.g. in a host. The reference genome of *B. pseudomallei* reveals genes coding for pyochelin, membrane-bound iron transporters and a haem uptake locus (*hmu*) (Holden *et al.*, 2009; Ong *et al.*, 2004; Tuanyok *et al.*, 2005). *B. pseudomallei* also produces malleobactin, a hydroxamate-type siderophore that removes iron from lactoferrin and transferrin (Alice *et al.*, 2006; Yang *et al.*, 1991, 1993). Whilst siderophore and other iron uptake systems have been considered as virulence factors in other bacteria like *P. aeruginosa* (pyoverdine), *Acinetobacter baumannii* (acinetobactin) (Peek *et al.*, 2012; Penwell *et al.*, 2012), and *B. cenocepacia* ornibactin, the role of these iron uptake systems in *B. pseudomallei* virulence remains unclear.

A recently published paper claims that the known siderophores and haemin uptake system in *B. pseudomallei* are dispensable for full virulence in a murine model of infection (Kvitko *et al.*, 2012). The authors provided the following evidence to corroborate their findings: (i) a clinical isolate naturally lacking the entire malleobactin gene cluster in its genome is fully virulent in a murine melioidosis model (Trunck *et al.*, 2009); (ii) deletion of the major siderophore malleobactin and pyochelin gene clusters revealed a residual siderophore activity. Interestingly, this residual siderophore activity appears to be unrelated to other known *Burkholderia* siderophores such as cepabactin and cepaciachelin. Finally, a quadruple mutant defective in production of malleobactin, pyochelin and haemin acquisition does not exhibit significant reduction in virulence in an acute mouse model. It appears that the pathogen is able to utilize ferritin as an iron source in the host environment and that a novel ferritin-iron acquisition pathway exists in *B. pseudomallei* to enable survival during infection. The association of this novel pathway to the metabolite malleilactone remains to be investigated.

Antibiotic resistance

Both *B. pseudomallei* and *B. mallei* are intrinsically resistant to commonly used antibiotics including aminoglycosides, polymyxins and beta-lactams. The *B. mallei* genome harbours more than 30 genes that are likely to play a role in drug resistance (Nierman *et al.*, 2004). The larger *B. pseudomallei* genome contains at least 10 operons coding for resistance-nodulation-cell-division (RND) efflux pump components, including AmrAB-OprA, a multidrug efflux pump conferring resistance to macrolide and aminoglycoside antibiotics (Moore *et al.*, 1999). This pump is absent in *B. mallei*, accounting for the sensitivity of the pathogen to these classes of antibiotics.

RND efflux pumps are recognized as major causes of multi-drug resistance in many Gram-negative bacteria and some of them have been characterized in *B. pseudomallei*. These include drug efflux pumps such as MexAB-OprM, MexXY-OprM, MexCD-OprJ, MexEF-OprN and AmrAB-OprM. The expression of several *B. pseudomallei* RND pumps can also be induced by exposure to toxic compounds (Kumar *et al.*, 2008). The efflux pumps are highly homologous and investigation of homologous efflux pumps in *B. thailandensis* reveals that some of them are able to play compensatory roles when other efflux pumps are inactivated, a form of adaptive resistance mechanism (Biot *et al.*, 2011). Whilst the AmrAB-OprA efflux pump appears to be required for full virulence in a murine model of pneumonic melioidosis (Jeddeloh *et al.*, 2003), there has been conflicting reports on the BpeAB-OprB efflux pump, possibly dependent on strain differences; it is associated with virulence in *B. pseudomallei* strain KHW but not in strain 1026b (Chan and Chua, 2005; Mima and Schweizer, 2010). In any case, the high levels of intrinsic resistance exhibited by this pathogen as a result of the armoury of efflux pumps it harbours, enhances its persistence and survival in unfavourable conditions, including during drug treatment in hosts, indirectly contributing to its pathogenesis.

The recent emergence of ceftazidime-resistant *B. pseudomallei* strains is of great concern since this antibiotic is currently the first-line of therapy for melioidosis in Asia. Several mechanisms appear to contribute to this resistance to ceftazadime. Tribuddharat *et al.* (2003) reported that acquired resistance in clinical *B. pseudomallei* strains to ceftazidime and amoxicillin-clavulanate was associated with single mutations in *penA* gene encoding a class A β-lactamase (Tribuddharat *et al.*, 2003). These mutations could be mapped to position 167, which is located within a conserved motif of the catalytic site of the *penA* gene (Pro to Ser mutation) (Ho *et al.*, 2002; Tribuddharat *et al.*, 2003) or to residue 69, leading to a Cys > Tyr substitution, which is critical for determining substrate specificity of the enzyme (Sam *et al.*, 2009). It appears that as a result of positive selection, the pathogen is able to evolve within a host undergoing prolonged drug treatment, acquiring this resistance. In *B. pseudomallei* clinical strains isolated in Thailand, the deletion of penicillin-binding protein 3 (*BPSS1219*) was sufficient to result in new resistance against ceftazidime (Chantratita *et al.*, 2011). More recently, the molecular mechanisms for ceftazidime resistance in *B. pseudomallei* strains from two Australian melioidosis patients who temporally developed resistant *B. pseudomallei* strains during ceftazidime therapy was characterized. The selective pressure during treatment resulted in two single-nucleotide polymorphisms that directly increased ceftazidime hydrolysis. In addition to the mutation observed previously (Cys69Tyr), the authors identified a novel mutation within the *penA* promoter region. Similar to previous studies, the amino acid substitution altered substrate specificity to other β-lactams, suggesting a potential fitness cost associated with this mutation (Sarovich *et al.*, 2012).

Future directions and conclusions

With increased interest and research funding into biowarfare agents, the list of virulence determinants characterized in *B. pseudomallei* and *B. mallei* continues to grow. However, there is still much to be discovered, especially for *B. mallei*, which is less studied than *B. pseudomallei*. These multipartite pathogens, with large genomes, are good models to unravel new and redundant pathways involved in environmental adaptation and virulence. The increasingly ease of sequencing techniques and accessibility to sequencing platforms will result in the availability of more genomes, a highly useful resource for researchers.

As both *B. pseudomallei* and *B. mallei* are classified as bio-threat agents, laboratory procedures involving life cultures will require highly regulated biosafety level 3 containment, making them difficult to work with. The lack of vaccine for both pathogens continues to push researchers on to search for potential candidates, and the race between man and microbes in novel drug treatment continues. Increasingly, the development of various platforms, including better tools for genetic manipulation, tiling microarrays customized for global *B. pseudomallei* gene expression studies, establishment of proteome and metabolome pipelines, and even optimized crystallography research on *B. pseudomallei*, adds on to the list of established work tools (*in vitro* cultures, animal models, etc.) already available to better dissect the pathogenesis of these highly infectious microbes. With better understanding of their biology, adaptation to niches and virulence mechanisms, the light might just be in sight soon.

References

Alice, A.F., Lopez, C.S., Lowe, C.A., Ledesma, M.A., and Crosa, J.H. (2006). Genetic and transcriptional analysis of the siderophore malleobactin biosynthesis and transport genes in the human pathogen *Burkholderia pseudomallei* K96243. J. Bacteriol. *188*, 1551–1566.

Arjcharoen, S., Wikraiphat, C., Pudla, M., Limposuwan, K., Woods, D.E., Sirisinha, S., and Utaisincharoen, P. (2007). Fate of a *Burkholderia pseudomallei* lipopolysaccharide mutant in the mouse macrophage cell line RAW 264.7: possible role for the O-antigenic polysaccharide moiety of lipopolysaccharide in internalization and intracellular survival. Infect. Immun. *75*, 4298–4304.

Atkins, T., Prior, R., Mack, K., Russell, P., Nelson, M., Prior, J., Ellis, J., Oyston, P.C., Dougan, G., and Titball, R.W. (2002). Characterisation of an acapsular mutant of *Burkholderia pseudomallei* identified by signature tagged mutagenesis. J. Med. Microbiol. *51*, 539–547.

Bateman, A., Birney, E., Cerruti, L., Durbin, R., Etwiller, L., Eddy, S.R., Griffiths-Jones, S., Howe, K.L., Marshall, M., and Sonnhammer, E.L. (2002). The Pfam protein families database. Nucleic Acids Res. *30*, 276–280.

Begley, M., Hill, C., and Gahan, C.G. (2006). Bile salt hydrolase activity in probiotics. Appl. Environ. Microbiol. *72*, 1729–1738.

Biggins, J.B., Liu, X., Feng, Z., and Brady, S.F. (2011). Metabolites from the induced expression of cryptic single operons found in the genome of *Burkholderia pseudomallei*. J. Am. Chem. Soc. *133*, 1638–1641.

Biggins, J.B., Ternei, M.A., and Brady, S.F. (2012). Malleilactone, a Polyketide Synthase-Derived Virulence Factor Encoded by the Cryptic Secondary Metabolome of *Burkholderia pseudomallei* Group Pathogens. J. Am. Chem. Soc. *134*, 13192–13195.

Biot, F.V., Valade, E., Garnotel, E., Chevalier, J., Villard, C., Thibault, F.M., Vidal, D.R., and Pages, J.M. (2011). Involvement of the efflux pumps in chloramphenicol selected strains of *Burkholderia thailandensis*: proteomic and mechanistic evidence. PLoS One *6*, e16892.

Biscola, F.T., Abe, C.M., and Guth, B.E. (2011). Determination of adhesin gene sequences in, and biofilm formation by, O157 and non-O157 Shiga toxin-producing Escherichia coli strains isolated from different sources. Appl. Environ. Microbiol. *77*, 2201–2208.

Brown, N.F., Logue, C.A., Boddey, J.A., Scott, R., Hirst, R.G., and Beacham, I.R. (2004). Identification of a novel two-partner secretion system from *Burkholderia pseudomallei*. Mol. Genet. Genomics *272*, 204–215.

Burtnick, M.N., DeShazer, D., Nair, V., Gherardini, F.C., and Brett, P.J. (2010). *Burkholderia mallei* cluster 1 type VI secretion mutants exhibit growth and actin polymerization defects in RAW 264.7 murine macrophages. Infect. Immun. *78*, 88–99.

Chan, Y.Y., and Chua, K.L. (2005). The *Burkholderia pseudomallei* BpeAB-OprB efflux pump: expression and impact on quorum sensing and virulence. J. Bacteriol. *187*, 4707–4719.

Chan, Y.Y., Bian, H.S., Tan, T.M., Mattmann, M.E., Geske, G.D., Igarashi, J., Hatano, T., Suga, H., Blackwell, H.E., and Chua, K.L. (2007). Control of quorum sensing by a *Burkholderia pseudomallei* multidrug efflux pump. J. Bacteriol. *189*, 4320–4324.

Chandler, J.R., Duerkop, B.A., Hinz, A., West, T.E., Herman, J.P., Churchill, M.E., Skerrett, S.J., and Greenberg, E.P. (2009). Mutational analysis of *Burkholderia thailandensis* quorum sensing and self-aggregation. J. Bacteriol. *191*, 5901–5909.

Chantratita, N., Wuthiekanun, V., Boonbumrung, K., Tiyawisutsri, R., Vesaratchavest, M., Limmathurotsakul, D., Chierakul, W., Wongratanacheewin, S., Pukritiyakamee, S., White, N.J., *et al.* (2007). Biological relevance of colony morphology and phenotypic switching by *Burkholderia pseudomallei*. J. Bacteriol. *189*, 807–817.

Chantratita, N., Rholl, D.A., Sim, B., Wuthiekanun, V., Limmathurotsakul, D., Amornchai, P., Thanwisai, A., Chua, H.H., Ooi, W.F., Holden, M.T., *et al.* (2011). Antimicrobial resistance to ceftazidime involving loss of penicillin-binding protein 3 in *Burkholderia pseudomallei*. Proc. Natl. Acad. Sci. U.S.A. *108*, 17165–17170.

Cheng, A.C. (2010). Melioidosis: advances in diagnosis and treatment. Curr. Opin. Infect. Dis. *23*, 554–559.

Cheng, A.C., and Currie, B.J. (2005). Melioidosis: epidemiology, pathophysiology, and management. Clin. Microbiol. Rev. *18*, 383–416.

Chua, K.L., Chan, Y.Y., and Gan, Y.H. (2003). Flagella are virulence determinants of *Burkholderia pseudomallei*. Infect. Immun. *71*, 1622–1629.

Chuaygud, T., Tungpradabkul, S., Sirisinha, S., Chua, K.L., and Utaisincharoen, P. (2008). A role of *Burkholderia pseudomallei* flagella as a virulent factor. Trans. R. Soc. Trop. Med. Hyg. *102 (Suppl 1)*, S140–144.

Cruz-Migoni, A., Hautbergue, G.M., Artymiuk, P.J., Baker, P.J., Bokori-Brown, M., Chang, C.T., Dickman, M.J., Essex-Lopresti, A., Harding, S.V., Mahadi, N.M., *et al.* (2011). A *Burkholderia pseudomallei* toxin inhibits helicase activity of translation factor eIF4A. Science *334*, 821–824.

Cuccui, J., Easton, A., Chu, K.K., Bancroft, G.J., Oyston, P.C., Titball, R.W., and Wren, B.W. (2007). Development of signature-tagged mutagenesis in *Burkholderia pseudomallei* to identify genes important in survival and pathogenesis. Infect. Immun. 75, 1186–1195.

Cui, J., Yao, Q., Li, S., Ding, X., Lu, Q., Mao, H., Liu, L., Zheng, N., Chen, S., and Shao, F. (2010). Glutamine deamidation and dysfunction of ubiquitin/NEDD8 induced by a bacterial effector family. Science *329*, 1215–1218.

D'Cruze, T., Gong, L., Treerat, P., Ramm, G., Boyce, J.D., Prescott, M., Adler, B., and Devenish, R.J. (2011). Role for the *Burkholderia pseudomallei* type three secretion system cluster 1 bpscN gene in virulence. Infect. Immun. *79*, 3659–3664.

DeShazer, D., Brett, P.J., Carlyon, R., and Woods, D.E. (1997). Mutagenesis of *Burkholderia pseudomallei* with Tn5-OT182: isolation of motility mutants and molecular characterization of the flagellin structural gene. J. Bacteriol. *179*, 2116–2125.

DeShazer, D., Brett, P.J., and Woods, D.E. (1998). The type II O-antigenic polysaccharide moiety of *Burkholderia pseudomallei* lipopolysaccharide is required for serum resistance and virulence. Mol. Microbiol. *30*, 1081–1100.

DeShazer, D., Brett, P.J., Burtnick, M.N., and Woods, D.E. (1999). Molecular characterization of genetic loci required for secretion of exoproducts in *Burkholderia pseudomallei*. J. Bacteriol. *181*, 4661–4664.

DeShazer, D., Waag, D.M., Fritz, D.L., and Woods, D.E. (2001). Identification of a *Burkholderia mallei* polysaccharide gene cluster by subtractive hybridization and demonstration that the encoded capsule is an essential virulence determinant. Microb. Pathog. *30*, 253–269.

Dobrindt, U., Hochhut, B., Hentschel, U., and Hacker, J. (2004). Genomic islands in pathogenic and environmental microorganisms. Nat. Rev. Microbiol. *2*, 414–424.

Douzi, B., Filloux, A., and Voulhoux, R. (2012). On the path to uncover the bacterial type II secretion system. Philos. Trans. R. Soc. Lond. B. Biol. Sci. *367*, 1059–1072.

Duerkop, B.A., Ulrich, R.L., and Greenberg, E.P. (2007). Octanoyl-homoserine lactone is the cognate signal for *Burkholderia mallei* BmaR1-BmaI1 quorum sensing. J. Bacteriol. *189*, 5034–5040.

Duerkop, B.A., Herman, J.P., Ulrich, R.L., Churchill, M.E., and Greenberg, E.P. (2008). The *Burkholderia mallei* BmaR3-BmaI3 quorum-sensing system produces and responds to N-3-hydroxy-octanoyl homoserine lactone. J. Bacteriol. *190*, 5137–5141.

Dussurget, O., Cabanes, D., Dehoux, P., Lecuit, M., Buchrieser, C., Glaser, P., and Cossart, P. (2002). Listeria monocytogenes bile salt hydrolase is a PrfA-regulated virulence factor involved in the intestinal and hepatic phases of listeriosis. Mol. Microbiol. *45*, 1095–1106.

Eberl, L. (2006). Quorum sensing in the genus *Burkholderia*. Int. J. Med. Microbiol. *296*, 103–110.

Essex-Lopresti, A.E., Boddey, J.A., Thomas, R., Smith, M.P., Hartley, M.G., Atkins, T., Brown, N.F., Tsang, C.H., Peak, I.R., Hill, J., *et al.* (2005). A type IV pilin, PilA, Contributes To Adherence of *Burkholderia pseudomallei* and virulence *in vivo*. Infect. Immun. *73*, 1260–1264.

Estes, D.M., Dow, S.W., Schweizer, H.P., and Torres, A.G. (2010). Present and future therapeutic strategies for melioidosis and glanders. Expert Rev. Anti. Infect. Ther. *8*, 325–338.

Fernandes, P.J., Guo, Q., Waag, D.M., and Donnenberg, M.S. (2007). The type IV pilin of *Burkholderia mallei* is highly immunogenic but fails to protect against lethal aerosol challenge in a murine model. Infect. Immun. *75*, 3027–3032.

Galyov, E.E., Brett, P.J., and DeShazer, D. (2010). Molecular insights into *Burkholderia pseudomallei* and *Burkholderia mallei* pathogenesis. Annu. Rev. Microbiol. *64*, 495–517.

Godoy, D., Randle, G., Simpson, A.J., Aanensen, D.M., Pitt, T.L., Kinoshita, R., and Spratt, B.G. (2003). Multilocus sequence typing and evolutionary relationships among the causative agents of melioidosis and glanders, *Burkholderia pseudomallei* and *Burkholderia mallei*. J. Clin. Microbiol. *41*, 2068–2079.

Ham, J.H., Melanson, R.A., and Rush, M.C. (2010). Burkholderia glumae: next major pathogen of rice? Mol. Plant Pathol. *12*, 329–339.

Hautbergue, G.M., and Wilson, S.A. (2012). BLF1, the first *Burkholderia pseudomallei* toxin, connects inhibition of host protein synthesis with melioidosis. Biochem. Soc. Trans. *40*, 842–845.

Hill, C. (2012). Virulence or Niche Factors: What's in a Name? J. Bacteriol. *194*, 5725–5727.

Ho, P.L., Cheung, T.K., Yam, W.C., and Yuen, K.Y. (2002). Characterization of a laboratory-generated variant of BPS beta-lactamase from *Burkholderia pseudomallei* that hydrolyses ceftazidime. J. Antimicrob. Chemother. *50*, 723–726.

Holden, M.T., Titball, R.W., Peacock, S.J., Cerdeno-Tarraga, A.M., Atkins, T., Crossman, L.C., Pitt, T., Churcher, C., Mungall, K., Bentley, S.D., *et al.* (2004). Genomic plasticity of the causative agent of melioidosis, *Burkholderia pseudomallei*. Proc. Natl. Acad. Sci. U.S.A. *101*, 14240–14245.

Holden, M.T., Seth-Smith, H.M., Crossman, L.C., Sebaihia, M., Bentley, S.D., Cerdeno-Tarraga, A.M., Thomson, N.R., Bason, N., Quail, M.A., Sharp, S., *et al.* (2009). The genome of Burkholderia cenocepacia J2315, an epidemic pathogen of cystic fibrosis patients. J. Bacteriol. *191*, 261–277.

Isshiki, Y., Matsuura, M., Dejsirilert, S., Ezaki, T., and Kawahara, K. (2001). Separation of 6-deoxy-heptan [correction of 6-deoxy-heptane] from a smooth-type lipopolysaccharide preparation of *Burkholderia pseudomallei*. FEMS Microbiol. Lett. *199*, 21–25.

Jeddeloh, J.A., Fritz, D.L., Waag, D.M., Hartings, J.M., and Andrews, G.P. (2003). Biodefense-driven murine model of pneumonic melioidosis. Infect. Immun. *71*, 584–587.

Kespichayawattana, W., Rattanachetkul, S., Wanun, T., Utaisincharoen, P., and Sirisinha, S. (2000). *Burkholderia pseudomallei* induces cell fusion and actin-associated membrane protrusion: a possible mechanism for cell-to-cell spreading. Infect. Immun. *68*, 5377–5384.

Khaki, P., Mosavari, N., Khajeh, N.S., Emam, M., Ahouran, M., Hashemi, S., Taheri, M.M., Jahanpeyma, D., and Nikkhah, S. (2012). Glanders outbreak at Tehran Zoo, Iran. Iran J. Microbiol. *4*, 3–7.

Khan, I., Wieler, L.H., Melzer, F., Elschner, M.C., Muhammad, G., Ali, S., Sprague, L.D., Neubauer, H., and Saqib, M. (2012). Glanders in Animals: A Review on Epidemiology, Clinical Presentation, Diagnosis and Countermeasures. Transbound Emerg. Dis. *60* (3), 204–22.

Kim, H.S., Schell, M.A., Yu, Y., Ulrich, R.L., Sarria, S.H., Nierman, W.C., and DeShazer, D. (2005). Bacterial genome adaptation to niches: divergence of the potential virulence genes in three Burkholderia species of different survival strategies. BMC Genomics *6*, 174.

Kiratisin, P., and Sanmee, S. (2008). Roles and interactions of *Burkholderia pseudomallei* BpsIR quorum-sensing system determinants. J. Bacteriol. *190*, 7291–7297.

Korotkov, K.V., Sandkvist, M., and Hol, W.G. (2012). The type II secretion system: biogenesis, molecular architecture and mechanism. Nat. Rev. Microbiol. *10*, 336–351.

Kulasakara, H., Lee, V., Brencic, A., Liberati, N., Urbach, J., Miyata, S., Lee, D.G., Neely, A.N., Hyodo, M., Hayakawa, Y., *et al.* (2006). Analysis of *Pseudomonas aeruginosa* diguanylate cyclases and phosphodiesterases reveals a role for bis-(3′-5′)-cyclic-GMP in virulence. Proc. Natl. Acad. Sci. U.S.A. *103*, 2839–2844.

Kumar, A., Mayo, M., Trunck, L.A., Cheng, A.C., Currie, B.J., and Schweizer, H.P. (2008). Expression of resistance-nodulation-cell-division efflux pumps in commonly used *Burkholderia pseudomallei* strains and clinical isolates from northern Australia. Trans. R. Soc. Trop. Med. Hyg. *102 (Suppl 1)*, S145–151.

Kvitko, B.H., Goodyear, A., Propst, K.L., Dow, S.W., and Schweizer, H.P. (2012). *Burkholderia pseudomallei* known siderophores and hemin uptake are dispensable for lethal murine melioidosis. PLoS Negl. Trop. Dis. *6*, e1715.

Lazar Adler, N.R., Stevens, J.M., Stevens, M.P., and Galyov, E.E. (2012). Autotransporters and Their Role in the Virulence of *Burkholderia pseudomallei* and *Burkholderia mallei*. Front. Microbiol. *2*, 151.

Lee, H.S., Gu, F., Ching, S.M., Lam, Y., and Chua, K.L. (2010a). CdpA is a *Burkholderia pseudomallei* cyclic di-GMP phosphodiesterase involved in autoaggregation, flagellum synthesis, motility, biofilm formation, cell invasion, and cytotoxicity. Infect. Immun. *78*, 1832–1840.

Lee, Y.H., Chen, Y., Ouyang, X., and Gan, Y.H. (2010b). Identification of tomato plant as a novel host model for *Burkholderia pseudomallei*. BMC Microbiol. *10*, 28.

Lin, C.H., Bourque, G., and Tan, P. (2008). A comparative synteny map of *Burkholderia* species links large-scale genome rearrangements to fine-scale nucleotide variation in prokaryotes. Mol. Biol. Evol. *25*, 549–558.

Lipscomb, L., and Schell, M.A. (2011). Elucidation of the regulon and cis-acting regulatory element of HrpB, the AraC-type regulator of a plant pathogen-like type III secretion system in *Burkholderia pseudomallei*. J. Bacteriol. *193*, 1991–2001.

Losada, L., Ronning, C.M., DeShazer, D., Woods, D., Fedorova, N., Kim, H.S., Shabalina, S.A., Pearson, T.R., Brinkac, L., Tan, P., *et al.* (2010). Continuing evolution of *Burkholderia mallei* through genome reduction and large-scale rearrangements. Genome Biol. Evol. *2*, 102–116.

Lumjiaktase, P., Diggle, S.P., Loprasert, S., Tungpradabkul, S., Daykin, M., Camara, M., Williams, P., and Kunakorn, M. (2006). Quorum sensing regulates dpsA and the oxidative stress response in *Burkholderia pseudomallei*. Microbiology *152*, 3651–3659.

Madigan, M.T., Martinko, J.M., Dunlap, P.V., and Clark, D.P., eds. (2009). Brock biology of microorganisms, 12th edition (Upper Saddle River, NJ, Pearson Education Inc).

Mima, T., and Schweizer, H.P. (2010). The BpeAB-OprB efflux pump of *Burkholderia pseudomallei* 1026b does not play a role in quorum sensing, virulence factor production, or extrusion of aminoglycosides but is a broad-spectrum drug efflux system. Antimicrob. Agents Chemother. *54*, 3113–3120.

Moore, R.A., DeShazer, D., Reckseidler, S., Weissman, A., and Woods, D.E. (1999). Efflux-mediated aminoglycoside and macrolide resistance in *Burkholderia pseudomallei*. Antimicrob. Agents Chemother. *43*, 465–470.

Muangman, S., Korbsrisate, S., Muangsombut, V., Srinon, V., Adler, N.L., Schroeder, G.N., Frankel, G., and Galyov, E.E. (2012). BopC is a type III secreted effector protein of *Burkholderia pseudomallei*. FEMS Microbiol. Lett. *323*, 75–82.

Muller, C.M., Conejero, L., Spink, N., Wand, M.E., Bancroft, G.J., and Titball, R.W. (2012). Role of RelA and SpoT in *Burkholderia pseudomallei* Virulence and Immunity. Infect. Immun. *80*, 3247–3255.

Nandi, T., Ong, C., Singh, A.P., Boddey, J., Atkins, T., Sarkar-Tyson, M., Essex-Lopresti, A.E., Chua, H.H., Pearson, T., Kreisberg, J.F., *et al.* (2010). A genomic survey of positive selection in *Burkholderia pseudomallei* provides insights into the evolution of accidental virulence. PLoS Pathog. *6*, e1000845.

Ngauy, V., Lemeshev, Y., Sadkowski, L., and Crawford, G. (2005). Cutaneous melioidosis in a man who was taken as a prisoner of war by the Japanese during World War II. J. Clin. Microbiol. *43*, 970–972.

Nierman, W.C., DeShazer, D., Kim, H.S., Tettelin, H., Nelson, K.E., Feldblyum, T., Ulrich, R.L., Ronning, C.M., Brinkac, L.M., Daugherty, S.C., *et al.* (2004). Structural flexibility in the *Burkholderia mallei* genome. Proc. Natl. Acad. Sci. U.S.A. *101*, 14246–14251.

Ong, C., Ooi, C.H., Wang, D., Chong, H., Ng, K.C., Rodrigues, F., Lee, M.A., and Tan, P. (2004). Patterns of large-scale genomic variation in virulent and avirulent Burkholderia species. Genome Res. *14*, 2295–2307.

Ooi, W.F., Ong, C., Nandi, T., Kreisberg, J.F., Chua, H.H., Sun, G., Chen, Y., Mueller, C., Conejero, L., Eshagi, M., *et al.* (2013). The condition-dependent transcriptional landscape of *Burkholderia pseudomallei*. PLoS Genet. In press.

Pal, M., Erskine, P.T., Gill, R.S., Wood, S.P., and Cooper, J.B. (2012). Near-atomic resolution analysis of BipD, a component of the type III secretion system of *Burkholderia pseudomallei*. Acta Crystallogr. Sect. F Struct. Biol. Cryst. Commun. *66*, 990–993.

Peacock, S.J., Limmathurotsakul, D., Lubell, Y., Koh, G.C., White, L.J., Day, N.P., and Titball, R.W. (2012). Melioidosis vaccines: a systematic review and appraisal of the potential to exploit biodefense vaccines for public health purposes. PLoS Negl. Trop. Dis. *6*, e1488.

Peek, M.E., Bhatnagar, A., McCarty, N.A., and Zughaier, S.M. (2012). Pyoverdine, the major siderophore in *Pseudomonas aeruginosa*, Evades NGAL Recognition. Interdiscip. Perspect. Infect. Dis. *2012*, 843509.

Penwell, W.F., Arivett, B.A., and Actis, L.A. (2012). The *Acinetobacter baumannii* entA gene located outside the acinetobactin cluster is critical for siderophore production, iron acquisition and virulence. PLoS One *7*, e36493.

Perry, M.B., MacLean, L.L., Schollaardt, T., Bryan, L.E., and Ho, M. (1995). Structural characterization of the lipopolysaccharide O antigens of *Burkholderia pseudomallei*. Infect. Immun. *63*, 3348–3352.

Reckseidler-Zenteno, S.L., DeVinney, R., and Woods, D.E. (2005). The capsular polysaccharide of *Burkholderia pseudomallei* contributes to survival in serum by reducing complement factor C3b deposition. Infect. Immun. *73*, 1106–1115.

Rolim, D.B., Rocha, M.F., Brilhante, R.S., Cordeiro, R.A., Leitao, N.P., Jr., Inglis, T.J., and Sidrim, J.J. (2009). Environmental isolates of *Burkholderia pseudomallei* in Ceara State, northeastern Brazil. Appl. Environ. Microbiol. *75*, 1215–1218.

Sam, I.C., See, K.H., and Puthucheary, S.D. (2009). Variations in ceftazidime and amoxicillin-clavulanate susceptibilities within a clonal infection of *Burkholderia pseudomallei*. J. Clin. Microbiol. *47*, 1556–1558.

Sarkar-Tyson, M., Thwaite, J.E., Harding, S.V., Smither, S.J., Oyston, P.C., Atkins, T.P., and Titball, R.W. (2007). Polysaccharides and virulence of *Burkholderia pseudomallei*. J. Med. Microbiol. *56*, 1005–1010.

Sarovich, D.S., Price, E.P., Von Schulze, A.T., Cook, J.M., Mayo, M., Watson, L.M., Richardson, L., Seymour, M.L., Tuanyok, A., Engelthaler, D.M., *et al.* (2012). Characterization of ceftazidime resistance mechanisms in clinical isolates of *Burkholderia pseudomallei* from Australia. PLoS One *7*, e30789.

Schell, M.A., Ulrich, R.L., Ribot, W.J., Brueggemann, E.E., Hines, H.B., Chen, D., Lipscomb, L., Kim, H.S., Mrazek, J., Nierman, W.C., *et al.* (2007). Type VI secretion is a major virulence determinant in *Burkholderia mallei*. Mol. Microbiol. *64*, 1466–1485.

Schubert, S., Rakin, A., and Heesemann, J. (2004). The *Yersinia* high-pathogenicity island (HPI): evolutionary and functional aspects. Int. J. Med. Microbiol. *294*, 83–94.

Scott, A.E., Twine, S.M., Fulton, K.M., Titball, R.W., Essex-Lopresti, A.E., Atkins, T.P., and Prior, J.L. (2011). Flagellar glycosylation in *Burkholderia pseudomallei* and *Burkholderia thailandensis*. J. Bacteriol. *193*, 3577–3587.

Shalom, G., Shaw, J.G., and Thomas, M.S. (2007). *In vivo* expression technology identifies a type VI secretion system locus in *Burkholderia pseudomallei* that is induced upon invasion of macrophages. Microbiology *153*, 2689–2699.

Sim, B.M., Chantratita, N., Ooi, W.F., Nandi, T., Tewhey, R., Wuthiekanun, V., Thaipadungpanit, J., Tumapa, S., Ariyaratne, P., Sung, W.K., *et al.* (2010). Genomic acquisition of a capsular polysaccharide virulence cluster by non-pathogenic *Burkholderia* isolates. Genome Biol. *11*, R89.

Simm, R., Morr, M., Kader, A., Nimtz, M., and Romling, U. (2004). GGDEF and EAL domains inversely regulate cyclic di-GMP levels and transition from sessility to motility. Mol. Microbiol. *53*, 1123–1134.

Sim, S.H., Yu, Y., Lin, C.H., Karuturi, R.K., Wuthiekanun, V., Tuanyok, A., Chua, H.H., Ong, C., Paramalingam, S.S., Tan, G., *et al.* (2008). The core and accessory genomes of *Burkholderia pseudomallei*: implications for human melioidosis. PLoS Pathog. *4*, e1000178.

Song, Y., Xie, C., Ong, Y.M., Gan, Y.H., and Chua, K.L. (2005). The BpsIR quorum-sensing system of *Burkholderia pseudomallei*. J. Bacteriol. *187*, 785–790.

Srinivasan, A., Kraus, C.N., DeShazer, D., Becker, P.M., Dick, J.D., Spacek, L., Bartlett, J.G., Byrne, W.R., and Thomas, D.L. (2001). Glanders in a military research microbiologist. N. Engl. J. Med. *345*, 256–258.

Stevens, M.P., Stevens, J.M., Jeng, R.L., Taylor, L.A., Wood, M.W., Hawes, P., Monaghan, P., Welch, M.D., and Galyov, E.E. (2005). Identification of a bacterial factor required for actin-based motility of *Burkholderia pseudomallei*. Mol. Microbiol. *56*, 40–53.

Sun, G.W., and Gan, Y.H. (2010). Unraveling type III secretion systems in the highly versatile *Burkholderia pseudomallei*. Trends Microbiol. *18*, 561–568.

Sun, G.W., Chen, Y., Liu, Y., Tan, G.Y., Ong, C., Tan, P., and Gan, Y.H. (2010). Identification of a regulatory cascade controlling Type III Secretion System 3 gene expression in *Burkholderia pseudomallei*. Mol. Microbiol. *76*, 677–689.

Suparak, S., Kespichayawattana, W., Haque, A., Easton, A., Damnin, S., Lertmemongkolchai, G., Bancroft, G.J., and Korbsrisate, S. (2005). Multinucleated giant cell formation and apoptosis in infected host cells is mediated by *Burkholderia pseudomallei* type III secretion protein BipB. J. Bacteriol. *187*, 6556–6560.

Tandhavanant, S., Thanwisai, A., Limmathurotsakul, D., Korbsrisate, S., Day, N.P., Peacock, S.J., and Chantratita, N. (2010). Effect of colony morphology variation of *Burkholderia pseudomallei* on intracellular survival and resistance to antimicrobial environments in human macrophages *in vitro*. BMC Microbiol. *10*, 303.

Tribuddharat, C., Moore, R.A., Baker, P., and Woods, D.E. (2003). *Burkholderia pseudomallei* class a beta-lactamase mutations that confer selective resistance against ceftazidime or clavulanic acid inhibition. Antimicrob. Agents Chemother. *47*, 2082–2087.

Trunck, L.A., Propst, K.L., Wuthiekanun, V., Tuanyok, A., Beckstrom-Sternberg, S.M., Beckstrom-Sternberg, J.S., Peacock, S.J., Keim, P., Dow, S.W., and Schweizer, H.P. (2009). Molecular basis of rare aminoglycoside susceptibility and pathogenesis of *Burkholderia pseudomallei* clinical isolates from Thailand. PLoS Negl. Trop. Dis. *3*, e519.

Tuanyok, A., Kim, H.S., Nierman, W.C., Yu, Y., Dunbar, J., Moore, R.A., Baker, P., Tom, M., Ling, J.M., and Woods, D.E. (2005). Genome-wide expression analysis of iron regulation in *Burkholderia pseudomallei* and *Burkholderia mallei* using DNA microarrays. FEMS Microbiol. Lett. *252*, 327–335.

Tuanyok, A., Leadem, B.R., Auerbach, R.K., Beckstrom-Sternberg, S.M., Beckstrom-Sternberg, J.S., Mayo, M., Wuthiekanun, V., Brettin, T.S., Nierman, W.C., Peacock, S.J., *et al.* (2008). Genomic islands from five strains of *Burkholderia pseudomallei*. BMC Genomics *9*, 566.

Ulrich, R.L., and DeShazer, D. (2004). Type III secretion: a virulence factor delivery system essential for the pathogenicity of *Burkholderia mallei*. Infect. Immun. *72*, 1150–1154.

Ulrich, R.L., Deshazer, D., Brueggemann, E.E., Hines, H.B., Oyston, P.C., and Jeddeloh, J.A. (2004a). Role of quorum sensing in the pathogenicity of *Burkholderia pseudomallei*. J. Med. Microbiol. *53*, 1053–1064.

Ulrich, R.L., Deshazer, D., Hines, H.B., and Jeddeloh, J.A. (2004b). Quorum sensing: a transcriptional regulatory system involved in the pathogenicity of *Burkholderia mallei*. Infect. Immun. *72*, 6589–6596.

Ulrich, R.L., Hines, H.B., Parthasarathy, N., and Jeddeloh, J.A. (2004c). Mutational analysis and biochemical characterization of the *Burkholderia thailandensis* DW503 quorum-sensing network. J. Bacteriol. *186*, 4350–4360.

Ussery, D.W., Kiil, K., Lagesen, K., Sicheritz-Ponten, T., Bohlin, J., and Wassenaar, T.M. (2009). The genus burkholderia: analysis of 56 genomic sequences. Genome Dyn. *6*, 140–157.

Valade, E., Thibault, F.M., Gauthier, Y.P., Palencia, M., Popoff, M.Y., and Vidal, D.R. (2004). The PmlI-PmlR quorum-sensing system in *Burkholderia pseudomallei* plays a key role in virulence and modulates production of the MprA protease. J. Bacteriol. *186*, 2288–2294.

Valdez, Y., Ferreira, R.B., and Finlay, B.B. (2009). Molecular mechanisms of Salmonella virulence and host resistance. Curr. Top. Microbiol. Immunol. *337*, 93–127.

Warawa, J., and Woods, D.E. (2005). Type III secretion system cluster 3 is required for maximal virulence of *Burkholderia pseudomallei* in a hamster infection model. FEMS Microbiol. Lett. *242*, 101–108.

Warawa, J.M., Long, D., Rosenke, R., Gardner, D., and Gherardini, F.C. (2009). Role for the *Burkholderia pseudomallei* capsular polysaccharide encoded by the wcb operon in acute disseminated melioidosis. Infect. Immun. *77*, 5252–5261.

Warner, J.M., Pelowa, D.B., Currie, B.J., and Hirst, R.G. (2007). Melioidosis in a rural community of Western Province, Papua New Guinea. Trans. R. Soc. Trop. Med. Hyg. *101*, 809–813.

Warner, J.M., Pelowa, D.B., Gal, D., Rai, G., Mayo, M., Currie, B.J., Govan, B., Skerratt, L.F., and Hirst, R.G. (2008). The epidemiology of melioidosis in the Balimo region of Papua New Guinea. Epidemiol. Infect. *136*, 965–971.

Wiersinga, W.J., van der Poll, T., White, N.J., Day, N.P., and Peacock, S.J. (2006). Melioidosis: insights into the pathogenicity of *Burkholderia pseudomallei*. Nat. Rev. Microbiol. *4*, 272–282.

Wikraiphat, C., Charoensap, J., Utaisincharoen, P., Wongratanacheewin, S., Taweechaisupapong, S., Woods, D.E., Bolscher, J.G., and Sirisinha, S. (2009). Comparative *in vivo* and *in vitro* analyses of putative virulence factors of *Burkholderia pseudomallei* using lipopolysaccharide, capsule and flagellin mutants. FEMS Immunol. Med. Microbiol. *56*, 253–259.

Wong, K.T., Puthucheary, S.D., and Vadivelu, J. (1995). The histopathology of human melioidosis. Histopathology *26*, 51–55.

Wongtrakoongate, P., Tumapa, S., and Tungpradabkul, S. (2012). Regulation of a quorum sensing system by stationary phase sigma factor RpoS and their co-regulation of target genes in *Burkholderia pseudomallei*. Microbiol. Immunol. *56*, 281–294.

Wuthiekanun, V., Mayxay, M., Chierakul, W., Phetsouvanh, R., Cheng, A.C., White, N.J., Day, N.P., and Peacock, S.J. (2005). Detection of *Burkholderia pseudomallei* in soil within the Lao People's Democratic Republic. J. Clin. Microbiol. *43*, 923–924.

Yang, H.M., Chaowagul, W., and Sokol, P.A. (1991). Siderophore production by *Pseudomonas pseudomallei*. Infect. Immun. *59*, 776–780.

Yang, H., Kooi, C.D., and Sokol, P.A. (1993). Ability of *Pseudomonas pseudomallei* malleobactin to acquire transferrin-bound, lactoferrin-bound, and cell-derived iron. Infect. Immun. *61*, 656–662.

Yao, Q., Cui, J., Zhu, Y., Wang, G., Hu, L., Long, C., Cao, R., Liu, X., Huang, N., Chen, S., *et al.* (2009). A bacterial type III effector family uses the papain-like hydrolytic activity to arrest the host cell cycle. Proc. Natl. Acad. Sci. U.S.A. *106*, 3716–3721.

Yu, Y., Kim, H.S., Chua, H.H., Lin, C.H., Sim, S.H., Lin, D., Derr, A., Engels, R., DeShazer, D., Birren, B., *et al.* (2006). Genomic patterns of pathogen evolution revealed by comparison of *Burkholderia pseudomallei*, the causative agent of melioidosis, to avirulent *Burkholderia thailandensis*. BMC Microbiol. *6*, 46.

Molecular Mechanisms of *Burkholderia cepacia* Complex Intracellular Survival in Phagocytic Cells

7

Miguel A. Valvano, Roberto Rosales-Reyes, Crystal L. Schmerk and Hanna Ostapska

Abstract

Burkholderia cepacia complex (Bcc) species are a group of Gram-negative opportunistic pathogens that chronically infect the airways of cystic fibrosis patients, but they can also infect patients with various types of immunosuppressive disorders. Bcc members are multidrug resistant bacteria that have the ability to persist in the infected host and also elicit robust inflammatory responses. Studies using macrophages, neutrophils and dendritic cells, combined with dramatic advances in the ability to genetically manipulate these microorganisms have contributed to increase our understanding of the molecular mechanisms of virulence in these pathogens and the molecular details of the cell host responses triggering inflammation. This chapter reviews our understanding of the pathogenic mechanisms used by Bcc to establish an intracellular niche in phagocytic cells and modulate host cell responses that ultimately end up in cell death and a proinflammatory response.

Introduction

The *Burkholderia cepacia* complex (Bcc) comprises a subset of *Burkholderia* species that have been found to be associated with respiratory infection in patients with cystic fibrosis (CF) (Coenye and Vandamme, 2003; Vandamme and Dawyndt, 2011). Bcc bacteria are opportunistic pathogens typically acquired from the environment. These bacteria infect and survive in various eukaryotic cells including human respiratory epithelial cells, neutrophils, macrophages, and amoebae (Loutet and Valvano, 2010; Saldías and Valvano, 2009). Bcc infections trigger robust inflammatory responses causing rapid deterioration of lung function in CF patients and in some cases leading to a deadly necrotizing pneumonia called *cepacia* syndrome. Uncontrolled inflammatory responses originate from the activation of the inflammasome as a consequence of bacterial intracellular infection of macrophages. In this chapter we will review our current understanding of the molecular mechanisms of intracellular survival of Bcc bacteria and the response of infected macrophages that leads to inflammasome activation, macrophage cell death, and inflammation. For other aspects of the pathogenesis of Bcc bacteria not covered in this chapter the reader should consult excellent recent reviews (Bazzini *et al.*, 2011; Drevinek and Mahenthiralingam, 2010; Ganesan and Sajjan, 2011; Leitão *et al.*, 2010; McClean and Callaghan, 2009; Vial *et al.*, 2011).

Bcc bacteria survive intracellularly in macrophages and free-living amoebae

Bcc strains have the ability to survive in free-living amoebae, as demonstrated by various groups using co-cultures with several species of *Acanthamoeba* (Anacarso *et al.*, 2010; Landers *et al.*, 2000; Marolda *et al.*, 1999). Detailed analysis of the intracellular survival in *Acanthamoeba* demonstrated that amoeba cells engulfing bacteria develop profuse cytoplasmic vacuolization (Marolda *et al.*, 1999), which was absent in cells infected with heat-killed Bcc bacteria or live

Escherichia coli. These experiments indicated that metabolically active intracellular bacteria are required for the formation of vacuoles. However, intracellular Bcc did not replicate to any significant degree at least up to 96 h post infection, while extracellular bacteria multiplied efficiently on by-products released by amoebae (Marolda *et al.*, 1999). The findings suggest that infected amoebae may be a reservoir for Bcc and perhaps may even provide a vehicle for transmission of these opportunistic pathogens. This conclusion was based on the observation that infected amoebae release small bacteria-containing vacuoles of about 5 μm in diameter, which could reach the lower respiratory tract (Marolda *et al.*, 1999).

The survival of Bcc strains in amoebae was investigated in more detail using *B. vietnamiensis* CEP040 and *B. cenocepacia* H111, two clinical isolates recovered from CF patients. Both strains survived but did not replicate within vacuoles of *A. polyphaga* (Lamothe *et al.*, 2004). Survival was demonstrated using bacteria fluorescently labelled with chloromethylfluorescein diacetate, which remained fluorescent and therefore viable for at least 24 h post infection. Furthermore, intracellular bacteria retained green fluorescent protein within the bacterial cytoplasm, while this protein rapidly escaped from engulfed, heat-killed bacteria into the vacuolar lumen (Lamothe *et al.*, 2004). The *B. cepacia*-containing vacuoles (BcCVs) accumulated the fluid phase marker Lysosensor™ Blue, suggesting they had low pH. However, it is difficult to conclude that bacteria indeed survive in a low pH vacuole because pH measurements were not performed using ratiometric techniques based on pH-dependent and -independent fluorophores. Transmission electron microscopy experiments revealed that the BcCVs did not accumulate cationized ferritin, a compound that localizes within the lysosome (Wells and Rikihisa, 1988), suggesting that Bcc bacteria survive in amoebae within a vacuole that is not fused to the lysosomal compartment.

Like amoebae, mammalian macrophages can also support intracellular survival of Bcc bacteria within membrane-bound vacuoles. In one study, bacterial entry stimulated the production of tumour necrosis factor alpha and the release of reactive oxygen radicals (ROS) in a lipopolysaccharide (LPS)-dependent manner (Saini *et al.*, 1999). Also, infected murine macrophages primed with interferon gamma produced less nitric oxide than uninfected cells, suggesting that intracellular Bcc not only can survive intracellularly, but can also resist microbicidal mechanisms. Intracellular survival was demonstrated in several murine (Lamothe *et al.*, 2007; Lamothe and Valvano, 2008; Saini *et al.*, 1999) and human (Martin and Mohr, 2000) macrophage cell lines. Intracellular survival was also described in a human type II pneumocyte cell line, A549 (Burns *et al.*, 1996), and in various bronchiolar epithelial cell lines (Duff *et al.*, 2006; Keig *et al.*, 2002; Moura *et al.*, 2008; Sajjan *et al.*, 2006; Schwab *et al.*, 2002; Taylor *et al.*, 2010). As in macrophages and amoebae, intracellular bacteria in epithelial cells survive in vacuoles that show a maturation arrest (Sajjan *et al.*, 2006).

The maturation of the BcCV in murine macrophages was investigated in detail using the *B. cenocepacia* strains J2315 and K56–2 (Flannagan *et al.*, 2012; Huynh *et al.*, 2010; Keith *et al.*, 2009; Lamothe *et al.*, 2007; Maloney and Valvano, 2006; Rosales-Reyes *et al.*, 2012b). Both strains are clonal and J2315 is the prototypic isolate of the widespread and transmissible ET12 clone (Johnson *et al.*, 1994). Phagosomes containing *B. cenocepacia* strain J2315, transiently recruit early endosome autoantigen (EEA1) and Rab5, and synthesize phosphatidylinositol-3-phosphate (Fig. 7.1), indicating that BcCVs progress normally to the early phagosomal stage (Huynh *et al.*, 2010; Lamothe *et al.*, 2007). However, unlike heat-inactivated bacteria, which reach lysosomes shortly after internalization, vacuoles containing live J2315 bacteria exhibit a significant delay in the accumulation of the late endosome/lysosome marker LAMP-1 and rarely fuse with lysosomes (Lamothe *et al.*, 2007). In addition to LAMP-1, BcCVs acquire the late phagosomal markers CD63 and Rab7 (Huynh *et al.*, 2010). During their arrested maturation, the BcCVs remain in contact with newly formed endosomes and maintain a luminal pH around 6.4, in stark contrast with vacuoles containing heat-inactivated J2315, which rapidly reached an average pH of

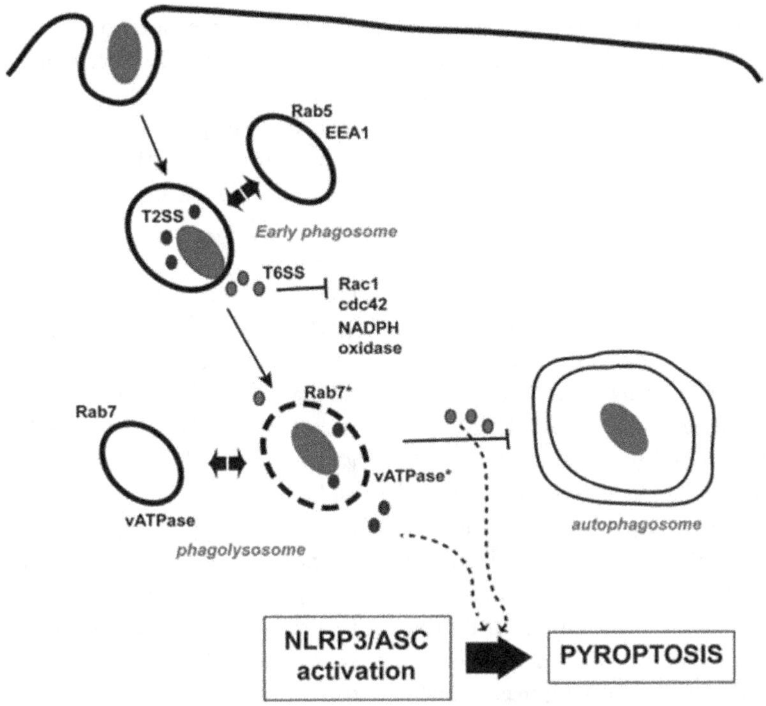

Figure 7.1 Diagram summarizing current knowledge of the intramacrophage survival mechanisms of *B. cenocepacia*. Upon engulfment the BcCV matures normally as an early phagosome by recruiting Rab5 and EEA1. However, further maturation is halted resulting in delayed acidification and recruitment of components of the NADPH oxidase and v-ATPase complexes. Delayed recruitment of the NADPH oxidase complex depends on a functional T6SS, which presumably secretes one or more effectors (in red) that inactivate Rac1 and Cdc42, also leading to defects in actin polymerization, phagocytosis, and cell migration. At later stages of maturation the BcCV accumulates Rab7 but this protein is inactivated by an unknown mechanism that most likely involves one or more effectors (in blue) secreted by the T2SS, which escape from the BcCV to the host cell's cytoplasm. The BcCV interacts with components of the autophagosome, but the full maturation of the autophagosome is also halted. BcCV membrane damage also triggers the assembly and activation of the NLRP3/ASC inflammasome resulting in macrophage cell death by pyroptosis. Asterisks indicate non-functional vATPase and Rab7, respectively. A colour version of this figure is located in the plate section at the back of the book.

4.8 and fused with lysosomes shortly after engulfment (Lamothe *et al.*, 2007). Similar results were obtained with formalin-fixed bacteria (Huynh *et al.*, 2010). Delay in acidification correlates with delayed recruitment of the 16 kDa subunit of the phagosomal vATPase onto the BcCV membrane (Rosales-Reyes *et al.*, 2012b). Fluorescence recovery after photobleaching and use of a probe that detects Rab7-guanosine triphosphate indicated that the Rab7 recruited to the BcCVs is inactive (Huynh *et al.*, 2010), providing a hint to a potential mechanism accounting for the inability of these vacuoles to fuse with lysosomes (Fig. 7.1).

Difficulties in determining *Burkholderia cenocepacia* intracellular survival are overcome using aminoglycoside-sensitive mutants

Several evidences indicate that intracellular Bcc in amoebae and macrophages are viable and metabolically active. However, it has been very difficult to reliably quantify the intracellular infection due to the extreme resistance of Bcc bacteria to antibiotics (Saini *et al.*, 1999). The gentamicin protection assay, traditionally employed to quantify intracellular bacteria (Elsinghorst, 1994; Isberg and Falkow, 1985), relies on the inability of aminoglycoside antibiotics to permeate the eukaryotic cell. However, antibiotic treatment of Bcc-infected macrophages requires large doses of kanamycin in combination with ceftazidime, which do not efficiently kill all extracellular bacteria (Burns *et al.*, 1996; Saini *et al.*, 1999), thus preventing a proper quantification of the intracellular bacterial load by colony counting. The development of an efficient protocol for constructing markerless gene deletion in *B. cenocepacia* (Flannagan *et al.*, 2008) allowed the construction of gentamicin-sensitive strains of *B. cenocepacia* in which the genes encoding an AmrAB-OprA-like efflux pump were deleted (Hamad *et al.*, 2010). *B. cenocepacia* strains carrying this deletion are hypersensitive to gentamicin, allowing the use of low concentrations of gentamicin to effectively kill extracellular bacteria. The gentamicin-sensitive strains can be used to reliably quantify the intracellular infection by *B. cenocepacia*, demonstrating that engulfed bacteria can replicate slowly over 24–48 h post infection (Hamad *et al.*, 2010). Importantly, these mutants also exhibit the same delay in phagolysosome fusion as their parental counterparts, indicating that the deleted efflux pump has no effect on the ability of intracellular bacteria to arrest phagosomal maturation.

Intramacrophage trafficking of *B. multivorans*

B. cenocepacia and *B. multivorans* are the Bcc species most commonly isolated from CF patients (Drevinek and Mahenthiralingam, 2010). Over the past decade, the majority of research groups have focused on the study of *B. cenocepacia*, largely due to the epidemic nature of these isolates that spread from patient to patient. However, the incidence of *B. multivorans* infections is steadily increasing. Currently, *B. multivorans* is the most prevalent Bcc strain infecting CF patients in North America and also the most common species infecting patients with chronic granulomatous disease (Zelazny *et al.*, 2009). *B. multivorans* can infect human lung epithelial cells (Duff *et al.*, 2006; Moura *et al.*, 2008), monocytes (Zelazny *et al.*, 2009), and dendritic cells (Macdonald and Speert, 2008). The invasion of human epithelial cells resulted in comparable levels of cytokine release and induction of apoptosis by *B. multivorans* and *B. cenocepacia* strains (Moura *et al.*, 2008). In contrast, in dendritic cells, intracellular *B. cenocepacia* led to impaired cell function and necrosis, even though *B. multivorans* elicited a similar release of cytokines (Macdonald and Speert, 2008).

A recent study by Schmerk and Valvano (2013) has revealed the details of *B. multivorans* interaction with macrophages during infection. Gentamicin sensitive derivatives of an environmental strain (ATCC 17616) and two CF clinical isolate strains (C5568 and C0514) of *B. multivorans* were used to infect both murine RAW264.7 and human THP-1 macrophages. The survival and replication of these strains was compared to that of a gentamicin-sensitive derivative of *B. cenocepacia*, which has been previously shown to replicate within RAW264.7 macrophages (Hamad *et al.*, 2010). The *B. multivorans* strains were able to replicate and survive within RAW264.7 macrophages in a manner similar to *B. cenocepacia*. Intramacrophage replication occurred during the first 24 h post infection with strains replicating ~ 1–2 $\log_{10}$ units. In contrast, the *B. multivorans* and *B. cenocepacia* strains do not replicate within human THP-1 macrophages and their numbers diminished marginally after 48 h of infection.

Both species of *Burkholderia* survived and replicated in a similar manner within macrophages, but this does not preclude differences in their trafficking properties after engulfment. *B. cenocepacia* does not traffic normally within RAW264.7 macrophages (Huynh *et al.*, 2010; Lamothe *et al.*, 2007), and can delay fusion of BcCVs with lysosomes by as much as 6 h (Lamothe *et al.*, 2007). It is generally thought that this delay in BcCV maturation is accomplished to allow *B. cenocepacia* adequate time to activate genes that are required for survival in the phagolysosome environment (Tolman and Valvano, 2012). However, the endocytic trafficking pattern of *B. multivorans* strains in murine RAW264.7 macrophages is different than observed in macrophages infected with *B. cenocepacia*. *B. multivorans* containing vacuoles (BmCVs) appeared to interact normally with the endocytic pathway, acquiring the late endosomal marker LAMP-1 and lysosomal marker dextran within 2 h of infection (Schmerk and Valvano,

2013). Thus, it would appear that the arrest observed in BcCV maturation is not required for the intramacrophage survival of *B. multivorans* strains. Despite that BmCVs exhibited a normal progression through the endocytic pathway, intracellular bacteria could survive and replicate within macrophages in a manner similar to *B. cenocepacia*. Together, these data reveal that *B. cenocepacia* and *B. multivorans* are capable of surviving within immune cells using different strategies.

Bacterial determinants for entry and survival

Very little is known about bacterial factors involved in entry and intramacrophage bacterial survival of *B. cenocepacia*. Recent evidence supports the notion that the O-antigen moiety of the lipopolysaccharide is associated with reduced bacterial engulfment by macrophages (Saldías *et al.*, 2009; Schmerk and Valvano, 2013). However, the macrophage receptor involved with bacterial recognition remains unknown. Using the rat agar bead model of chronic lung infection revealed that a *B. cenocepacia* transposon mutant carrying an insertion in the *mgtC* gene was dramatically attenuated (Hunt *et al.*, 2004), and a subsequent study demonstrated that this mutant fails to delay phagosomal maturation resulting in rapid targeting of the BcCV to the lysosomal compartment and a compromise in bacterial survival (Maloney and Valvano, 2006). MgtC is essential for the intramacrophage survival of several intracellular bacteria but its biochemical function has not been elucidated (Alix and Blanc-Potard, 2007). Furthermore, two sigma factors controlling the regulation of multiple genes, RpoN and RpoE, are required for the intracellular survival of *B. cenocepacia* (Flannagan and Valvano, 2008; Saldías *et al.*, 2008), but the specific gene determinants involved in survival have not been elucidated.

Factors involved in resistance to oxidative damage play a role in intracellular survival of *B. cenocepacia*. For example, some strains produce a brown melanin-like pigment that can scavenge free radicals, resulting in the attenuation of the host cell oxidative burst. The synthesis of this pigment requires a homogentisate (HGA) precursor. Disruption of the *hppD* gene (BCAL0207) encoding a 4-hydroxyphenylpyruvic acid dioxygenase abolishes pigment production. The *hppD* mutant becomes sensitive to hydrogen peroxide and to extracellularly generated superoxide anions, but it is resistant to endogenous superoxide (Keith *et al.*, 2007). Upon engulfment by macrophages, non-pigmented bacteria colocalize in a dextran-positive vacuole, but this process can be halted in the presence of the NADPH oxidase inhibitor diphenyleneiodonium, and also the inducible nitric oxide inhibitor amino guanidine (Keith *et al.*, 2007). Therefore, pigment production by *B. cenocepacia* protects intracellular bacteria from oxidative damage.

In phagocytes, the NADPH oxidase plays a crucial role in host defence. The enzyme is usually dormant in resting phagocytes. However, after activation by phagocytosis, the complex is assembled and becomes active to generate large quantities of superoxide anions at the plasma membrane and phagosomal membranes (Segal, 2005; Vignais, 2002). Also, the delay of the BcCV in the incorporation and activation of the NADPH oxidase components suggests that *B. cenocepacia* must escape oxidative damage to overcome the host cell defensive mechanisms. This is underscored by elegant experiments performed by Bylund *et al.* (2006) showing that many Bcc isolates produce large amounts of exopolysaccharide that interfere with the function of human neutrophils *in vitro*. These interactions lead to inhibition of neutrophil chemotaxis and production of reactive oxygen species (ROS), both essential components of innate neutrophil-mediated host defences. These authors provided direct evidence that exopolysaccharides inhibit enzymatic generation of ROS in cell-free systems, indicating that they have the ability to scavenge these bactericidal molecules (Bylund *et al.*, 2006).

B. cenocepacia also produces two Zn^{2+}-dependent metalloproteases, ZmpA and ZmpB, whose expression is controlled by quorum sensing and the AtsR hybrid sensor kinase (Aubert *et al.*, 2013; Corbett *et al.*, 2003; Gingues *et al.*, 2005; O'Grady *et al.*, 2012). These proteases can cleave and inactivate antimicrobial peptides involved in innate immunity including β-defensin 1, cathelicidin LL-37, elafin, secretory leucocyte inhibitor, and protamine (Kooi and Sokol, 2009). *In vivo*, ZmpA and ZmpB are required for the maturation arrest of the BcCVs and bacterial intra-macrophage survival (Rosales-Reyes *et al.*, 2012a), suggesting these proteases play a significant role in the ability of *B. cenocepacia* to establish an intracellular niche.

The *B. cenocepacia* Type VI secretion system mediates a proinflammatory macrophage response leading to cell death by pyroptosis

Bacterial intracellular survival typically relies on one or more specialized secretory systems responsible for injecting bacterial molecules into the cytoplasm of eukaryotic cells. These molecules target host-signalling pathways, allowing bacteria to establish an intracellular niche (Records, 2011; van der Heijden and Finlay, 2012; Voth *et al.*, 2012). *B. cenocepacia* expresses one Type III (T3SS), two Type IV (T4SS-1 and T4SS-2), and one Type VI (T6SS) secretion systems. A report by Sajjan *et al.* (2008) concluded that the plasmid encoded T4SS-1 was required for intracellular survival and replication in macrophages. Using markers of the classical endocytic pathway in colocalization experiments, these authors showed that mutant and parental strains reside transiently in early endosomes, but a greater proportion of the mutant bacteria are targeted for lysosomal degradation. In our laboratory, we have constructed deletion mutants defective in each of the secretion systems singly and in combination, including a mutant defective in all the secretion systems (Tolman and Valvano, unpublished data). Using these mutants, we could not find differences in intracellular survival in murine macrophages (Tolman and Valvano, unpublished data). Therefore, the role of the T4SS and the other secretion systems in the biology of *B. cenocepacia* intracellular infection is controversial and requires further research.

T6SS affects the normal function of small Rho GTPases

T6SS is a recently recognized secretion system widely spread in Gram-negative pathogenic (Cascales, 2008; Pukatzki *et al.*, 2006) and non-pathogenic bacteria (Jani and Cotter, 2010). The *B. cenocepacia* T6SS is essential for virulence *in vivo*, as a library of mutants attenuated for virulence in the rat agar bead model of lung infection (Hunt *et al.*, 2004) contained three independent transposon mutants in a region that was subsequently identified as a T6SS gene cluster (Aubert *et al.*, 2008). The T6SS expression is negatively regulated by a global regulator designated AtsR (Aubert *et al.*, 2013). Macrophage infection assays showed that an *atsR* mutant induces the formation of actin-mediated cell projections in a process that is T6SS dependent (Aubert *et al.*, 2008). Surprisingly, the T6SS is not required for the maturation arrest of the BcCV (Rosales-Reyes *et al.*, 2012b). Further research established that intracellular *B. cenocepacia* expressing the T6SS alter the activation of Rac1 and Cdc42 GTPases by reducing the cellular pool of GTP-bound Rac1 and Cdc42 (Fig. 7.1) (Flannagan *et al.*, 2012; Rosales-Reyes *et al.*, 2012b). GTPase inactivation requires internalization of viable bacteria and also inhibits macropinocytosis and phagocytosis (Flannagan *et al.*, 2012). The T6SS also increases the cellular pool of GTP-bound RhoA and decreases cofilin activity (Rosales-Reyes *et al.*, 2012b). These effects lead to abnormal actin polymerization causing collapse of lamellipodia and failure to retract the uropod. The T6SS also prevents the recruitment of soluble subunits of the NADPH oxidase complex including Rac1 to the BcCV membrane, but is not involved in causing the BcCV maturation arrest (Rosales-Reyes *et al.*, 2012b).

T6SS mediates damage of the BcCV resulting in an increased proinflammatory response by macrophages

Bcc infections are often persistent and lead to exacerbated inflammatory responses (Downey *et al.*, 2007; Moura *et al.*, 2008; Ventura *et al.*, 2009). The *B. cenocepacia* T6SS induces a proinflammatory response in macrophages, which depends on the activation of the NLRP3 inflammasome and results in the stimulation of IL-1β secretion in a caspase-1-dependent manner that ultimately leads to cell death by pyroptosis (Rosales-Reyes *et al.*, 2012a). Furthermore, the T6SS mediates the disruption of the BcCV membrane, which allows the escape of proteins secreted by the T2SS into the macrophage cytoplasm. This was demonstrated by measuring the cAMP concentration in the cytosol of macrophages infected with bacteria expressing fusion derivatives of the T2SS-secreted metalloproteases ZmpA and ZmpB with the calmodulin-activated adenylcyclase (Rosales-Reyes *et al.*, 2012a). Calmodulin-activated adenylcyclase can only be enzymatically active when present in the cytosol (Sory and Cornelis, 1994) supporting a model whereby two bacterial secretory pathways cooperate, resulting in the release of bacterial products that are not normally secreted into eukaryotic cells directly into the cytosol of the host cell. This is further underscored by the observation that ZmpA and ZmpB are required for efficient IL-1β secretion in a T6SS-dependent fashion (Rosales-Reyes *et al.*, 2012a).

Murine macrophages use the NLRP3 inflammasome and the host adapter molecule ASC to detect intracellular *B. cenocepacia* (Fig. 7.1) (Rosales-Reyes *et al.*, 2012a). In contrast, detection of intracellular *B. cenocepacia* in human macrophages requires pyrin as well as ASC (Gavrilin *et al.*, 2012). This was demonstrated in human monocytes and THP-1 cells stably expressing either small interfering RNA against pyrin or YFP-pyrin and YFP-ASC. Upon engulfment of *B. cenocepacia*, suppression of pyrin levels dramatically reduced caspase-1 activation and IL-1β, while overexpression of pyrin or ASC induced a robust IL-1β response in *B. cenocepacia* infected macrophages. The activation of the NLRP3/ASC and pyrin/ASC inflammasomes in murine and human macrophages, respectively, also required a functional T6SS (Gavrilin *et al.*, 2012; Rosales-Reyes *et al.*, 2012a). Therefore, a model is emerging from all of these studies in murine and human macrophages indicating the expression of T6SS plays a proinflammatory role, resulting in damage of the BcCV, release of bacterial mediators of inflammation accumulated in the lumen of the BcCV, which ultimately results in host cell death. This correlates well with clinical observations documenting the proinflammatory nature of the respiratory infection by *B. cenocepacia* in CF patients.

Role of the CF transmembrane conductance regulator (CFTR) in *B. cenocepacia* intracellular survival

One of the most remarkable observations in the biology of the *B. cenocepacia* infection in macrophages was the finding that the maturation arrest of the BcCVs is much more exaggerated in CFTR-defective macrophages than in CFTR-normal macrophages (Lamothe and Valvano, 2008). The CFTR-associated phagosomal maturation defect was absent in macrophages exposed to heat-inactivated *B. cenocepacia* and macrophages infected with a non-CF pathogen such as *Salmonella enterica*, an intracellular pathogen that once internalized rapidly traffics to acidic compartments that acquire lysosomal markers (Lamothe and Valvano, 2008). These results suggest that not only defective CFTR, but also viable *B. cenocepacia* are required for the altered trafficking phenotype. Therefore, CFTR may play a role in the mechanism of clearance of the intracellular infection, as *B. cenocepacia* bacterial cells that localized to the lysosome lost cell envelope integrity (Lamothe *et al.*, 2007). BcCVs in *B. cenocepacia*-infected CFTR-defective macrophages also displayed a more pronounced delay in the assembly of the NADPH oxidase than that seen in infected CFTR-normal macrophages (Keith *et al.*, 2009). During phagosomal maturation, the V-type ATPase (V-ATPase) shuttles H^+ into the maturing phagosome, lowering the pH from somewhat neutral (~pH 6.0) to acidic (~pH 4.5) (Vieira *et al.*, 2002). It is believed that Cl^- ions contribute to efficient phagosomal acidification by preventing the rapid influx of H^+ by the vATPase from establishing an electrochemical gradient that would lower membrane conductance and thus halt any significant change to intraphagosomal pH (Sonawane and Verkman, 2003), although others have recently proposed that a cation counterflux is also important for lysosomal acidification (Steinberg *et al.*, 2010). Cl^- may be provided via any of several channels, including members of the chloride channel (Cl) family (Jentsch, 2007) and CFTR (Di *et al.*, 2006). The suggestion that CFTR is the Cl^- ion channel essential for phagosomal acidification (Di *et al.*, 2006) has been disproved by several groups (Barriere *et al.*, 2009; Haggie and Verkman, 2007; Lamothe and Valvano, 2008; Painter *et al.*, 2008, 2010) and therefore cannot explain the delayed maturation arrest of BcCVs in CFTR-defective macrophages.

An alternative role for CFTR-mediated Cl^- anions is the activation of the myeloperoxidase (MPO) to produce hypochlorous acid (HOCl) in neutrophils (Painter and Wang, 2006). Cl^- availability to PMNs affects their HOCl production and microbicidal capacity, which is reduced in CFTR-defective cells (Painter *et al.*, 2008) as a consequence of a defective traffic of Cl^- (Painter *et al.*, 2010). Others have shown that zebra fish neutrophils with a defective CFTR also have a similarly reduced microbicidal activity (Phennicie *et al.*, 2010). However, myeloperoxidase is not typically expressed in macrophages and monocytes. Unpublished research in our laboratory using *B. cenocepacia* labelled at the cell surface with Cl^--sensitive and -insensitive fluorophores allowed us to measure the intraphagosomal Cl^- concentration in CFTR-normal and CFTR-defective murine macrophages upon bacterial engulfment (Ostapska and Valvano, unpublished data). No significant difference in the Cl^- ion levels was found between the two cell types at various times post infection ranging from 1 to 10 h. There was also no significant difference in the survival rate of *B. cenocepacia* in CFTR-normal and CFTR-defective macrophages that were exposed to Cl^-. As expected, HOCl production was not detected in CFTR-normal and CFTR-defective macrophage lysates, as well as in human THP1 monocyte lysates suggesting that this particular oxidative bactericidal mechanism employed ubiquitously in neutrophils is indeed absent in the macrophages employed for these studies.

One alternative possibility to explain a role for CFTR in microbicidal clearance is accumulation of misfolded protein in aggregates, which has been observed in CF airway epithelial cells (Okiyoneda *et al.*, 2010). Defective CFTR results in defective autophagy and decreases the clearance of protein aggregates, which in turns leads to inflammation (Luciani *et al.*, 2010, 2011, 2012). Conceivably, defective autophagy might alter the ability of macrophages to clear the intracellular infection. Indeed, several intracellular pathogens reside in vacuoles with autophagosome features (Deretic and Levine, 2009). In this scenario, the CFTR defect and Bcc infection could have a synergistic effect down-regulating autophagy, which would be consistent with the pronounced delay in the BcCV maturation under a functionally impaired CFTR. At least in the case of the CFTR-ΔF508 mutation, the defective CFTR protein cannot progress to the apical cell membrane and abnormal proteins accumulate in a macromolecular complex called the aggresome, which sequesters components of the autophagy pathway.

An autophagosome is a compartment that engulfs non-functional organelles and parts of the cytoplasm, and subsequently delivers these components to the lysosome for degradation and recovery of essential nutrients during periods of cell starvation or stress. Recent results have shown that in CFTR-ΔF508 murine macrophages, *B. cenocepacia* persists in vacuoles that rarely fuse with lysosomes. These infected cells produce higher levels IL-1β than infected CFTR-normal macrophages (Abdulrahman *et al.*, 2011). It was previously suggested that intracellular *B. cenocepacia* reside in autophagosomes that fuse with lysosomes at later stages of infection (Sajjan *et al.*, 2008). A comparison between infected CFTR-ΔF508 and CFTR-normal macrophages revealed that the autophagy genes are down-regulated. However, the CFTR-ΔF508 cells contain more bacteria than the CFTR-normal macrophages. Treatment of CFTR-ΔF508 macrophages with the autophagy stimulating agent rapamycin markedly decreases *B. cenocepacia* infection *in vitro* by enhancing the clearance of *B. cenocepacia via* induced autophagy (Abdulrahman *et al.*, 2011). More importantly, rapamycin decreased bacterial burden and inflammation in the lungs of CF mice, revealing that autophagy can control the fate of the *B. cepacia* intracellular infection and the ensuing inflammatory response.

The mechanism for the link between *B. cenocepacia* and autophagy in CFTR-ΔF508 macrophages was uncovered in a recent study (Abdulrahman *et al.*, 2012), and involves the adaptor protein SQSTM1/p62. This is an ubiquitin binding adaptor protein that is required for the delivery of ubiquitinated cargo to the autophagosome. In

CFTR-normal macrophages, p62 depletion and overexpression cause increased and decreased bacterial intracellular survival, respectively. However, depletion of p62 in CFTR-ΔF508 macrophages results in decreased bacterial survival, whereas overexpression of p62 leads to increased *B. cenocepacia* intracellular multiplication. Depletion of p62 from CFTR-ΔF508 macrophages results in the release of the autophagy molecule beclin 1 from the mutant CFTR aggregates and allows p62 redistribution and recruitment to the BcCV, mediating the recruitment of the autophagy marker LC3 to the BcCV and bacterial clearance via autophagy. Therefore, the studies discussed above provide evidence that a strong link exists between *B. cenocepacia* infection and the CFTR defect, and suggest an avenue to investigate novel targets to reduce the bacterial load and proinflammatory responses in the CF patients.

Conclusions

The survival of Bcc bacteria, in particular *B. cenocepacia* and *B. multivorans* in macrophages provides a model system to investigate the interactions of opportunistic pathogens with host cells responsible for innate immune recognition and beneficial or pathological inflammatory responses at the molecular level. The elucidation of the details of the survival, especially the ability of *B. cenocepacia* to synergize with the CFTR defect and its consequences on the mechanism of autophagy, provides new avenues to explore novel therapeutic approaches to ameliorate and control the profuse inflammatory responses triggered by infection and eradicate the bacterial load.

Acknowledgements

Research in the Valvano laboratory has been supported by grants from Cystic Fibrosis Canada, Marie Curie Actions, the Cystic Fibrosis Trust, the Natural Sciences and Engineering Research Council and the Canadian Institutes of Health Research. C.L.S. was supported by a Cystic Fibrosis Canada fellowship. M.A.V. was supported by a Canada Research Chair in Microbial Pathogenesis and Infectious Diseases.

References

Abdulrahman, B.A., Khweek, A.A., Akhter, A., Caution, K., Kotrange, S., Abdelaziz, D.H., Newland, C., Rosales-Reyes, R., Kopp, B., McCoy, K., *et al.* (2011). Autophagy stimulation by rapamycin suppresses lung inflammation and infection by *Burkholderia cenocepacia* in a model of cystic fibrosis. Autophagy *7*, 1359–1370.

Abdulrahman, B.A., Abu Khweek, A., Akhter, A., Caution, K., Tazi, M., Hassan, H., Zhang, Y., Rowland, P.D., Malhotra, S., Aeffner, F., *et al.* (2012). Depletion of the ubiquitin binding adaptor molecule SQSTM1/p62 from macrophages harboring *cftr* ΔF508 mutation improves the delivery of *Burkholderia cenocepacia* to the autophagic machinery. J. Biol. Chem. *288*, 2049–2058.

Alix, E., and Blanc-Potard, A.B. (2007). MgtC: a key player in intramacrophage survival. Trends Microbiol. *15*, 252–256.

Anacarso, I., Guerrieri, E., Bondi, M., de Niederhausern, S., Iseppi, R., Sabia, C., Contri, M., Borella, P., and Messi, P. (2010). Influence of *Legionella pneumophila* and other water bacteria on the survival and growth of *Acanthamoeba polyphaga*. Arch. Microbiol. *192*, 877–882.

Aubert, D.F., Flannagan, R.S., and Valvano, M.A. (2008). A novel sensor kinase-response regulator hybrid controls biofilm formation and virulence in *Burkholderia cenocepacia*. Infect. Immun. *76*, 1979–1991.

Aubert, D.F., O'Grady, E.P., Hamad, M.A., Sokol, P.A., and Valvano, M.A. (2013). The *Burkholderia cenocepacia* sensor kinase hybrid AtsR is a global regulator modulating quorum-sensing signalling. Environ. Microbiol. *15*, 372–385.

Barriere, H., Bagdany, M., Bossard, F., Okiyoneda, T., Wojewodka, G., Gruenert, D.C., Radzioch, D., and Lukacs, G.L. (2009). Revisiting the role of cystic fibrosis transmembrane conductance regulator and counterion permeability in the pH regulation of endocytic organelles. Mol. Biol. Cell *20*, 3125–3141.

Bazzini, S., Udine, C., and Riccardi, G. (2011). Molecular approaches to pathogenesis study of *Burkholderia cenocepacia*, an important cystic fibrosis opportunistic bacterium. Appl. Microbiol. Biotechnol. *92*, 887–895.

Burns, J.L., Jonas, M., Chi, E.Y., Clark, D.K., Berger, A., and Griffith, A. (1996). Invasion of respiratory epithelial cells by *Burkholderia (Pseudomonas) cepacia*. Infect. Immun. *64*, 4054–4059.

Bylund, J., Burgess, L.A., Cescutti, P., Ernst, R.K., and Speert, D.P. (2006). Exopolysaccharides from *Burkholderia cenocepacia* inhibit neutrophil chemotaxis and scavenge reactive oxygen species. J. Biol. Chem. *281*, 2526–2532.

Cascales, E. (2008). The type VI secretion toolkit. EMBO Rep. *9*, 735–741.

Coenye, T., and Vandamme, P. (2003). Diversity and significance of *Burkholderia* species occupying diverse ecological niches. Environ. Microbiol. *5*, 719–729.

Corbett, C.R., Burtnick, M.N., Kooi, C., Woods, D.E., and Sokol, P.A. (2003). An extracellular zinc metalloprotease gene of *Burkholderia cepacia*. Microbiology *149*, 2263–2271.

Deretic, V., and Levine, B. (2009). Autophagy, immunity, and microbial adaptations. Cell Host Microbe *5*, 527–549.

Di, A., Brown, M.E., Deriy, L.V., Li, C., Szeto, F.L., Chen, Y., Huang, P., Tong, J., Naren, A.P., Bindokas, V., *et al.* (2006). CFTR regulates phagosome acidification in macrophages and alters bactericidal activity. Nat. Cell Biol. *8*, 933–944.

Downey, D.G., Martin, S.L., Dempster, M., Moore, J.E., Keogan, M.T., Starcher, B., Edgar, J., Bilton, D., and Elborn, J.S. (2007). The relationship of clinical and inflammatory markers to outcome in stable patients with cystic fibrosis. Pediatr. Pulmonol. *42*, 216–220.

Drevinek, P., and Mahenthiralingam, E. (2010). *Burkholderia cenocepacia* in cystic fibrosis: epidemiology and molecular mechanisms of virulence. Clin. Microbiol. Infect. *16*, 821–830.

Duff, C., Murphy, P.G., Callaghan, M., and McClean, S. (2006). Differences in invasion and translocation of *Burkholderia cepacia* complex species in polarised lung epithelial cells *in vitro*. Microb. Pathog. *41*, 183–192.

Elsinghorst, E.A. (1994). Measurement of invasion by gentamicin resistance. Methods Enzymol. *236*, 405–420.

Flannagan, R.S., and Valvano, M.A. (2008). *Burkholderia cenocepacia* requires RpoE for growth under stress conditions and delay of phagolysosomal fusion in macrophages. Microbiology *154*, 643–653.

Flannagan, R.S., Linn, T., and Valvano, M.A. (2008). A system for the construction of targeted unmarked gene deletions in the genus *Burkholderia*. Environ. Microbiol. *10*, 1652–1660.

Flannagan, R.S., Jaumouillé, V., Huynh, K.K., Plumb, J.D., Downey, G.P., Valvano, M.A., and Grinstein, S. (2012). *Burkholderia cenocepacia* disrupts host cell actin cytoskeleton by inactivating Rac and Cdc42. Cell Microbiol. *14*, 239–254.

Ganesan, S., and Sajjan, U.S. (2011). Host Evasion by *Burkholderia cenocepacia*. Front Cell Infect. Microbiol. *1*, 25.

Gavrilin, M.A., Abdelaziz, D.H., Mostafa, M., Abdulrahman, B.A., Grandhi, J., Akhter, A., Abu Khweek, A., Aubert, D.F., Valvano, M.A., Wewers, M.D., *et al.* (2012). Activation of the pyrin inflammasome by intracellular *Burkholderia cenocepacia*. J. Immunol. *188*, 3469–3477.

Gingues, S., Kooi, C., Visser, M.B., Subsin, B., and Sokol, P.A. (2005). Distribution and expression of the ZmpA metalloprotease in the *Burkholderia cepacia* complex. J. Bacteriol. *187*, 8247–8255.

Haggie, P.M., and Verkman, A.S. (2007). Cystic fibrosis transmembrane conductance regulator-independent phagosomal acidification in macrophages. J. Biol. Chem. *282*, 31422–31428.

Hamad, M.A., Skeldon, A.M., and Valvano, M.A. (2010). Construction of aminoglycoside-sensitive *Burkholderia cenocepacia* strains for use in studies of intracellular bacteria with the gentamicin protection assay. Appl. Environ. Microbiol. *76*, 3170–3176.

Hunt, T.A., Kooi, C., Sokol, P.A., and Valvano, M.A. (2004). Identification of *Burkholderia cenocepacia* (formerly *Burkholderia cepacia* genomovar III) genes required for bacterial survival *in vivo*. Infect. Immun. *72*, 4010–4022.

Huynh, K.K., Plumb, J.D., Downey, G.P., Valvano, M.A., and Grinstein, S. (2010). Inactivation of macrophage Rab7 by *Burkholderia cenocepacia*. J. Innate Immun. *2*, 522–533.

Isberg, R.R., and Falkow, S. (1985). A single genetic locus encoded by *Yersinia pseudotuberculosis* permits invasion of cultured animal cells by *Escherichia coli*K-12. Nature (London) *317*, 262–264.

Jani, A.J., and Cotter, P.A. (2010). Type VI secretion: not just for pathogenesis anymore. Cell Host Microbe. *8*, 2–6.

Jentsch, T.J. (2007). Chloride and the endosomal–lysosomal pathway: emerging roles of CLC chloride transporters. J. Physiol. *578*, 633–640.

Johnson, W.M., Tyler, S.D., and Rozee, K.R. (1994). Linkage analysis of geographic and clinical clusters in *Pseudomonas cepacia* infections by multilocus enzyme electrophoresis and ribotyping. J. Clin. Microbiol. *32*, 924–930.

Keig, P.M., Ingham, E., Vandamme, P.A., and Kerr, K.G. (2002). Differential invasion of respiratory epithelial cells by members of the *Burkholderia cepacia* complex. Clin. Microbiol. Infect. *8*, 47–49.

Keith, K.E., Killip, L., He, P., Moran, G.H., and Valvano, M.A. (2007). *Burkholderia cenocepacia* C5424 produces a pigment with antioxidant properties using a homogentisate intermediate. J. Bacteriol. *189*, 9057–9065.

Keith, K.E., Hynes, D.W., Sholdice, J.E., and Valvano, M.A. (2009). Delayed association of the NADPH oxidase complex with macrophage vacuoles containing the opportunistic pathogen *Burkholderia cenocepacia*. Microbiology *155*, 1004–1015.

Kooi, C., and Sokol, P.A. (2009). *Burkholderia cenocepacia* zinc metalloproteases influence resistance to antimicrobial peptides. Microbiology *155*, 2818–2825.

Lamothe, J., Thyssen, S., and Valvano, M.A. (2004). *Burkholderia cepacia* complex isolates survive intracellularly without replication within acidic vacuoles of *Acanthamoeba polyphaga*. Cell Microbiol. *6*, 1127–1138.

Lamothe, J., Huynh, K.K., Grinstein, S., and Valvano, M.A. (2007). Intracellular survival of *Burkholderia cenocepacia* in macrophages is associated with a delay in the maturation of bacteria-containing vacuoles. Cell Microbiol. *9*, 40–53.

Lamothe, J., and Valvano, M.A. (2008). *Burkholderia cenocepacia*-induced delay of acidification and phagolysosomal fusion in cystic fibrosis transmembrane conductance regulator (CFTR)-defective macrophages. Microbiology *154*, 3825–3834.

Landers, P., Kerr, K.G., Rowbotham, T.J., Tipper, J.L., Keig, P.M., Ingham, E., and Denton, M. (2000). Survival and growth of *Burkholderia cepacia* within the free-living amoeba *Acanthamoeba polyphaga*. Eur. J. Clin. Microbiol. Infect. Dis. *19*, 121–123.

Leitão, J.H., Sousa, S.A., Ferreira, A.S., Ramos, C.G., Silva, I.N., and Moreira, L.M. (2010). Pathogenicity, virulence factors, and strategies to fight against *Burkholderia cepacia* complex pathogens and related species. Appl. Microbiol. Biotechnol. *87*, 31–40.

Loutet, S.A., and Valvano, M.A. (2010). A decade of *Burkholderia cenocepacia* virulence determinant research. Infect. Immun. *78*, 4088–4100.

Luciani, A., Villella, V.R., Esposito, S., Brunetti-Pierri, N., Medina, D., Settembre, C., Gavina, M., Pulze, L., Giardino, I., Pettoello-Mantovani, M., *et al.* (2010). Defective CFTR induces aggresome formation and lung inflammation in cystic fibrosis through ROS-mediated autophagy inhibition. Nat. Cell Biol. *12*, 863–875.

Luciani, A., Villella, V.R., Esposito, S., Brunetti-Pierri, N., Medina, D.L., Settembre, C., Gavina, M., Raia, V., Ballabio, A., and Maiuri, L. (2011). Cystic fibrosis: a disorder with defective autophagy. Autophagy *7*, 104–106.

Luciani, A., Villella, V.R., Esposito, S., Gavina, M., Russo, I., Silano, M., Guido, S., Pettoello-Mantovani, M., Carnuccio, R., Scholte, B., *et al.* (2012). Targeting autophagy as a novel strategy for facilitating the therapeutic action of potentiators on ΔF508 cystic fibrosis transmembrane conductance regulator. Autophagy *8*, 1657–1672.

McClean, S., and Callaghan, M. (2009). *Burkholderia cepacia* complex: epithelial cell-pathogen confrontations and potential for therapeutic intervention. J. Med. Microbiol. *58*, 1–12.

Macdonald, K.L., and Speert, D.P. (2008). Differential modulation of innate immune cell functions by the *Burkholderia cepacia* complex: *Burkholderia cenocepacia* but not *Burkholderia multivorans* disrupts maturation and induces necrosis in human dendritic cells. Cell Microbiol. *10*, 2138–2149.

Maloney, K.E., and Valvano, M.A. (2006). The *mgtC* gene of *Burkholderia cenocepacia* is required for growth under magnesium limitation conditions and intracellular survival in macrophages. Infect. Immun. *74*, 5477–5486.

Marolda, C.L., Hauröder, B., John, M.A., Michel, R., and Valvano, M.A. (1999). Intracellular survival and saprophytic growth of isolates from the *Burkholderia cepacia* complex in free-living amoebae. Microbiology *145*, 1509–1517.

Martin, D.W., and Mohr, C.D. (2000). Invasion and intracellular survival of *Burkholderia cepacia*. Infect. Immun. *68*, 24–29.

Moura, J.A., Cristina de Assis, M., Ventura, G.C., Saliba, A.M., Gonzaga, L., Si-Tahar, M., Marques Ede, A., and Plotkowski, M.C. (2008). Differential interaction of bacterial species from the *Burkholderia cepacia* complex with human airway epithelial cells. Microbes Infect. *10*, 52–59.

O'Grady, E.P., Viteri, D.F., and Sokol, P.A. (2012). A unique regulator contributes to quorum sensing and virulence in *Burkholderia cenocepacia*. PLoS ONE *7*, e37611.

Okiyoneda, T., Barrière, H., Bagdány, M., Rabeh, W.M., Du, K., Höhfeld, J., Young, J.C., and Lukacs, G.L. (2010). Peripheral protein quality control removes unfolded CFTR from the plasma membrane. Science *329*, 805–810.

Painter, R.G., and Wang, G. (2006). Direct measurement of free chloride concentrations in the phagolysosomes of human neutrophils. Anal. Chem. *78*, 3133–3137.

Painter, R.G., Bonvillain, R.W., Valentine, V.G., Lombard, G.A., LaPlace, S.G., Nauseef, W.M., and Wang, G. (2008). The role of chloride anion and CFTR in killing of *Pseudomonas aeruginosa* by normal and CF neutrophils. J Leukoc Biol *83*, 1345–1353.

Painter, R.G., Marrero, L., Lombard, G.A., Valentine, V.G., Nauseef, W.M., and Wang, G. (2010). CFTR-mediated halide transport in phagosomes of human neutrophils. J. Leukoc. Biol. *87*, 933–942.

Phennicie, R.T., Sullivan, M.J., Singer, J.T., Yoder, J.A., and Kim, C.H. (2010). Specific resistance to *Pseudomonas aeruginosa* infection in zebrafish is mediated by the Cystic Fibrosis transmembrane conductance regulator. Infect. Immun. 1–36.

Pukatzki, S., Ma, A.T., Sturtevant, D., Krastins, B., Sarracino, D., Nelson, W.C., Heidelberg, J.F., and Mekalanos, J.J. (2006). Identification of a conserved bacterial protein secretion system in *Vibrio cholerae* using the *Dictyostelium* host model system. Proc. Natl. Acad. Sci. U.S.A. *103*, 1528–1533.

Records, A.R. (2011). The type VI secretion system: a multipurpose delivery system with a phage-like machinery. Mol. Plant Microbe. Interact. *24*, 751–757.

Rosales-Reyes, R., Aubert, D.F., Tolman, J.S., Amer, A.O., and Valvano, M.A. (2012a). *Burkholderia cenocepacia* type VI secretion system mediates escape of type II secreted proteins into the cytoplasm of infected macrophages. PLoS ONE *7*, e41726.

Rosales-Reyes, R., Skeldon, A.M., Aubert, D.F., and Valvano, M.A. (2012b). The Type VI secretion system of *Burkholderia cenocepacia* targets multiple Rho family GTPases disrupting the actin cytoskeleton and the assembly of NADPH oxidase complex in macrophages. Cell Microbiol. *14*, 255–273.

Saini, L., Galsworthy, S., John, M., and Valvano, M.A. (1999). Intracellular survival of *Burkholderia cepacia* complex isolates in the presence of macrophage cell activation. Microbiology *145*, 3465–3475.

Sajjan, U.S., Yang, J.H., Hershenson, M.B., and LiPuma, J.J. (2006). Intracellular trafficking and replication of *Burkholderia cenocepacia* in human cystic fibrosis airway epithelial cells. Cell Microbiol. *8*, 1456–1466.

Sajjan, S.U., Carmody, L.A., Gonzalez, C.F., and LiPuma, J.J. (2008). A type IV secretion system contributes to intracellular survival and replication of *Burkholderia cenocepacia*. Infect. Immun. *76*, 5447–5455.

Saldías, M.S., and Valvano, M.A. (2009). Interactions of *Burkholderia cenocepacia* and other *Burkholderia cepacia* complex bacteria with epithelial and phagocytic cells. Microbiology *155*, 2809–2817.

Saldías, M.S., Ortega, X., and Valvano, M.A. (2009). *Burkholderia cenocepacia* O antigen lipopolysaccharide prevents phagocytosis by macrophages and adhesion to epithelial cells. J. Med. Microbiol. *58*, 1542–1548.

Saldías, M.S., Lamothe, J., Wu, R., and Valvano, M.A. (2008). *Burkholderia cenocepacia* requires the RpoN sigma factor for biofilm formation and intracellular trafficking within macrophages. Infect. Immun. *76*, 1059–1067.

Schmerk, C.L., and Valvano, M.A. (2013). *Burkholderia multivorans* survival and trafficking within macrophages. J. Med. Microbiol. *62*, 173–184.

Schwab, U., Leigh, M., Ribeiro, C., Yankaskas, J., Burns, K., Gilligan, P., Sokol, P., and Boucher, R. (2002). Patterns of epithelial cell invasion by different species of the *Burkholderia cepacia* complex in well-differentiated human airway epithelia. Infect. Immun. *70*, 4547–4555.

Segal, A.W. (2005). How neutrophils kill microbes. Annu. Rev. Immunol. *23*, 197–223.

Sonawane, N.D., and Verkman, A.S. (2003). Determinants of [Cl^-] in recycling and late endosomes and Golgi complex measured using fluorescent ligands. J. Cell Biol. *160*, 1129–1138.

Sory, M.P., and Cornelis, G.R. (1994). Translocation of a hybrid YopE-adenylate cyclase from *Yersinia enterocolitica* into HeLa cells. Mol. Microbiol. *14*, 583–594.

Steinberg, B.E., Huynh, K.K., Brodovitch, A., Jabs, S., Stauber, T., Jentsch, T.J., and Grinstein, S. (2010). A cation counterflux supports lysosomal acidification. J. Cell Biol. *189*, 1171–1186.

Taylor, J.B., Hogue, L.A., LiPuma, J.J., Walter, M.J., Brody, S.L., and Cannon, C.L. (2010). Entry of *Burkholderia* organisms into respiratory epithelium: CFTR, microfilament and microtubule dependence. J. Cyst. Fibros. *9*, 36–43.

Tolman, J.S., and Valvano, M.A. (2012). Global changes in gene expression by the opportunistic pathogen *Burkholderia cenocepacia* in response to internalization by murine macrophages. BMC Genomics *13(1):63.*

van der Heijden, J., and Finlay, B.B. (2012). Type III effector-mediated processes in *Salmonella* infection. Future Microbiol. *7*, 685–703.

Vandamme, P., and Dawyndt, P. (2011). Classification and identification of the *Burkholderia cepacia* complex: Past, present and future. Syst. Appl. Microbiol. *34*, 87–95.

Ventura, G.M., Balloy, V., Ramphal, R., Khun, H., Huerre, M., Ryffel, B., Plotkowski, M.C., Chignard, M., and Si-Tahar, M. (2009). Lack of MyD88 protects the immunodeficient host against fatal lung inflammation triggered by the opportunistic bacteria *Burkholderia cenocepacia*. J. Immunol. *183*, 670–676.

Vial, L., Chapalain, A., Groleau, M.C., and Deziel, E. (2011). The various lifestyles of the *Burkholderia cepacia* complex species: a tribute to adaptation. Environ. Microbiol. *13*, 1–12.

Vieira, O.V., Botelho, R.J., and Grinstein, S. (2002). Phagosome maturation: aging gracefully. Biochem. J. *366*, 689–704.

Vignais, P.V. (2002). The superoxide-generating NADPH oxidase: structural aspects and activation mechanism. Cell Mol. Life Sci. *59*, 1428–1459.

Voth, D.E., Broederdorf, L.J., and Graham, J.G. (2012). Bacterial Type IV secretion systems: versatile virulence machines. Future Microbiol. *7*, 241–257.

Wells, M.Y., and Rikihisa, Y. (1988). Lack of lysosomal fusion with phagosomes containing *Ehrlichia risticii* in P388D1 cells: abrogation of inhibition with oxytetracycline. Infect. Immun. *56*, 3209–3215.

Zelazny, A.M., Ding, L., Elloumi, H.Z., Brinster, L.R., Benedetti, F., Czapiga, M., Ulrich, R.L., Ballentine, S.J., Goldberg, J.B., Sampaio, E.P., *et al.* (2009). Virulence and cellular interactions of *Burkholderia multivorans* in chronic granulomatous disease. Infect. Immun. *77*, 4337–4344.

The Phytopathogenic *Burkholderia*

8

Bruna G. Coutinho, Mayukh Das, Zulma R. Suárez-Moreno, Carlos F. Gonzalez and Vittorio Venturi

Abstract

Historically, species of the genus *Burkholderia* have been known to be pathogenic to plants, and currently they are known as the etiological agents of several diseases of a wide variety of plants, causing wilt, rot, blight and canker. This group includes: *B. andropogonis*, *B. caryophilii*, *B. cepacia*, *B. cenocepacia*, *B. gladioli*, *B. glumae*, *B. plantarii*, *B. seminalis* and *B. rhizoxinica*. These species have evolved a spectrum of extracellular factors, including phytotoxins, which allow them to cause disease in a variety of hosts. Some of the species presented here may produce these extracellular factors to obtain nutrients by affecting cell membranes, whereas others are typical vascular pathogens invading plant host by wounding the plant surface.

In this chapter we describe the preferred plant hosts of each phytopathogenic *Burkholderia* species, and their virulence factors, as well as the mechanisms involved in disease development.

Introduction

The genus *Burkholderia* has been phylogenetically well defined, consisting of species that are functionally diverse (Compant *et al.*, 2008). In fact, *Burkholderia* species are recurrently isolated from many different environmental niches, including soil and water, plants, animals and humans. Interestingly, a growing number of *Burkholderia* strains and species have also been reported as plant-associated bacteria (Balandreau *et al.*, 2001; Caballero-Mellado *et al.*, 2004; Sessitsch *et al.*, 2005; Barrett and Parker, 2006; Bontemps *et al.*, 2010). This latter group of plant associated *Burkholderia* spp. can be free-living in the rhizosphere as well as epiphytic and endophytic, including obligate endosymbionts and phytopathogens. The phytopathogenic *Burkholderia* species are the causal agent of disease in economically important crops, such as rice, onion, carnation and maize.

To address *Burkholderia* as plant pathogens it is important to highlight that plant diseases are a result of complex interactions between pathogen and host. In fact, phyllosphere and rhizosphere play major roles in determining the health and productivity of crops, and may serve as major avenues of interaction with pathogens that can lead to infection and culminate in disease. For example, some foliar pathogens exhibit an epiphytic phase on plants and can maintain populations in healthy plants (Leveau and Lindow, 2001; Waffa, 2002). In addition, recent studies have also demonstrated that there is a very important connection between avirulent resident bacteria and incoming pathogens resulting in a marked effect on the host immune response and, thus on disease development (Venturi and Passos, 2012).

The dynamics of plant disease are often described in terms of the interaction of three major components: humans, vectors and time. Plant diseases occur when pathogens infect susceptible plants under environmental conditions that favour infection and growth of the pathogen, which explains why small variations on these conditions may reduce or enhance bacterial pathogenicity.

The introduction of new crops or cultivars into planting regimes by humans can also introduce susceptible hosts. In addition, changes in weather patterns are known to affect crop physiology and, therefore their susceptibility to infection, as is the case for the emergence of bacterial panicle blight in rice in Latin America and the United States, caused by *Burkholderia glumae.* In this particular case, disease emergence was dramatically enhanced by changes in temperature and moisture patterns (Zeigler and Alvarez, 1987; Nandakumar *et al.*, 2007, 2009).

This chapter describes the phytopathogenic species belonging to the genus *Burkholderia,* some of which are also pathogens of humans. The *Burkholderia* group of plant pathogens includes *B. andropogonis, B. caryophylli, B. cepacia, B. gladioli, B. cenocepacia, B. glumae, B. plantarii* and *B. seminalis,* which are the etiological agents of disease for a variety of plants. We are also including the endosymbiotic *B. rhizoxinica,* which resides inside *Rhizopus* spp. known to be a rice pathogen. The most relevant pathogenicity factors are presented and their interaction with the plant host is also explored. A summary of the main features of each phytopathogenic *Burkholderia* spp. is presented in Table 8.1.

General features of the diseases and their aetiological agents

B. andropogonis

[*Bacterium*] *andropogonis* was described by Smith (1911) as the causative agent of stripe disease of sorghum (*Andropogon* sp.) and leaf spot of velvet bean (*Stizolobium deeringianum*). In the same publication, [*Bacterium*] *woodsii* was described as an important pathogen of carnation (*Dianthus caryophyllus*) (Smith, 1911). After more than a decade, both species were transferred to the genus *Pseudomonas* (Stevens, 1925; Stapp, 1928). Later it was shown that *P. andropogonis* and *P. woodsii* both belonged to the solanacearum rRNA branch of the pseudomonads (rRNA group II) (Palleroni, 1984; De Vos *et al.*, 1985). Based on DNA–rRNA hybridizations, *P. andropogonis* was transferred to the genus *Burkholderia* by Gillis *et al.* (1995) and the same authors stated that *P. woodsii* and *B. andropogonis* could be considered as synonyms because of the nearly identical phenotypic characteristics of both microorganisms (Gillis *et al.*, 1995). More recently, DNA–DNA hybridization, 16S rDNA sequence, protein pattern, and fatty acid analyses confirmed the previously published phenotypic data (Coenye *et al.*, 2001).

Table 8.1 Main characteristics of the plant pathogenic *Burkholderia* spp.

Species	Plant host	Virulence factors
B. andropogonis	Sorghum, carnation and others	Rhizobitoxine
B. caryophylli	Carnation	Polygalacturonase*, cellulase*
B. cenocepacia	Banana	Lipase*, protease*, ptw T4SS, quorum sensing
B. cepacia	Onion	Polygalacturonase, lipase*, protease*, ptw T4SS, quorum sensing
B. gladioli	*Gladiolus* spp., rice and others	Polygalacturonase*, toxoflavin
B. glumae	Rice and others	Polygalacturonase*, lipase, T2SS, Hrp T3SS, quorum sensing
B. plantarii	Rice	Tropolone, quorum sensing
B. seminalis	Apricot fruit	ND
B. rhizoxinica (endosymbiont of *Rhizopus* spp.)	Rice	Rhizoxin

ND, not determined; ptw T4SS, plant tissue watersoaking type IV secretion system; T2SS, type II secretion system; Hrp T3SS, hypersensitive response and pathogenicity type III secretion system. *Produced by the pathogen and important for the virulence of other phytopathogenic bacteria, but without any experimental data on its importance for this particular species.

Figure 8.1 Symptoms of bacterial leaf stripe, caused by *Burkholderia andropogonis*, on sorghum. (Photo courtesy Anne K. Vidaver, University of Nebraska-Lincoln.)

B. andropogonis has a broad host range and geographical occurrence. Its hosts include 52 species of 15 families of unrelated monocot and dicot plants such as white clover, carnation, bougainvillea, and other ornamental plants (Cother *et al.*, 2004). *B. andropogonis* has also been recovered from lesions on crops such as sweet and field maize (*Zea mays*), blueberry (*Vaccinium* species), sorghum (*Sorghum bicolor*), carnation (*Dianthus caryophyllus*), coffee (*Coffea arabica*), statice (*Limonium* sp.), rye (*Secale cereale*), and clover (*Trifolium* species) (Bradbury, 1986). *B. andropogonis*, the causal agent of bacterial leaf stripe, is one of the three major bacterial diseases of sorghum, and strict quarantine regulations against importation of *B. andropogonis*-infested sorghum feed grains and seeds are imposed by numerous countries (Claflin, 2000).

Bacterial leaf stripe is characterized by red streaks and blotches on the leaves and sheaths (Fig. 8.1). Lesions at first appear as watersoaked, dark green specks. Single lesions are only a few millimetres in diameter, but they may extend to appear later as long red streaks between the veins. When lesions merge, they may cover a large part of a leaf blade (Tarr, 1962; Claflin, 2000). Bacterial stripe is most severe under warm wet conditions and can survive on weeds such as Johnsongrass and Sudangrass. Resistant varieties in combination with weed control can reduce disease incidence (Claflin, 2000). More recently, *B. andropogonis* has been reported as the causal agent of blight of golden cane palms in Australia (Young *et al.*, 2007), bacterial leaf spot on *Ruscus* sp. in Brazil (Almeida *et al.*, 2009) and brown leaf spot of citrus in the United States (Duan *et al.*, 2009).

B. caryophylli

B. caryophylli is primarily described as the causal agent of bacterial wilt of carnations; however, *B. caryophylli* has also been shown to infect *Dianthus barbatus* (Sweet William) and *D. allwoodii* (Border Pinks) through artificial inoculation (OEPP/EPPO, 1997) as well as rot onions (Ballard *et al.*, 1970; Palleroni, 1984) *B. caryophylli* is widely distributed throughout North and South America as well as Europe (OEPP/EPPO, 1997).

B. caryophylli can only enter plants through wounds, and subsequently colonizes the vascular system of the stem and roots (Dowson, 1929; Hellmers, 1958; Garibaldi, 1967). The natural spread of the pathogen is very slow and over extremely short distances. The primary infection source is infected cuttings taken from mother plants with a latent infection. Bacteria can pass from one cutting to another in the water of the propagating bed or, if the cuttings are held in water, before planting out (Dimcock, 1950). The observed slow, scattered spread of the disease indicates that spread occurs only from one root system to another. Bacterial 'slime' can be observed when stems crack, and can then serve as inoculum from one plant to another (Dimcock, 1950). Symptoms may take two to three years to manifest themselves, particularly when cuttings are mildly infected and maintained at relatively low temperatures (< 20°C). This enables the dissemination of infected cuttings, which can be devastating to the floral industry. Foliage becomes grayish-green, later yellowing and wilting (Dimcock, 1950; Saddler, 1994).

In infected plants, at soil temperatures below 17°C, longitudinal, internodal stem cracks appear, usually at the base of the plant, and later develop into deep cankers (Fig. 8.2). At 20–25°C, cankers are less frequent and wilting is the common symptom. Visual observation of peeled stems reveals sticky, brownish-yellow, narrow or broad, longitudinal stripes in the vascular tissue; in cross section, these appear as irregular brownish spots with a watersoaked margin (Dimcock, 1950; Saddler, 1994). Roots of infected plants, once wilting

Figure 8.2 Symptoms of stem canker, caused by *Burkholderia caryophylli*, on carnation. (Photo courtesy the Istituto Sperimentale Per La Floricoltura, Italy.)

occurs, are more or less rotten, the plants being easily pulled out of the soil and, on cutting, roots show discontinuous brown spots (Saddler, 1994).

B. cepacia

In 1942 W.H. Burkholder reported a scale rot of onion bulbs that was common to New York State and the pathogen was given the name *P. alliicola* (Burkholder, 1942). In 1950, he reported on another rot of onion known among growers as 'sour skin' and 'slippery skin' that did not have the watery and glassy appearance of the rot caused by *P. alliicola*, but was slimy and yellow (Burkholder, 1950). Burkholder studied the pathogen and determined that young growing plants did not show disease symptoms, whereas inoculated bulbs did. He concluded that infection probably took place when the onion tops were cut off at harvest time and that the increased use of high-speed mechanical toppers may have contributed to the incidence of disease. Host range studies showed that the pathogen did not cause disease on potato, narcissus bulbs, or calla lily rhizomes. After conducting morphological and biochemical testing Burkholder (1950) concluded 'that the pathogen and the disease it causes appear to be new to literature' and reported the new bacterium as belonging to the genus *Pseudomonas* with the species Latin name, 'cepacia' meaning 'of or like onion'.

B. cepacia is the causal agent of sour skin, which is primarily a disease of onion (Burkholder, 1950), and has been reported from onion-growing areas all over the world (Kawamoto and Lorbeer, 1974). Primary symptoms of *B. cepacia* infection on onions include a slimy (but initially firm), pale yellow to light brown decay and breakdown of one or a few inner bulb scales (Fig. 8.3). Adjacent outer scales and the centre of the bulb may remain firm. Externally, bulbs appear sound, but the neck region may soften after leaves have collapsed. In advanced stages, healthy scales can slip off during handling. Young leaves sometimes die back, starting at the tips (Kawamoto and Lorbeer, 1972, 1974).

Disease losses due to sour skin may occur in storage, but infection generally occurs in the field after bulb formation. The pathogen, present in soil or irrigation water, invades bulb tissue through

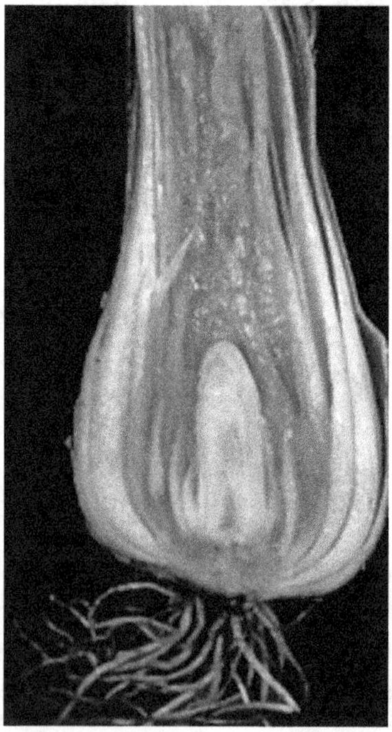

Figure 8.3 Symptoms of sour skin of onion caused by *Burkholderia cepacia*. (Photo courtesy the American Phytopathological Society.)

wounds made by removal of the tops at harvest. The disease is most damaging at temperatures above 30°C (Parke and Gurian-Sherman, 2001). Young leaves are much more susceptible than mature leaves, which are usually symptom-free. Infection can advance into the bulb via an infected leaf and its corresponding scale with infection not moving into adjacent scales (Kawamoto and Lorbeer, 1972, 1974). Infection with *B. cepacia* has been associated with the splashing of contaminated furrow or overhead irrigation water (Teviotdale *et al.*, 1989).

Strains of *B. cepacia* produce a single endopolygalacturonase at low pH which is responsible for the maceration of onion tissue (Ulrich, 1975; Gonzalez and Vidaver, 1979; Gonzalez *et al.*, 1997). Gonzalez and co-workers (1997) have cloned and sequenced the plasmid-borne *pehA* (pectic enzyme hydrolase) gene from the plant pathogenic type strain of *B. cepacia*. Peh-negative derivatives remain capable of causing a plant tissue water soaking (ptw) phenotype that results from loss of cell membrane integrity and the accumulation of fluids in the intracellular spaces of plant tissue (Gonzalez *et al.*, 1997). Broad-spectrum antibacterial and antifungal metabolites are also produced by *B. cepacia in vitro* and may predispose plant tissue to infection or contribute to their competitive survival in soil (Hu and Young, 1998).

B. cenocepacia

B. cenocepacia was proposed by Vandamme and co-workers, in 2003, after the reclassification of *B. cepacia* genomovar III (Vandamme *et al.*, 2003). Strains of this species have been isolated from numerous plant rhizospheres, sputum of cystic fibrosis patients, pharmaceutical and industrial products, soil and surface waters (LiPuma *et al.*, 2002; Ramette *et al.*, 2005; Dalmastri *et al.*, 2007; Jacobs *et al.*, 2008; Mahenthiralingam *et al.*, 2008). In addition, several strains of *B. cenocepacia* were isolated from the rhizosphere of onions and were shown to be pathogenic in a detached onion bulb scale assay, causing symptoms of water soaking, maceration and/or necrosis, which are similar to the ones caused by *B. cepacia* (Jacobs *et al.*, 2008). The onion pathogenicity of *B. cenocepacia* may be explained by its ability to cause the ptw phenotype (Engledow *et al.*, 2004).

Moreover, *B. cenocepacia* has been identified as the causal agent of banana finger-tip rot in Taiwan. Diseased fingers are recognized externally by their small and distorted shape with the pulp presenting brown discoloration and gelatinous consistency (Lee *et al.*, 2003; Lee and Chan, 2007). However, losses from the disease are not severe, as only a few fingers from the youngest hands are affected (Lee *et al.*, 2004). Studies have shown that the stigma of the banana finger is an infection court for *B. cenocepacia*, as 8% of stigmata from symptomless fingers were infested with bacteria. The pathogen could be disseminated in rain water, moving from infected tissues to healthy ones. When in the stigma, the bacteria may multiply, because of the high relative humidity of the environment, and then penetrate into banana fingers through natural openings of stigmata. The contamination of the pulp leads to discoloration and softening of the fruit; however, the bacteria never spread beyond the vascular system within the floral stem (Lee *et al.*, 2004).

B. gladioli

B. gladioli was originally described as a phytopathogen on *Gladiolus* species, causing rot of gladiolus corms (Severini, 1913). This species also includes the formerly known species *P. marginata* (McCulloch, 1921) and *P. alliicola* (Burkholder, 1942). However, other hosts include onion, iris and freesia (Tsuchiya and Muko, 1963). *B. gladioli* has also been reported as a pathogen of *Dendrobium* orchids in Hawaii (Uchida, 1995). The most common symptoms caused by infection by *B. gladioli* are infected leaves containing dark-green to brown lesions surrounded by water soaking (Fig. 8.4). When infected plants are exposed to environmental conditions which include rain or high humidity, the watersoaked area can expand rapidly to cause soft decay over the entire leaf, which can result in leaf drop off (Keith *et al.*, 2005).

In a survey to determine rice seed-associated bacterial populations, Cottyn *et al.* (2001) found

Figure 8.4 Symptoms of an orchid artificially inoculated with *Burkholderia gladioli* (Keith *et al.*, 2005). (Photo courtesy the American Phytopathological Society.)

that, out of 428 isolates tested, two were identified as *B. gladioli* and five as *B. glumae.* They reported the *B. gladioli* as being less aggressive on greenhouse inoculated plants than *B. glumae* (Cottyn *et al.*, 2001). The *B. gladioli* isolates caused brown necrotic lesions on the sheath and chlorosis of the second and third leaves within three days post inoculation; with eventual collapse of the plant. In an earlier study Miyagawa *et al.* (2000) reported that proliferation of *B. glumae* and *B. plantarii* is suppressed in rice seeds infected with *B. gladioli* (Miyagawa, 2000). More recently, Maeda and co-workers (2006) suggested that the ecology and pathogenicity mechanism of *B. gladioli* on rice plants remains unclear. Moreover, they suggest that the antagonistic activity of *B. gladioli* against *B. glumae* and *B. plantarii* in rice plants as reported by Miyagawa *et al.* (2000) shows that an understanding of the ecology of *B. gladioli* on rice plants would facilitate not only the development of control systems for *B. gladioli*-induced disease, but also an understanding of the ecology of *B. glumae* and *B. plantarii* on rice plants. The co-association of *B. glumae* and *B. gladioli* in rice plants showing panicle blight symptoms was further supported by the reports from Panama and the southern United States (Nandakumar *et al.*, 2007, 2009).

B. glumae

B. glumae causes rot of rice grains and seedlings (panicle blight), and has emerged during the past two decades as the most important bacterial pathogen of rice in Japan, Korea, Taiwan, Indonesia, India, Vietnam, Philippines and the southern United States (Goto and Ohata, 1956; Uematsu *et al.*, 1976; Maeda *et al.*, 2006; Sayler *et al.*, 2006; Lim *et al.*, 2009; Nandakumar *et al.*, 2009). Additionally, it has been reported that *B. glumae* can cause wilting symptoms in sunflower, tomato, sesame, perilla, aubergine, and hot pepper (Jeong *et al.*, 2003).

Panicle blight, caused by infection from *B. glumae,* is characterized by brown or straw-coloured discoloration of florets on a panicle branch (Fig. 8.5). Lesions are not apparent below the grain, and panicle branches remain green. However, in some cases, there is browning of the sheath, which may be confused with sheath rot caused by *Sarocladium oryzae.* The grain stops developing and the florets turn grey. It is more severe in late-planted rice where the rice is maturing and late night temperatures are high. Rice crops planted with infected seeds will suffer severe losses during summers with unusually high nighttime temperatures (Shahjahan *et al.*, 2000b).

The pathogen, *B. glumae,* is seed borne (Goto and Ohata, 1956; Uematsu *et al.*, 1976) and grows epiphytically on the plant through the booting stage (Goto and Ohata, 1956). The bacterium multiplies rapidly on the surface of emerging panicles and infects flowers just after emergence (Goto and Ohata, 1956). Symptoms include small (1 to 5 mm) tan lesions with brown margins on the leaves and spikelets, resulting in abortion of infected kernels before complete fill. Infected florets may exhibit dark grey or brown bases, but the rachis of the panicle remains green. A prominent sheath lesion several centimetres long with a tan centre and reddish brown margin may form on individual flag leaf sheaths of infected tillers. The panicles often remain erect, due to a loss of grain weight, instead of bending over. The disease

Figure 8.5 Symptoms of panicle blight showing unfilled glumes on rice caused by *Burkholderia glumae*. (Photo courtesy Joseph P. Krausz, Texas A&M University-College Station.)

causes spikelet sterility as well as discoloration of the developing grains and significant yield losses. Glumes become bleached or discoloured with the tip of the glume being a lighter colour than the base. The bacterium produces toxoflavin (1,6-dimethylpyrimidio[5,4-e]–1,2,4-triazine-5,7 (1H,6H)-dione), which is essential for pathogenicity (Sato *et al.*, 1989; Iiyama *et al.*, 1995; Furuya *et al.*, 1997; Suzuki *et al.*, 1998; Nagamatsu, 2001).

In Japan, *B. glumae* causes yield losses of up to 34% (Yoshimura *et al.*, 1987; Tsushima *et al.*, 1996). Although panicle blight has been reported in Japan, Korea and Taiwan (Goto and Ohata, 1956; Uematsu *et al.*, 1976; Urakami *et al.*, 1994) it was not until 2007, that a survey of 623 symptomless rice seed samples reported that two originating from Hainan province contained *B. glumae* (Luo *et al.*, 2007). The authors indicated that this survey confirmed that the pathogen was already in the mainland of China. They noted that this raised the potential risk of seedling rot outbreaks in large-scale production operations that grow rice seedlings for transplantation. In Korea, the occurrence of rice grain rot is becoming more common, because recent summers in the southern Korean peninsula have been very hot and humid (Jeong *et al.*, 2003). Furthermore, the observed occurrence of disease caused by *B. glumae* on non-traditional crops such as sunflower, tomato, sesame, perilla, aubergine, and hot pepper in Korea are postulated to occur when field crops and rice are cultivated in close proximity in the same growing season. The authors speculate that since this disease is highly dependent on weather conditions, the ecological aspects of their occurrence need further investigation (Jeong *et al.*, 2003), and we concur that abiotic factors may indeed play a role in the infection of non-traditional hosts by *B. glumae*.

Panicle blight of rice is also one of the most important new diseases in rice production areas of the southern United States. In 1995, an epidemic of panicle blight occurred across Arkansas, Louisiana and Texas (Shahjahan *et al.*, 2000b). Its occurrence has been increasing since 1995 and a survey conducted in 1999, in Louisiana, confirmed the association of panicle blight of rice with *B. glumae*. In field trials artificially inoculated plants at full boot-heading stages showed as much as 80% yield loss (Shahjahan *et al.*, 2000a). In Louisiana it has been noted that little disease developed in fields until the weather turned very hot in July and August and more than 90% of cultivars planted in the rice growing area were susceptible (Shahjahan *et al.*, 2000b). Rice pathologists in Texas have noted that new introduced cultivars appear to be more susceptible to panicle blight (Joseph Krausz, personal communication). *B. glumae* isolates from Arkansas showed little or no genetic diversity, with more than 99.5% identity, by 16S–23S rDNA intergenic transcribed spacer (ITS) sequence analysis (Sayler *et al.*, 2006). The genetic similarity of isolates from Arkansas was not surprising since the same ITS region of *B. glumae* isolates from different regions of Japan were also found to be identical (Takeuchi *et al.*, 1997). Nandakumar and co-workers (2009) reported that 76 and 5% of the strains isolated from rice fields in Louisiana, Arkansas, Texas and

Mississippi were confirmed as *B. glumae* and *B. gladioli*, respectively.

B. plantarii

B. plantarii causes a seed-borne disease of rice and is considered to persist in the basal parts of the rice plant in the field (Iguchi, 1991; Azegami, 1994). It is speculated that the bacteria then move to upper parts of plants and infect rice seeds (Azegami, 1999). When the rice seeds carrying the bacterium are sown in nursery boxes, the disease occurs under high temperature and high humidity. Seedling blight occurs only in nursery boxes used for mechanical transplanting. Outbreaks sometimes occur at commercial rice seedling centres and can cause severe damage (Fig. 8.6). The toxin tropolone is associated with the symptoms in plants (Azegami *et al.*, 1987; Urakami *et al.*, 1994).

B. plantarii has been isolated from several graminaceous weeds (Satou and Matuda, 1998), one pteridophytous weed, and three dicotyledonous weeds (Tanaka *et al.*, 1994) growing on the levees of paddy fields. The mode of transmission from the weeds to rice seeds has not yet been clarified; however, an alternative route of infection has been suggested by Miyagawa and Inoue (2002). *B. plantarii* latently inhabits graminaceous weeds on levees, probably around the basal parts of the weeds (Satou and Matuda, 1998). When the level of paddy water is high enough to cover these basal parts or probably during rainfall, *B. plantarii* is released into paddy water. *B. plantarii* disperses by rain splash and wind, latently infecting rice panicles (Miyagawa and Inoue, 2002).

Figure 8.6 (A) Uninoculated rice plant. (B) Rice plant inoculated with *Burkholderia plantarii*, exhibiting symptoms of seedling blight (Iwai *et al.*, 2002). (Photo courtesy the American Phytopathological Society.)

Taxonomically Coenye *et al.*, (1999) have proposed that the single *B. vandii* isolate described by Urakami *et al.* (1994) represents the same species as *B. plantarii*, and that *B. vandii* should be considered a junior synonym of *B. plantarii*. The proposal is based on the high level of DNA–DNA hybridization data that are in agreement with protein electrophoretic similarity despite some differences in biochemical characteristics. They propose that biochemical differences observed indicate the presence of biochemical subgroups within this species and may warrant the designation of different biotypes or even subspecies within *B. plantarii* (Coenye *et al.*, 1999).

B. seminalis

The species *B. seminalis* was proposed by Vanlaere and co-workers in 2008 after the characterization of several *Burkholderia cepacia* complex (Bcc) isolates through polyphasic taxonomic approaches. Strains from this species had been isolated from sputum of cystic fibrosis patients and rice seed surface (Vanlaere *et al.*, 2008). However, in 2009, this bacterium was identified as the causal agent of fruit rot disease of apricot in China (Fang *et al.*, 2009; Li *et al.*, 2010).

The first stage of the disease is the appearance of water-soaked lesions on the surface of pre-harvested fruits. As these lesions expand and elongate, the diseased tissue becomes soft and watery. The progress of the disease results in soft rot of the entire fruit, which reduces the quality and marketability of the apricot. Studies suggest that the bacteria infect the fruit tissue via wounds, as the pathogen was only able to cause disease when inoculated by needle puncture or infiltration, but not by spraying (Fang *et al.*, 2009).

B. rhizoxinica

Several species of phytopathogenic fungi belonging to the genus *Rhizopus* are known for causing rice seedling blight, which starts with an abnormal swelling of the seedling roots without any sign of infection by the pathogen (Ibaragi, 1973; Furuya *et al.*, 1974; Gho *et al.*, 1978; Noda *et al.*, 1980). This typical symptom is caused by a macrocyclic polyketide metabolite, designated rhizoxin that was believed to be produced by the fungus (Iwasaki *et al.*, 1984; Iwasaki *et al.*, 1986). However, studies have shown that the phytotoxin is, in fact, produced by endosymbiotic bacteria living within the fungal cytosol. This endosymbiont was designated *Burkholderia rhizoxinica* (Partida-Martinez and Hertweck, 2005; Partida-Martinez *et al.*, 2007a). In this way the fungus harbours bacterial endosymbionts for the production of a toxin able to weakens or even kill the plant, releasing nutrients for both. Besides that, bacteria also benefit from the 'safe' niche provided by the interior of the fungus (Partida-Martinez and Hertweck, 2005).

B. rhizoxinica was shown to be vertically transmitted through spores of its fungal host, which is a more efficient way to ensure stability of symbiosis (Partida-Martinez *et al.*, 2007b). Moreover, fungal spores are only formed when symbionts are present, meaning that the vegetative reproduction of the fungus is dependent on the endosymbiont. The loss of the ability to produce endogenous sporulation factors, probably due to evolution processes, led the fungus reliant on the endobacteria for reproduction, which is another mechanism for the maintenance of the bacterial–fungal association (Partida-Martinez *et al.*, 2007b). Interestingly, when the endobacteria is isolated, it loses the ability to produce rhizoxin in pure culture over time. Partida-Martinez and Hertweck (2005) suggest that this fact may be due to genomic instability or down-regulation of biosynthesis genes. However, when the symbiont is re-introduced into a symbiont-free *Rhizopus* strain, the production of rhizoxin is activated again, which indicates that the interaction between fungus and bacteria provide signals that trigger and/or maintain rhizoxin biosynthesis (Partida-Martinez and Hertweck, 2005).

Virulence associated factors and their molecular features

Extracellular enzymes

Polygalacturonase

The plant cell wall is composed of approximately 90% polysaccharides and 10% proteins. The polysaccharides are divided in cellulose, hemicelluloses and pectin and the proportion of each one may vary among species (McNeil *et al.*, 1984). Phytopathogenic bacteria are able to secrete several types of cell-wall-degrading enzymes to breakdown these polysaccharides and invade the plant (Ward and Moo-Young, 1989; Massa *et al.*, 2007b) and first produced are the pectin degrading enzymes (English *et al.*, 1971; Jones *et al.*, 1972; Tomassini *et al.*, 2009). This class of enzymes not only weakens the plant cell wall, but also exposes the other polymers to degradation by hemicellulases and cellulases, being important virulence factors for many plant pathogenic bacteria, including soft rot-causing bacteria (Kotoujansky, 1987; Rodriguez-Palenzuela *et al.*, 1991; Massa *et al.*, 2007b).

Polygalacturonases are one of the most studied microbial pectic enzymes and are responsible for the hydrolysis of the pectic glycosidic bonds. These enzymes may attack the polymer in a random or in a terminal way, being called endo-polygalacturonase or exo-polygalacturonase, respectively. An endo-polygalacturonase activity will produce a mixture of oligomers of various sizes, whereas an exo-polygalacturonase will generate only one product. The production of polygalacturonases has been previously reported in *B. cepacia*, *B. caryophylli*, *B. gladioli* and *B. glumae* (Gross and Cody, 1985); however, only the polygalacturonases produced by *B. cepacia* ATCC 25416 and *B. glumae* ATCC 33617 have been characterized (Ulrich, 1975; Gonzalez *et al.*, 1997; Massa *et al.*, 2007a,b; Degrassi *et al.*, 2008). The purification and characterization of the pectolytic enzyme from *B. cepacia* ATCC 25416 revealed the presence of an endo-polygalacturonase, named PehA, showing significant low degree of amino acid identity and similarity with known fungal and bacterial endo-polygalacturonases. The production of PehA is dependent on the presence of polygalacturonic acid and the enzyme has optimal activity at pH 3.5. Such acidic optimal pH value can be explained by the fact that the enzyme is produced during the infection of the onion tissue, and the onion juice leaking out from the site of infection has a pH value of approximately 4.0 (Gonzalez *et al.*, 1997; Massa *et al.*, 2007b). Moreover, the *pehA* gene of *B. cepacia* ATCC 25416 is present on a plasmid and its expression is regulated by quorum sensing (QS) (Gonzalez *et al.*, 1997; Aguilar *et al.*, 2003). The activity of PehA is essential for maceration of onion tissue (Gonzalez *et al.*, 1997).

Two endo-polygalacturonases of *B. glumae* ATCC 33617, PehA and PehB were previously purified and characterized (Degrassi *et al.*, 2008). These two enzymes display very high identity (over 70%) with each other and with PehA from *B. cepacia* ATCC 25416. Moreover, they have the same primary structures and similar biochemical properties as the PehA of *B. cepacia*, indicating that these three endo-polygalacturonases constitute a subfamily of these enzymes. However, the deletion of either one of these genes, *pehA* or *pehB*, from *B. glumae* AU6208 did not abolish the ability of the pathogen to cause panicle blight in rice. These results suggest either a functional redundancy between the two isoenzymes or they are not involved in the initial stages of the disease process of *B. glumae*. Moreover, the studies of Degrassi and co-workers (2008) have shown that the expression of PehB, but not PehA, is regulated by QS.

Interestingly, a further study with 54 clinical and agricultural isolates of *B. cenocepacia* reported a positive correlation between the presence of *pehA* gene and the pathogenicity towards onion (Springman *et al.*, 2009; Gonzalez and LiPuma, unpublished). These results suggest that the production of polygalacturonase is an important virulence factor for *B. cenocepacia* as it was also shown for *B. cepacia*.

Cellulases

Cellulases have a role in breaking down the cellulose present in primary and secondary cell walls of the plant. Usually, this type of enzyme is not essential for pathogenicity, but can act in synergy with other cell-wall degrading enzymes to attack the plant host (Toth *et al.*, 2003). *B. caryophylli* was reported many years ago to produce cellulases (Gehring, 1962) and, although no correlation between virulence and cellulase production has been shown as yet, secretion of the enzyme by plant pathogens may facilitate the entry of pathogens and, thus have a synergistic role in virulence. The sequencing of *B. glumae* genomes has revealed the presence of genes related to the production of cellulases, however, their role in virulence is currently unknown.

Lipases

Lipases are produced by many different bacterial species and they catalyse the hydrolysis and synthesis of esters of glycerol with long-chain fatty acids. These microbial lipases have numerous applications in different industrial sectors since they are able to catalyse a large variety of reactions. The lipase of *B. glumae* for example is currently used in the detergent industry (Frenken *et al.*, 1993; Rosenau and Jaeger, 2000; Boekema *et al.*, 2007).

A previous study from Devescovi and co-workers (2007) reported the ability of the clinical strain *B. glumae* AU6208, isolated from an infant with chronic granulomatous disease, to cause severe panicle blight on rice. Moreover, it was shown that a mutant of this strain defective in *lipA*, which encodes LipA lipase, showed a dramatic decrease in its ability to cause disease in rice when compared to the wild type strain, suggesting that this enzyme is an important virulence factor (Devescovi *et al.*, 2007). This suggests that LipA lipase may be able to act together with other cell wall-degrading enzymes to break down components of plant cell walls. Lipases may be involved in the hydrolysis of ester bonds from the polysaccharides forming plant walls, promoting their degradation. Another possible role of these enzymes is in the degradation of epicuticular waxes, which cover and protect the host epidermal cells. It is worth noting that a relevant insight from the study of Devescovi and co-workers (2007) was that the lipase production by *B. glumae* is also regulated by QS.

B. cepacia and *B. cenocepacia* have been reported as producing lipases, however this phenotype was mainly observed in clinical strains (Lonon *et al.*, 1988; Jorgensen *et al.*, 1991; Carvalho *et al.*, 2007), consequently no information is available regarding their importance in plant pathogenicity in these species.

Proteases

As the plant cell wall is composed of only 10% proteins, the importance of proteases for plant–pathogen interactions is limited; however stress induced wall proteins such as extensins have been reported to have a role in plant defence (Bassler *et al.*, 1993). In contrast, there have been extensive studies on the proteases of pathogens of animals, whose intercellular matrix is mainly protein (Walton, 1994).

Strains of *B. cepacia* and *B. cenocepacia* produce two zinc metalloproteases ZmpA and ZmpB, which are important for the virulence in animal models (Corbett *et al.*, 2003; Gingues *et al.*, 2005; Kooi *et al.*, 2006). The expression of both proteases was shown to be regulated by QS (Sokol *et al.*, 2003; Malott *et al.*, 2005; Kooi *et al.*, 2006). The production of ZmpA was also reported for the onion pathogen *B. cepacia* ATCC 25416, however its importance to virulence in a plant system is yet to be determined (Aguilar *et al.*, 2003; Gingues *et al.*, 2005).

The importance of proteases in few other plant pathogenic systems is well established. For instance, it was reported that the two major extracellular proteases (PRT1 and PRT2) produced by *Xanthomonas campestris* pv. *campestris* are important virulence factors, since protease-deficient mutants show reduced symptoms (Dow *et al.*, 1990). This finding is not applicable for *Erwinia* spp., as their proteases appear to play only a minor role in pathogenesis of plants (Marits *et al.*, 1999). The *Erwinia* proteases may act either to provide amino acids for biosynthesis of microbial proteins or degradation of resistance host proteins (Heilbronn and Lyon, 1990; Kyostio *et al.*, 1991).

Toxins

Phytopathogenic bacteria produce a wide variety of phytotoxins that are toxic to plant cells and influence the appearance of symptoms along plant disease. In some cases, the toxins are not essential for pathogenicity, but they may act as virulence factors affecting disease severity.

Toxoflavin

B. glumae is the major cause of bacterial panicle blight and rice grain rot (Cottyn *et al.*, 1996; Suzuki *et al.*, 1998) and is able to cause bacterial wilt in many other field crops (Jeong *et al.*, 2003). This pathogen produces the phytotoxins toxoflavin, fervenulin and reumycin, which are very similar in their structure (Fig. 8.7) (Sato *et al.*, 1989; Kim *et al.*, 2004; Suzuki *et al.*, 2004). Toxoflavin is the most toxic to rice and is encoded by the polycistronic operon *toxABCDE* (Sato *et al.*, 1989; Suzuki *et al.*, 2004). Mutations in this operon cause impairment in the production of all of the three phytotoxins, which suggests that fervenulin and reumycin also need the toxoflavin biosynthesis pathway for their biosynthesis (Kim *et al.*, 2004; Suzuki *et al.*, 2004).

Toxoflavin acts as an active electron carrier between NADH and oxygen, and can produce hydrogen peroxide and bypass the cytochrome system (Latuasan and Berends, 1961). It causes a reduction on the elongation of sprouts and roots of rice seedlings and symptoms of bacterial grain rot on panicles (Iiyama *et al.*, 1995; Suzuki *et al.*, 1998; Yoneyama *et al.*, 1998). This toxin is also responsible for the bacterial wilt in crops other than rice (Jeong *et al.*, 2003) and is produced not only by *B. glumae*, but also by strains of *B. gladioli* (Suzuki *et al.*, 1998).

Toxoflavin is essential for the pathogenesis of *B. glumae*, as strains deficient in toxin production are avirulent (Yoneyama *et al.*, 1998; Jeong *et al.*, 2003; Kim *et al.*, 2004). In contrast, other studies using mutants disrupted in toxoflavin biosynthesis genes, demonstrated that they were still able to induce symptoms on rice panicles, even though these symptoms were milder than those caused by the parental strain (Suzuki *et al.*, 2004). These results suggest that, although toxoflavin is of great importance for the virulence of this pathogen, there are other virulence factors involved in this complicated process and that some of these may be strain specific.

Toxoflavin transport in *B. glumae* is dependent on the polycistronic operon *toxFGHI* and its expression and secretion is regulated by several different elements, such as ToxR, a LysR-type transcriptional activator, ToxJ, another transcriptional activator of toxoflavin biosynthesis and

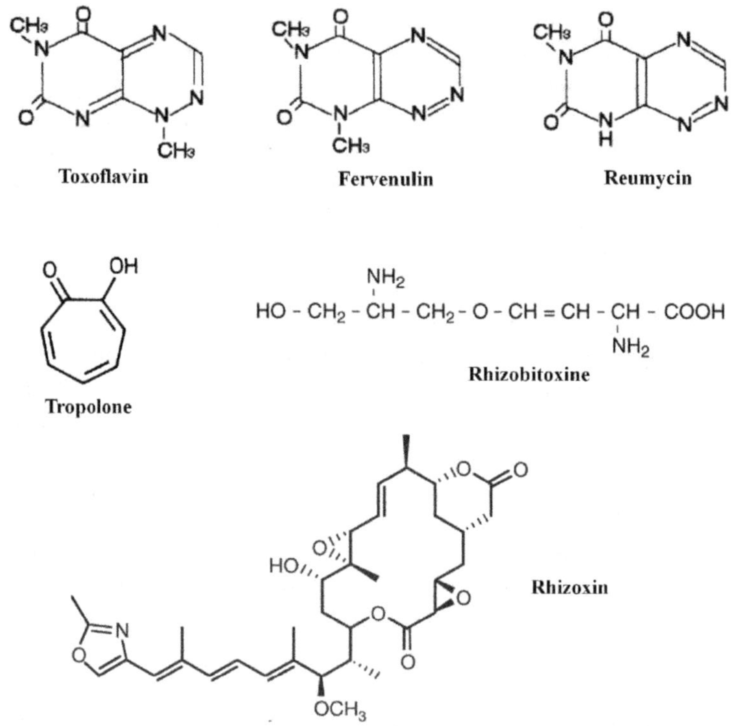

Figure 8.7 Structure of toxins produced by phytopathogenic *Burkholderia* spp. Toxoflavin, fervenulin and reumycin produced by *B. glumae*; Tropolone produced by *B. plantarii*; Rhizobitoxine produced by *B. andropogonis*; and Rhizoxin produced by *B. rhizoxinica*.

TofI/TofR, the QS system (Kim *et al.*, 2004; Suzuki *et al.*, 2004). A model for the regulatory cascade leading to the expression of the toxoflavin biosynthesis genes was proposed, in which TofI/TofR regulates the expression of ToxJ that, in turn, activates the transcription of *toxR*. Both ToxJ and ToxR activate the expression of the genes for the biosynthesis and transport of toxoflavin (Kim *et al.*, 2004; Kim *et al.*, 2009). In addition, toxoflavin was shown to act as a co-inducer for ToxR (Kim *et al.*, 2004), and its expression is also dependent on growth temperature, which is maximal at 37°C, while no detectable amount is produced at 25–28°C (Matsuda and Sato 1988).

Further studies with toxoflavin have revealed its antibacterial, antifungal and herbicidal activities (Nagamatsu, 2001). Moreover, this toxin was identified as a potential novel anticancer agent, since it has activity against Polo-like kinases that are involved in signal transduction (Goh *et al.* 2004).

Tropolone

B. plantarii produces tropolone, a non-benzoic aromatic compound with a seven-membered ring that has properties characteristic of phenols and acids (Fig. 8.7) (Azegami *et al.*, 1985). Tropolone is toxic to rice seedlings causing chlorosis, root growth inhibition and wilting, similar to the symptoms caused by the pathogen itself (Azegami *et al.*, 1985, 1987).

Tropolone shares many characteristics with siderophores, such as its chemical structure that is similar to phenolate-catecholate type of siderophores (Neilands, 1981). This toxin is also inactivated in the presence of iron (Azegami *et al.*, 1987) and is not produced at temperatures higher than 35°C (Azegami *et al.*, 1986), which is also a characteristic of siderophores (Weinberg, 1978).

Besides rice toxicity, tropolone also exhibits antimicrobial and antifungal activities (Trust, 1975; Morita *et al.*, 2003) and displays a strong insecticidal activity against *Tyrophagus putrescentiae* (mould mite) and *Dermatophagoides farina* (dust mite) (Morita *et al.*, 2003).

Rhizobitoxine

Rhizobitoxine is an enol-ether amino acid (2-amino-4-[2-amino-3-hydroxypropoxy]-trans-3-butenoic acid) produced by *B. andropogonis* and by the legume symbiont *Bradyrhizobium elkanii* (Fig. 8.7) (Mitchell *et al.*, 1986). This molecule induces foliar chlorosis of soybeans and is regarded as a phytotoxin since it is responsible for the chlorotic symptoms caused by *B. andropogonis* in maize and sorghum (Johnson *et al.*, 1959; Owens and Wright, 1965; Mitchell *et al.*, 1986; Xiong and Fuhrmann, 1996). Interestingly, rhizobitoxine plays a beneficial role in the establishment of the symbiosis between *B. elkanii* and its host (Duodu *et al.*, 1999).

Rhizobitoxine is an analogue of cystathionine, thus it acts as an inhibitor of β-cystathionase, which catalyses the conversion of cystathionine to homocysteine, pyruvate, and ammonium, in the methionine biosynthesis pathway (cystathionine β-lyase) of bacteria and plants (Owens *et al.*, 1968; Giovanelli *et al.*, 1973). Another role of rhizobitoxine is the inhibition of 1-aminocyclopropane-1-carboxylate (ACC) synthase in the ethylene biosynthesis pathway (Yasuta *et al.*, 1999). ACC synthase is the rate-limiting enzyme in ethylene biosynthesis in plants catalysing the conversion of *S*-adenosylmethionine to ACC. This ability to reduce plant ethylene production was shown to enhance the nodulation process and competitiveness of *B. elkanii* (Duodu *et al.*, 1999) and is probably a way to reduce plant defence reactions by *B. andropogonis* (Okazaki *et al.*, 2004).

Rhizoxin

Rhizoxin (Fig. 8.7) is an antimitotic polyketide macrolide responsible for rice seedling blight and, consequently, for severe losses in agriculture (Iwasaki *et al.*, 1984; Tsuruo *et al.*, 1986). For a long time, it was believed that this phytotoxin was produced by the saprotrophic fungus *Rhizopus microsporus*, however Partida-Martinez and Hertweck (2005) convincingly demonstrated that a bacterial endosymbiont, designated *B. rhizoxinica*, was responsible for the production of rhizoxin. Its toxicity is due to the efficient binding of β-tubulin thereby blocking cell division in the roots. This antimitotic activity of rhizoxin has also been observed in other eukaryotic cells, including human and murine tumour cells and their resistant sublines (Iwasaki *et al.*, 1986; Tsuruo *et al.*, 1986; Tolcher *et al.*, 2000). For this reason, rhizoxin is being studied as a potential antitumour drug (Jordan *et al.*, 1998; Lafontaine *et al.*, 2003).

The gene cluster responsible for rhizoxin biosynthesis has been identified in the chromosome of *B. rhizoxinica*. It is composed of 10 open reading frames designated *rhiA-rhiJ* that encode a hybrid system of non-ribosomal peptide synthetase and polyketide synthase, and is flanked by two transposase genes, which may represent a potentially mobile region of the genome (Partida-Martinez and Hertweck, 2007; Lackner *et al.*, 2011).

Pathogen-associated secretion systems

To successfully infect a host plant, bacteria depend on protein secretion systems that are responsible for the secretion of proteins into the extracellular milieu or transport proteins and/or DNA directly into the host cell cytosol.

The type II secretion system (T2SS) is the most commonly used system for the transport of proteins from the bacterial periplasm to the extracellular milieu and is responsible for the secretion of enzymes such as proteases, lipases and cell wall-degrading enzymes that are important for plant–pathogen interactions (Johnson *et al.*, 2006). The exact contribution of each T2SS substrate to bacterial virulence and the identity of their cognate plant targets are not yet fully understood. However, it is likely that the enzymes secreted by this system have an important role in the degradation of the plant cell wall, which is a considerable barrier for phytopathogenic bacteria (McNeil *et al.*, 1984).

T2SS systems were revealed as virulence factors in some plant pathogenic bacteria such as *R. solanacearum*, *Erwinia* spp. and *Xanthomonas campestris* pv. *campestris*, in which mutants for this secretion system are non-pathogenic (Dums *et al.*, 1991; Lindeberg and Collmer, 1992; Cianciotto, 2005; Liu *et al.*, 2005). Two studies have demonstrated that when *B. glumae* is mutated in its general secretion pathway genes for the T2SS it has a lower colonization capability and conferred reduced disease severity in rice than the parent strain. This indicates that the T2SS-dependent extracellular proteins are important for full virulence of this pathogen (Kang *et al.*, 2008; Goo *et al.*, 2010). Until now, the importance of T2SS for pathogenicity has only been studied in *B. glumae*, but it would be expected that this secretion system is also important in other phytopathogenic *Burkholderia* spp., as mutants for this secretion system would lack transport of such virulence factors as polygalacturonases, lipases, cellulases or proteases.

Many phytopathogenic bacteria utilize a Hrp (hypersensitive response and pathogenicity) type III secretion system (T3SS) to inject virulence effector proteins into plant cells. For these bacteria the effector proteins transported by the Hrp T3SS directly interfere with the host defence response and suppress plant defences (Alfano and Collmer, 2004; Mudgett, 2005). In *R. solanacearum* the Hrp T3SS and associated 'Pop' (*Pseudomonas* outer protein) play an important role in pathogenicity (Alfano and Collmer, 2004). The study of Kang and co-workers (2008) demonstrated a reduced virulence of *B. glumae* Hrp T3SS-deficient mutants, as well as reduced growth in rice, indicating that these mutants are less pathogenic because their ability to colonize plant tissues is impaired.

As mentioned above, phytopathogenic strains of *B. cepacia* and *B. cenocepacia* produce the polygalacturonase PehA, which is important for maceration of onion tissue. Curing the PehA-encoding plasmid of *B. cepacia* ATCC 25416 makes it unable to macerate onion (Gonzalez *et al.*, 1997). However, PehA mutants remain capable of causing the plant tissue watersoaking (ptw) phenotype (Gonzalez *et al.*, 1997), which results from loss of cell membrane integrity and the accumulation of fluids in the intracellular spaces of plant tissue. In a survey of Bcc species it was found that all *B. cepacia* and *B. cenocepacia* strains tested, as well as a *B. vietnamiensis* strain, produced the ptw phenotype, suggesting that members of this genomic species may produce a cytotoxic factor that affects plant cell integrity (Engledow *et al.*, 2004). Although the nature of the cytotoxic factor is currently unknown, Engledow and co-workers (2004) have shown that a type IV secretion system (T4SS) is indispensable for the generation of ptw.

The T4SS from *Agrobacterium tumefaciens* was the first to be identified and it delivers oncogenic transfer DNA and effector proteins to plant cells during infection (Christie, 2001; Kumar and Das, 2002). This system has been divided into four subgroups depending on its specialized function, these are: effector molecule translocation into host target cells; conjugation of chromosomal and plasmid DNA; DNA uptake and transformation; and DNA release into the extracellular milieu (Backert and Meyer, 2006).

The ptw T4SS identified in *B. cenocepacia* K56-2 is encoded by a plasmid, is a chimera composed of VirB/D4 and F-specific subunits, and is responsible for the secretion of a plant cytotoxic protein(s) that causes the ptw phenotype. Mutations on the ptw T4SS cluster abolish the phenotype, which shows its importance for the pathogenicity of *B. cenocepacia* K56–2. It is believed that the role of the effector(s) is to provide bacterial cells with nutrients by causing leakage of the plant cell cytosol (Engledow *et al.*, 2004).

Quorum sensing

N-Acylhomoserine lactone (AHL) QS system is involved in the pathogenicity of several phytopathogenic bacteria through the regulation of biofilm formation, plasmid transfer, production of degradative enzymes and many other phenotypes (von Bodman *et al.*, 2003).

Members of the Bcc are known to possess a common and highly conserved AHL QS system designated CepI/CepR (Gotschlich *et al.*, 2001; Lutter *et al.*, 2001; Eberl, 2006). CepI is the responsible for the synthesis of *N*-octanoylhomoserine lactone (C8-AHL), which is able to interact with CepR to regulate gene expression. This QS system is involved in the regulation of proteases, chitinases, swarming motility, biofilm formation, siderophore and polygalacturonase production (Venturi *et al.*, 2004; Eberl, 2006). In the phytopathogenic species of the Bcc, *B. cenocepacia* and *B. cepacia*, QS was shown to be important for their virulence against alfalfa and onion, respectively (Aguilar *et al.*, 2003; Eberl, 2006). *B. cepacia cepI* mutants are less virulent in an onion rot model, as demonstrated by attenuated tissue maceration compared to the wild type strain. This attenuation was attributed to the decrease in secreted polygalacturonase activity by 40% in the QS mutant (Aguilar *et al.*, 2003).

The rice pathogen *B. glumae* has an AHL QS system, designated TofI/TofR, that is highly similar to the CepI/R system producing and responding to C8-AHL (Kim *et al.*, 2004). As mentioned above, the QS of *B. glumae* indirectly regulates the synthesis and secretion of toxoflavin, as it activates the transcriptional activator ToxJ (Kim *et al.*, 2004). Kim and co-workers (2004) suggested that since the production of toxoflavin is highly energy consuming, *B. glumae* cells must ensure that they have reached a critical cell density before start producing it. The production of lipase and flagella were also found to be regulated by QS, and both of them have been reported as important virulence factors of this pathogen, since mutants for either one of these traits are no longer pathogenic to rice (Devescovi *et al.*, 2007; Kim *et al.*, 2007). The results of these studies demonstrate that QS is essential for the pathogenicity of *B. glumae*.

B. plantarii, another rice pathogen, also regulates virulence factors in a cell-density manner. The QS sensing system of *B. plantarii*, PlaI/R, has been previously studied and characterized (Solis *et al.*, 2006). PlaI/R shares a high similarity with TofI/R of *B. glumae*, and it is also closely related to CepI/R of the Bcc (75% identity). The inactivation of *plaI* considerably delays the appearance of seedling blight in rice, which indicates the importance of QS for disease development. However, the virulence factor(s) regulated by QS in *B. plantarii* is (are) poorly understood (Solis *et al.*, 2006). Besides QS, the stationary phase alternative sigma factor RpoS, which is an important global regulator, has also been implicated in seedling blight (Solis *et al.*, 2006).

Concluding remarks

In this chapter, we have reviewed the current knowledge of the phytopathogenic *Burkholderia* spp. which cause a variety of symptoms in several plant hosts. Despite their importance as plant pathogens, *Burkholderia* phytopathogens have not been intensively investigated by scientists. The rice pathogen *B. glumae* is the most studied species, probably because rice grain rot epidemics are emerging in several areas around the world. In addition, *B. glumae* is related to *B. pseudomallei* and to members of the Bcc meaning that the virulence mechanisms evolved by this pathogen can be paralleled to the ones of opportunistic human pathogens.

Unlike many other genera of phytopathogens, *Burkholderia* displays ecological versatility and are able to compete and survive in many different environments. This is reflected by their multireplicon genomes carrying a large number of genes, which also allows them to utilize many different compounds as carbon sources and to withstand various stress conditions. So far, five complete genomes from phytopathogenic *Burkholderia* species are available. Those are *B. cenocepacia*, *B. cepacia*, *B. gladioli*, *B. glumae* and *B. rhizoxinica* (http://www.burkholderia.com/index.jsp) and genome projects are currently in progress for *B. seminalis* and *B. plantarii* (http://www.ncbi.nlm.nih.gov/genomes/lproks.cgi). Available data suggest that genomes from these group of phytopathogenic species range from 3.75 Mbp (*B. rhizoxinica*) to 9.05 (*B. gladioli*), comprised from one to three chromosomes. So far the presence of plasmids has only been demonstrated in *B. rhizoxinica*, but genome sequencing in multiple strains is required to understand the relevance of these elements in the niche adaptation and metabolic pathways of these phytopathogens. It remains unclear whether *Burkholderia* phytopathogens have a host specific interaction since they do not display the hypersensitive reaction typical of an incompatible response with plant hosts. More details of their genome content, plant, and specific niche adaptation will become clearer with the genome sequencing and comparative genomic analysis of more *Burkholderia* species and strains of diverse life style. In addition, the diversity and versatility of this group of species are largely unexplored as well as is the production of secondary metabolites.

The virulence associated factors employed by *Burkholderia* spp. are also diverse and often characterized by non-specific toxins that induce symptoms such as chlorosis and necrosis. Many pathogens are expected to utilize the type II secretion machinery to secrete a cocktail of hydrolytic enzymes, which are involved in affecting the structural integrity of plant tissue and, thus play an important role in obtaining nutrients for the invading bacteria. *B. glumae*, for instance, employs QS, phytotoxins, lipase production and motility in the development of virulence. Future research will highlight the general mechanisms of *Burkholderia* phytopathogens, bacteria which have not evolved a specialized host-interaction, but cause serious and emerging diseases.

References

Aguilar, C., Bertani, I., and Venturi, V. (2003). Quorum-sensing system and stationary-phase sigma factor (*rpoS*) of the onion pathogen *Burkholderia cepacia* genomovar I type strain, ATCC 25416. Appl. Environ. Microbiol. *69*(3), 1739–1747.

Alfano, J.R., and Collmer, A. (2004). Type III secretion system effector proteins: double agents in bacterial disease and plant defense. Annu. Rev. Phytopathol. *42*, 385–414.

Almeida, I.M.G., Beriam, L.O.S., Sannazzaro, A.M., and Rodrigues Neto, J. (2009). Mancha bacteriana em *Ruscus* sp. causada por *Burkholderia andropogonis* no Brasil. Trop. Plant Pathol. *34*, 339–342.

Azegami, K. (1994). Studies on bacterial seedling blight of rice. Bull. Natl. Inst. Agro-Environ. Sci. *11*, 1–80.

Azegami, K. (1999). Bacterial seedling blight. Ecology and Control of Seed Borne Diseases. Ohata, K., Kuniyasu, K., Takahashi, H., Tochihara, H., and Nagao, N. Japan Plant Protection Association, Tokyo: 133–135.

Azegami, K., Nishiyama, K., Watanabe, Y., Suzuki, T., Yoshida, M., Nose, K., and Toda, S. (1985). Tropolone as a root growth inhibitor produced by a pant pathogenic *Pseudomonas* sp. causing seedling blight of rice. Annu. Phytopathol. Soc. Jpn. *51*, 315–317.

Azegami, K., Nishiyama, K., and Watanabe, Y. (1986). Effects of temperature and iron on the occurrence of rice seedling blight. Annu. Phytopathol. Soc. Jpn. *52*, 501.

Azegami, K., Nishiyama, K., Watanabe, Y., Kadota, I., Ohuchi, A., and Fukazawa, C. (1987). *Pseudomonas plantarii* sp. nov., the causal agent of rice seedling blight. Int. J. Syst. Bacteriol. *37*, 144–152.

Backert, S., and Meyer, T.F. (2006). Type IV secretion systems and their effectors in bacterial pathogenesis. Curr. Opin. Microbiol. *9*, 207–217.

Balandreau, J., Viallard, V., Cournoyer, B., Coenye, T., Laevens, S., and Vandamme, P. (2001). *Burkholderia cepacia* genomovar III is a common plant-associated bacterium. Appl. Environ. Microbiol. *67*(2), 982–985.

Ballard, R.W., Palleroni, N.J., Doudoroff, M., Stanier, R.Y., and Mandel, M. (1970). Taxonomy of the aerobic pseudomonads: *Pseudomonas cepacia, P. marginata, P. alliicola* and *P. caryophylli*. J. Gen. Microbiol. *60*(2), 199–214.

Barrett, C.F., and Parker, M.A. (2006). Coexistence of *Burkholderia, Cupriavidus*, and *Rhizobium* sp. nodule bacteria on two *Mimosa* spp. in Costa Rica. Appl. Environ. Microbiol. 72, 1198–1206.

Bassler, B.L., Wright, M., Showalter, R.E., and Silverman, M.R. (1993). Intercellular signalling in *Vibrio harveyi*: sequence and function of genes regulating expression of luminescence. Mol. Microbiol. *9*, 773–786.

Boekema, B.K., Beselin, A., Breuer, M., Hauer, B., Koster, M., Rosenau, F., Jaeger, K.E., and Tommassen, J. (2007). Hexadecane and Tween 80 stimulate lipase production in *Burkholderia glumae* by different mechanisms. Appl. Environ. Microbiol. *73*, 3838–3844.

Bontemps, C., Elliott, G.N., Simon, M.F., Dos Reis Junior, F.B., Gross, E., Lawton, R.C., Neto, N.E., de Fatima Loureiro, M., De Faria, S.M., Sprent, J.I., *et al.* (2010). *Burkholderia* species are ancient symbionts of legumes. Mol. Ecol. *19*(1), 44–52.

Bradbury, J.F. (1986). Guide to plant pathogenic bacteria. Kew, Surrey, UK: CAB International Mycological Institute.

Burkholder, W.H. (1942). Three bacterial plant pathogens. *Phytomonas caryophylli* sp. n., *Phytomonas alliicola* sp. n. and *Phytomonas manihotis* (Artaud, Berthet and Bondar) Viégas. Phytopathology *32*, 141–149.

Burkholder, W.H. (1950). Sour skin, a bacterial rot of onion bulbs. Phytopathology *40*, 115–117.

Caballero-Mellado, J., Martinez-Aguilar, L., Paredes-Valdez, G., and Santos, P.E. (2004). *Burkholderia unamae* sp. nov., an N2-fixing rhizospheric and endophytic species. Int. J. Syst. Evol. Microbiol. *54*(Pt 4), 1165–1172.

Carvalho, A.P., Ventura, G.M., Pereira, C.B., Leao, R.S., Folescu, T.W., Higa, L., Teixeira, L.M., Plotkowski, M.C., Merquior, V.L., Albano, R.M., and Marques, E.A. (2007). *Burkholderia cenocepacia, B. multivorans, B. ambifaria* and *B. vietnamiensis* isolates from cystic fibrosis patients have different profiles of exoenzyme production. APMIS *115*, 311–318.

Christie, P.J. (2001). Type IV secretion: intercellular transfer of macromolecules by systems ancestrally related to conjugation machines. Mol. Microbiol. *40*(2), 294–305.

Cianciotto, N.P. (2005). Type II secretion: a protein secretion system for all seasons. Trends Microbiol. *13*(12), 581–588.

Claflin, L.E. (2000). Diseases caused by prokaryotes. Compendium of sorghum diseases. Frederiksen, R.A., and Odvody, G.N., eds. St. Paul, MN: American Phytopathological Society: 5–7.

Coenye, T., Holmes, B., Kersters, K., Govan, J.R., and Vandamme, P. (1999). *Burkholderia cocovenenans* (van Damme *et al.*, 1960) Gillis *et al.*, 1995 and *Burkholderia vandii* Urakami *et al.*, 1994 are junior synonyms of *Burkholderia gladioli* (Severini, 1913) Yabuuchi *et al.*, 1993 and *Burkholderia plantarii* (Azegami *et al.*, 1987) Urakami *et al.*, 1994, respectively. Int. J. Syst. Bacteriol. *49* (Pt 1), 37–42.

Coenye, T., Vandamme, P., Govan, J.R., and LiPuma, J.J. (2001). Taxonomy and identification of the

Burkholderia cepacia complex. J. Clin. Microbiol. *39*(10), 3427–3436.

Compant, S., Nowak, J., Coenye, T., Clement, C., and Ait Barka, E. (2008). Diversity and occurrence of *Burkholderia* spp. in the natural environment. FEMS Microbiol. Rev. *32*(4), 607–626.

Corbett, C.R., Burtnick, M.N., Kooi, C., Woods, D.E., and Sokol, P.A. (2003). An extracellular zinc metalloprotease gene of *Burkholderia cepacia*. Microbiol. *149*(Pt 8), 2263–2271.

Cother, E.J., Noble, D., Peters, B.J., Albiston, A., and Ash, G.J. (2004). A new bacterial disease of jojoba (*Simmondsia chinensis*) caused by *Burkholderia andropogonis*. Plant Pathol. *53*, 129–135.

Cottyn, B., Van Outryve, M.F., Cerez, M.T., De Cleene, M., Swings, J., and Mew, T.W. (1996). Bacterila Diseases of Rice. II. Characterization of pathogenic bacteria associated with sheath rot complex and grain discoloration of rice in the Philippines. Plant Dis. *80*, 438–445.

Cottyn, B., Regalado, E., Lanoot, B., De Cleene, M., Mew, T.W., and Swings, J. (2001). Bacterial populations associated with rice seed in the tropical environment. Phytopathol. *91*, 282–292.

Dalmastri, C., Baldwin, A., Tabacchioni, S., Bevivino, A., Mahenthiralingam, E., Chiarini, L., and Dowson, C. (2007). Investigating *Burkholderia cepacia* complex populations recovered from Italian maize rhizosphere by multilocus sequence typing. Environ. Microbiol. *9*, 1632–1639.

De Vos, P., Goor, M., Gillis, M., and De Ley, J. (1985). Ribosomal Ribonucleic Acid Cistron Similarities of Phytopathogenic *Pseudomonas* Species. Int. J. Syst. Bacteriol. *35*(2), 169–184.

Degrassi, G., Devescovi, G., Kim, J., Hwang, I., and Venturi, V. (2008). Identification, characterization and regulation of two secreted polygalacturonases of the emerging rice pathogen *Burkholderia glumae*. FEMS Microbiol. Ecol. *65*(2), 251–262.

Devescovi, G., Bigirimana, J., Degrassi, G., Cabrio, L., LiPuma, J.J., Kim, J., Hwang, I., and Venturi, V. (2007). Involvement of a quorum-sensing-regulated lipase secreted by a clinical isolate of *Burkholderia glumae* in severe disease symptoms in rice. Appl. Environ. Microbiol. *73*, 4950–4958.

Dimcock, A.W. (1950). Major carnation diseases and their spread. Florists Exchange Horticulture, Trade World *114*, 21–72.

Dow, J.M., Clarke, B.R., Milligan, D.E., Tang, J.L., and Daniels, M.J. (1990). Extracellular proteases from *Xanthomonas campestris* pv. *campestris*, the black rot pathogen. Appl. Environ. Microbiol. *56*(10), 2994–2998.

Dowson, W.J. (1929). On the stem rot or wilt disease of carnations. Ann. Appl. Biol. *16*(2), 261–280.

Duan, Y.P., Sun, X., Zhou, L.J., Gabriel, D.W., Benyon, L.S., and Gottwald, T. (2009). Bacterial Brown Leaf Spot of Citrus, a New Disease Caused by *Burkholderia andropogonis*. Plant Dis. *93*(6), 607–614.

Dums, F., Dow, J.M., and Daniels, M.J. (1991). Structural characterization of protein secretion genes of the bacterial phytopathogen *Xanthomonas campestris* pathovar *campestris*: relatedness to secretion systems of other Gram-negative bacteria. Mol. Gen. Genet. *229*(3), 357–364.

Duodu, S., Bhuvaneswari, T.V., Stokkermans, T.J., and Peters, N.K. (1999). A positive role for rhizobitoxine in *Rhizobium*–legume symbiosis. Mol. Plant Microbe Interact. *12*, 1082–1089.

Eberl, L. (2006). Quorum sensing in the genus *Burkholderia*. Int. J. Med. Microbiol. *296*(2–3), 103–110.

Engledow, A.S., Medrano, E.G., Mahenthiralingam, E., LiPuma, J.J., and Gonzalez, C.F. (2004). Involvement of a plasmid-encoded type IV secretion system in the plant tissue watersoaking phenotype of *Burkholderia cenocepacia*. J. Bacteriol. *186*(18), 6015–6024.

English, P.D., Jurale, J.B., and Albersheim, P. (1971). Host-Pathogen Interactions: II. Parameters Affecting Polysaccharide-degrading Enzyme Secretion by *Colletotrichum lindemuthianum* Grown in Culture. Plant Physiol. *47*(1), 1–6.

Fang, Y., Li, B., Wang, F., Liu, B.P., Wu, Z.Y., Su, T., Qiu, W., and Xie, G.L. (2009). Bacterial fruit rot of apricot caused by *Burkholderia cepacia* in China. Plant Pathol. J. *25*, 429–432.

Frenken, L.G., Egmond, M.R., Batenburg, A.M., and Verrips, C.T. (1993). *Pseudomonas glumae* lipase: increased proteolytic stability by protein engineering. Protein Eng. *6*, 637–642.

Furuya, N., Iiyama, K., Shiozaki, N., and Matsuyama, N. (1997). Phytotoxin produced by *Burkholderia gladioli*. J. Fac. Agric. Kyushu Univ. *42*, 33–37.

Furuya, S., Kurata, M., and Saito, T. (1974). Studies on *Rhizopus* sp. causing growth injury of young rice seedlings and its chemical control. Proc. Assoc. Plant Protect. *9*, 49–55.

Garibaldi, A. (1967). A new bacterial disease of carnation in Italy. Inform. Fitopatol. *17*, 9–11, 79–80.

Gehring, F. (1962). Untersuchungen uber den Infektionsverlauf einer durch *Pectobacterium parthenii* (Starr) Hellmers var. dianthicola Hellmers verursachten Nelkenbakteriose sowie über enzymatische Eigenschaften dieses Bakteriums im Vergleich mit *Pseudomonas caryophylli* (Burkholder) Starr et Burkholder und einigen typischen Hassfauleerregern. Phytopathol. Z. *43*, 383–407.

Gho, N., Sato, Z., Yaoita, T., and Aoyagi, K. (1978). Studies on the control of *Rhizopus* in the nursery cases of rice seedlings. 5. Influence of a phytotoxic substance produced by *Rhizopus* on growht of rice. Proc. Assoc. Plant Protect. *26*, 90–94.

Gillis, M., Van Van, T., Bardin, R., Goor, M., Hebbar, P., Willems, A., Segers, P., Kersters, K., Heulin, T., and Fernandez, M.P. (1995). Polyphasic Taxonomy in the Genus *Burkholderia* Leading to an Emended Description of the Genus and Proposition of *Burkholderia vietnamiensis* sp. nov. for N2-Fixing Isolates from Rice in Vietnam. Int. J. Syst. Bacteriol. *45*(2), 274–289.

Gingues, S., Kooi, C., Visser, M.B., Subsin, B., and Sokol, P.A. (2005). Distribution and expression of the ZmpA metalloprotease in the *Burkholderia cepacia* complex. J. Bacteriol. *187*(24), 8247–8255.

Giovanelli, J., Owens, L.D., and Mudd, S.H. (1973). Beta-cystathionase *in vivo* inactivation by rhizobitoxine and role of the enzyme in methionine biosynthesis in corn seedlings. Plant Physiol. *51*, 492–503.

Goh, K.C., Wang, H., Yu, N., Zhou, Y., Zheng, Y., Lim, Z., Sangthongpitag, K., Fang, L., Du, M., Wang, X., *et al.* (2004). PLK1 as a potential drug target in cancer therapy. Drug Dev. Res. *62*, 349–361.

Gonzalez, C.F., and Vidaver, A.K. (1979). Bacteriocin, plasmid and pectolytic diversity in *Pseudomonas cepacia* of clinical and plant origin. J. Gen. Microbiol. *110*(1), 161–170.

Gonzalez, C.F., Pettit, E.A., Valadez, V.A., and Provin, E.M. (1997). Mobilization, cloning, and sequence determination of a plasmid-encoded polygalacturonase from a phytopathogenic *Burkholderia* (*Pseudomonas*) *cepacia*. Mol. Plant Microbe Interact. *10*(7), 840–851.

Goo, E., Kang, Y., Kim, H., and Hwang, I. (2010). Proteomic analysis of quorum sensing-dependent proteins in *Burkholderia glumae*. J. Proteome Res. *9*(6), 3184–3199.

Goto, K., and Ohata, K. (1956). New bacterial diseases of rice (brown stripe and grain rot). Annu. Phytopathol. Soc. Jpn. *21*, 46–47.

Gotschlich, A., Huber, B., Geisenberger, O., Tögl, A., Steidle, A., Riedel, K., Hill, P., Tümmler, B., Vandamme, P., Middleton, B., *et al.* (2001). Synthesis of multiple *N*-acyl-homoserine lactones is wide-spread among the members of the *Burkholderia cepacia* complex. Syst. Appl. Microbiol. *24*, 1–14.

Gross, D.C., and Cody, Y.S. (1985). Mechanisms of plant pathogenesis by *Pseudomonas* species. Can. J. Microbiol. *31*, 403–410.

Heilbronn, J., and Lyon, G.D. (1990). The ineffectuality of potato protease inhibitor on the extracellular protease from *Erwinia carotovora* subsp. *carotovora*. J. Appl. Microbiol. *69*, 25–29.

Hellmers, E. (1958). Four wilt diseases of perpetual flowering carnation in Denmark. Dansk Botanisk Arkiv *18*, 1–200.

Hu, F.P., and Young, J.M. (1998). Biocidal activity in plant pathogenic *Acidovorax*, *Burkholderia*, *Herbaspirillum*, *Ralstonia* and *Xanthomonas* spp. J. Appl. Microbiol. *84*, 263–271.

Ibaragi, T. (1973). Studies on rice seedling blight. I. Growth injury caused by *Rhizopus* sp. under high temperature. Ann. Phytopathol. Soc. Jpn. *39*, 141–144.

Iguchi, K. (1991). Growth development of rice plant suffering from bacterial seedling blight after transplanting and the population changes of the causal agent, *Pseudomonas plantarii* in rice plant. Proc. Kanto P1. Prot. Soc. *38*, 23–24.

Iiyama, K., Furuya, N., Takanami, Y., and Matsuyama, N. (1995). A role of phytotoxin in virulence of *Pseudomonas glumae*. Annu. Phytopathol. Soc. Jpn. *61*, 470–476.

Iwai, T., Kaku, H., Honkura, R., Nakamura, S., Ochiai, H., Sasaki, T., and Ohashi, Y. (2002). Enhanced resistance to seed-transmitted bacterial diseases in transgenic rice plants overproducing an oat cell-wall-bound thionin. Mol. Plant Microbe Interact. *15*(6), 515–521.

Iwasaki, S., Kobayashi, H., Furukawa, J., Namikoshi, M., Okuda, S., Sato, Z., Matsuda, I., and Noda, T. (1984). Studies on macrocyclic lactone antibiotics. VII. Structure of a phytotoxin 'rhizoxin' produced by *Rhizopus chinensis*. J. Antibiot. (Tokyo) 37(4), 354–362.

Iwasaki, S., Namikoshi, M., Kobayashi, H., Furukawa, J., Okuda, S., Itai, A., Kasuya, A., Iitaka, Y., and Sato, Z. (1986). Studies on macrocyclic lactone antibiotics. VIII. Absolute structures of rhizoxin and a related compound. J. Antibiot. (Tokyo) 39(3), 424–429.

Jacobs, J.L., Fasi, A.C., Ramette, A., Smith, J.J., Hammerschmidt, R., and Sundin, G.W. (2008). Identification and onion pathogenicity of *Burkholderia cepacia* complex isolates from the onion rhizosphere and onion field soil. Appl. Environ. Microbiol. *74*(10), 3121–3129.

Jeong, Y., Kim, J., Kim, S., Kang, Y., Nagamatsu, T., and Hwang, I. (2003). Toxoflavin produced by *Burkholderia glumae* causing rice grain rot is responsible for inducing bacterial wilt in many field crops. Plant Dis. *87*, 890–895.

Johnson, H.W., Means, U.M., and Clark, F.E. (1959). Responses of Seedlings to Extracts of Soybean Nodules Bearing Selected Strains of *Rhizobium Japonicum*. Nature *183*(4657), 308–309.

Johnson, T.L., Abendroth, J., Hol, W.G., and Sandkvist, M. (2006). Type II secretion: from structure to function. FEMS Microbiol. Lett. *255*(2), 175–186.

Jones, T.M., Anderson, A.J., and Albersheim, P. (1972). Host–pathogen interactions IV. Studies on the polysaccharide-degrading enzymes secreted by *Fusarium oxysporum* f. sp. *lycopersici*. Physiol. Plant Pathol. *2*, 153–166.

Jordan, A., Hadfield, J.A., Lawrence, N.J., and McGown, A.T. (1998). Tubulin as a target for anticancer drugs: agents which interact with the mitotic spindle. Med. Res. Rev. *18*(, 259–296.

Jorgensen, S., Skov, K.W., and Diderichsen, B. (1991). Cloning, sequence, and expression of a lipase gene from *Pseudomonas cepacia*: lipase production in heterologous hosts requires two *Pseudomonas* genes. J. Bacteriol. *173*, 559–567.

Kang, Y., Kim, J., Kim, S., Kim, H., Lim, J.Y., Kim, M., Kwak, J., Moon, J.S., and Hwang, I. (2008). Proteomic analysis of the proteins regulated by HrpB from the plant pathogenic bacterium *Burkholderia glumae*. Proteomics *8*, 106–121.

Kawamoto, S.O., and Lorbeer, J.W. (1972). Multiplication of *Pseudomonas cepacia* in onion leaves. Phytopathology *62*, 1263–1265.

Kawamoto, S.O., and Lorbeer, J.W. (1974). Infection of onion leaves by *Pseudomonas cepacia*. Phytopathol. *64*, 1440–1445.

Keith, L.M., Sewake, K.T., and Zee, F.T. (2005). Isolation and characterization of *Burkholderia gladioli* from orchids in Hawaii. Plant Dis. *89*, 1273–1278.

Kim, J., Kim, J.G., Kang, Y., Jang, J.Y., Jog, G.J., Lim, J.Y., Kim, S., Suga, H., Nagamatsu, T., and Hwang, I. (2004).

Quorum sensing and the LysR-type transcriptional activator ToxR regulate toxoflavin biosynthesis and transport in *Burkholderia glumae*. Mol. Microbiol. *54*(4), 921–934.

Kim, J., Kang, Y., Choi, O., Jeong, Y., Jeong, J.E., Lim, J.Y., Kim, M., Moon, J.S., Suga, H., and Hwang, I. (2007). Regulation of polar flagellum genes is mediated by quorum sensing and FlhDC in *Burkholderia glumae*. Mol. Microbiol. *64*, 165–179.

Kim, J., Oh, J., Choi, O., Kang, Y., Kim, H., Goo, E., Ma, J., Nagamatsu, T., Moon, J.S., and Hwang, I. (2009). Biochemical evidence for ToxR and ToxJ binding to the *tox* operons of *Burkholderia glumae* and mutational analysis of ToxR. J. Bacteriol. *191*, 4870–4878.

Kooi, C., Subsin, B., Chen, R., Pohorelic, B., and Sokol, P.A. (2006). *Burkholderia cenocepacia* ZmpB is a broad-specificity zinc metalloprotease involved in virulence. Infect. Immun. *74*, 4083–4093.

Kotoujansky, A. (1987). Molecular genetics of pathogenesis by soft-rot erwinias. Annu. Rev. Phytopathol. *25*, 405–430.

Kumar, R.B., and Das, A. (2002). Polar location and functional domains of the *Agrobacterium tumefaciens* DNA transfer protein VirD4. Mol. Microbiol. *43*, 1523–1532.

Kyostio, S.R., Cramer, C.L., and Lacy, G.H. (1991). *Erwinia carotovora* subsp. *carotovora* extracellular protease: characterization and nucleotide sequence of the gene. J. Bacteriol. *173*, 6537–6546.

Lackner, G., Moebius, N., Partida-Martinez, L., and Hertweck, C. (2011). Complete genome sequence of *Burkholderia rhizoxinica*, an Endosymbiont of *Rhizopus microsporus*. J. Bacteriol. *193*, 783–784.

Lafontaine, J.A., Provencal, D.P., Gardelli, C., and Leahy, J.W. (2003). Enantioselective Total Synthesis of the Antitumor Macrolide Rhizoxin D. J. Org. Chem. *68*, 4215–4234.

Latuasan, H.E., and Berends, W. (1961). On the origin of the toxicity of toxoflavin. Biochim. Biophy. Acta *52*, 502–508.

Lee, Y.-A., Shiao, Y.-Y., and Chao, C.-P. (2003). First report of *Burkholderia cepacia* as a pathogen of banana finger-tip rot in Taiwan. Plant Dis. *87*, 601.

Lee, Y.-A., Chan, C.-W., Shiao, Y.-Y., and Chao, C.-P. (2004). Determination of genomovar status of *Burkholderia cepacia* that causes banana finger-tip rot and development of a simple semi-selective medium for the isolation of the pathogen. Plant Pathol. Bull. (Taiwan) *13*, 177–184.

Lee, Y.A., and Chan, C.W. (2007). Molecular Typing and Presence of Genetic Markers Among Strains of Banana Finger-Tip Rot Pathogen, *Burkholderia cenocepacia*, in Taiwan. Phytopathology 97, 195–201.

Leveau, J.H., and Lindow, S.E. (2001). Quantitative monitoring of bacterial sugar consumption in the phyllosphere. Proc. Natl. Acad. Sci. U.S.A. *98*, 3446–3453.

Li, B., Fang, Y., Zhang, G.Q., Yu, R.R., Lou, M.M., Xie, G.L., Wang, Y.L., and Sun, G.C. (2010). Molecular Characterization of *Burkholderia cepacia* Complex Isolates Causing Bacterial Fruit Rot of Apricot. Plant Pathol. J. *26*, 223–230.

Lim, J., Lee, T.H., Nahm, B.H., Choi, Y.D., Kim, M., and Hwang, I. (2009). Complete genome sequence of *Burkholderia glumae* BGR1. J. Bacteriol. *191*, 3758–3759.

Lindeberg, M., and Collmer, A. (1992). Analysis of eight out genes in a cluster required for pectic enzyme secretion by *Erwinia chrysanthemi*: sequence comparison with secretion genes from other Gram-negative bacteria. J. Bacteriol. *174*(22), 7385–7397.

LiPuma, J.J., Spilker, T., Coenye, T., and Gonzalez, C.F. (2002). An epidemic *Burkholderia cepacia* complex strain identified in soil. Lancet *359*, 2002–2003.

Liu, H., Zhang, S., Schell, M.A., and Denny, T.P. (2005). Pyramiding unmarked deletions in *Ralstonia solanacearum* shows that secreted proteins in addition to plant cell-wall-degrading enzymes contribute to virulence. Mol. Plant Microbe Interact. *18*, 1296–1305.

Lonon, M.K., Woods, D.E., and Straus, D.C. (1988). Production of lipase by clinical isolates of *Pseudomonas cepacia*. J. Clin. Microbiol. *26*(5), 979–984.

Luo, J., Xie, G., Li, B., and Lihui, X. (2007). First Report of *Burkholderia glumae* Isolated from Symptomless Rice Seeds in China. Plant Dis. *91*(10), 1363.

Lutter, E., Lewenza, S., Dennis, J.J., Visser, M.B., and Sokol, P.A. (2001). Distribution of quorum-sensing genes in the *Burkholderia cepacia* complex. Infect. Immun. *69*, 4661–4666.

McCulloch, L. (1921). A bacterial disease of gladiolus. Science *54*, 115–116.

McNeil, M., Darvill, A.G., Fry, S.C., and Albersheim, P. (1984). Structure and function of the primary cell walls of plants. Annu. Rev. Biochem. *53*, 625–663.

Maeda, Y., Shinohara, H., Kiba, A., Ohnishi, K., Furuya, N., Kawamura, Y., Ezaki, T., Vandamme, P., Tsushima, S., and Hikichi, Y. (2006). Phylogenetic study and multiplex PCR-based detection of *Burkholderia plantarii*, *Burkholderia glumae* and *Burkholderia gladioli* using *gyrB* and *rpoD* sequences. Int. J. Syst. Evol. Microbiol. *56*(Pt 5), 1031–1038.

Mahenthiralingam, E., Baldwin, A., and Dowson, C.G. (2008). *Burkholderia cepacia* complex bacteria: opportunistic pathogens with important natural biology. J. Appl. Microbiol. *104*, 1539–1551.

Malott, R.J., Baldwin, A., Mahenthiralingam, E., and Sokol, P.A. (2005). Characterization of the *cciIR* quorum-sensing system in *Burkholderia cenocepacia*. Infect. Immun. 73, 4982–4992.

Marits, R., Koiv, V., Laasik, E., and Mae, A. (1999). Isolation of an extracellular protease gene of *Erwinia carotovora* subsp. *carotovora* strain SCC3193 by transposon mutagenesis and the role of protease in phytopathogenicity. Microbiology*145*, 1959–1966.

Massa, C., Clausen, M.H., Stojan, J., Lamba, D., and Campa, C. (2007a). Study of the mode of action of a polygalacturonase from the phytopathogen *Burkholderia cepacia*. Biochem. J. *407*, 207–217.

Massa, C., Degrassi, G., Devescovi, G., Venturi, V., and Lamba, D. (2007b). Isolation, heterologous expression and characterization of an endo-polygalacturonase produced by the phytopathogen *Burkholderia cepacia*. Protein Expr. Purif. *54*(2), 300–308.

Matsuda, I., and Sato, Z. (1988). Regulation between pathogenicity and pigment productivity in the causal agent of bacterial grain rot of rice. Ann. Phytopathol. Soc. Jpn. *54*, 378.

Mitchell, R.E., Frey, E.J., and Benn, M.K. (1986). Rhizobitoxine and 1-threo-hydroxythreonine production by the plant pathogen *Pseudomonas andropogonis*. Phytochem. *25*, 2711–2715.

Miyagawa, H. (2000). Biocontrol of bacterial seedling blight of rice caused by *Burkholderia gladioli* using with its avirulent isolate. Jpn. J. Phytopathol. *66*, 232–238.

Miyagawa, H., and Inoue, H. (2002). Alternative route of infection for bacterial seedling blight of rice caused by *Burkholderia plantarii*. J. Gen. Plant Pathol. *68*, 356–362.

Morita, Y., Matsumura, E., Okabe, T., Shibata, M., Sugiura, M., Ohe, T., Tsujibo, H., Ishida, N., and Inamori, Y. (2003). Biological activity of tropolone. Biol. Pharm. Bull. *26*(10), 1487–1490.

Mudgett, M.B. (2005). New insights to the function of phytopathogenic bacterial type III effectors in plants. Annu. Rev. Plant Biol. *56*, 509–531.

Nagamatsu, T. (2001). Syntheses, Transformation, and Biological Activities of 7-Azapteridine Antibiotics: Toxoflavin, Fervenulin, Reumycin and Their Analogues. Recent Res. Devel. Org. Bioorg. Chem. *4*, 97–121.

Nandakumar, R., Rush, M.C., and Correa, F. (2007). Association of *Burkholderia glumae* and *B. gladioli* with Panicle Blight Symptoms on Rice in Panama. Plant Dis. *91*, 767–767.

Nandakumar, R., Shahjahan, A.K.M., Yuan, X.L., Dickstein, E.R., Groth, D.E., Clark, C.A., Cartwright, R.D., and Rush, M.C. (2009). *Burkholderia glumae* and *B. gladioli* cause bacterial panicle blight in rice in the Southern United States. Plant Dis. *93*, 896–905.

Neilands, J.B. (1981). Microbial iron compounds. Annu. Rev. Biochem. *50*(1), 715–731.

Noda, T., Hashiba, T., and Sato, Z. (1980). The structural changes in young swollen roots of rice seedlings infected with *Rhizopus chinensis* Saito. Ann. Phytopathol. Soc. Jpn *46*, 40–45.

O.E.P.P/E.P.P.O. (1997). Data sheets on quarantine pests, *Burkholderia caryophylli*. Available at: http://www.eppo.org/QUARANTINE/bacteria/Burkholderia_caryophilli/PSDMCA_ds.pdf.

Okazaki, S., Nukui, N., Sugawara, M., and Minamisawa, K. (2004). Rhizobial strategies to enhance symbiotic interaction: Rhizobitoxine and 1-aminocyclopropane-1-carboxylate deaminase. Microbes Environ. *19*, 99–111.

Owens, L.D., and Wright, D.A. (1965). Rhizobial-induced chlorosis in soybeans: isolation, production in nodules, and varietal specificity of the toxin. Plant Physiol. *40*, 927–930.

Owens, L.D., Guggenheim, S., and Hilton, J.L. (1968). *Rhizobium*-synthesized phytotoxin: An inhibitor of β-cystathionase in *Salmonella typhimurium*. Biochim. Biophys. Acta *158*, 219–225.

Palleroni, N.J. (1984). Genus I *Pseudomonas* Migula *1894*, 237AL. Bergey's Manual of Systematic Bacteriology. Krieg, N.R.H., Baltimore, J.G., eds.: Williams & Wilkins. 1, pp. 141–199.

Parke, J.L., and Gurian-Sherman, D. (2001). Diversity of the *Burkholderia cepacia* complex and implications for risk assessment of biological control strains. Annu Rev Phytopathol *39*, 225–258.

Partida-Martinez, L.P., and Hertweck, C. (2005). Pathogenic fungus harbours endosymbiotic bacteria for toxin production. Nature *437*(7060), 884–888.

Partida-Martinez, L.P., and Hertweck, C. (2007). A gene cluster encoding rhizoxin biosynthesis in '*Burkholderia rhizoxina*', the bacterial endosymbiont of the fungus *Rhizopus microsporus*. Chembiochemistry *8*, 41–45.

Partida-Martinez, L.P., Groth, I., Schmitt, I., Richter, W., Roth, M., and Hertweck, C. (2007a). *Burkholderia rhizoxinica* sp. nov. and *Burkholderia endofungorum* sp. nov., bacterial endosymbionts of the plant-pathogenic fungus *Rhizopus microsporus*. Int. J. Syst. Evol. Microbiol. *57*(Pt 11), 2583–2590.

Partida-Martinez, L.P., Monajembashi, S., Greulich, K.O., and Hertweck, C. (2007b). Endosymbiont-dependent host reproduction maintains bacterial-fungal mutualism. Curr. Biol. *17*, 773–777.

Ramette, A., LiPuma, J.J., and Tiedje, J.M. (2005). Species Abundance and Diversity of *Burkholderia cepacia* Complex in the Environment. Appl. Environ. Microbiol. *71*, 1193–1201.

Rodriguez-Palenzuela, P., Burr, T.J., and Collmer, A. (1991). Polygalacturonase is a virulence factor in *Agrobacterium tumefaciens* biovar 3. J. Bacteriol. *173*, 6547–6552.

Rosenau, F., and Jaeger, K. (2000). Bacterial lipases from *Pseudomonas*: regulation of gene expression and mechanisms of secretion. Biochimie *82*, 1023–1032.

Saddler, G.S. (1994). *Burkholderia caryophylli*. IMI Descriptions of Pathogenic Fungi and Bacteria No. 1215. CAB International, Wallingford, UK.

Sato, Z., Koiso, Y., Iwasaki, S., and Matsuda, I. (1989). Toxins produced by *Pseudomonas glumae*. Annu. Phytopathol. Soc. Jpn. *55*, 356–359.

Satou, Y., and Matuda, I. (1998). Isolation of *Pseudomonas plantarii*, pathogen of bacterial seedling blight of rice, from *Graminiae* weeds. Proc. Assoc. P1. Prot. Hokuriku *46*, 49–52.

Sayler, R.J., Cartwright, R.D., and Yang, Y. (2006). Genetic characterization and real-time PCR detection of *Burkholderia glumae*, a newly emerging bacterial pathogen of rice in the United States. Plant Dis. *90*, 603–610.

Sessitsch, A., Coenye, T., Sturz, A.V., Vandamme, P., Barka, E.A., Salles, J.F., Van Elsas, J.D., Faure, D., Reiter, B., Glick, B.R., *et al.* (2005). *Burkholderia phytofirmans* sp. nov., a novel plant-associated bacterium with plant-beneficial properties. Int. J. Syst. Evol. Microbiol. *55*(Pt 3), 1187–1192.

Severini, G. (1913). Una bacteriosi dell'Ixia maculata e del *Gladiolus coluilli*. Annu. Bot. (Roma) *11*, 413–424.

Shahjahan, A.K.M., Groth, D.E., Clark, C.A., Linscombe, S.D., and Rush, M.C. (2000a). Epidemiological studies on panicle blight of rice: critical stage of infection and the effect of infected seeds on disease development and yield of rice. Proc. Rice Tech. Work. Group 28:77. (Biloxi, MS: Feb. 27– 1 March 2000).

Shahjahan, A.K.M., Rush, M.C., Groth, D., and Clark, C. (2000b). Panicle blight: Recent research points to a bacterial cause. Rice J. *103*, 26–28.

Smith, E.F. (1911). Bacteria in relation to plant diseases. Carnegie Inst. Washington Publ. *2*, 1–368.

Sokol, P.A., Sajjan, U., Visser, M.B., Gingues, S., Forstner, J., and Kooi, C. (2003). The CepIR quorum-sensing system contributes to the virulence of *Burkholderia cenocepacia* respiratory infections. Microbiology *149*(Pt 12), 3649–3658.

Solis, R., Bertani, I., Degrassi, G., Devescovi, G., and Venturi, V. (2006). Involvement of quorum sensing and RpoS in rice seedling blight caused by *Burkholderia plantarii*. FEMS Microbiol. Lett. *259*(1), 106–112.

Springman, A.C., Jacobs, J.L., Somvanshi, V.S., Sundin, G.W., Mulks, M.H., Whittam, T.S., Viswanathan, P., Gray, R.L., LiPuma, J.J., and Ciche, T.A. (2009). Genetic diversity and multihost pathogenicity of clinical and environmental strains of *Burkholderia cenocepacia*. Appl. Environ. Microbiol. *75*, 5250–5260.

Stapp, C. (1928). Schizomycetes (Spaltpilze oder bakterien). Handbuch der Planzenkrankheiten, Sorauer, P., ed. (Berlin: Paul Parey), pp.1–25.

Stevens, F.L. (1925). Plant Disease Fungi. (New York: Macmillan).

Suzuki, F., Zhu, Y., Sawada, H., and Matsuda, I. (1998). Identification of proteins involved in toxin production by *Pseudomonas glumae*. Annu. Phytopathol. Soc. Jpn. *64*, 75–79.

Suzuki, F., Sawada, H., Azegami, K., and Tsuchiya, K. (2004). Molecular characterization of the *tox* operon involved in toxoflavin biosynthesis of *Burkholderia glumae*. J. Gen. Plant Pathol. *70*, 97–107.

Takeuchi, T., Sawada, H., Suzuki, F., and Matsuda, I. (1997). Specific detection for *Burkholderia plantarii* and *B. glumae* by PCR using primers selected from the 16S-23S rDNA spacer regions. Annu. Phytopathol. Soc. Jpn. *63*, 455–462.

Tanaka, T., Katou, T., and Fujita, Y. (1994). Isolation of *Pseudomonas plantarii*, pathogen of bacterial seedling blight of rice, from the weeds around the paddy field. Annu. Phytopathol. Soc. Jpn. *60*, 640–643.

Tarr, S.A.J. (1962). Diseases of sorghum, sudan grass and broom corn. Kew, Surrey, UK: Commonwealth Mycological Institute.

Teviotdale, B.L., Davis, R.M., Guerard, J.P., and Harper, D.H. (1989). Effect of irrigation management on sour skin of onion. Plant Dis. 73, 819–822.

Tolcher, A.W., Aylesworth, C., Rizzo, J., Izbicka, E., Campbell, E., Kuhn, J., Weiss, G., Von Hoff, D.D., and Rowinsky, E.K. (2000). A phase I study of rhizoxin (NSC 332598) by 72-hour continuous intravenous infusion in patients with advanced solid tumors. Ann. Oncol. *11*, 333–338.

Tomassini, A., Sella, L., Raiola, A., D'Ovidio, R., and Favaron, F. (2009). Characterization and expression of *Fusarium graminearum* endo-polygalacturonases *in vitro* and during wheat infection. Plant Pathol. *58*(3), 556–564.

Toth, I.K., Bell, K.S., Holeva, M.C., and Birch, P.R. (2003). Soft rot erwiniae: from genes to genomes. Mol. Plant Pathol. *4*(1), 17–30.

Trust, T.J. (1975). Antibacterial activity of tropolone. Antimicrob. Agents Chemother. *7*(5), 500–506.

Tsuchiya, Y., and Muko, H. (1963). Occurrence of neck rot disease of *Freesia* sp. in Japan. Chukan Hokoku Plant Pathol Sect. NIAS *16*, 82–84.

Tsuruo, T., Oh-hara, T., Iida, H., Tsukagoshi, S., Sato, Z., Matsuda, I., Iwasaki, S., Okuda, S., Shimizu, F., Sasagawa, K., *et al.* (1986). Rhizoxin, a macrocyclic lactone antibiotic, as a new antitumor agent against human and murine tumor cells and their vincristine-resistant sublines. Cancer Res. *46*, 381–385.

Tsushima, S., Naito, H., and M. Koitabashi (1996). Population dynamics of *Pseudomonas glumae*, the causal agent of bacterial grain rot of rice, on leaf sheaths of rice plants in relation to disease development in the field. Ann. Phytopathol. Soc. Jpn. *62*, 108–113.

Uchida, J. (1995). Bacterial diseases of *Dendrobium*. Hawaii Inst. Trop. Agric. Hum. Resour. Brief *158*.

Uematsu, T., Yoshimura, D., Nishiyama, K., Ibaraki, T., and Fuji, H. (1976). Occurrence of bacterial seedling rot in nursery flat, caused by grain rot bacterium *Pseudomonas glumae*. Annu. Phytopathol. Soc. Jpn. *42*, 310–312.

Ulrich, J.M. (1975). Pectic enzymes of *Pseudomonas cepacia* and penetration of polygalacturonase into cells. Physiol. Plant Pathol. *5*(1), 37–44.

Urakami, T., Ito-Yoshida, C., Araki, H., Kijima, T., Suzuki, K.-I., and Komagata, K. (1994). Transfer of *Pseudomonas plantarii* and *Pseudomonas glumae* to *Burkholderia* as *Burkholderia* spp. and Description of *Burkholderia vandii* sp. nov. Int. J. Syst. Bacteriol. *44*, 235–245.

Vandamme, P., Holmes, B., Coenye, T., Goris, J., Mahenthiralingam, E., LiPuma, J.J., and Govan, J.R. (2003). *Burkholderia cenocepacia* sp. nov.--a new twist to an old story. Res. Microbiol. *154*(2), 91–96.

Vanlaere, E., LiPuma, J.J., Baldwin, A., Henry, D., De Brandt, E., Mahenthiralingam, E., Speert, D., Dowson, C., and Vandamme, P. (2008). *Burkholderia latens* sp. nov., *Burkholderia diffusa* sp. nov., *Burkholderia arboris* sp. nov., *Burkholderia seminalis* sp. nov. and *Burkholderia metallica* sp. nov., novel species within the *Burkholderia cepacia* complex. Int. J. Syst. Evol. Microbiol. *58*, 1580–1590.

Venturi, V., and Passos, D. (2012). Incoming pathogens team up with harmless 'resident' bacteria. Trends in Microbiology *20*(4), 160–164.

Venturi, V., Friscina, A., Bertani, I., Devescovi, G., and Aguilar, C. (2004). Quorum sensing in the *Burkholderia cepacia* complex. Res. Microbiol. *155*(4), 238–244.

Von Bodman, S.B., Bauer, W.D., and Coplin, D.L. (2003). Quorum sensing in plant-pathogenic bacteria. Annu. Rev. Phytopathol. *41*, 455–482.

Waffa, M.H. (2002). Sustainable Agriculture Management of Plant Disease. Online Journal of Biological Sciences 2, 280–284.

Walton, J.D. (1994). Deconstructing the Cell Wall. Plant Physiol. *104*(4), 1113–1118.

Ward, O.P., and Moo-Young, M. (1989). Enzymatic degradation of cell wall and related plant polysaccharides. Crit. Rev. Biotechnol. *8*(4), 237–274.

Weinberg, E.D. (1978). Iron and infection. Microbiol. Rev. *42*, 45–66.

Xiong, K., and J.J. Fuhrmann (1996). Comparison of rhizobitoxine-induced inhibition of β-cystathione from different bradyrhizobia and soybean genotypes. Plant Soil *186*, 53–61.

Yasuta, T., Satoh, S., and Minamisawa, K. (1999). New Assay for Rhizobitoxine Based on Inhibition of 1-Aminocyclopropane-1-Carboxylate Synthase. Appl. Environ. Microbiol. *65*(2), 849–852.

Yoneyama, K., Kono, Y., Yamaguchi, I., Horikoshi, M., and Hirooka, T. (1998). Toxoflavin is an essential factor for virulence of *Burkholderia glumae* causing rice seedling rot disease. Ann. Phytopathol. Soc. Jpn. *64*, 91–96.

Yoshimura, D., Yoshida, K., and Otofuji, M. (1987). Effect of meteorological factors on the occurrence of rice bacterial grain rot. Proc. Assoc. Plant Prot. Kyushu *33*, 9–12.

Young, A.J., Grice, K.R.E., Trevorrow, P.R., and Vawdrey, L.L. (2007). *Burkholderia andropogonis* blight of golden cane palms in north Queensland. Australasian Plant Disease Notes 2, 131–132.

Zeigler, R.S., and Alvarez, E. (1987). Grain discoloration of rice caused by Pseudomonas glumae in Latin America. Plant Disease 73, 368.

Molecular Basis for Antibiotic Resistance in the Genus *Burkholderia*

9

Silvia Bazzini, Silvia Buroni, Claudia Udine, Maria R. Pasca and Giovanna Riccardi

Abstract

Bacteria belonging to the *Burkholderia cepacia* complex (Bcc), as well as other *Burkholderia* spp. such as *Burkholderia pseudomallei* and *Burkholderia mallei*, are able to cause infections in humans. Their eradication is difficult due to the high level of intrinsic resistance to several antibiotics, but the efficacy of the antibiotic treatment would likely benefit from a better knowledge on the drug resistance mechanisms. Different resistant mechanisms have been identified in these bacteria, including alteration of drug target site, enzymatic drug inactivation or modification, reduced membrane permeability, and/or activation of efflux systems. To circumvent these mechanisms, a wide range of antimicrobial compounds have been tested and new ones are being discovered in order to improve the clinical approach. Among current therapies, tobramycin, ciprofloxacin, or meropenem/ceftazidime in combination with chloramphenicol, trimethoprim–sulphamethoxazole or aztreonam are mostly recommended.

Burkholderia and the burden of antibiotic resistance

Burkholderia species are Gram-negative bacteria which are present in a wide range of ecological niches, such as soil, water, plants, animals and humans, by adopting different lifestyles. This biochemical adaptability is also reflected in the host–pathogen interactions: beneficial associations can be established as well as pathogenic ones (Valvano, 2006). Among the pathogenic species, *Burkholderia cenocepacia* is often responsible for serious infections in immunocompromised patients, especially those affected by cystic fibrosis (CF). It is an opportunistic microorganism, belonging to the *Burkholderia cepacia* complex (Bcc), a group of genetically closely related species, containing *Burkholderia ambifaria*, *Burkholderia anthina*, *Burkholderia arboris*, *B. cenocepacia*, *Burkholderia cepacia*, *Burkholderia contaminans*, *Burkholderia diffusa*, *Burkholderia dolosa*, *Burkholderia lata*, *Burkholderia latens*, *Burkholderia metallica*, *Burkholderia multivorans*, *Burkholderia pyrrocinia*, *Burkholderia seminalis*, *Burkholderia stabilis*, *Burkholderia ubonensis* and *Burkholderia vietnamiensis* (Compant *et al.*, 2008; Vanlaere *et al.*, 2008; Vanlaere *et al.*, 2009).

B. cenocepacia emerged as an important CF pathogen about 30 years ago, and infections with this organism are associated with increased morbidity and mortality. This bacterium is able to infect the respiratory tract of CF patients and can lead to a wide range of clinical outcomes. Indeed, *B. cenocepacia* colonization can lead to no apparent deterioration, but the '*cepacia* syndrome', a progressive necrotizing pneumonia, often leads to a raid decline in lung function and is associated with increased mortality (Jones *et al.*, 2001). Another prevalent species in CF patients is *B. multivorans*, accounting for 85–97% of all Bcc infections together with *B. cenocepacia* (Drevinek and Mahenthiralingam, 2010). This microorganism was identified in an outbreak of infection in the early 1990s and cases of fatal epidemics similar to those associated with *B. cenocepacia* have been reported (Blackburn *et al.*, 2004; LiPuma, 2010).

Other *Burkholderia* spp. which are important pathogens, but do not belong to the Bcc, are *Burkholderia pseudomallei* and *Burkholderia mallei*; the former being the aetiological agent of melioidosis, the latter causing glanders in animals and humans (Galyov *et al.*, 2010). In addition to these, *Burkholderia gladioli* can cause disease in both plants and humans, where it can be responsible for pulmonary infections, while *Burkholderia thailandensis* is considered an opportunistic pathogen, since only a few cases of infection in immunocompromised patients have been reported (Brett *et al.*, 1998; Glass *et al.*, 2006; Murray *et al.*, 2008).

The major problem related to the clinical therapy of these infections is the antibiotic resistance, thus influencing the choice, use and future of chemotherapy. The treatment of *B. cenocepacia*, for example, is very complicated due to the high level of intrinsic resistance to clinically relevant antibiotics such as cephalosporins, aminoglycosides, polymyxins and β–lactams. This pathogen can display both intrinsic and inducible resistance and for this reason its eradication is difficult to achieve (Drevinek and Mahenthiralingam, 2010). *B. cenocepacia* isolates can also show a Multidrug Resistance (MDR) phenotype, which is defined as resistance to all of the agents belonging to at least two out of three classes of antibiotics, such as quinolones, aminoglycosides, and β-lactam agents, including monobactams and carbapenems (Saiman and Siegel, 2003). Furthermore, cross-resistance among unrelated antibiotics can often develop, as observed, for example, in *B. pseudomallei*, *B. mallei* and *B. thailandensis*, making clinical treatment difficult (Viktorov *et al.*, 2008).

Currently, the drug resistance mechanisms which have been identified in *Burkholderia* spp. are comparable to those identified in other Gram-negative bacteria. Indeed, antibiotic resistance can be due to different mechanisms: alteration of the drug target site, increase in target production, enzymatic inactivation or modification of the drug, reduced membrane permeability, and/or activation of efflux systems. Moreover, the MDR of Bcc species has been attributed to reduced outer membrane permeability (Moore and Hancock, 1986; Parr *et al.*, 1987; Aronoff, 1988a), production of modifying enzymes such as β-lactamases (Trépanier *et al.*, 1997), and alteration of antibiotic targets (Burns *et al.*, 1989). The efflux systems seem to be involved as well: it is noteworthy that a single species of multidrug efflux transporter can produce a MDR phenotype. Although these pumps are able to confer low-level intrinsic resistance, the presence of antimicrobial compounds or mutations in the regulatory genes can significantly increase the expression of the transporters (Misra and Bavro, 2009). All these data suggest that the high level of antibiotic resistance in *Burkholderia* spp. is due to many different interacting factors, leading to infections which are very difficult to treat.

Drugs and mechanisms of action

In order to fight fatal infections, CF patients are treated with different antibiotics. Antibiotics are prescribed in three different clinical settings: firstly when the patient is colonized by *Pseudomonas aeruginosa*, then when other pathogens are detected in the sputum and finally during periodic exacerbations to restore lung function. However, in order to prescribe an adequate therapy, the pathogen causing the infection must be identified and a susceptibility profile must be determined. Here we provide an overview of the drugs which can be used for the treatment of Bcc. In general, the use of meropenem plus ceftazidime, chloramphenicol, trimethoprim–sulphamethoxazole or aztreonam is recommended (Gibson *et al.*, 2003; Waters, 2012). As regarding the *Burkholderia* species which do not belong to the Bcc, *B. pseudomallei* is of particular importance, being the causative agent of melioidosis. The drugs typically used to treat *B. pseudomallei* infections are doxycycline, chloramphenicol, sulphamethoxazole and trimethoprim (Dance *et al.*, 1989).

The following antibiotics have been investigated in relation to *Burkholderia* antimicrobial resistance or are commonly used to treat infections. Among the drugs targeting the bacterial cell wall are:

Amoxicillin/clavulanic acid is a combination antibiotic composed of amoxicillin trihydrate, a β-lactam antibiotic, and potassium clavulanate, a β-lactamase inhibitor. This combination is used in order to increase the spectrum of action of amoxicillin against β-lactamase positive organisms. Amoxicillin acts by binding to penicillin-binding protein 1A, preventing the formation of a cross-link of two linear peptidoglycan strands, inhibiting the bacterial cell wall synthesis.

Aztreonam is a monocyclic β-lactam isolated from *Chromobacterium violaceum*. Preliminary results on its activity were reported in 1981 by Acar *et al.* It blocks peptidoglycan crosslinking thus impairing the biosynthesis of the bacterial cell wall. In particular, it has a very high affinity for penicillin-binding protein 3 of Gram-negative bacteria. Aztreonam is bactericidal as cell lysis, mediated by autolysins, occurs due to its action. It is resistant to β-lactamases (such as penicillinases and cephalosporinases) due to its molecular structure. In fact it contains a single monocyclic β-lactam nucleus which makes it structurally different from other β-lactams. In particular, the aminothiazolyl oxime side chain at position 3 and a methyl group at position 4 confer specificity and β-lactamase stability to aztreonam. In addition, it does not induce β-lactamase activity (Livermore and Williams, 1981). Aztreonam is used for the treatment of Bcc infections that are multi-resistant to other antibiotics, such as cephalosporins, penicillin, and aminoglycosides.

Ceftazidime is a semisynthetic third-generation aminothiazolyl cephalosporin (Wise *et al.*, 1980), a β-lactam antibiotic, with a broad-spectrum activity. Its mechanism of action is the inhibition of cell wall synthesis, which results in a bactericidal activity. In particular, ceftazidime is able to inhibit penicillin binding protein 3. It is widely used against many clinically important pathogens (in particular Gram-negative) which are resistant to ampicillin and other cephalosporins due to the production of β-lactamases of plasmid or chromosomal-origin.

Imipenem is a broad spectrum β-lactam carbapenem (Kesado *et al.*, 1980). It inhibits cell wall synthesis of various Gram-positive and Gram-negative bacteria by binding to penicillin binding proteins. It is also resistant to β-lactamase activity (Braveny, 1984).

Meropenem was first described in 1989 by Edwards and collaborators who demonstrated its activity against a broad spectrum of aerobes, including *Staphylococcus aureus*, β-haemolytic streptococci, *Streptococcus pneumoniae*, *Haemophilus influenzae*, *Neisseria* spp., members of the family *Enterobacteriaceae*, *Pseudomonas* spp., and Gram-positive and Gram-negative anaerobes (Edwards *et al.*, 1989). It is a broad spectrum injectable antibiotic used to treat a variety of infections. It is a β-lactam belonging to the subgroup of carbapenems. Meropenem is a bactericidal antibiotic which inhibits bacterial wall synthesis by binding to penicillin-binding proteins, in particular PBPs 2, 3 and 4. Like ceftazidime, it is resistant to degradation by β-lactamases or cephalosporinases.

Piperacillin is a semisynthetic, broad-spectrum antibiotic derived from ampicillin (Jones *et al.*, 1977). It acts by binding to specific penicillin-binding proteins inside the bacterial cell wall. Piperacillin inhibits the final stage of bacterial cell wall synthesis. Piperacillin is stable against the activity of many broad-spectrum β-lactamases (Eliopoulos and Moellering, 1982).

Temocillin is a semisynthetic 6-α-methoxy β-lactam which exhibits exceptional stability to a wide range of bacterial β-lactamases (Slocombe *et al.*, 1981). It is a narrow spectrum penicillin. Van Acker and collaborators showed that temocillin has good *in vitro* activity against planktonic and 4-h-old Bcc biofilms, but not against 24-h-old biofilms (Van Acker *et al.*, 2010).

Antibiotics that target the protein biosynthesis comprise:

Amikacin is a semi-synthetic aminoglycoside derived from kanamycin A (Price *et al.*, 1974). It targets protein synthesis by binding to four nucleotides of the 16S rRNA in the 30S ribosome. This interferes with the formation of an initiation complex with mRNA, resulting in misreading and as a consequence incorrect amino acids are inserted into the polypeptide leading to non-functional peptides. Aminoglycosides are mostly used against aerobic, Gram-negative bacteria, some mycobacteria, and sometimes against Gram-positive bacteria. Amikacin has a broad spectrum of activity due to the fact that it has been shown to be a poor substrate for most enzymes that inactivate other aminoglycosides through O-phosphorylation, O-adenylylation, or N-acetylation (Price *et al.*, 1974). However, amikacin is nephrotoxic and ototoxic.

Chloramphenicol (originally known as chloromycetin) is a bacteriostatic antimicrobial isolated in 1948 by David Gottlieb from the bacterium *Streptomyces venezuelae* (Gottlieb *et al.*, 1948). It was introduced into medical practice the following year. It is a broad-spectrum antibiotic, being effective against both Gram-positive and Gram-negative bacteria. It blocks bacterial growth by inhibiting protein synthesis. In particular, it prevents protein chain elongation by binding to A2451 and A2452 residues in the 23S rRNA of the 50S ribosomal subunit, thus inhibiting the peptidyl transferase activity of the bacterial ribosome (Pestka, 1969). Bone marrow toxicity is one of the most serious side effects associated with the use of chloramphenicol.

Doxycycline is a lipophilic synthetic tetracycline derivative. Doxycycline reversibly binds to the 30S ribosomal subunits, blocking the binding of aminoacyl tRNA to the mRNA and inhibiting protein synthesis.

Minocycline is chemically derived from tetracycline (Redin, 1966). It is a broad-spectrum bacteriostatic antibiotic. It binds to the 30S ribosomal subunit, preventing the binding of tRNA to the mRNA–ribosome complex, thus interfering with protein synthesis. Diarrhoea and central nervous system-related side effects are common.

Tobramycin is an aminoglycoside, effective against Gram-negative bacteria, and is produced by *Streptomyces tenebrarius* (Meyer *et al.*, 1971). In order to be active it must be taken up by the cell by active transport. Tobramycin acts by inhibiting protein synthesis, binding irreversibly to one of two binding sites on the 30S ribosomal subunit, thus preventing formation of the 70S complex. In this way, mRNA cannot be translated and cell death occurs. Like other aminoglycosides, tobramycin is ototoxic (Brummett *et al.*, 1971) and also nephrotoxic.

Among the drugs which are able to block biosynthetic pathways, we find:

Trimethoprim is a bactericidal, pyrimidine analogue. It acts by binding to dihydrofolate reductase thus inhibiting the reduction of dihydrofolic acid to tetrahydrofolic acid, an essential precursor in the thymidine synthesis pathway. As bacteria are unable to take up folic acid from the environment, this leads to shortage of nucleotides and subsequent inhibition of DNA synthesis. Trimethoprim is usually used in combination with sulphonamides, in particular with sulphamethoxazole (co-trimoxazole) (Reisberg *et al.*, 1966).

Sulphamethoxazole was introduced in therapy in 1961 (Holloman *et al.*, 1961). It inhibits dihydrofolate synthetase, an enzyme involved in the same pathway as the target of trimethoprim, but further upstream. In particular, sulphonamides inhibit the enzymatic conversion of pteridine and *p*-aminobenzoic acid to dihydropteroic acid by competing with *p*-aminobenzoic acid for binding to dihydrofolate synthetase. Sulphamethoxazole is bacteriostatic. The fact that trimethoprim and sulphamethoxazole are used in combination reduces the development of resistance. As regarding side effects, the interference with folic acid metabolism may affect haematopoiesis.

Finally, antibiotics targeting DNA replication include:

Ciprofloxacin is a synthetic second-generation fluoroquinolone. It is a broad-spectrum antimicrobial which acts on DNA replication. The bactericidal action of ciprofloxacin results from inhibition of the enzymes DNA gyrase and topoisomerase IV. Its use as antimicrobial was first reported in 1983 (Wise *et al.*, 1983), while in 1992 Bhakta and collaborators described its activity against *Burkholderia* isolates (Bhakta *et al.*, 1992).

Combination therapy

Since Bcc bacteria show a high level of intrinsic resistance, combinations of antibiotics must be used. Unfortunately, it is not possible to predict if a drug combination would be effective (Aaron *et al.*, 2000). So, *in vitro* testing is always recommended. Synergy testing is performed in order to find the most effective combinations that can be administered. Several antibiotic combinations have been shown to have good activity against *Burkholderia*. For example, chloramphenicol and minocycline or chloramphenicol plus ceftazidime, were found to be active combinations (Burns and Saiman, 1999). In other studies, different combinations of antibiotics were tested (Aaron *et al.*, 2000), with the most effective combinations including meropenem plus minocycline, amikacin, ceftazidime, chloramphenicol, or trimethoprim/sulphamethoxazole. Zhou and collaborators observed that minocycline, meropenem, and ceftazidime were the most active after testing over 2000 Bcc CF strains for drug susceptibility and synergy (23 combinations) (Zhou *et al.*, 2007).

Potential novel antimicrobial agents

New therapeutic combinations are being investigated and a few approaches appear promising. One example is amiloride, a sodium channel blocker, in association with tobramycin aerosols (Middleton *et al.*, 2005).

As regarding the non Bcc species, a novel compound to be used in clinical therapy is cethromycin (Mima *et al.*, 2011). Cethromycin is a ketolide molecule with broad-spectrum activity which, in contrast to macrolides, is poorly extruded by the efflux systems AmrAB-OprA and BpeAB-OprB, making it promising to eradicate *B. pseudomallei* infections (MIC range 4–64 mg/l for clinical isolates and 4–32 mg/l for environmental isolates) (Mima *et al.*, 2011). Its mechanism of action has been described in *Streptococcus pneumoniae* (Capobianco *et al.*, 2000). It acts by binding at two sites of the bacterial ribosome. To date, only *in vitro* studies have been performed and future *in vivo* experimentation is needed to confirm if it will have clinical efficacy.

As the problem of resistance due to the presence of β-lactamases is of particular interest in Gram-negative bacteria, combinations of β-lactamases inhibitors have been developed (Page *et al.*, 2011). For example, BAL30376 is a combination of two β-lactamase inhibitors (BAL29880 and clavulanic acid) plus BAL19764 which is a siderophore conjugated monobactam. This combination has been tested against *Burkholderia* spp. and its MIC_{90} is 16 μg/ml (Page *et al.*, 2011). A different field to be explored for non-Bcc bacteria is the use of old drugs, as in the case of co-trimoxazole. *B. pseudomallei* infections can be treated using this drug, mostly in combination with ceftazidime, even if controversial results have been shown (Goldberg and Bishara, 2012).

Potential future therapy could include 4-amino-4-deoxy-L-arabinose (Ara4N) pathway inhibitors as potentiators of antimicrobial peptides activity against most of *Burkholderia* spp. (Kline *et al.*, 2008). Indeed, the specific features of lipopolysaccharide are essential to these bacteria and the disruption of such pathways could improve current clinical approach.

A study from Cassidy *et al.* (2012) aimed to determine if photodynamic antimicrobial chemotherapy (PACT) was effective in the treatment of Bcc infection alone or in combination with antibiotics. In particular, they saw that when strains were grown in biofilm, antibiotic treatment combined with PACT had a bactericidal activity of 92% (against 47% of the antibiotic treatment alone). So the authors conclude that PACT could be a good treatment option, either alone or in combination with antibiotics for treatment of Bcc pulmonary infection (Cassidy *et al.*, 2012).

The need for new drugs made researchers worldwide look for alternative sources of antimicrobial compounds. For example, Romoli and collaborators described the use of microorganisms from Antarctica in order to find molecules able to inhibit Bcc bacteria (Romoli *et al.*, 2011). In particular, they used bacteria isolated from Antarctic sponges and found a set of bacteria belonging to the genera *Pseudoalteromonas*, *Arthrobacter* and *Psychrobacter* able to do this. This ability is due to the production of small volatile organic compounds (Romoli *et al.*, 2011). In addition to these non-*Burkholderia* species producing antibiotics, it has been recently found that Bcc bacteria themselves produce antibiotics that kill other *Burkholderia* as well as multidrug-resistant Gram-positive and -negative pathogens (Mahenthiralingam *et al.*, 2011). *B. ambifaria* strains were found to produce a very potent polyketide antibiotic, enacyloxin IIa, which targeted a range of drug resistant Gram-negatives including *B. multivorans* (Mahenthiralingam *et al.*, 2011). These early studies which show that *Burkholderia* can produce polyketide antibiotics such as those normally found in *Streptomyces* bacteria are very promising and suggest that *Burkholderia* can serve as novel source of antibiotics.

Resistance mechanisms

Gram-negative bacteria have five main resistance mechanisms: alteration of drug target site, increase in target production, enzymatic inactivation or modification of drug, reduced membrane permeability, and/or activation of efflux systems. Alteration of targets and enzymatic modification are usually related to a single class of drugs, while the reduction of membrane permeability and active efflux, both widespread in *Burkholderia* pathogens, can determine broad-spectrum drug resistance.

Modification of drug target site

Drug targets of commonly used antibiotics include: penicillin binding proteins for β-lactams; ribosomes for aminoglycosides, macrolides, tetracyclines, and chloramphenicol; DNA gyrase for fluoroquinolones; and folic acid biosynthesis for trimethoprim and sulphonamides. In the Bcc and related species the modification of the antibiotic target was not considered the main resistance mechanism, but an increasing number of reports on mutations associated with resistance have been published recently.

One of the first examples of resistance characterized in Bcc bacteria was modification of the enzyme dihydrofolate reductase (DHFR). Burns *et al.* (1989) analysed DHFR enzyme isolated from susceptible and resistant Bcc strains, and suggested that the difference between the two enzymes is due to a mutation in the active site, probably determining the distinct resistance phenotype in association with a decrease in permeability. In *Burkholderia glumae* and *B. vietnamiensis*, mutations in the gyrase-encoding gene (*gyrA*) have been associated with quinolone resistance (Maeda *et al.*, 2004). In *Burkholderia cepacia* ciprofloxacin-resistant mutants, mutations in *gyrA* conferred 12- to 64-fold increases in MIC, while an additional *parC* mutation conferred a >256-fold increase in MIC (Pope *et al.*, 2008). Amino acid substitutions in *gyrA* have also been detected in *B. pseudomallei* and *B. mallei*, conferring a high level of resistance (Viktorov *et al.*, 2008).

A different mechanism, responsible for β-lactam resistance, has been described by Chantratita and colleagues (Chantratita *et al.*, 2011). They observed that in *B. pseudomallei*, ceftazidime resistance is mediated by the deletion of penicillin-binding protein 3 gene, and that this event was associated with a larger genomic loss of at least 49 genes (Chantratita *et al.*, 2011). In *B. pseudomallei* PenA (a class A β-lactamase) modifications seem to contribute to β–lactam resistance. Indeed, an up-regulation of the expression of the *penA* gene results in increased resistance, whereas *penA* deletion confers susceptibility to various β–lactams (Rholl *et al.*, 2011). In addition, mutations in PenA located near the active site can alter the substrate profile by changing the chemical properties of the binding pocket. To date, only meropenem activity seems not to be affected by either the over-expression adaptions nor the mutations in *penA* gene. Furthermore, Sarovich and colleagues revealed the molecular mechanisms of resistance in clinical isolates of *B. pseudomallei* through the identification of two single-nucleotide polymorphisms (SNPs) located within the *penA* promoter region and near the active site of PenA, respectively. These two SNPs are associated with antibiotic resistance, altering the substrate specificity to β-lactams (Sarovich *et al.*, 2012).

It is known that in Gram-negative bacteria, β-lactamases are able to expand their substrate spectrum through simple mutations (Bush and Fisher, 2011). Yi *et al.* (2012) have recently identified in *B. thailandensis* that single amino acid substitutions, mostly located in the active-site pocket, are able to expand the substrate specificity of its PenA β-lactamase. Their results support the hypothesis that PenA could be a major determinant for ceftazidime resistance in *B. thailandensis* and likely other *Burkholderia* spp., since mutations located in *penA* homologues in *B. cepacia* and *B. pseudomallei* are associated with the same resistance, even if not all the mutations and their localization have been confirmed in the various species. (Poirel *et al.*, 2009; Rholl *et al.*, 2011; Yi *et al.*, 2012). Nonetheless, mutations in genes different from *penA* have been identified in presence of an increased antibiotic exposure, suggesting that other resistance determinants can be found in these bacteria (Yi *et al.*, 2012).

Enzymatic drug inactivation

Resistance to β-lactams and aminoglycosides is commonly associated with enzymatic drug inactivation, although in *Burkholderia* spp. this process is rarely the only resistance mechanism. β-lactams are inactivated by periplasmatic β-lactamases through the hydrolysis of the β–lactam ring, thus restoring the formation of the peptidoglycan

(Bush and Fisher, 2011). Aminoglycoside-inactivating enzymes are able to inhibit the activity of the corresponding antibiotic, which affects ribosomal protein synthesis, through N-acetylation, O-nucleotidylation, and O-phosphorylation, by attacking a specific amino or hydroxyl group (Kondo and Hotta, 1999).

In Bcc species β-lactamases are the most frequently reported enzymes and they have already been well characterized (Aronoff, 1988b; Trepanier *et al.*, 1997). For example, mutant strains over-expressing β–lactamases have been observed to exhibit a high level of resistance (Aronoff, 1988b; Niumsup and Wuthiekanun, 2002). Moreover, in the *B. cenocepacia* J2315 genome, these enzymes containing β–lactamase Pfam domains (PF00144) have been hypothesized to have antimicrobial resistance functions (Holden *et al.*, 2009). In *B. cenocepacia* both Class A and Class C β-lactamases encoding genes have been shown to be up-regulated in response to meropenem in both the wild-type at sub MIC exposure and constitutively in the meropenem adapted strain (Sass *et al.*, 2012). These enzymes have been also identified in *B. pseudomallei* (Cheung *et al.*, 2002; Dance *et al.*, 1991; Keith *et al.*, 2005), *B. thailandensis* (Keith *et al.*, 2005) and *B. gladioli* (Wagner *et al.*, 2002), and have been shown to be involved in β-lactam resistance. Most of these enzymes are inducible and the resistance phenotype is often associated with a defect in permeability. Indeed, in *B. pseudomallei* the antibiotic selective pressure can rapidly induce modifications in the enzyme, and distinct resistance patterns have been reported to emerge (Godfrey *et al.*, 1991).

Reduced membrane permeability

One of the most common resistance mechanism among *Burkholderia* spp. is associated with cell permeability, which is able to impair antibiotics uptake. However, it is noteworthy that efflux, drug target alteration and inactivating enzymes can act synergistically with reduced permeability, thus increasing the antibiotic resistance level of these bacteria (Woodruff and Hancock, 1988). Antibiotic permeability can be affected by LPS structure, porines, and the presence of a bacterial capsule. The LPS structure impacts the permeability to aminoglycosides, polymyxins and cationic peptides of the immune system through the modification of the divalent cation binding sites on LPS itself. Indeed, the core oligosaccharide of the LPS in Bcc strains has less phosphate and 3-deoxy-D-manno-octo-2-ulosonic acid, which causes the neutralization of the negative charges on the bacterial membrane (Cox and Wilkinson, 1991). Moreover, changes in the size of LPS can confer antibiotic resistance to trimethoprim and quinolones, due to the lack of high molecular weight O-antigen (Rajyaguru and Muszynski, 1997).

The biosynthesis of hopanoids, which are involved in the stability of the outer membrane, can affect the permeability and contribute to intrinsic resistance to polymyxins in *B. multivorans* (Malott *et al.*, 2012). The impairment in the self-promoted uptake of these cationic antimicrobials has also been observed in *B. pseudomallei* and *B. cenocepacia*, complementary to other features such as LPS biosynthesis and composition (Burtnick and Woods, 1999; Loutet *et al.*, 2011; Schmerk *et al.*, 2011). Indeed, the presence of 4-amino-4-deoxy-L-arabinose (Ara4N) sugar in the lipid A component of LPS is able to increase polymyxin B resistance also in other Gram-negative bacteria, even though its precise contribution to resistance has not yet been determined (Loutet and Valvano, 2011).

In *B. cenocepacia* the stability of the core oligosaccharide of LPS is able to confer resistance to polymyxins and antimicrobial peptides since the core truncation leads to the loss of different sugars in the LPS chain and results in increased sensitivity (Ortega *et al.*, 2009). Another resistance determinant identified in this species is the alternative sigma factor RpoE, which regulates the expression of genes for extracytoplasmic stress response and is involved in polymyxin B resistance (Loutet *et al.*, 2011). All these data led Loutet *et al.* (2011) to hypothesize a two-tier model in which the outer membrane impermeability and the modified LPS molecule are responsible for the majority of antimicrobial peptide resistance, while other minor determinants play a small role, but contribute to the high global level of resistance in *B. cenocepacia* (Loutet *et al.*, 2011). Nevertheless, it has been observed that in *B. vietnamiensis* aminoglycoside resistance is not determined by the LPS modifications, since Ara4N residues are important for the viability of strains of this species (Jassem *et al.*, 2011). RND efflux pumps seem to be involved in the adaptive resistance of this species and their role in aminoglycosides export has been recently suggested also in *B. cenocepacia* and *B. pseudomallei* (Buroni *et al.*, 2009; Trunck *et al.*, 2009).

In addition, protein channels located in the outer membrane can mediate the decrease in membrane permeability that has been described in most *Burkholderia* spp. (Hancock *et al.*, 1983; Parr *et al.*, 1987; Siritapetawee *et al.*, 2004). Indeed, the number, size and conductance of porins can regulate the entry of hydrophilic molecules such as chloramphenicol, trimethoprim, ciprofloxacin or β-lactams into the bacterial cell (Aronoff, 1988a; Burns and Clark, 1992). In 1992, Burns and Clark observed a decrease in porin-mediated outer membrane permeability in the Bcc when grown in presence of salicylate, and suggested that the OpcS protein could function as a salicylate-suppressed porin (Burns and Clark, 1992). This protein seems to have a selective substrate spectrum, since the membrane penetration and the susceptibility to chloramphenicol, trimethoprim and ciprofloxacin is decreased if salicylate is added, while ceftazidime susceptibility remained unchanged (Burns and Clark, 1992).

Efflux

The efflux-mediated extrusion of antibiotics is suggested to be the main resistance mechanism in Gram-negative bacteria often acting synergistically with permeability barriers (Nikaido, 2001; Ryan *et al.*, 2001). Active efflux can be mediated by five families of pumps: the major facilitator superfamily (MFS), the ATP-binding cassette family (ABC), the small multidrug resistance family (SMR), the multidrug and toxic compound extrusion family (MATE) and the resistance–nodulation–division family (RND) (Poole, 2005).

The transporters belonging to the RND family are able to catalyse the active efflux of many different antibiotics and chemotherapeutic agents, and can mediate MDR in Gram-negative bacteria including the Bcc (Nikaido and Takatsuka, 2009). They comprise a transporter protein (RND) located in the cytoplasmic membrane, a membrane adaptor protein (MFP) in the periplasm, and an outer membrane channel protein (OMP). In *B. cenocepacia* several RND pumps have been identified, sharing 50–71% homology with Mex pumps of *P. aeruginosa* (Buroni *et al.*, 2009; Guglierame *et al.*, 2006). Among them, CeoAB-OpcM, also called RND-10, is able to confer resistance to chloramphenicol, trimethoprim and ciprofloxacin (Nair *et al.*, 2004) and RND-4 seems to be involved in the extrusion of different compounds, since the deletion mutant demonstrated an increase in drug susceptibility to several antibiotics (Bazzini *et al.*, 2011; Buroni *et al.*, 2009).

In *B. vietnamiensis*, the gene encoding NorM efflux pump can complement AcrAB from *Escherichia coli* and is involved in the protection of the cell from quinolones and cationic peptides (Fehlner-Gardiner and Valvano, 2002). This gene also appears to be conserved in at least six Bcc species. In addition, *B. pseudomallei* possesses several RND efflux systems: AmrAB-OprA conferring resistance to aminoglycosides and macrolides, BpeAB-OprB to fluoroquinolones, tetracyclines, acriflavine and macrolides and BpeEF-OprC to chloramphenicol and trimethoprim (Chan and Chua, 2005; Kumar *et al.*, 2006; Mima and Schweizer, 2010; Moore *et al.*, 1999). It is noteworthy that *B. pseudomallei*, *B. mallei* and *B. thailandensis* can have an increase in the expression of genes encoding transporters belonging to RND and MFS efflux families, conferring cross-resistance to fluoroquinolones and cephalosporins, as suggested in other Gram-negative bacteria by previous reports (Poole, 2000; Viktorov *et al.*, 2008). In *B. thailandensis*, for example, chloramphenicol-induced MDR has been observed: in this microorganism resistance to fluoroquinolones, tetracyclines, trimethoprim and some β-lactams is associated with the induction of two different RND efflux pumps (Biot *et al.*, 2011). The first one is a BpeAB-OprB homologue, while the second one is homologous to BpeEF-OprC, (although only the outer membrane channel component has been found) (Biot *et al.*, 2011; Kumar *et al.*, 2006; Mima and Schweizer, 2010).

Also high-level resistance can be mediated by the overexpression of efflux systems. This is the case for cethromycin and AmrAB-OprA up-regulation in *B. pseudomallei*, even if other minor resistance mechanisms appear to be involved (Mima *et al.*, 2011). An analogous process has been described in *B. cenocepacia*, where the increase in *fsr* gene (fosmidomycin resistance gene) expression results in fosmidomycin efflux and related resistant phenotype (Messiaen *et al.*, 2011).

All these data suggest that bacteria are able to coordinate the efflux systems to overcome antibiotic stress and the selective pressure induced by one antimicrobial compound is able to generate resistance to different antibiotic classes.

Biofilm-related drug resistance

Biofilm-growing cells have been reported to proliferate in chronic infections in CF airway, developing further mechanisms of drug resistance, such as: limited antibiotic diffusion (Szomolay *et al.*, 2005), matrix-mediated drug inactivation, reduced susceptibility of the bacterial cells due to nutrients depletion and metabolic slowdown (Hill *et al.*, 2005), phenotypic differentiation through the formation of persister cells (Roberts and Stewart, 2005), and/or up-regulation of biofilm-specific resistance genes (Gillis *et al.*, 2005). In *Burkholderia* spp. the molecular correlation between these mechanisms and virulence or resistance has not been demonstrated yet (Cunha *et al.*, 2004). However, conventional antibiotics are often ineffective at eradicating Bcc infections, especially when associated with biofilms (Brackman *et al.*, 2011).

As QS is known to play an important role in biofilm resistance and following the deletion of QS systems in *B. cenocepacia*, a significant reduction in the resistance to ceftazidime and meropenem was observed, and tobramycin showed higher efficacy on sessile cells. However, a comprehensive description of the underlying molecular mechanism is still lacking (Udine *et al.*, 2013). A first hypothesis on how QS can control downstream processes has been proposed by Schmid *et al.* (2012) and Deng *et al.* (2012). According to the authors, the QS-dependent transcriptional regulation cascade involves a c-di-GMP receptor protein or an alternative unknown common regulator (Schmid *et al.*, 2012). In conclusion, it would be crucial to gain a deeper knowledge on this sophisticated process in order to find new inhibitors to support drug therapy. QS inhibitors (QSI) are hypothesized to improve bacterial eradication. *B. multivorans* and *B. cenocepacia* biofilms treated with QSI appeared to be impaired in their structure and stability (Brackman *et al.*, 2009). In 2011 Brackman and collaborators investigated the use of QSI in combination with conventional antimicrobial agents to check if they were able to enhance the susceptibility of bacterial biofilms (Brackman *et al.*, 2011). They demonstrated that biofilm-grown *B. cenocepacia* strains were more susceptible to the combination QSI-antibiotic than to the treatment with an antibiotic alone. (Brackman *et al.*, 2011).

Conclusion

The major problem related to the clinical therapy of infections due to *Burkholderia* species is the antibiotic resistance. *B. cenocepacia* isolates often show a MDR phenotype. Antibiotic resistance can be due to different mechanisms: alteration of the drug target, increase in target production, enzymatic inactivation or modification of the drug, reduced membrane permeability, activation of efflux systems and biofilm formation. Moreover, the transcriptomic analysis of *B. cenocepacia* exposed to different antibiotics conducted by Sass and collaborators (2011) revealed that resistance in these bacteria is multifactorial. In fact this microorganism is able to express different genes and activate many pathways at a time once exposed to antibiotics and these changes are stably maintained (Sass *et al.*, 2011). In this way, multiple antibiotics must be administered in order to achieve successful therapies. We provided an overview of the drugs which can be used for the treatment of Bcc, being combination therapy the prevalent way to follow. In addition to classical resistance mechanisms, *Burkholderia* also seems to alter the central metabolism as means to adapt to high level resistance (Sass *et al.*, 2011). In this way the treatment of infections due to *Burkholderia* species is very problematic and new drugs and new targets are urgently needed to improve CF patients' lives.

References

Aaron, S.D., Ferris, W., Henry, D.A., Speert, D.P., and Macdonald, N.E. (2000). Multiple combination bactericidal antibiotic testing for patients with cystic fibrosis infected with *Burkholderia cepacia*. Am. J. Respi.r Crit. Care Med. *161*, 1206–1212.

Acar, J.F., Kitzis, M.D., and Goldstein, F.W. (1981). *In-vitro* activity of SQ 26,776 against multiply-resistant enterobacteria--preliminary results. J. Antimicrob. Chemother. *8 (Suppl E)*, 97–101.

Aronoff, S.C. (1988a) Outer membrane permeability in *Pseudomonas cepacia*: diminished porin content in a beta-lactam-resistant mutant and in resistant cystic fibrosis isolates. Antimicrob. Agents Chemother. *32*, 1636–1639.

Aronoff, S.C. (1988b) Derepressed beta-lactamase production as a mediator of high-level beta-lactam resistance in *Pseudomonas cepacia*. Pediatr. Pulmonol. *4*, 72–77.

Bazzini, S., Udine, C., Sass, A., Pasca, M.R., Longo, F., Emiliani, G., Fondi, M., Perrin, E., Decorosi, F., Viti, C., *et al.* (2011). Deciphering the role of RND efflux transporters in *Burkholderia cenocepacia*. PLoS One *6*, e18902.

Bhakta, D.R., Leader, I., Jacobson, R., Robinson-Dunn, B., Honicky, R.E., and Kumar, A. (1992). Antibacterial properties of investigational, new, and commonly used antibiotics against isolates of *Pseudomonas cepacia* in Michigan. Chemotherapy *38*, 319–323.

Biot, F.V., Valade, E., Garnotel, E., Chevalier, J., Villard, C., Thibault, F.M., Vidal, D.R., and Pagès, J.M. (2011). Involvement of the efflux pumps in chloramphenicol selected strains of *Burkholderia thailandensis*: proteomic and mechanistic evidence. PLoS One *6*, e16892.

Blackburn, L., Brownlee, K., Conway, S., and Denton, M. (2004). 'Cepacia syndrome' with *Burkholderia multivorans*, 9 years after initial colonization. J. Cyst. Fibros. *3*, 133–134.

Brackman, G., Hillaert, U., Van Calenbergh, S., Nelis, H.J., and Coenye, T. (2009). Use of quorum sensing inhibitors to interfere with biofilm formation and development in *Burkholderia multivorans* and *Burkholderia cenocepacia*. Res. Microbiol. *160*, 144–151.

Brackman, G., Cos, P., Mae,s L., Nelis, H.J., and Coenye, T. (2011). Quorum sensing inhibitors increase the susceptibility of bacterial biofilms to antibiotics *in vitro* and *in vivo*. Antimicrob. Agents Chemother. *55*, 2655–2661.

Braveny, I. (1984). *In vitro* activity of imipenem--a review. Eur. J. Clin. Microbiol. *3*, 456–462.

Brett, P.J., DeShazer, D., and Woods, D.E. (1998). *Burkholderia thailandensis* sp. nov., a *Burkholderia pseudomallei*-like species. Int. J. Syst. Bacteriol. *1*, 317–320.

Brizendine, K.D., Baddley, J.W., Pappas, P.G., Leon, K.J., and Rodriguez, J.M. (2012). Fatal *Burkholderia gladioli* infection misidentified as *Empedobacter brevis* in a lung transplant recipient with cystic fibrosis. Transpl. Infect. Dis. *14*, E13–18.

Brummett, R.E., Vernon, J.A., and Meikle, M.M. (1971). Ototoxicity of tobramycin in guinea pigs. Arch. Otolaryngol. *94*, 59–63.

Burns, J.L., and Clark, D.K. (1992). Salicylate-inducible antibiotic resistance in *Pseudomonas cepacia* associated with absence of a pore-forming outer membrane protein. Antimicrob. Agents Chemother. *36*, 2280–2285.

Burns, J., and Saiman, L. (1999). *Burkholderia cepacia* infections in cystic fibrosis. Pediatr. Infect. Dis. J. *18*, 155–156.

Burns, J.L., Lien, D.M., and Hedin, L.A. (1989). Isolation and characterization of dihydrofolate reductase from trimethoprim-susceptible and trimethoprim-resistant *Pseudomonas cepacia*. Antimicrob. Agents Chemother. *33*, 1247–1251.

Buroni, S., Pasca, M.R., Flannagan, R.S., Bazzini, S., Milano, A., Bertani, I., Venturi, V., Valvano, M.A., and Riccardi, G. (2009). Assessment of three Resistance-Nodulation-Cell Division drug efflux transporters of *Burkholderia cenocepacia* in intrinsic antibiotic resistance. BMC Microbiol. *9*, 200.

Burtnick, M.N., and Woods, D.E. (1999). Isolation of polymyxin B-susceptible mutants of *Burkholderia pseudomallei* and molecular characterization of genetic loci involved in polymyxin B resistance. Antimicrob. Agents Chemother. *43*, 2648–2656.

Bush K., and Fisher, J.F. (2011). Epidemiological expansion, structural studies, and clinical challenges of new β–lactamases from Gram-negative bacteria. Annu. Rev. Microbiol. *65*, 455–478.

Capobianco, J.O., Cao, Z., Shortridge, V.D., Ma, Z., Flamm, R.K., and Zhong, P. (2000). Studies of the novel ketolide ABT-773: transport, binding to ribosomes, and inhibition of protein synthesis in *Streptococcus pneumoniae*. Antimicrob. Agents Chemother. *44*, 1562–1567.

Cassidy, C.M., Donnelly, R.F., Elborn, J.S., Magee, N.D., and Tunney, M.M. (2012). Photodynamic Antimicrobial Chemotherapy (PACT) in combination with antibiotics for treatment of *Burkholderia cepacia* complex infection. J. Photochem. Photobiol. B. *106*, 95–100.

Chan, Y.Y., and Chua, K.L. (2005). The *Burkholderia pseudomallei* BpeAB-OprB efflux pump: expression and impact on quorum sensing and virulence. J. Bacteriol. *187*, 4707–4719.

Chantratita, N., Rholl, D.A., Sim, B., Wuthiekanun, V., Limmathurotsakul, D., Amornchai, P., Thanwisai, A., Chua, H.H., Ooi, W.F., Holden, M.T., *et al.* (2011). Antimicrobial resistance to ceftazidime involving loss of penicillin-binding protein 3 in *Burkholderia pseudomallei*. Proc. Natl. Acad. Sci. U.S.A. *108*, 17165–17170.

Cheung, T.K., Ho, P.L., Woo, P.C., Yuen, K.Y., and Chau, P.Y. (2002). Cloning and expression of class A beta-lactamase gene *blaA*(BPS) in *Burkholderia pseudomallei*. Antimicrob. Agents Chemother. *46*, 1132–1135.

Compant, S., Nowak, J., Coenye, T., Clement, C., and Ait Barka, E. (2008). Diversity and occurrence of

Burkholderia spp. in the natural environment. FEMS Microbiol. Rev. *32*, 607–626.

Cox, A.D., and Wilkinson, S.G. (1991). Ionizing groups in lipopolysaccharides of *Pseudomonas cepacia* in relation to antibiotic resistance. Mol. Microbiol. *5*, 641–646.

Cunha, M.V., Sousa, S.A., Leitão, J.H., Moreira, L.M., Videira, P.A., and Sá-Correia, I. (2004). Studies on the involvement of the exopolysaccharide produced by cystic fibrosis-associated isolates of the *Burkholderia cepacia* complex in biofilm formation and in persistence of respiratory infections. J. Clin. Microbiol. *42*, 3052–3058.

Dance, D.A., Wuthiekanun, V., Chaowagul, W., and White, N.J. (1989). The antimicrobial susceptibility of *Pseudomonas pseudomallei*. Emergence of resistance *in vitro* and during treatment. J. Antimicrob. Chemother. *24*, 295–309.

Dance, D.A., Wuthiekanun, V., Chaowagul, W., Suputtamongkol, Y., and White, N.J. (1991). Development of resistance to ceftazidime and co-amoxiclav in *Pseudomonas pseudomallei*. J. Antimicrob. Chemother. *28*, 321–324.

Deng, Y., Schmid, N., Wang, C., Wang, J., Pessi, G., Wu, D., Lee, J., Aguilar, C., Ahrens, C.H., Chang, C., *et al.* (2012). Cis-2-dodecenoic acid receptor RpfR links quorum-sensing signal perception with regulation of virulence throughcyclic dimeric guanosine monophosphate turnover. Proc. Natl. Acad. Sci. U.S.A. *109*, 15479–15484.

Drevinek, P., and Mahenthiralingam, E. (2010). *Burkholderia cenocepacia* in cystic fibrosis: epidemiology and molecular mechanisms of virulence. Clin. Microbiol. Infect. *16*, 821–830.

Edwards, J.R., Turner, P.J., Wannop, C., Withnell, E.S., Grindey, A.J., and Nairn, K. (1989). *In vitro* antibacterial activity of SM-7338, a carbapenem antibiotic with stability to dehydropeptidase I. Antimicrob. Agents Chemother. *33*, 215–222.

Eliopoulos, G.M., and Moellering, R.C. Jr. (1982). Azlocillin, mezlocillin, and piperacillin: new broad-spectrum penicillins. Ann. Intern. Med. *97*, 755–760.

Fehlner-Gardiner, C.C., and Valvano, M.A. (2002). Cloning and characterization of the *Burkholderia vietnamiensis norM* gene encoding a multi-drug efflux protein. FEMS Microbiol. Lett. *215*, 279–283.

Galyov, E.E., Brett, P.J., and DeShazer, D. (2010). Molecular insights into *Burkholderia pseudomallei* and *Burkholderia mallei* pathogenesis. Annu. Rev. Microbiol. *64*, 495–517.

Gibson, R.L., Burns, J.L., and Ramsey, B.W. (2003). Pathophysiology and management of pulmonary infections in cystic fibrosis. Am. J. Respir. Crit. Care Med. *168*, 918–951.

Gillis, R.J., White, K.G., Choi, K.H., Wagner, V.E., Schweizer, H.P., and Iglewski, B.H. (2005). Molecular basis of azithromycin-resistant *Pseudomonas aeruginosa* biofilms. Antimicrob. Agents Chemother. *49*, 3858–3867.

Glass, M.B., Gee, J.E., Steigerwalt, A.G., Cavuoti, D., Barton, T., Hardy, R.D., Godoy, D., Spratt, B.G., Clarck, T.A., and Wilkins, P.P. (2006). Pneumonia and septicemia caused by *Burkholderia thailandensis* in the United States. J. Clin. Microbiol. *44*, 4601–4604.

Godfrey, A.J., Wong, S., Dance, D.A.B., Chaowagul, W., and Bryan, L.E. (1991). *Pseudomonas pseudomallei* resistance to β-lactam antibiotics due to alterations in the chromosomally encoded β-lactamase. Antimicrob. Agents Chemother. *35*, 1635–1640.

Goldberg, E., and Bishara, J. (2012). Contemporary unconventional clinical use of co-trimoxazole. Clin. Microbiol. Infect. *18*, 8–17.

Gottlieb, D., Bhattacharyya, P.K., Anderson, H.W., and Carter, H.E. (1948). Some properties of an antibiotic obtained from a species of *Streptomyces*. J. Bacteriol. *55*, 409–417.

Graves, M., Robin, T., Chipman, A.M., Wong, J., Khashe, S., and Janda, J.M. (1997). Four additional cases of *Burkholderia gladioli* infection with microbiological correlates and review. Clin. Infect. Dis. *25*, 838–842.

Guglierame, P., Pasca, M.R., De Rossi, E., Buroni, S., Arrigo, P., Manina, G., and Riccardi, G. (2006). Efflux pump genes of the resistance-nodulation-division family in *Burkholderia cenocepacia* genome. BMC Microbiol. *6*, 66.

Hancock, R.E., Poole, K., Gimple, M., and Benz, R. (1983). Modification of the conductance, selectivity and concentration-dependent saturation of *Pseudomonas aeruginosa* protein P channels by chemical acetylation. Biochim. Biophys. Acta. *735*, 137–144.

Hill, D., Rose, B., Pajkos, A., Robinson, M., Bye, P., Bell, S., Elkins, M., Thompson, B., Macleod, C., Aaron, S.D., and Harbour, C. (2005). Antibiotic susceptibilities of *Pseudomonas aeruginosa* isolates derived from patients with cystic fibrosis under aerobic, anaerobic, and biofilm conditions. J. Clin. Microbiol. *43*, 5085–5090.

Holden, M.T., Seth-Smith, H.M., Crossman, L.C., Sebaihia, M., Bentley, S.D., Cerdeño-Tárraga, A.M., Thomson, N.R., Bason, N., Quail, M.A., Sharp, S., *et al.* (2009). The genome of *Burkholderia cenocepacia* J2315, an epidemic pathogen of cystic fibrosis patients. J. Bacteriol. *191*, 261–277.

Holloman, J.L.Jr., Fullilove, R.E.Sr., Yergan, C.M., and Barouk, E. (1961). Clinical experience with a new sulfonamide sulphamethoxazole. Appl. Ther. 3, 775–779.

Jassem, A.N., Zlosnik, J.E., Henry, D.A., Hancock, R.E., Ernst, R.K., and Speert, D.P. (2011). *In vitro* susceptibility of *Burkholderia vietnamiensis* to aminoglycosides. Antimicrob. Agents Chemother. *55*, 2256–2264.

Jones, A.M., Dodd, M.E., and Webb, A.K. (2001). *Burkholderia cepacia*: current clinical issues, environmental controversies and ethical dilemmas. Eur. Respir. J. *17*, 295–301.

Jones, R.N., Thornsberry, C., Barry, A.L., Fuchs, P.C., Gavan, T.L., and Gerlach, E.H. (1977). Piperacillin (T-1220), a new semisynthetic penicillin: *in vitro* antimicrobial activity comparison with carbenicillin, ticarcillin, ampicillin, cephalothin, cefamandole and cefoxitin. J. Antibiot. (Tokyo). *30*, 1107–1114.

Keith, K.E., Oyston, P.C., Crossett, B., Fairweather, N.F., Titball, R.W., Walsh, T.R., and Brown, K.A. (2005).

Functional characterization of OXA-57, a class D beta-lactamase from *Burkholderia pseudomallei*. Antimicrob. Agents Chemother. *49*, 1639–1641.

Kesado, T., Hashizume, T., and Asahi, Y. (1980). Antibacterial activities of a new stabilized thienamycin, N-formimidoyl thienamycin, in comparison with other antibiotics. Antimicrob. Agents Chemother. *17*, 912–917.

Kline, T., Trent, M.S., Stead, C.M., Lee, M.S., Sousa, M.C., Felise, H.B., Nguyen, H.V., and Miller, S.I. (2008). Synthesis of and evaluation of lipid A modification by 4-substituted 4-deoxy arabinose analogs as potential inhibitors of bacterial polymyxin resistance. Bioorg. Med. Chem. Lett. *18*, 1507–1510.

Kondo, S., and Hotta, K. (1999). Semisynthetic aminoglycoside antibiotics: development and enzymatic modifications. J. Infect. Chemother. *5*, 1–9.

Kumar, A., Chua, K.L., and Schweizer, H.P. (2006). Method for regulated expression of single copy efflux pump genes in a surrogate *Pseudomonas aeruginosa* strain: identification of the BpeF-OprC chloramphenicol and trimethoprim efflux pump of *Burkholderia pseudomallei* 1026b. Antimicrob. Agents Chemother. *50*, 3460–3463.

LiPuma, J.J. (2010). The changing microbial epidemiology in cystic fibrosis. Clin. Microbiol. Rev. *23*, 299–323.

Livermore, D.M., and Williams, J.D. (1981). *In-vitro* activity of the monobactam, SQ 26,776, against Gram-negative bacteria and its stability to their beta-lactamases. J. Antimicrob. Chemother. *8 (Suppl E)*, 29–37.

Loutet, S.A., and Valvano, M.A. (2011). Extreme antimicrobial peptide and polymyxin B resistance in the genus *Burkholderia*. Front. Microbiol. *2*, 159.

Loutet, S.A., Mussen, L.E., Flannagan, R.S., and Valvano, M.A. (2011). A two-tier model of polymyxin B resistance in *Burkholderia cenocepacia*. Environ. Microb. Reports. *3*, 278–285.

Maeda, Y., Kiba, A., Ohnishi, K., and Hikichi, Y. (2004). Implications of amino acid substitutions in GyrA at position 83 in terms of oxolinic acid resistance in field isolates of *Burkholderia glumae*, a causal agent of bacterial seedling rot and grain rot of rice. Appl. Environ. Microbiol. *70*, 5613–5620.

Mahenthiralingam, E., Song, L., Sass, A., White, J., Wilmot, C., Marchbank, A., Boaisha, O., Paine, J., Knight, D., and Challis, G.L. (2011). Enacyloxins are products of an unusual hybrid modular polyketide synthase encoded by a cryptic *Burkholderia ambifaria* Genomic Island. Chem. Biol. *18*, 665–677.

Malott, R.J., Steen-Kinnaird, B.R., Lee, T.D., and Speert, D.P. (2011). Identification of hopanoid biosynthesis genes involved in polymyxin resistance in *Burkholderia multivorans*. Antimicrob. Agents Chemother. *56*, 464–471.

Messiaen, A.S., Verbrugghen, T., Declerck, C., Ortmann, R., Schlitzer, M., Nelis, H., Van Calenbergh, S., and Coenye, T. (2011). Resistance of the *Burkholderia cepacia* complex to fosmidomycin and fosmidomycin derivatives. Int. J. Antimicrob. Agents. *38*, 261–264.

Meyer, R.D., Young, L.S., and Armstrong, D. (1971). Tobramycin (nebramycin factor 6): *in vitro* activity against *Pseudomonas aeruginosa*. Appl. Microbiol. *22*, 1147–1151.

Middleton, P.G., Kidd, T.J., and Williams, B. (2005). Combination aerosol therapy to treat *Burkholderia cepacia* complex. Eur. Respir. J. *26*, 305–308.

Mima, T., and Schweizer, H.P. (2010). The BpeAB-OprB efflux pump of *Burkholderia pseudomallei* 1026b does not play a role in quorum sensing, virulence factor production, or extrusion of aminoglycosides but is a broad-spectrum drug efflux system. Antimicrob. Agents Chemother. *54*, 3113–3120.

Mima, T., Schweizer, H.P., and Xu, Z.Q. (2011). *In vitro* activity of cethromycin against *Burkhoderia pseudomallei* and investigation of mechanism of resistance. J. Antimicrob. Chemother. *66*, 73–78.

Misra, R., and Bavro, V.N. (2009). Assembly and transport mechanism of tripartite drug efflux systems. Biochim. Biophys. Acta. *1794*, 817–825.

Moore, R.A., and Hancock, R.E. (1986). Involvement of outer membrane of *Pseudomonas cepacia* in aminoglycoside and polymyxin resistance. Antimicrob. Agents Chemother. *30*, 923–926.

Moore, R.A., DeShazer, D., Reckseidler, S., Weissman, A., and Woods, D.E. (1999). Efflux-mediated aminoglycoside and macrolide resistance in *Burkholderia pseudomallei*. Antimicrob. Agents Chemother. *43*, 465–470.

Murray, S., Charbeneau, J., Marshall, B.C., and LiPuma, J.J. (2008). Impact of *Burkholderia* infection on lung transplantation in cystic fibrosis. Am. J. Respir. Crit. Care Med. *178*, 363–371.

Nair, B.M., Cheung, K.J. Jr, Griffith, A., and Burns, J.L. (2004). Salicylate induces an antibiotic efflux pump in *Burkholderia cepacia* complex genomovar III (*B. cenocepacia*). J. Clin. Invest. *113*, 464–4.

Ni, N., Li, M., Wang, J., and Wang, B. (2009). Inhibitors and antagonists of bacterial quorum sensing. Med. Res. Rev. *29*, 65–124.

Nikaido, H. (2001). Preventing drug access to targets: cell surface permeability barriers and active efflux in bacteria. Semin. Cell Dev. Biol. *12*, 215–223.

Nikaido, H., and Takatsuka, Y. (2009). Mechanisms of RND multidrug efflux pumps. Biochim. Biophys. Acta *1794*, 769–781.

Niumsup, P., and Wuthiekanun, V. (2002). Cloning of the class D β-lactamase gene from *Burkholderia pseudomallei* and studies on its expression in ceftazidime-susceptible and -resistant strains. J. Antimicrob. Chemother. *50*, 445–455.

Ortega, X.P., Silipo, A., Saldías, M.S., Bates, C.C., Molinaro, A., and Valvano, M.A. (2009). Biosynthesis and structure of the *Burkholderia cenocepacia* K56-2 lipopolysaccharide core oligosaccharide: truncation of the coreoligosaccharide leads to increased binding and sensitivity to polymyxin B. J. Biol. Chem. *284*, 21738–21751.

Page, M.G.P., Dantier, C., Desarbre, E., Gaucher, B., Gebhardt, K., and Schmitt-Hoffmann, A. (2011). *In vitro* and *in vivo* properties of BAL30376, a_β-lactam and dual β-lactamase inhibitor combination with enhanced activity against Gram-negative bacilli that express multiple β-lactamases. Antimicrob. Agents Chemother. *55*, 1510–1519.

Parr, T.R. Jr, Moore, R.A., Moore, L.V., and Hancock, R.E. (1987). Role of porins in intrinsic antibiotic resistance of *Pseudomonas cepacia*. Antimicrob. Agents Chemother. *31*, 121–123.

Peeters, E., Nelis, H.J., and Coenye, T. (2009). *In vitro* activity of ceftazidime, ciprofloxacin, meropenem, minocycline, tobramycin and trimethoprim/sulfamethoxazole against planktonic and sessile *Burkholderia cepacia* complex bacteria. J. Antimicrob. Chemother. *64*, 801–809.

Pestka, S. (1969). Studies on the formation of transfer ribonucleic acid–ribosome complexes. X. Phenylalanyl-oligonucleotide binding to ribosomes and the mechanism of chloramphenicol action. Biochem. Biophys. Res. Commun. *36*, 589–595.

Poirel, L., Rodriguez-Martinez, J.M., Plesiat, P., and Nordmann, P. (2009). Naturally occurring class A beta-lactamases from the *Burkholderia cepacia* complex. Antimicrob. Agents Chemother. *53*, 876–882.

Poole, K. (2000). Efflux-mediated resistance to fluoroquinolones in Gram-negative bacteria. Antimicrob. Agents Chemother. *44*, 2233–2241.

Poole, K. (2005). Efflux-mediated antimicrobial resistance. J. Antimicrob. Chemother. *56*, 20–51.

Pope, C.F., Gillespie, S.H., Pratten, J.R., and McHugh, T.D. (2008). Fluoroquinolone-resistant mutants of *Burkholderia cepacia*. Antimicrob. Agents Chemother. *52*, 1201–1203.

Price, K.E., Pursiano, T.A., and DeFuria, M.D. (1974). Activity of BB-K8 (amikacin) against clinical isolates resistant to one or more aminoglycoside antibiotics. Antimicrob. Agents Chemother. *5*, 143–152.

Rajyaguru, J.M., and Muszynski, M.J. (1997). Association of resistance to trimethoprim/sulphamethoxazole, chloramphenicol and quinolones with changes in major outer membrane proteins and lipopolysaccharide in *Burkholderia cepacia*. J. Antimicrob. Chemother. *40*, 803–809.

Redin, G.S. (1966). Antibacterial activity in mice of minocycline, a new tetracycline. Antimicrob. Agents Chemother. (Bethesda) *6*, 371–376.

Reisberg, B., Herzog, J., and Weinstein, L. (1966). *In vitro* antibacterial activity of trimethoprim alone and combined with sulfonamides. Antimicrob. Agents Chemother. (Bethesda) *6*, 424–429.

Rholl, D.A., Papp-Wallace, K.M., Tomaras, A.P., Vasil, M.L., Bonomo, R.A., and Schweizer, H.P. (2011). Molecular investigations of PenA-mediated β–lactam resistance in *Burkholderia pseudomallei*. Front. Microbiol. *2*, 139.

Roberts, M.E., and Stewart, P.S. (2005). Modelling protection from antimicrobial agents in biofilms through the formation of persister cells. Microbiology *151*, 75–80.

Romoli, R., Papaleo, M.C., de Pascale, D., Tutino, M.L., Michaud, L., LoGiudice, A., Fani, R., and Bartolucci, G. (2011). Characterization of the volatile profile of Antarctic bacteria by using solid-phase microextraction-gas chromatography–mass spectrometry. J. Mass Spectrom. *46*, 1051–1059.

Ross, J.P., Holland, S.M., Gill, V.J., DeCarlo, E.S., and Gallin, J.I. (1995). Severe *Burkholderia (Pseudomonas) gladioli* infection in chronic granulomatous disease: report of two successfully treated cases. Clin. Infect. Dis. *21*, 1291–1293.

Ryan, B.M., Dougherty, T.J., Beaulieu, D., Chuang, J., Dougherty, B.A., and Barrett, J.F. (2001). Efflux in bacteria: what do we really know about it? Expert Opin. Investig. Drugs *10*, 1409–1422.

Saiman, L., Siegel, J., and the Cystic Fibrosis Foundation. (2003). Infection control recommendations for patients with cystic fibrosis: microbiology, important pathogens, and infection control practices to prevent patient-to-patient transmission. Infect. Control Hosp. Epidemiol. *24*, S6–52.

Sarovich, D.S., Price, E.P., Von Schulze, A.T., Cook, J.M., Mayo, M., Watson, L.M., Richardson, L., Seymour, M.L., Tuanyok, A., Engelthaler, D.M., *et al.* (2012). Characterization of ceftazidime resistance mechanisms in clinical isolates of *Burkholderia pseudomallei* from Australia. PLoS One *7*, e30789.

Sass, A., Marchbank, A., Tullis, E., Lipuma, J.J., and Mahenthiralingam, E. (2011). Spontaneous and evolutionary changes in the antibiotic resistance of *Burkholderia cenocepacia* observed by global gene expression analysis. BMC Genomics *12*, 373.

Schmid, N., Pessi, G., Deng, Y., Aguilar, C., Carlier, A.L., Grunau, A., Omasits, U., Zhang, L., Ahrens, C.H., and Eberl, L. (2012). The AHL- and BDSF-dependent quorum sensing systems control specific and overlapping sets of genes in *Burkholderia cenocepacia* H111. PLoS ONE *7*, e49966.

Schmerk, C.L., Bernards, M.A., and Valvano, M.A. (2011). Hopanoid production is required for low pH tolerance, antimicrobial resistance, and motility in *Burkholderia cenocepacia*. J. Bacteriol. *193*, 6712–6723.

Siritapetawee, J., Prinz, H., Samosornsuk, W., Ashley, R.H., and Suginta, W. (2004). Functional reconstitution, gene isolation and topology modelling of porins from *Burkholderia pseudomallei* and *Burkholderia thailandensis*. Biochem. J. 377, 579–587.

Slocombe, B., Basker, M.J., Bentley, P.H., Clayton, J.P., Cole, M., Comber, K.R., Dixon, R.A., Edmondson, R.A., Jackson, D., Merrikin, D.J., and Sutherland, R. (1981). BRL 17421, a novel beta-lactam antibiotic, highly resistant to beta-lactamases, giving high and prolonged serum levels in humans. Antimicrob, Agents Chemother. *20*, 38–46.

Szomolay, B., Klapper, I., Dockery, J., and Stewart, P.S. (2005). Adaptive responses to antimicrobial agents in biofilms. Environ. Microbiol. 7, 1186–1191.

Trépanier, S., Prince, A., and Huletsky, A. (1997). Characterization of the *penA* and *penR* genes of *Burkholderia cepacia* 249 which encode the chromosomal class A penicillinase and its LysR-type transcriptional regulator. Antimicrob. Agents Chemother. *41*, 2399–2405.

Trunck, L.A., Propst, K.L., Wuthiekanun, V., Tuanyok, A., Beckstrom-Sternberg, S.M., Beckstrom-Sternberg, J.S., Peacock, S.J., Keim, P., Dow, S.W., and Schweizer, H.P. (2009). Molecular basis of rare aminoglycoside susceptibility and pathogenesis of *Burkholderia pseudomallei* clinical isolates from Thailand. PLoS Negl. Trop. Dis. *3*, e519.

Udine, C., Brackman, G., Bazzini, S., Buroni, S., Van Acker, H., Pasca, M.R., Riccardi, G., and Coenye, T. (2013). Phenotypic and genotypic characterisation of *Burkholderia cenocepacia* J2315 mutants affected in homoserine lactone and diffusible signal factor-based quorum sensing systems suggests interplay between both types of systems. PLoS ONE *8*, e55112.

Valvano, M.A. (2006). Infections by *Burkholderia* spp.: the psychodramatic life of an opportunistic pathogen. Future Microbiol. *1*, 145–149.

Van Acker, H., Van Snick, E., Nelis, H.J., and Coenye, T. (2010). *In vitro* activity of temocillin against planktonic and sessile *Burkholderia cepacia* complex bacteria. J, Cyst, Fibros. *9*, 450–454.

Vanlaere. E., LiPuma. J.J., Baldwin. A., Henry. D., De Brandt. E., Mahenthiralingam. E., Speert. D., Dowson. C., and Vandamme. P. (2008). *Burkholderia latens* sp. nov., *Burkholderia diffusa* sp. nov., *Burkholderia arboris* sp. nov., *Burkholderia seminalis* sp. nov. and *Burkholderia metallica* sp. nov., novel species within the *Burkholderia cepacia* complex. Int. J. Syst. Evol. Microbiol. *58*, 1580–1590.

Vanlaere, E., Baldwin, A., Gevers, D., Henry, D., De Brandt, E., LiPuma, J.J., Mahenthiralingam, E., Speert, D.P., Dowson, C., and Vandamme, P. (2009). Taxon K, a complex within the *Burkholderia cepacia* complex, comprises at least two novel species, *Burkholderia contaminans* sp. nov. and *Burkholderia lata* sp. nov. Int. J. Syst. Evol. Microbiol. *59*, 102–111.

Viktorov, D.V., Zakharova, I.B., Podshivalova, M.V., Kalinkina, E.V., Merinova, O.A., Ageeva, N.P., Antonov, V.A., Merinova, L.K., and Alekseev, V.V. (2008). High-level resistance to fluoroquinolones and cephalosporins in *Burkholderia pseudomallei* and closely related species. Trans. R. Soc. Trop. Med. Hyg. *102* (Suppl. 1), S103–110.

Wagner, U.G., Petersen, E.I., Schwab, H., and Kratky, C. (2002). EstB from *Burholderia gladioli*: a novel esterase with a beta-lactamase fold reveals steric factors to discriminate between esterolytic and beta-lactam cleaving activity. Protein Sci. *11*, 467–478.

Waters, V. (2012). New treatments for emerging cystic fibrosis pathogens other than *Pseudomonas*. Curr. Pharm. Des. *18*, 696–725.

Wise, R., Andrews, J.M., and Bedford, K.A. (1980). Comparison of *in vitro* activity of GR 20263, a novel cephalosporin derivative, with activities of other beta-lactam compounds. Antimicrob, Agents Chemother. *17*, 884–889.

Wise. R., Andrews. J.M., and Edwards. L.J. (1983). *In vitro* activity of Bay 09867, a new quinoline derivative, compared with those of other antimicrobial agents. Antimicrob. Agents Chemother. *23*, 559–564.

Woodruff, W.A., and Hancock, R.E. (1988). Construction and characterization of *Pseudomonas aeruginosa* protein F-deficient mutants after *in vitro* and *in vivo* insertion mutagenesis of the cloned gene. J. Bacteriol. *170*, 2592–2598.

Yi, H., Cho, K.H., Cho, Y.S., Kim, K., Nierman, W.C., and Kim, H.S. (2012). Twelve positions in a β-lactamase that can expand its substrate spectrum with a single amino acid substitution. PLoS One *7*, e37585.

Zhou, J., Chen, Y., Tabibi, S., Alba, L., Garber, E., and Saiman, L. (2007). Antimicrobial susceptibility and synergy studies of *Burkholderia cepacia* complex isolated from patients with cystic fibrosis. Antimicrob. Agents Chemother. *51*, 1085–1088.

Cell-to-cell Communication and Biofilm Formation of Members of the Genus *Burkholderia*: A Story of Multilingually Talented Bacteria

10

Nadine Schmid, Gabriella Pessi, Claudio Aguilar and Leo Eberl

Abstract

Members of the genus *Burkholderia* appear to be particularly loquacious. Evidence has accumulated over the past few years that these bacteria are capable of utilizing at least three different chemical languages: *N*-acylhomoserine lactones, *cis*-2-unsaturated fatty acids and quinolones. Here we summarize the current knowledge of the underlying molecular architectures of these communication systems and show that they are involved in the control of some highly conserved functions, including the production of extracellular proteolytic enzymes, motility, antifungal activity, expression of virulence factors and biofilm formation. Particular attention is paid to the role of these communication systems in the formation of surface-associated consortia and the genes that are required for the sessile lifestyle of this group of bacteria. The new emerging role of the intracellular secondary messenger cyclic diguanylate (c-di-GMP) in biofilm formation and especially as a downstream regulatory element of the fatty acid signalling cascade is discussed. Finally, given that these communication systems are required for both biofilm formation and pathogenicity, they are suggested to represent highly valuable targets for the development of novel antibacterial drugs.

Members of the genus *Burkholderia* speak multiple languages

N-Acylhomoserine lactone (AHL)-based QS systems in *Burkholderia* sp.

AHL-based QS systems typically rely on two proteins: an AHL synthase and an AHL receptor protein. At low cell densities the expression of the AHL synthase gene, coding for a protein of the LuxI family, occurs at a basal level such that the signal molecules slowly accumulate in the culture medium. However, upon reaching a critical threshold concentration, the AHL molecule binds to its cognate receptor, a protein of the LuxR family of transcriptional regulators. This complex in turn binds to specific sites in the promoter regions of target genes, activating or repressing their transcription (Williams, 2007). First indications that a *Burkholderia* species produces AHLs were obtained in a study that showed that the culture supernatant of a *B. cepacia* strain was able to activate different AHL biosensors (McKenney *et al.*, 1995). A genetic screen for mutants with altered siderophore production in *Burkholderia cenocepacia* K56-2 eventually lead to the identification of the CepI/R system (Lewenza *et al.*, 1999). The AHL synthase, CepI, was shown to direct the synthesis of *N*-octanoyl-homoserine lactone (C8-HSL) and minor amounts of *N*-hexanoylhomoserine lactone (C6-HSL) (Fig. 10.1). Subsequent studies revealed that the CepR/AHL complex binds as a dimer to an imperfect palindromic sequence that often partially overlaps the

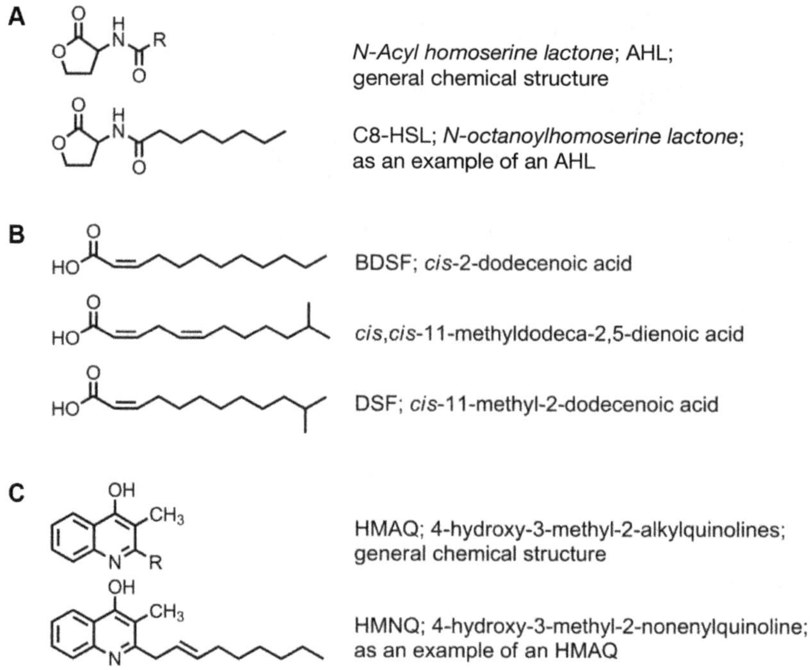

Figure 10.1 The different QS signalling molecules used by *Burkholderia* species. (A) Signalling molecules in AHL-based QS. (B) Signalling molecules in fatty acid based QS. (C) Signalling molecules in quinolone-based QS.

−35 regions of target promoters, thereby initiating transcription of downstream genes (Chambers *et al.*, 2006; Wei *et al.*, 2011). This recognition sequence, referred to as *cep* box, is also present in the promoter region of *cepI*. Thus, once the QS system is triggered a positive feedback loop that leads to a rapid increase in AHL levels and consequently to target gene expression, is established. In addition, the CepR/AHL complex also negatively controls its own expression (Lewenza and Sokol, 2001).

The CepI/R system appears to be present in all 17 *Burkholderia cepacia* complex (Bcc) species, although direct experimental evidence is missing for a few of them, namely *B. arboris, B. diffusa, B. latens, B. metallica,* and *B. seminalis* (Gotschlich *et al.*, 2001; Venturi *et al.*, 2004; Wopperer *et al.*, 2006). CepI/R orthologues have also been identified in all members of the pseudomallei group, consisting of *B. pseudomallei, B. mallei, B. thailandensis* and *B. oklahomensis,* as well as in in some species of the recently defined plant-beneficial-environmental *Burkholderia* group (PBE), although it remains to be seen whether they are true orthologues or belong to independent evolutionary lineages (Suárez-Moreno *et al.*, 2010). Recent work has also provided evidence that many members of the PBE group harbour another highly conserved LuxIR pair, designated BraI/R, that relies on the AHL signal 3-oxo-C14-HSL (Suárez-Moreno *et al.*, 2008). Finally, many *Burkholderia* strains have been demonstrated to harbour more than one AHL-dependent QS system and produce multiple AHL signal molecules (Riedel and Eberl, 2007). It is beyond the scope of this chapter to summarize all these systems and we will rather focus on *B. cenocepacia*, which is the best-studied model organism.

In addition to the CepI/R system, *B. cenocepacia* strains belonging to the ET12 lineage carry the CciI/R QS system on the *B. cenocepacia* island (cci), a 31.7 kb low GC content pathogenicity island encoding 35 putative ORFs (Baldwin *et al.*, 2004). This QS system differs from the CepI/R system in that *cciI* and *cciR* form a transcriptional unit, whereas *cepI* and *cepR* are divergently transcribed (Malott *et al.*, 2005). Furthermore, the main product of CciI is C6-HSL, while C8-HSL, the main product of CepI, is only produced in minor amounts (Malott *et al.*, 2005). The CepI/R QS system and the CciI/R QS system do not act independently: CepR is required for transcription of the *cciIR* operon, whereas CciR negatively controls the expression of CepI, but not of CepR (Malott *et al.*, 2005). Results from a transcriptomic analysis suggested a reciprocal regulation of target genes by the two QS systems, which ensures the fine-tuned expression of target genes (O'Grady *et al.*, 2009).

Over the past few years it has become evident that the *B. cenocepacia* QS circuitry is far more complex than initially anticipated, as various higher-level regulatory elements as well as downstream regulators were identified. An additional regulatory element influencing the activity of the CepI/R QS system is ShvR, a LysR type transcriptional regulator that was initially shown to be responsible for shiny colony variants of *B. cenocepacia* K56-2 (Bernier *et al.*, 2008). In a recent study it has been shown that ShvR controls the expression of more than 1000 genes (O'Grady *et al.*, 2011), including *cciI/R* and *cepI/R*, although the effect on the latter system was only weak. Consequently, ShvR affects AHL kinetics: maximal AHL production peaked earlier in a *shvR* mutant compared to wild type (O'Grady *et al.*, 2011). However, the regulons of ShvR and CepR/CciR only partially overlap and phenotypic analyses of mutants defective in *shvR*, *cepR* or *cciR* suggest that ShvR not only influences expression of CepR- and CciR-regulated genes, but also of a set of genes which is not under direct QS control (O'Grady *et al.*, 2011).

In addition to ShvR, another global regulator, AtsR, was shown to modulate QS signalling. This response regulator was first identified as a repressor of biofilm formation and was shown to control expression of a type VI secretion system (Aubert *et al.*, 2008). A recent study demonstrated that deletion of *atsR* led to increased expression of *cepI/R* and *cciI/R*, which in turn led to premature and increased AHL production. This finding suggests that AtsR plays a role in controlling the timing and fine-tuning of AHL-dependent gene expression. However, AtsR is also able to repress QS-regulated genes independently of AHL production, indicating that AtsR may operate via a novel yet uncharacterized mechanism (Aubert *et al.*, 2012).

In contrast to ShvR and AtsR, which downregulate both QS systems, a recent study identified a regulator that promotes transcription of *cepI/R* and *cciI/R* in *B. cenocepacia* K56–2 (O'Grady *et al.*, 2012). This regulator, BCAM1871, is cotranscribed with *cepI* and acts as an enhancer of AHL production (positive feedback). Mutants in BCAM1871 showed reduced transcription levels of *cciI*, *cciR* and of *shvR*. However, and similar to AtsR, some phenotypes were affected independently of AHL levels or CepI/R function, suggesting that BCAM1871 may operate through an AHL-independent mechanism (O'Grady *et al.*, 2012).

As many other Proteobacteria, *B. cenocepacia* harbours an orphan or solo *luxR* gene, i.e. a *luxR* gene not associated with a cognate *luxI* AHL synthase gene (for a review see (Subramoni and Venturi, 2009). This LuxR solo, designated CepR2, is a 237 amino-acid orthologue of CepR and contains all the conserved residues of LuxR transcriptional regulators (Malott *et al.*, 2009). In *B. cenocepacia* K56–2, CepR2 is negatively regulated by CciR and by itself, and it does not affect expression of the CepI/R or the CciI/R system. CepR2 was shown to be both a negative and a positive regulator of gene expression and does not require AHL to exert its function (Malott *et al.*, 2009). Interestingly, the set of genes regulated by CepR2 partially overlaps with those of the CepI/R and CciI/R regulons, including the genes encoding the metalloproteases ZmpA and ZmpB and the nematocidal factor AidA.

Fatty acid signalling in *Burkholderia* sp.

In 2008, Boon *et al.* discovered that some *Burkholderia* strains produce a long chain fatty acid signal molecule, the structure of which was identified as *cis*-2-dodecenoic acid, referred to as BDSF (*Burkholderia* diffusible signal factor). BDSF is a structural and functional analogue of *cis*-11-methyl-2-dodecenoic, the diffusible signal factor

(DSF) of *Xanthomonas campestris* pv. *campestris* (Xcc), which is a well characterized QS signal in this organism. Analyses of structural derivatives of DSF revealed that the key features of this type of QS signal molecules in Xcc are the double bond between C1 and C2, as well as its *cis*-conformation (Wang *et al.*, 2003), both features are also present in BDSF. Genetic analysis revealed that in *B. cenocepacia* J2315, the gene responsible for BDSF synthesis is BCAM0581. This gene codes for an enoyl-CoA hydratase that shares 37% identity at the amino acid level with RpfF, the main DSF synthase of Xcc. *In trans* expression of BCAM0581 in a DSF-deficient Xcc strain can rescue the defects of the mutant, including biofilm formation and extracellular polysaccharide production (Boon *et al.*, 2008). BCAM0581 has been recently renamed to $rpfF_{Bc}$ and shown to encode a bifunctional enzyme that catalyses the dehydration of 3-hydroxydodecanoyl-ACP to *cis*-2-dodecenoyl-ACP and cleaves the thioester bond to give BDSF (Bi *et al.*, 2012). Besides BDSF, some *Burkholderia* strains also produce additional DSF-family signals like DSF or *cis,cis*-11-methyldodeca-2,5-dienoic acid (DSF with an additional double bond in *cis*-configuration between C5 and C6) (Deng *et al.*, 2010). Examination of nine species from the Bcc (*B. lata, B. multivorans, B. cenocepacia, B. stabilis, B. vietnamiensis, B. dolosa, B. ambifaria, B. anthina* and *B. pyrrocinia*) showed that all produced BDSF, with four of them (*B. multivorans, B. stabilis, B. anthina,* and *B. pyrrocinia*) additionally producing *cis,cis*-11-methyldodeca-2,5-dienoic acid and *B. multivorans* producing DSF (Deng *et al.*, 2010). Interestingly, the observed diversity of DSF-family signal molecules in Bcc species is not related to variations found in the $RpfF_{Bc}$ primary structure, but is likely due to the availability of different precursors in the different Bcc species (Deng *et al.*, 2010). Blast analyses revealed orthologues of $RpfF_{Bc}$ from *B. cenocepacia* J2315 also in members of the PBE cluster (*B. bryophila, B. graminis, B. kururiensis, B. phenoliruptrix, B. phymatum, B. phytofirmans, B. terrae and B. xenovorans*) (Winsor *et al.*, 2008; Deng *et al.*, 2012).

Although Xcc is using a similar signal molecule as *B. cenocepacia*, the signal receptor as well as the signalling cascade is completely different. The DSF receptor of Xcc, RpfC, is a hybrid sensor kinase that phosphorylates its cognate response regulator RpfG. In addition to a receiver domain RpfG contains a HD-GYP domain, which was shown to be responsible for the c-di-GMP phosphodiesterase activity of the protein thereby affecting the cellular c-di-GMP level (Ryan and Dow, 2010). However, neither *rpfC* nor *rpfG* orthologues could be identified in any of the sequenced Bcc genomes. The fact that the gene next to $rpfF_{Bc}$ (BCAM0580, named *rpfR*) is required for biofilm formation (Huber *et al.*, 2002), a phenotype also known to be affected by BDSF (Ryan *et al.*, 2009; Deng *et al.*, 2010), prompted a more detailed study that revealed that BCAM0580 encodes a novel fatty acid signal receptor (Deng *et al.*, 2012). RpfR is a protein that comprises a PAS, a GGDEF and an EAL domain. The latter two are known to be responsible for c-di-GMP synthesis and degradation, respectively. A *B. cenocepacia* H111 *rpfR* mutant phenocopies a BDSF-deficient mutant, including an increased intracellular c-di-GMP level, reduced motility, impaired biofilm formation, and lowered proteolytic activity. Biochemical characterization of the purified RpfR protein demonstrated that BDSF binds to the PAS domain with high affinity and stimulates the phosphodiesterase activity of the protein through induction of allosteric conformational changes. RpfR represents the first example of a c-di-GMP metabolic enzyme that is directly activated by cell–cell communication signals via protein–ligand interaction. A bioinformatic analysis revealed that the $rpfF_{Bc}$/*rpfR* gene cluster is not only highly conserved in 11 *Burkholderia* species (*B. cenocepacia, B. cepacia, B. multivorans, B. dolosa, B. ambifaria, B. vietnamiensis, B. ubonensis, B. phymatum, B. xenovorans, B. graminis* and *B. phytofirmans*), but also in members of other genera, including *Achromobacter, Yersinia, Serratia, Enterobacter, Pantoea* and *Cronobacter* (Deng *et al.*, 2012).

Quinolone-based QS in *Burkholderia* sp.

In addition to AHL- and BDSF-based QS systems, a third potential QS system, which relies on quinolone signal molecules, was identified in some *Burkholderia* strains. The archetypal quinolone-based QS system is the one of *Pseudomonas aeruginosa,* which utilizes

2-heptyl-3-hydroxy-4(1*H*)-quinolone (also known as the Pseudomonas quinolone signal, PQS) to control expression of many virulence genes. PQS is also considered an integral component of the AHL-dependent QS network of this organism (Diggle *et al.*, 2006). The biosynthesis of PQS relies on the *pqsABCD* gene products, which direct the synthesis of 2-heptyl-4-quinolone (HHQ), the precursor of PQS. HHQ is thought to be an extracellular messenger that is released from, and taken up by, bacterial cells, where it is converted into PQS. Using LC-MS/MS analysis it was shown that several *B. pseudomallei* strains, *B. cenocepacia*, *B. thailandensis* and *B. ambifaria* produce, in addition to HHQ, other 4-hydroxy-2-alkylquinolone derivatives and also 4-hydroxy-3-methyl-2-alkylquinolones but not PQS (Diggle *et al.*, 2006; Vial *et al.*, 2008). Mutants of *B. ambifaria* defective in quinolone signalling were shown to have an increased production of AHLs and, consequently, slightly up-regulated expression of AHL-dependent phenotypes (Vial *et al.*, 2008). In *B. pseudomallei* mutation of the *pqsA* homologue not only resulted in the loss of HHQ production but also in altered colony morphology and increased elastase production (Diggle *et al.*, 2006). Additional work will be required to elucidate the molecular mechanisms underlying the quinolone signalling cascade interaction with the AHL- and BDSF-dependent QS circuitries.

QS-regulated phenotypes and regulons

The impact of QS on the physiology of *Burkholderia* species has been assessed by phenotypic characterization as well as by global profiling approaches (transcriptomes or proteomes) using defined QS mutants. The majority of these studies focused on pathogenic species, including Bcc species, the pseudomallei group and the rice pathogen *B. glumae*. In Table 10.1 we show QS-regulated phenotypes of different *Burkholderia* species. Some of the listed functions, including biofilm formation, proteolytic activity, motility, antifungal activity and pathogenicity are highly conserved, suggesting that they represent ancestral QS-regulated phenotypic traits.

QS-regulated functions were also mapped by functional genomics approaches. A transcriptome analysis of CepR-regulated genes in *B. cenocepacia* H111 identified 57 genes that were more than threefold up- or down-regulated in a *cepR* mutant relative to the wild-type strain (Inhülsen *et al.*, 2012). The same custom *B. cenocepacia* oligonucleotide microarray was also used to identify the overlap between the CepI/R and CciI/R systems in *B. cenocepacia* K56–2 (O'Grady *et al.*, 2009). When compared with the wild type, 860 genes were found to be more than twofold up- or down-regulated in a K56-2 *cepR* mutant, 595 genes in a *cciR* mutant and 489 genes in a *cepR cciR* double mutant (O'Grady *et al.*, 2009). Among those genes, 196 genes were regulated by both CepR and CciR, with the vast majority (167) being positively regulated by CepR and negatively regulated by CciR. With only very few exceptions all 57 CepR-regulated genes identified in strain H111 (Inhülsen *et al.*, 2012) were also among the CepR-regulated genes of strain K56-2. Both studies showed that *aidA*, which encodes a factor required for nematode pathogenicity, was among the most stringently CepR-activated genes. Interestingly, CciR had a negative effect on *aidA* transcription in strain K56-2. The dual regulation of *aidA* in K56–2 provides an explanation for the observation that the *cepI* mutants of H111 and K56-2 but not the K56-2 *cciI* mutant are attenuated in virulence in the *C. elegans* infection model (Uehlinger *et al.*, 2009). Other AHL-regulated genes that were identified in both studies include several proteases, an operon encoding three lectins, a type I pilus, a non-ribosomal peptide synthetase gene cluster, and genes required for swimming motility (Inhülsen *et al.*, 2012; O'Grady *et al.*, 2009).

Global proteomic analyses have been used to study the QS regulon of *B. cenocepacia* H111. An early study by Riedel *et al.* (2003) employed two-dimensional gel-electrophoresis (2-DE) to compare the proteomes of the H111 wild-type and its isogenic *cepI* mutant grown in presence or absence of AHLs. Fifty five of the 985 detected protein spots were found to be differentially regulated. Due to limited genome sequence data at that time, only 19 of the 55 QS-regulated proteins could be identified. In good agreement with more recent transcriptomic data (Inhülsen *et al.*,

Table 10.1 Experimentally determined phenotypes of QS mutants in *Burkholderia* species

Species[a]	Mutated QS system	Biofilm[c]	Protease	Chitinase	Lipase	Siderophore	Swarming motility	Swimming motility	Antifungal activity
B. ambifaria	*cepI/R*		↓[d]						↓
	HMAQ	↑				↑			↑
B. anthina	*cepI/R*	↓	↓ (1), – (1)				↓		
B. cenocepacia with cciIR	*cepI/R*	↓	↓		↓	↑	↓	↓	
	cciI/R	↓ (cciR), ↔ (cciI)	↑ (cciI), ↓ (cciR)			↔	↓ (cciI), ↔ (cciR)	↑	
	rpfF/R	↓					↓		
B. cenocepacia without cciIR	*cepI/R*	↓	↓	↓	↔	↑	↓	↔	↓
	rpfF/R	↓	↓				↓		
B. cepacia	*cepI/R*	↓	↓	↔	↔	↔			↓
B. dolosa	*cepI/R*	↓	–						
B. lata	*cepI/R*					↔			↓
B. multivorans	*cepI/R*	↓	–						
B. pyrrocinia	*cepI/R*	↓(1), ↑(1)	↓						↓
B. stabilis	*cepI/R*	↓	↓				↓		↓
B. vietnamiensis	*cepI/R*	↓(3), ↑ (2), ↔ (1)	–		–	↔	–		↓
	bviI/R	↔	–		–	↔	–		
B. pseudomallei	*pmlI/R* or *bpsI/R*	↓	↑, ↔[e]		↔	↑			
B. thailandensis	*btaI/R*		↔		↓, ↑	↔↑			

Pathogenicity: *C. elegans*	Pathogenicity: *Galleria*	Pathogenicity: mouse/rat/ hamster	Pathogenicity: zebrafish	Pathogenicity: alfalfa	Other QS-regulated phenotypes	Reference
↓	↓					Wopperer *et al*. (2006), Uehlinger *et al*. (2009)
						Vial *et al*. (2008)
↓						Wopperer *et al*. (2006)
↓	↔	↓	↓	↔		Lewenza *et al*. (1999), Sokol (2003), Malott *et al*. (2005), Tomlin *et al*. (2005), O'Grady *et al*. (2009), Uehlinger *et al*. (2009), Vergunst *et al*. (2010)
↔	↔	↓		↔		Malott *et al*. (2005), Tomlin *et al*. (2005), O'Grady *et al*. (2009), Uehlinger *et al*. (2009)
	↓	↓	↓		Germtube formation in *Candida albicans*:↓; adherence to porcine mucin: ↓, cytotoxicity: ↓	Boon *et al*. (2008), Deng *et al*. (2009), Ryan *et al*. (2009), McCarthy *et al*. (2010)
↓	↓ cepI, ↔ cepR	↓, ↔[f]				Huber *et al*. (2001), Köthe *et al*. (2003), Sousa *et al*. (2007), Schmidt *et al*. (2009), Uehlinger *et al*. (2009)
↓						Deng *et al*. (2012)
↓					Onion pathogenicity: ↓	Aguilar *et al*., (003), Wopperer *et al*. (2006), Schmidt *et al*. (2009)
↓ (1), – (1)						Wopperer *et al*. (2006)
						Schmidt *et al*. (2009)
– (6), ↓(1)						Wopperer *et al*. (2006)
↓						Wopperer *et al*. (2006), Schmidt *et al*. (2009)
↓						Wopperer *et al*. (2006), Schmidt *et al*. (2009)
↓ (3), – (2)	↓			–		Conway and Greenberg (2002), Wopperer *et al*. (2006), Schmidt *et al*. (2009), Uehlinger *et al*. (2009)
				–		Malott and Sokol (2007)
↓		↓			Sensitivity to oxidative stress: ↑	Ulrich *et al*. (2004a), Valade *et al*. (2004), Song *et al*. (2005), Lumjiaktase *et al*. (2006), Gamage *et al*. (2011)
		↔			Bactobilin synthesis ↓, self aggregation ↓	Ulrich *et al*. (2004c), Chandler *et al*. (2009), Duerkop *et al*. (2009), Seyedsayamdost *et al*. (2010)

Table 10.1 continued

Species[a]	Mutated QS system	Biofilm[c]	Protease	Chitinase	Lipase	Siderophore	Swarming motility	Swimming motility	Antifungal activity
B. mallei	*bmaI/R*		↔		↔				
B. glumae	*tofI/R*				↓		↓	↓	
B. plantarii	*plaI/R*								
B. kururiensis	*braI/R*	↑	↔		↔	↔	↔	↔	
B. xenovorans	*braI/R*	↔	↔		↔	↔	↔	↔	
B. unamae	*braI/R*	↑							
B. caryophylii	AHL[b]								↓
B. phenazinium	AHL[b]								↓
B. bryophila	AHL[b]								↓
B. megapolitana	AHL[b]								↓

[a]*Burkholderia* species, in which a QS regulated phenotype has been observed.
[b]The phenotypic testing has been performed with an AHL-degrading approach.
[c]Number in parentheses indicates the number of strains in which the according regulation has been observed if a phenotype is strain-dependent.
[d]↓ down-regulated in QS mutant, ↑ up-regulated in QS mutant, ↔ not regulated in QS mutant (experimentally verified), – phenotype not observed in wild-type strain.
[e]Opposing observations in different studies.
[f]Reciprocally by the different QS systems: btaI/R1 and btaI/R3 ↑, btaI/R2.

Pathogenicity: *C. elegans*	Pathogenicity: *Galleria*	Pathogenicity: mouse/rat/ hamster	Pathogenicity: zebrafish	Pathogenicity: alfalfa	Other QS-regulated phenotypes	Reference
		↓				Ulrich *et al.* (2004b)
					Catalase: ↓, toxoflavin synthesis: ↓, heat shock resistance ↓, rice pathogenicity: ↓,	Kim *et al.* (2004, 2007, 2012), Devescovi *et al.* (2007), Chun *et al.* (2009)
					rice seedling blight ↓	Solis *et al.* (2006)
					EPS production: ↓	Suárez-Moreno *et al.* (2008, 2010)
					EPS production: ↓	Suárez-Moreno *et al.* (2008, 2010)
					EPS production: ↓, phenol degradation: ↓	Suárez-Moreno *et al.* (2008, 2010)
						Schmidt *et al.* (2009)
						Schmidt *et al.* (2009)
						Schmidt *et al.* (2009)
						Schmidt *et al.* (2009)

2012; O'Grady *et al.*, 2009), the proteins included the ZmpB metalloprotease, the FimA protein, the BclA lectin and AidA. A more recent study, employing a gel-free proteomics methodology, identified 22 proteins that were down-regulated in a *cepI* as well as in a *cepR* mutant of *B. cenocepacia* H111 (Inhülsen *et al.*, 2012). This analysis not only provided confirmatory results but also showed that many of the QS-regulated proteins were not among the QS-regulated genes identified by the microarray analyses, suggesting that expression of many of the identified proteins may be controlled by QS at the posttranscriptional level.

2-DE was also employed to identify AHL-regulated proteins in *B. glumae* BGR1 (Goo *et al.*, 2010). Among 79 differentially expressed proteins identified, 59 were extracellular proteins, such as flagellin, lipases, and proteases, proteins involved in anti-oxidation and membrane proteins. The high proportion of extracellular proteins among the QS-regulated proteins was elegantly explained by the finding that expression of the type II secretion system, which is required for secretion of many extracellular proteins, is controlled by QS (Goo *et al.*, 2010). Likewise, 2-DE was used to map the proteins controlled by one (the *bpsI/R* system) of the three AHL-dependent QS systems in *B. pseudomallei* (Wongtrakoongate *et al.*, 2012). Forty-five of the 65 identified genes were found to be also regulated by the stationary sigma factor RpoS, most likely because RpoS is involved in the regulation of *bpsI* expression.

In a recent study the BDSF stimulon of *B. cenocepacia* H111 was mapped by RNA-Seq and shotgun proteomics (Schmid *et al.*, 2012). These investigations provided evidence that many of the BDSF-regulated genes or proteins are also controlled by the CepI/R system, suggesting that the two regulons partially overlap. This is in accordance with previous studies that have shown that AHLs and BDSF control similar phenotypic traits, including biofilm formation, proteolytic activity and pathogenicity (Deng *et al.*, 2009, 2012, Fig. 10.2). The detailed analysis of two mutually regulated operons, one encoding three lectins and the other one encoding the large surface protein BapA and its type I secretion machinery, revealed that both AHLs and BDSF are required for full expression, suggesting that the two signalling systems operate in parallel (Fig. 10.3). In accordance with this, it was demonstrated that both AHLs and BDSF are required for biofilm formation and protease production. Even though AHL production is reduced by 50% in the *rpfF*$_{Bc}$ mutant, the two QS systems were found to regulate the tested phenotypes and genes independently, suggesting that they are not hierarchically arranged.

Molecular mechanisms and factors important for biofilm formation of *Burkholderia* sp.

As pointed out above, one of the phenotypes regulated by both the AHL- and BDSF-dependent QS system (no data on the role of quinolone signalling are available) is the formation of

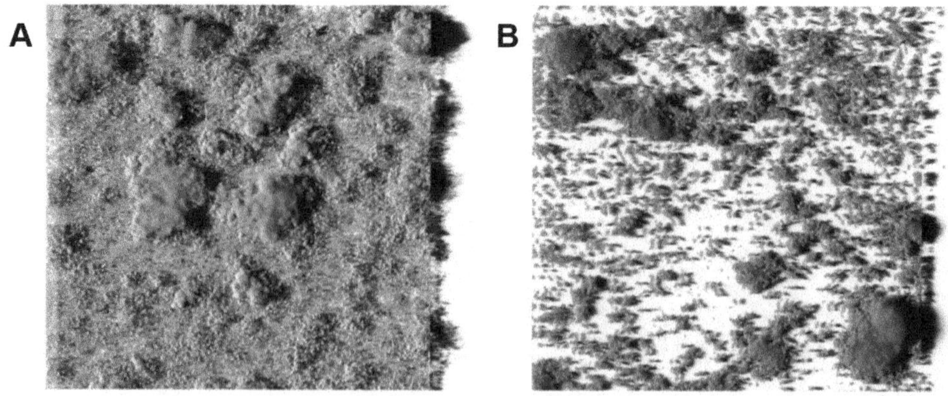

Figure 10.2 Simulated 3-D view of a 48h old biofilm of *B. cenocepacia* H111. (A) wild type. (B) isogenic double mutant defective in both *cepI* (AHL synthase) and *rpfF*$_{Bc}$ (BDSF synthase).

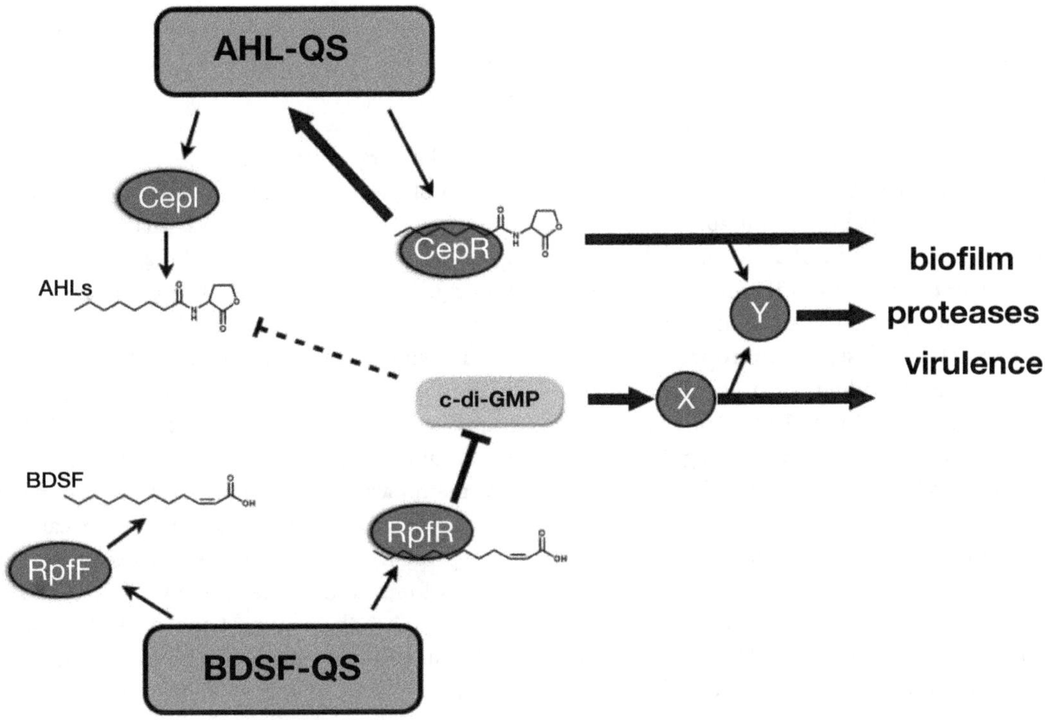

Figure 10.3 Model for the possible interactions of the AHL- and BDSF-dependent QS systems of *B. cenocepacia* H111 QS. The CepI/R system consists of the AHL synthase CepI directing the synthesis of C8-HSL, and of the transcriptional regulator CepR. The RpfF/RpfR system consists of RpfF which directs the synthesis of BDSF, and of its cognate receptor RpfR, which upon binding of BDSF degrades c-di-GMP. The two QS systems operate in parallel and control specific as well as overlapping sets of genes. We propose the existence of a c-di-GMP receptor protein X that stimulates transcription of target genes independent of CepR. However, the two QS cascades could also converge and control the expression or the activity status of an unknown common regulator Y, which in turn regulates expression of target genes. C-di-GMP has a negative regulatory effect on AHL levels via an unknown mechanism (depicted by the dashed line).

surface-associated communities, referred to as biofilms. In nature bacteria are thought to exist predominantly as biofilms on both biotic and abiotic surfaces. Biofilms are also of eminent clinical importance, as sessile cells are up to 1000-fold more resistant to antibiotics than their planktonic counterparts. In the following we will discuss factors that have been reported to play a role in biofilm formation of *Burkholderia* sp.

Pili

The adhesion to surfaces is often mediated by specialized proteinaceous surface appendages such as fimbriae or pili or in some bacteria also flagella (Watnick and Kolter, 2000). *Burkholderia* sp. expresses flagella and various types of pili (Goldstein *et al.*, 1995). Using transmission electron microscopy (TEM), five types of pili were identified in Bcc strains: mesh (Msh), filamentous (Fil), spine (Spn), spike (Spk), and cable (Cbl) pili. The expression of these different pili types has been linked to the ecological niches from which the respective strains have been isolated from. Msh pili have been shown to be produced by many clinical and environmental strains, whereas Cbl pili were only expressed by epidemic CF isolates. Fil pili were found to be associated with non-epidemic CF isolates, Spn pili with non-CF clinical isolates, and Spk pili were exclusively produced by environmental strains. The cable pilus Cbl has been demonstrated to be important for adhesion of Bcc strains to epithelial cells and is therefore thought to be of particular clinical relevance (Sajjan *et al.*, 2002). It is noteworthy that only a few Bcc strains harbour the gene coding for the Cbl pilus and not all stains possessing the gene produce pili (Sajjan *et al.*, 2002).

Many *Burkholderia* strains produce a homologue of the *E. coli* FimA protein, the major subunit of the type I pilus. In *E. coli* this pilus is critical for initial attachment to biotic and abiotic surfaces (Pratt and Kolter, 1998). A FimA homologue in *B. cenocepacia* H111 was identified among the CepR-regulated genes (Riedel *et al.*, 2003). *fimA* (BCAL1677) is the first gene of an operon that also contains genes encoding for a chaperone-usher secretion apparatus (BCAL1678–1681) (Holden *et al.*, 2009; Inhülsen *et al.*, 2012). Regulatory elements like homologues of FimB and FimE, which control phase variation in *E. coli*, or a tip fibrillum gene are missing in *B. cenocepacia* H111. In spite of the fact that type I fimbriae are crucially important in some organisms for adhesion to surfaces, a *fimA* mutant of *B. cenocepacia* H111 was neither defective in biofilm biomass nor in biofilm architecture (Inhülsen *et al.*, 2012). As the analysis of the H111 genome indicated the presence of at least two additional chaperon-usher-type pili, two Flp-type pili and a type IV pilus, it is conceivable that these additional surface appendages may mask the biofilm defects of the *fimA* mutant. In *B. pseudomallei*, the type IV pilin-encoding gene *pilA* was shown to be essential for microcolony development, and thus for optimal association with eukaryotic cells, but is not required for direct adherence to cultured human cells (Boddey *et al.*, 2006).

Genes required for biofilm maturation

Using a simple microtitre dish based assay Huber *et al.* (2002) screened a random transposon insertion library of *B. cenocepacia* H111. Given the relatively long incubation time (48 h), mutants defective in the late steps of biofilm formation were enriched in this screen. Of 5000 transposon insertion mutants screened, 13 exhibited defects in biofilm formation on a polystyrene surface without being impaired in growth. However, all the mutants were defective in the development of the typical three-dimensional biofilm structure of *B. cenocepacia* H111. The 13 mutants were shown to carry the transposon in genes encoding surface proteins, proteins involved in the biogenesis and maintenance of an integral outer membrane, regulators or proteins involved in QS (Huber *et al.*, 2002).

Bacterial surface proteins

In two of the transposon mutants a gene encoding a large surface protein was inactivated. This gene was named *bapA* (biofilm-associated protein A). Various members of this protein family were shown to be involved in the colonization of diverse substrates: Mus-20 (mutants unattached to seeds) in the adhesion of *P. putida* KT2440 to maize seeds (Espinosa-Urgel *et al.*, 2000), Bap of *Staphylococcus aureus* in biofilm formation on mammary gland epithelium of ruminants (Cucarella *et al.*, 2001), Esp (Enterococcal surface protein) of *Enterococcus faecalis* in human skin colonization (Toledo-Arana *et al.*, 2001), *lapA* (large adhesion protein) of *P. fluorescens* in biofilm formation on abiotic surfaces (Hinsa *et al.*, 2003), and BapA of *Salmonella enterica* in air–broth interface pellicle and biofilm formation (Latasa *et al.*, 2005). These large proteins are usually secreted via a type I secretion system and are believed to remain loosely associated with the cell surface (Hinsa *et al.*, 2003; Latasa *et al.*, 2005). In a more recent study, a *bapA* mutant was found to be severely hampered in biofilm formation on an abiotic surface and showed defects in terms of attached biofilm mass as well as biofilm architecture (Inhülsen *et al.*, 2012). BapA appears to be exported via a type I secretion system. The three genes encoding this export machinery are located downstream of *bapA* and are co-transcribed with *bapA* as a single transcription unit. A mutant with a mutated type I transporter exhibited the same biofilm defect as a *bapA* mutant and microscopic inspection of a strain expressing a BapA–mCherry fusion protein suggested that BapA is no longer exported (Inhülsen *et al.*, 2012).

Although not identified in the mutant screen, an operon encoding three lectins (*bclACB*, BCAM0186 to BCAM0184) was found to influence biofilm structural development (Inhülsen *et al.*, 2012). All three lectins share an almost identical PA-IIL-like C-terminal domain; BclB and BclC have additional N-terminal domains. PA-IIL is a soluble lectin from *P. aeruginosa* with a strong affinity to fucose. A *P. aeruginosa* mutant not expressing PA-IIL is impaired in biofilm formation on glass slides when compared to the wild-type strain (Winzer *et al.*, 2000; Tielker *et al.*, 2005). Two of the three lectins of *B. cenocepacia*, BclA (BCM0186) and BclC (BCAM0185), were the subject of recent studies (Lameignere *et al.*, 2008, 2010; Sulák *et al.*, 2010, 2011; Marchetti *et al.*, 2012). These investigations revealed that BclA forms homodimers and displays a strict specificity for oligomannose-type oligosaccharides that are present on human glycoproteins (Lameignere *et al.*, 2008; Marchetti *et al.*, 2012). Moreover, it was shown that BclA binds to bacterial surfaces and to biofilms (Marchetti *et al.*, 2012). BclC, which also contains the C-terminal PA-IIL domain, possesses in addition an N-terminal domain comprising a TNF-α-like fold with fucose binding properties. BclC forms hexamers and, due to its two different lectin domains, is a super lectin with dual carbohydrate specificity (Sulák *et al.*, 2010, 2011). BclB is the least characterized of the three lectins encoded by *B. cenocepacia*. In a recent study it was shown that BclB is associated with the bacterial cell surface (Inhülsen *et al.*, 2012). This study also revealed that the structure of the biofilm formed by a *bclACB* mutant is different from the one of the wild type. The mutant biofilm exhibited a very characteristic biofilm architecture structure with hollow microcolonies, which could not be observed with the wild type. This change in biofilm morphology could only be rescued when the complementation was done with the intact *bclACB* operon, suggesting that the three lectins are not redundant and that all three lectins are needed for biofilm structural development. The authors speculated that the surface-exposed BclACB lectins may mediate contact to neighbouring cells within the biofilm or with the biofilm matrix.

Proteins involved in the biogenesis and maintenance of an integral outer membrane

Two genes essential for maintaining cell shape and outer membrane composition, *tolA* and *rodA*, were found to be required for biofilm formation. In *P. aeruginosa*, the homologue of *tolA* affects LPS structure and its expression was shown to be induced during biofilm formation (Whiteley *et al.*, 2001). In addition, *tolA* is involved in the biogenesis and maintenance of an integral outer membrane in both *E. coli* and *P. putida* (Lazzaroni *et al.*, 1999; Llamas *et al.*, 2000). In agreement with this finding, *tolA* mutants of *B. cenocepacia* H111 were elongated and grew in chains, forming flat and unstructured biofilms (Huber *et al.*, 2002). In contrast to *tolA* mutants, *rodA* mutants are coccoid and form hyper-structured biofilms with big microcolonies. Taken together, these data suggests that cell shape and cell surface composition are important factors for biofilm development.

Exopolysaccharide production

Exopolysaccharides (EPS) are considered to be an essential constituent of the biofilm matrix, even though the genes required for EPS biosynthesis were not identified in the mutant screen of Huber *et al.* (2002). The majority of Bcc strains produce large amounts of EPS and so far five different EPS molecules have been identified (Herasimenka *et al.*, 2007). Among them, cepacian, a polysaccharide with a branched heptasaccharide repeating unit, is the most abundant EPS produced by Bcc species (Herasimenka *et al.*, 2007). It has been shown that cepacian plays a role in biofilm maturation, specifically in the establishment of thick biofilms, but is not required for the initiation of biofilm formation (Cunha *et al.*, 2004). The enzymes required for cepacian biosynthesis are encoded by two separated gene clusters, *bce-I* and *bce-II* (Moreira *et al.*, 2003; Ferreira *et al.*, 2010). Interestingly, due to a frame shift mutation in the *bceB* gene (in the *bce-I* cluster), the ET12 lineage strain *B. cenocepacia* J2315, which is used as model strain in many laboratories, does not produce cepacian (Holden *et al.*, 2009). Despite the fact that bacteria growing in biofilms have been shown to be more resistant to both antibiotics and phagocyte killing than their planktonic counterparts (Høiby *et al.*, 2010), no clear correlation between cepacian production and the clinical outcome of the infection could be established when CF isolates of *B. cepacia*, *B. multivorans*, *B. cenocepacia* and *B. stabilis* were analysed (Cunha *et al.*, 2004).

Zlosnik *et al.* (2008) reported an inverse correlation between the quantity of mucoid exopolysaccharide production by Bcc bacteria and rate of decline in CF lung function, which is opposite to what is known from *P. aeruginosa*. It has been suggested that non-mucoid isolates are associated with increased disease severity while the mucoid phenotype may be associated with bacterial persistence. In agreement with this hypothesis, it was shown that EPS from a clinical *B. cenocepacia* isolate inhibited the chemotaxis of neutrophils and scavenges reactive oxygen species, both essential components of innate neutrophil-mediated host defences (Bylund *et al.*, 2006). In addition, it is possible that the loss of EPS production provides the bacteria with a competitive advantage in the lung and/or allows for increased virulence factor production (Zlosnik *et al.*, 2008).

The role of QS in biofilm formation

A role for AHL-mediated QS in regulation of biofilm formation was first reported by Davies *et al.* (1998) for *P. aeruginosa*. Specifically, it was shown that a QS mutant formed flat and undifferentiated biofilms when compared to the wild-type biofilm. Various QS-regulated functions, including the biosynthesis of rhamnolipids, the production of the biofilm matrix polysaccharide Pel, anaerobic denitrification, and *P. aeruginosa* quinolone signal (PQS)-dependent DNA release, have been identified as obvious links between biofilm structural development and cell-to-cell communication (for reviews, see Aguilar *et al.*, 2009; de Kievit, 2009). Recent studies have shown that QS is necessary for the formation of the cap portion of the mushroom-shaped microcolonies, which are typical of *P. aeruginosa* biofilms (Yang *et al.*, 2007, 2009; Patriquin *et al.*, 2008). However, it is noteworthy that in studies using slightly changed experimental settings or hydrodynamic conditions, no significant differences between biofilms of the wild-type and those formed by QS-negative mutants were observed (Stoodley *et al.*, 1999; Heydorn *et al.*, 2002; Schaber *et al.*, 2007), suggesting that different experimental settings have a strong influence on expression of QS-regulated genes.

B. cenocepacia H111 was the second example of a bacterium that requires a functional AHL-dependent QS system for the formation of differentiated biofilms (Huber *et al.*, 2001). While both wild type and mutants defective in the CepI/R QS system form characteristic microcolonies after initial attachment to an abiotic surface, striking differences can be observed in the later stages of biofilm development: the wild-type biofilm rapidly matures whereas the QS-defective strains are arrested at the microcolony stage and thus form a thinner and less structured biofilm. The central importance of the CepI/R system in biofilm formation was also confirmed in a subsequent study, in which three mutants were identified that had insertions in *yciR*, *suhB* and *yciL*. As these mutants showed reduced levels of AHL, it was suggested that their biofilm defect is due to impaired expression of CepI/R (Huber *et al.*, 2001). In *B. cenocepacia* K56–2, the influence of the CepI/R and CciI/R QS systems on biofilm formation is rather complex: mutations in either *cepI* or *cepR* led to reduced amounts of biofilm; however, mutation of *cciI* or *cepI cciI* did not affect the ability to form a biofilm, while a *cciR* as well as a *cepR cciR* double mutant were deficient in biofilm formation (Tomlin *et al.*, 2005). As with strain H111, higher-level regulators appear to fine-tune QS-dependent biofilm development of strain K56–2. The sensor kinase hybrid AtsR, which is an important regulator of type VI secretion system and virulence factor expression, was shown to affect the timing of AHL production in this organism. This effect is thought to be the reason for the biofilm defect observed with an *atsR* mutant (Aubert *et al.*, 2008).

The influence of QS on biofilm formation was confirmed for several *Burkholderia* species. A survey by Wopperer *et al.*, (2006) has shown that the large majority of Bcc strains, belonging to nine species, formed biofilms in an AHL-dependent manner. In addition, mature biofilms of *B. cenocepacia* and *B. multivorans* were reduced upon treatment with QS-inhibitory compounds, reaffirming the importance of QS in the late stages of biofilm development (Brackman *et al.*, 2009). Likewise, it has been shown that biofilm formation of *B. pseudomallei* is dependent on AHL signalling (Gamage *et al.*, 2011).

In a recent study, three AHL-regulated candidate gene clusters were tested for their role in biofilm formation of *B. cenocepacia* H111 (Inhülsen *et al.*, 2012). The three loci were (i) the *bclACB* lectin operon, (ii) the large surface protein *bapA* and its transporter and (iii) a gene cluster encoding a type I pilus. This analysis revealed that

BapA plays a major role in biofilm formation on abiotic surfaces while inactivation of the type I pilus showed little effect both in a static microtitre dish-based biofilm assay and in flow-through cells. Inactivation of the *bclACB* lectin genes gave rise to biofilms with an aberrant structure (see above).

The BDSF-dependent QS system of *B. cenocepacia* was also shown to have a strong effect on biofilm formation (Ryan *et al.*, 2009; Deng *et al.*, 2012). In fact, one of genes identified in the Huber *et al.* (2002) screen that are essential for biofilm formation, YciR, was recently identified as the main BDSF sensor and was renamed to RpfR (Deng *et al.*, 2012; see above). Consistent with the results obtained by Huber *et al.* (2002), the reconstructed RpfR mutant was deficient in biofilm formation and phenocopied the defect of the BDSF synthase mutant (Deng *et al.*, 2012). Most recent work has shown that the BDSF circuitry is also involved in the regulation of BapA expression (Schmid *et al.*, 2012), providing a rationale for the biofilm defects observed with the $rpfF_{Bc}$ or *rpfR* mutants. Expression of *bapA* in *B. cenocepacia* H111 is therefore regulated by both the AHL- and the BDSF-dependent QS system. However, additional work is required to fully understand how these two signalling systems orchestrate the expression of this key factor for biofilm formation.

Interspecies communication in mixed biofilms

Bcc strains and *P. aeruginosa* can form mixed biofilms in the lungs of CF patients. Given that both organisms control biofilm formation and expression of other virulence factors by QS, it appears likely that the two species are capable of communicating with each other. Initial evidence for interspecies signalling was reported by McKenney *et al.* (1995), who showed that the production of siderophores, lipases and proteases by *B. cepacia* can be stimulated by the addition of concentrated *P. aeruginosa* PAO1 culture supernatants. Circumstantial evidence for an interaction of the two pathogens was also obtained when the longitudinal AHL profiles of *P. aeruginosa* isolated from chronically infected CF patients were analysed (Geisenberger *et al.*, 2000). During the sampling period, two patients became co-infected with *B. cepacia*. While in one patient the observed AHL profile of *P. aeruginosa* isolates did not change, in the other case a dramatic reduction in the amount of AHL produced by *P. aeruginosa* was observed. Intriguingly, this reduction was only observed during the 6-month period of co-infection. After clearance from *B. cepacia*, the *P. aeruginosa* AHL profile was resumed (Geisenberger *et al.*, 2000). More direct evidence for AHL-mediated cross-talk was obtained by the use of fluorescent biosensors, which were employed to visualize communication in mixed biofilms of *B. cenocepacia* H111 and clinical isolates of *P. aeruginosa* (Riedel *et al.*, 2001). In the two biofilm model systems that were used, artificial flow chambers and alginate beads introduced into the lungs of mice, a *cepI* mutant of *B. cenocepacia* was able to respond to the AHLs released by *P. aeruginosa.* Interestingly, the opposite, namely *P. aeruginosa* responding to *B. cenocepacia,* was not observed (Riedel *et al.*, 2001). In another study, however, it was demonstrated that *P. aeruginosa* and *B. cenocepacia* K56–2 were both able to utilize the heterologous signalling molecule (Lewenza *et al.*, 2002), suggesting that communication between the two organisms is strain dependent.

Evidence is accumulating that *cis*-2-unsaturated fatty acids of the DFS family may also be used for interspecies communication (Ryan *et al.*, 2009). Many *Burkholderia* strains not only produce BDSF but some strains also synthesize DSF, the ancestral fatty acid signal isolated from Xcc, or other molecules of the DSF family (Deng *et al.*, 2010). It is noteworthy in this context that DSF was shown to strongly bind to the BDSF RpfR receptor of *B. cenocepacia* and to be able to rescue the phenotypic defects of a BDSF mutant (Deng *et al.*, 2012), suggesting that the two signals are interchangeable.

In a recent study Twomey *et al.* (2012) examined the role of DSF and BDSF in the regulation of virulence and persistence of *P. aeruginosa* in the cystic fibrosis airways. *P. aeruginosa* does not produce BDSF or DSF but the structurally related compound *cis*-2-decenoic acid, which was shown to induce the dispersal of biofilms (Davies and Marques, 2009). Twomey *et al.* (2012) showed that the presence of DSF and BDSF in sputa of CF patients was correlated with the colonization of the patient with *B. cenocepacia* and/or *Stenotrophomonas maltophilia*, which is known to produce DSF (Fouhy *et al.*, 2007). In the presence of DSF, persistence of *P. aeruginosa* was increased in a CF mouse model. Furthermore, polymyxin tolerance was enhanced when *P. aeruginosa* was grown as a biofilm in the presence of DSF on human airway epithelial cells (Twomey *et al.*, 2012). This study provided evidence that interspecies DSF-mediated bacterial interactions occur in the CF lung and it is tempting to speculate that this is also the case with BDSF, the main DSF family signal molecule produced by various *Burkholderia* species.

In addition to their roles in interspecies communication, DSF family signal molecules have also been implicated in inter-kingdom signalling. Many *cis*-2-unsaturated fatty acids, including DSF, BDSF and *cis,cis*-11-methyldodeca-2,5-dienoic acid, have been demonstrated to efficiently inhibit germ tube formation of the fungus *Candida albicans* (Boon *et al.*, 2008; Deng *et al.*, 2010).

QS as target for antibiofilm and antipathogenic drugs

QS as a target for the eradication of biofilms

Disruption of AHL-mediated QS systems has emerged as a promising strategy for the development of antimicrobial and antibiofilm drugs (Hentzer *et al.*, 2002; Sokol *et al.*, 2007). AHL signalling molecules can be degraded enzymatically, a process known as quorum quenching (Dong and Zhang, 2005; Dong *et al.*, 2007). AHL inactivation can occur through AHL-acylases or AHL-lactonases. AHL-acylases degrade AHL by removing the fatty acid side chain from the homoserine lactone ring of AHLs (Dong *et al.*, 2007) and have been reported to be produced by *Ralstonia* sp. (Lin *et al.*, 2003; Chen *et al.*, 2009), *Pseudomonas* spp. (Huang *et al.*, 2006; Sio *et al.*, 2006; Shepherd and Lindow, 2009; Wahjudi *et al.*, 2011), *Streptomyces* sp. (Park *et al.*, 2005), *Anabaena* sp. (Romero *et al.*, 2008), and the Gram-positive *Rhodococcus erythropolys* (Uroz *et al.*, 2005). Among these enzymes, there are some which prefer long-chain AHLs such as C_{10} to C_{14} (Huang *et al.*, 2006; Sio *et al.*, 2006). The second class of enzymes with AHL degrading activity includes AHL lactonases which have been identified in Gram-positive bacteria such as *Bacillus* sp. and *Arthrobacter* sp. (Dong *et al.*, 2001, 2002; Park, 2003) and Gram-negative bacteria such as *Klebsiella* sp. and *Agrobacterium* sp. (Zhang *et al.*, 2002; Park, 2003). Heterologous expression of the AHL lactonase *aiiA* from *Bacillus* spp. in *Pseudomonas* or *Burkholderia* resulted in a marked decrease in the amount of AHL signals produced (Dong *et al.*, 2000; Reimmann *et al.*, 2002). It was recently shown that *Bacillus* sp. isolates which show quorum quenching activity were able to impair biofilm growth in *B. pseudomallei* (Ramli *et al.*, 2012). Heterologous expression of the AiiA lactonase in different *B. cepacia* complex (Bcc) species was used to demonstrate that some QS-regulated functions, including protease production, biofilm formation and nematode pathogenicity, are highly conserved among Bcc strains (Wopperer *et al.*, 2006).

Another strategy to interrupt QS is by means of signal antagonists, the so-called QS inhibitors (QSIs) (Rasmussen and Givskov, 2006). To date, many QSIs have been identified, some of which have been isolated from natural products, such as halogenated furanones isolated from the marine alga *Delisea pulchra* (Givskov *et al.*, 1996) and 4-nitro-pyridine-*N*-oxide from garlic (Rasmussen *et al.*, 2005a). The furanone compound C-30 was shown to inhibit biofilm formation and promote a faster clearance of *P. aeruginosa* lung infection in mice (Hentzer *et al.*, 2002; Wu *et al.*, 2004). Furanones are thought to stimulate degradation of LuxR receptor proteins by cellular proteases (Manefield *et al.*, 2002) and to compete with AHL for binding to LuxR (Müh *et al.*, 2006). Other compounds such as a sulphur-rich molecule from garlic ajoene (Jakobsen *et al.*, 2012b), iberin from horseradish (Jakobsen *et al.*, 2012a), patulin and penicillic acid from fungi (Rasmussen *et al.*, 2005b) have been shown to have QSI activity and to exhibit antimicrobial activity against a number of Gram-positive and Gram-negative bacteria (Naganawa *et al.*, 1996).

A rational drug design approach in combination with a high-throughput screening using AHL biosensors was employed by Riedel *et al.* (2006) to develop QS blockers of the CepI/R QS system of *B. cenocepacia*. This approach lead to the identification of a compound that is entirely unrelated to AHLs and furanones but interferes with the expression of QS-regulated functions such as proteases, swarming motility and biofilm formation of *B. cenocepacia,* presumably via a competitive inhibition mechanism. Recent work confirmed the efficacy of this compound on biofilm formation of *B. cenocepacia* and *B. multivorans* (Brackman *et al.,* 2009). Although the clinical performance of QSI compounds remains to be seen, such drugs show great promise, either alone or in combination with conventional antimicrobial agents (Brackman *et al.,* 2011), for treatment of *Burkholderia* sp. infections.

Future directions

Members of the genus *Burkholderia* have emerged as multilingual talents that were demonstrated to master several bacterial languages. AHL-mediated communication is probably the most widespread and certainly the best investigated cell-to-cell signalling system operating in members of this genus. Work of the past few years has identified many AHL-controlled functions in diverse *Burkholderia* species. These investigations revealed a core set of highly conserved AHL-regulated functions that include biofilm formation, proteolytic activity, motility, antifungal activity and pathogenicity. The more recently discovered BDSF signalling system was also shown to be involved in the regulation of biofilm formation and virulence and evidence has been presented that the two QS regulons partly overlap (Schmid *et al.,* 2012). A main direction for future work will be a detailed analysis of connections between the two regulatory cascades. We have recently presented a working model of the QS circuitry of *B. cenocepacia* (Schmid *et al.,* 2012). In this model we assume a yet unidentified c-di-GMP receptor protein X, the activity of which is directly controlled by the phosphodiesterase activity of the BDSF sensor RpfR. Protein X is thought to be required for transcription of target genes either directly or via additional regulators. Although the two QS systems could operate independently of each other, we favour the idea that the two regulatory cascades converge and control the expression or the activity status of an unknown common regulator Y, which in turn regulates expression of target genes (Fig. 10.3). Additional work will be required to distinguish between the two possibilities and to identify the responsible c-di-GMP effector.

The intracellular secondary messenger c-di-GMP is considered to be a key signal for the bacterial lifestyle, with high levels being typical for sessile cells during chronic infections and low levels for planktonic cells and acute infections (Jonas *et al.,* 2009). Besides the initial investigations on the effects of the BDSF receptor RpfR on intracellular c-di-GMP levels very little work was done to investigate the role of this secondary messenger in virulence and biofilm formation in *Burkholderia* sp. In *B. pseudomallei* a cyclic di-GMP phosphodiesterase, CdpA, was identified and shown to be involved in autoaggregation, motility, biofilm formation, cell invasion, and cytotoxicity (Lee *et al.,* 2010). For *B. cenocepacia* it has been demonstrated that artificially elevated intracellular levels of c-di-GMP promoted wrinkly colony, pellicle and biofilm formation (Fazli *et al.,* 2011). In this study a transcriptional regulator, BCAM 1349, was identified that can bind c-di-GMP and is involved in the production of a number of biofilm matrix components, including cellulose and fimbriae. We believe that these anecdotal reports are just the tip of the iceberg of a new emerging role of c-di-GMP in controlling pathogenicity and the sessile lifestyle of members of the genus *Burkholderia.*

Acknowledgements

Work on quorum sensing and biofilm formation has been supported by the Swiss National Science Foundation (Projects 31003A-122013 and 143773).

References

Aguilar, C., Friscina, A., Devescovi, G., Kojic, M., and Venturi, V. (2003). Identification of quorum-sensing-regulated genes of *Burkholderia cepacia.* J. Bacteriol. *185,* 6456–6462.

Aguilar, C., Carlier, A.L., Riedel, K., and Eberl, L. (2009). Cell-to-cell communication in biofilms of Gram-negative bacteria. In Bacterial Signaling, Krämer, R., and Jung, K., eds. (Weinheim, Germany: Wiley-VCH Verlag), pp. 23–40.

Aubert, D.F., Flannagan, R.S., and Valvano, M.A. (2008). A novel sensor kinase-response regulator hybrid controls biofilm formation and type VI secretion system activity in *Burkholderia cenocepacia*. Infect. Immun. *76*, 1979–1991.

Aubert, D.F., O'Grady, E.P., Hamad, M.A., Sokol, P.A., and Valvano, M.A. (2013). The *Burkholderia cenocepacia* sensor kinase hybrid AtsR is a global regulator modulating quorum-sensing signalling. Environ. Microbiol. *15*, 372–385.

Baldwin, A., Sokol, P.A., Parkhill, J., and Mahenthiralingam, E. (2004). The *Burkholderia cepacia* epidemic strain marker is part of a novel genomic island encoding both virulence and metabolism-associated genes in *Burkholderia cenocepacia*. Infect. Immun. *72*, 1537–1547.

Bernier, S.P., Nguyen, D.T., and Sokol, P.A. (2008). A LysR-type transcriptional regulator in *Burkholderia cenocepacia* influences colony morphology and virulence. Infect. Immun. *76*, 38–47.

Bi, H., Christensen, Q.H., Feng, Y., Wang, H., and Cronan, J.E. (2012). The *Burkholderia cenocepacia* BDSF quorum sensing fatty acid is synthesized by a bifunctional crotonase homologue having both dehydratase and thioesterase activities. Mol. Microbiol. *83*, 840–855.

Boddey, J.A., Flegg, C.P., Day, C.J., Beacham, I.R., and Peak, I.R. (2006). Temperature-regulated microcolony formation by *Burkholderia pseudomallei* requires *pilA* and enhances association with cultured human cells. Infect. Immun. *74*, 5374–5381.

Boon, C., Deng, Y., Wang, L.-H., He, Y., Xu, J.-L., Fan, Y., Pan, S.Q., and Zhang, L.-H. (2008). A novel DSF-like signal from *Burkholderia cenocepacia* interferes with *Candida albicans* morphological transition. ISME J. *2*, 27–36.

Brackman, G., Hillaert, U., Van Calenbergh, S., Nelis, H.J., and Coenye, T. (2009). Use of quorum sensing inhibitors to interfere with biofilm formation and development in *Burkholderia multivorans* and *Burkholderia cenocepacia*. Res. Microbiol. *160*, 144–151.

Brackman, G., Cos, P., Maes, L., Nelis, H.J., and Coenye, T. (2011). Quorum sensing inhibitors increase the susceptibility of bacterial biofilms to antibiotics *in vitro* and *in vivo*. Antimicrob. Agents Chemother. *55*, 2655–2661.

Bylund, J., Burgess, L.-A., Cescutti, P., Ernst, R.K., and Speert, D.P. (2006). Exopolysaccharides from *Burkholderia cenocepacia* inhibit neutrophil chemotaxis and scavenge reactive oxygen species. J. Biol. Chem. *281*, 2526–2532.

Chambers, C.E., Lutter, E.I., Visser, M.B., Law, P.P.Y., and Sokol, P.A. (2006). Identification of potential CepR regulated genes using a cep box motif-based search of the *Burkholderia cenocepacia* genome. BMC Microbiol. *6*, 104.

Chandler, J.R., Duerkop, B.A., Hinz, A., West, T.E., Herman, J.P., Churchill, M.E.A., Skerrett, S.J., and Greenberg, E.P. (2009). Mutational analysis of *Burkholderia thailandensis* quorum sensing and self-aggregation. J. Bacteriol. *191*, 5901–5909.

Chen, C.-N., Chen, C.-J., Liao, C.-T., and Lee, C.-Y. (2009). A probable aculeacin A acylase from the *Ralstonia solanacearum* GMI1000 is *N*-acyl-homoserine lactone acylase with quorum-quenching activity. BMC Microbiol. *9*, 89.

Chun, H., Choi, O., Goo, E., Kim, N., Kim, H., Kang, Y., Kim, J., Moon, J.S., and Hwang, I. (2009). The quorum sensing-dependent gene *katG* of *Burkholderia glumae* is important for protection from visible light. J. Bacteriol. *191*, 4152–4157.

Conway, B.-A., and Greenberg, E.P. (2002). Quorum-sensing signals and quorum-sensing genes in *Burkholderia vietnamiensis*. J. Bacteriol. *184*, 1187–1191.

Cucarella, C., Solano, C., Valle, J., Amorena, B., Lasa, I., and Penadés, J.R. (2001). Bap, a *Staphylococcus aureus* surface protein involved in biofilm formation. J. Bacteriol. *183*, 2888–2896.

Cunha, M.V., Sousa, S.A., Leitão, J.H., Moreira, L.M., Videira, P.A., and Sá-Correia, I. (2004). Studies on the involvement of the exopolysaccharide produced by cystic fibrosis-associated isolates of the *Burkholderia cepacia* complex in biofilm formation and in persistence of respiratory infections. J. Clin. Microbiol. *42*, 3052–3058.

Davies, D.G., and Marques, C.N.H. (2009). A fatty acid messenger is responsible for inducing dispersion in microbial biofilms. J. Bacteriol. *191*, 1393–1403.

Davies, D.G., Parsek, M.R., Pearson, J.P., Iglewski, B.H., Costerton, J.W., and Greenberg, E.P. (1998). The involvement of cell-to-cell signals in the development of a bacterial biofilm. Science *280*, 295–298.

De Kievit, T.R. (2009). Quorum sensing in *Pseudomonas aeruginosa* biofilms. Environ. Microbiol. *11*, 279–288.

Deng, Y., Boon, C., Eberl, L., and Zhang, L.-H. (2009). Differential modulation of *Burkholderia cenocepacia* virulence and energy metabolism by the quorum-sensing signal BDSF and its synthase. J. Bacteriol. *191*, 7270–7278.

Deng, Y., Wu, J., Eberl, L., and Zhang, L.-H. (2010). Structural and functional characterization of diffusible signal factor family quorum-sensing signals produced by members of the *Burkholderia cepacia* complex. Appl. Environ. Microbiol. *76*, 4675–4683.

Deng, Y., Schmid, N., Wang, C., Wang, J., Pessi, G., Wu, D., Lee, J., Aguilar, C., Ahrens, C.H., Chang, C., *et al.* (2012). *Cis*-2-dodecenoic acid receptor RpfR links quorum-sensing signal perception with regulation of virulence through cyclic dimeric guanosine monophosphate turnover. Proc. Natl. Acad. Sci. U.S.A. *108*, 15479–15484.

Devescovi, G., Bigirimana, J., Degrassi, G., Cabrio, L., LiPuma, J.J., Kim, J., Hwang, I., and Venturi, V. (2007). Involvement of a quorum-sensing-regulated lipase secreted by a clinical isolate of *Burkholderia glumae* in severe disease symptoms in rice. Appl. Environ. Microbiol. *73*, 4950–4958.

Diggle, S.P., Lumjiaktase, P., Dipilato, F., Winzer, K., Kunakorn, M., Barrett, D.A., Chhabra, S.R., Cámara, M., and Williams, P. (2006). Functional genetic analysis reveals a 2-Alkyl-4-quinolone signaling system in the human pathogen *Burkholderia pseudomallei* and related bacteria. Chem. Biol. *13*, 701–710.

Dong, Y.-H., and Zhang, L.-H. (2005). Quorum sensing and quorum-quenching enzymes. J. Microbiol. *43*, 101–109.

Dong, Y.H., Xu, J.L., Li, X.Z., and Zhang, L.H. (2000). AiiA, an enzyme that inactivates the acylhomoserine lactone quorum-sensing signal and attenuates the virulence of *Erwinia carotovora*. Proc. Natl. Acad. Sci. U.S.A. *97*, 3526–3531.

Dong, Y.H., Wang, L.H., Xu, J.L., Zhang, H.B., Zhang, X.F., and Zhang, L.H. (2001). Quenching quorum-sensing-dependent bacterial infection by an *N*-acyl homoserine lactonase. Nature *411*, 813–817.

Dong, Y., Gusti, A.R., Zhang, Q., Xu, J., and Zhang, L. (2002). Identification of quorum-quenching *N*-acyl homoserine lactonases from *Bacillus* species. Appl. Environ. Microbiol. *68*, 1754–1759.

Dong, Y.-H., Wang, L., and Zhang, L.-H. (2007). Quorum-quenching microbial infections: mechanisms and implications. Philos. Trans. R. Soc. Lond. B Biol. Sci. *362*, 1201–1211.

Duerkop, B.A., Varga, J., Chandler, J.R., Peterson, S.B., Herman, J.P., Churchill, M.E.A., Parsek, M.R., Nierman, W.C., and Greenberg, E.P. (2009). Quorum-sensing control of antibiotic synthesis in *Burkholderia thailandensis*. J. Bacteriol. *191*, 3909–3918.

Espinosa-Urgel, M., Salido, A., and Ramos, J.L. (2000). Genetic analysis of functions involved in adhesion of *Pseudomonas putida* to seeds. J. Bacteriol. *182*, 2363–2369.

Fazli, M., O'Connell, A., Nilsson, M., Niehaus, K., Dow, J.M., Givskov, M., Ryan, R.P., and Tolker-Nielsen, T. (2011). The CRP/FNR family protein Bcam1349 is a c-di-GMP effector that regulates biofilm formation in the respiratory pathogen *Burkholderia cenocepacia*. Mol. Microbiol. *82*, 327–341.

Ferreira, A.S., Leitão, J.H., Silva, I.N., Pinheiro, P.F., Sousa, S.A., Ramos, C.G., and Moreira, L.M. (2010). Distribution of cepacian biosynthesis genes among environmental and clinical *Burkholderia* strains and role of cepacian exopolysaccharide in resistance to stress conditions. Appl. Environ. Microbiol. *76*, 441–450.

Fouhy, Y., Scanlon, K., Schouest, K., Spillane, C., Crossman, L., Avison, M.B., Ryan, R.P., and Dow, J.M. (2007). Diffusible signal factor-dependent cell–cell signaling and virulence in the nosocomial pathogen *Stenotrophomonas maltophilia*. J. Bacteriol. *189*, 4964–4968.

Gamage, A.M., Shui, G., Wenk, M.R., and Chua, K.L. (2011). *N*-Octanoylhomoserine lactone signalling mediated by the BpsI-BpsR quorum sensing system plays a major role in biofilm formation of *Burkholderia pseudomallei*. Microbiology *157*, 1176–1186.

Geisenberger, O., Givskov, M., Riedel, K., Høiby, N., Tümmler, B., and Eberl, L. (2000). Production of *N*-acyl-L-homoserine lactones by *P. aeruginosa* isolates from chronic lung infections associated with cystic fibrosis. FEMS Microbiol. Lett. *184*, 273–278.

Givskov, M., De Nys, R., Manefield, M., Gram, L., Maximilien, R., Eberl, L., Molin, S., Steinberg, P.D., and Kjelleberg, S. (1996). Eukaryotic interference with homoserine lactone-mediated prokaryotic signalling. J. Bacteriol. *178*, 6618–6622.

Goldstein, R., Sun, L., Jiang, R.Z., Sajjan, U., Forstner, J.F., and Campanelli, C. (1995). Structurally variant classes of pilus appendage fibers coexpressed from *Burkholderia (Pseudomonas) cepacia*. J. Bacteriol. *177*, 1039–1052.

Goo, E., Kang, Y., Kim, H., and Hwang, I. (2010). Proteomic analysis of quorum sensing-dependent proteins in *Burkholderia glumae*. J. Proteome Res. *9*, 3184–3199.

Gotschlich, A., Huber, B., Geisenberger, O., Tögl, A., Steidle, A., Riedel, K., Hill, P., Tümmler, B., Vandamme, P., Middleton, B., *et al.* (2001). Synthesis of multiple *N*-acylhomoserine lactones is wide-spread among the members of the *Burkholderia cepacia* complex. Syst. Appl. Microbiol. *24*, 1–14.

Hentzer, M., Riedel, K., Rasmussen, T.B., Heydorn, A., Andersen, J.B., Parsek, M.R., Rice, S.A., Eberl, L., Molin, S., Høiby, N., *et al.* (2002). Inhibition of quorum sensing in *Pseudomonas aeruginosa* biofilm bacteria by a halogenated furanone compound. Microbiology *148*, 87–102.

Herasimenka, Y., Cescutti, P., Impallomeni, G., Campana, S., Taccetti, G., Ravenni, N., Zanetti, F., and Rizzo, R. (2007). Exopolysaccharides produced by clinical strains belonging to the *Burkholderia cepacia* complex. J. Cyst. Fibros. *6*, 145–152.

Heydorn, A., Ersbøll, B., Kato, J., Hentzer, M., Parsek, M.R., Tolker-Nielsen, T., Givskov, M., and Molin, S. (2002). Statistical analysis of *Pseudomonas aeruginosa* biofilm development: impact of mutations in genes involved in twitching motility, cell-to-cell signaling, and stationary-phase sigma factor expression. Appl. Environ. Microbiol. *68*, 2008–2017.

Hinsa, S.M., Espinosa-Urgel, M., Ramos, J.L., and O'Toole, G.A. (2003). Transition from reversible to irreversible attachment during biofilm formation by *Pseudomonas fluorescens* WCS365 requires an ABC transporter and a large secreted protein. Mol. Microbiol. *49*, 905–918.

Holden, M.T.G., Seth-Smith, H.M.B., Crossman, L.C., Sebaihia, M., Bentley, S.D., Cerdeño-Tárraga, A.M., Thomson, N.R., Bason, N., Quail, M.A., Sharp, S., *et al.* (2009). The genome of *Burkholderia cenocepacia* J2315, an epidemic pathogen of cystic fibrosis patients. J. Bacteriol. *191*, 261–277.

Huang, J.J., Petersen, A., Whiteley, M., and Leadbetter, J.R. (2006). Identification of QuiP, the product of gene PA1032, as the second acyl-homoserine lactone acylase of *Pseudomonas aeruginosa* PAO1. Appl. Environ. Microbiol. *72*, 1190–1197.

Huber, B., Riedel, K., Hentzer, M., Heydorn, A., Gotschlich, A., Givskov, M., Molin, S., and Eberl, L. (2001). The *cep* quorum-sensing system of *Burkholderia cepacia* H111 controls biofilm formation and swarming motility. Microbiology *147*, 2517–2528.

Huber, B., Riedel, K., Köthe, M., Givskov, M., Molin, S., and Eberl, L. (2002). Genetic analysis of functions involved in the late stages of biofilm development in *Burkholderia cepacia* H111. Mol. Microbiol. *46*, 411–426.

Høiby, N., Bjarnsholt, T., Givskov, M., Molin, S., and Ciofu, O. (2010). Antibiotic resistance of bacterial biofilms. Int. J. Antimicrob. Agents *35*, 322–332.

Inhülsen, S., Aguilar, C., Schmid, N., Suppiger, A., Riedel, K., and Eberl, L. (2012). Identification of functions linking quorum sensing with biofilm formation in *Burkholderia cenocepacia* H111. Microbiol. Open *1*, 225–242.

Jakobsen, T.H., Bragason, S.K., Phipps, R.K., Christensen, L.D., Van Gennip, M., Alhede, M., Skindersoe, M., Larsen, T.O., Høiby, N., Bjarnsholt, T., *et al.* (2012a). Food as a source for quorum sensing inhibitors: iberin from horseradish revealed as a quorum sensing inhibitor of *Pseudomonas aeruginosa*. Appl. Environ. Microbiol. *78*, 2410–2421.

Jakobsen, T.H., Van Gennip, M., Phipps, R.K., Shanmugham, M.S., Christensen, L.D., Alhedȩ, M., Skindersoe, M.E., Rasmussen, T.B., Friedrich, K., Uthe, F., *et al.* (2012b). Ajoene, a sulfur-rich molecule from garlic, inhibits genes controlled by quorum sensing. Antimicrob. Agents Chemother. *56*, 2314–2325.

Jonas, K., Melefors, O., and Römling, U. (2009). Regulation of c-di-GMP metabolism in biofilms. Future Microbiol. *4*, 341–358.

Kim, J., Kim, J.-G., Kang, Y., Jang, J.Y., Jog, G.J., Lim, J.Y., Kim, S., Suga, H., Nagamatsu, T., and Hwang, I. (2004). Quorum sensing and the LysR-type transcriptional activator ToxR regulate toxoflavin biosynthesis and transport in *Burkholderia glumae*. Mol. Microbiol. *54*, 921–934.

Kim, J., Kang, Y., Choi, O., Jeong, Y., Jeong, J.-E., Lim, J.Y., Kim, M., Moon, J.S., Suga, H., and Hwang, I. (2007). Regulation of polar flagellum genes is mediated by quorum sensing and FlhDC in *Burkholderia glumae*. Mol. Microbiol. *64*, 165–179.

Kim, H., Goo, E., Kang, Y., Kim, J., and Hwang, I. (2012). Regulation of universal stress protein genes by quorum sensing and RpoS in *Burkholderia glumae*. J. Bacteriol. *194*, 982–992.

Köthe, M., Antl, M., Huber, B., Stoecker, K., Ebrecht, D., Steinmetz, I., and Eberl, L. (2003). Killing of *Caenorhabditis elegans* by *Burkholderia cepacia* is controlled by the *cep* quorum-sensing system. Cell Microbiol. *5*, 343–351.

Lameignere, E., Malinovská, L., Sláviková, M., Duchaud, E., Mitchell, E.P., Varrot, A., Sedo, O., Imberty, A., Wimmerová, M., Avikov, M.S.L., *et al.* (2008). Structural basis for mannose recognition by a lectin from opportunistic bacteria *Burkholderia cenocepacia*. Biochem. J. *411*, 307–318.

Lameignere, E., Shiao, T.C., Roy, R., Wimmerova, M., Dubreuil, F., Varrot, A., and Imberty, A. (2010). Structural basis of the affinity for oligomannosides and analogs displayed by BC2L-A, a *Burkholderia cenocepacia* soluble lectin. Glycobiology *20*, 87–98.

Latasa, C., Roux, A., Toledo-Arana, A., Ghigo, J.-M., Gamazo, C., Penadés, J.R., and Lasa, I. (2005). BapA, a large secreted protein required for biofilm formation and host colonization of *Salmonella enterica* serovar Enteritidis. Mol. Microbiol. *58*, 1322–1339.

Lazzaroni, J.C., Germon, P., Ray, M.C., and Vianney, A. (1999). The Tol proteins of *Escherichia coli* and their involvement in the uptake of biomolecules and outer membrane stability. FEMS Microbiol. Lett. *177*, 191–197.

Lee, H.S., Gu, F., Ching, S.M., Lam, Y., and Chua, K.L. (2010). CdpA is a *Burkholderia pseudomallei* cyclic di-GMP phosphodiesterase involved in autoaggregation, flagellum synthesis, motility, biofilm formation, cell invasion, and cytotoxicity. Infect. Immun. *78*, 1832–1840.

Lewenza, S., Conway, B., Greenberg, E.P., and Sokol, P.A. (1999). Quorum sensing in *Burkholderia cepacia*: identification of the LuxRI homologs CepRI. J. Bacteriol. *181*, 748–756.

Lewenza, S., and Sokol, P.A. (2001). Regulation of ornibactin biosynthesis and *N*-acyl-L-homoserine lactone production by CepR in *Burkholderia cepacia*. J. Bacteriol. *183*, 2212–2218.

Lewenza, S., Visser, M.B., and Sokol, P.A. (2002). Interspecies communication between *Burkholderia cepacia* and *Pseudomonas aeruginosa*. Can. J. Microbiol. *48*, 707–716.

Lin, Y.-H., Xu, J.-L., Hu, J., Wang, L.-H., Ong, S.L., Leadbetter, J.R., and Zhang, L.-H. (2003). Acyl-homoserine lactone acylase from *Ralstonia* strain XJ12B represents a novel and potent class of quorum-quenching enzymes. Mol. Microbiol. *47*, 849–860.

Llamas, M.A., Ramos, J.L., and Rodríguez-Herva, J.J. (2000). Mutations in each of the *tol* genes of *Pseudomonas putida* reveal that they are critical for maintenance of outer membrane stability. J. Bacteriol. *182*, 4764–4772.

Lumjiaktase, P., Diggle, S.P., Loprasert, S., Tungpradabkul, S., Daykin, M., Cámara, M., Williams, P., and Kunakorn, M. (2006). Quorum sensing regulates *dpsA* and the oxidative stress response in *Burkholderia pseudomallei*. Microbiology *152*, 3651–3659.

McCarthy, Y., Yang, L., Twomey, K.B., Sass, A., Tolker-Nielsen, T., Mahenthiralingam, E., Dow, J.M., and Ryan, R.P. (2010). A sensor kinase recognizing the cell–cell signal BDSF (*cis*-2-dodecenoic acid) regulates virulence in *Burkholderia cenocepacia*. Mol. Microbiol. *77*, 1220–1236.

McKenney, D., Brown, K.E., and Allison, D.G. (1995). Influence of *Pseudomonas aeruginosa* exoproducts on virulence factor production in *Burkholderia cepacia*: evidence of interspecies communication. J. Bacteriol. *177*, 6989–6992.

Malott, R.J., and Sokol, P.A. (2007). Expression of the *bviIR* and *cepIR* quorum-sensing systems of *Burkholderia vietnamiensis*. J. Bacteriol. *189*, 3006–3016.

Malott, R.J., Baldwin, A., Mahenthiralingam, E., and Sokol, P.A. (2005). Characterization of the *cciIR* quorum-sensing system in *Burkholderia cenocepacia*. Infect. Immun. *73*, 4982–4992.

Malott, R.J., O'Grady, E.P., Toller, J., Inhülsen, S., Eberl, L., and Sokol, P.A. (2009). A *Burkholderia cenocepacia* orphan LuxR homolog is involved in quorum-sensing regulation. J. Bacteriol. *191*, 2447–2460.

Manefield, M., Rasmussen, T.B., Henzter, M., Andersen, J.B., Steinberg, P., Kjelleberg, S., and Givskov, M. (2002). Halogenated furanones inhibit quorum sensing through accelerated LuxR turnover. Microbiology *148*, 1119–1127.

Marchetti, R., Malinovska, L., Lameignère, E., Adamova, L., De Castro, C., Cioci, G., Stanetty, C., Kosma, P., Molinaro, A., Wimmerova, M., *et al.* (2012). *Burkholderia cenocepacia* lectin A binding to heptoses from the bacterial lipopolysaccharide. Glycobiology *22*, 1387–1398.

Moreira, L.M., Videira, P.A., Sousa, S.A., Leitão, J.H., Cunha, M.V., and Sá-Correia, I. (2003). Identification and physical organization of the gene cluster involved in the biosynthesis of *Burkholderia cepacia* complex exopolysaccharide. Biochem. Biophys. Res. Commun. *312*, 323–333.

Müh, U., Schuster, M., Heim, R., Singh, A., Olson, E.R., and Greenberg, E.P. (2006). Novel *Pseudomonas aeruginosa* quorum-sensing inhibitors identified in an ultra-high-throughput screen. Antimicrob. Agents Chemother. *50*, 3674–3679.

Naganawa, R., Iwata, N., Ishikawa, K., Fukuda, H., Fujino, T., and Suzuki, A. (1996). Inhibition of microbial growth by ajoene, a sulfur-containing compound derived from garlic. Appl. Environ. Microbiol. *62*, 4238–4242.

O'Grady, E.P., Viteri, D.F., Malott, R.J., and Sokol, P.A. (2009). Reciprocal regulation by the CepIR and CciIR quorum sensing systems in *Burkholderia cenocepacia*. BMC Genomics *10*, 441.

O'Grady, E.P., Nguyen, D.T., Weisskopf, L., Eberl, L., and Sokol, P.A. (2011). The *Burkholderia cenocepacia* LysR-type transcriptional regulator ShvR influences expression of quorum-sensing, protease, type II secretion, and *afc* genes. J. Bacteriol. *193*, 163–176.

O'Grady, E.P., Viteri, D.F., and Sokol, P.A. (2012). A unique regulator contributes to quorum sensing and virulence in *Burkholderia cenocepacia*. PLoS One *7*, e37611.

Park, S.-Y. (2003). AhlD, an *N*-acylhomoserine lactonase in *Arthrobacter* sp., and predicted homologues in other bacteria. Microbiology *149*, 1541–1550.

Park, S., Kang, H., Jang, H., Lee, J., Koo, B., and Yum, D.-Y. (2005). Identification of extracellular *N*-acylhomoserine lactone acylase from a *Streptomyces* sp. and its application to quorum quenching. Appl. Environ. Microbiol. *71*, 2632–2641.

Patriquin, G.M., Banin, E., Gilmour, C., Tuchman, R., Greenberg, E.P., and Poole, K. (2008). Influence of quorum sensing and iron on twitching motility and biofilm formation in *Pseudomonas aeruginosa*. J. Bacteriol. *190*, 662–671.

Pratt, L.A., and Kolter, R. (1998). Genetic analysis of *Escherichia coli* biofilm formation: roles of flagella, motility, chemotaxis and type I pili. Mol. Microbiol. *30*, 285–293.

Ramli, N.S.K., Eng Guan, C., Nathan, S., and Vadivelu, J. (2012). The effect of environmental conditions on biofilm formation of *Burkholderia pseudomallei* clinical isolates. PLoS ONE 7, e44104.

Rasmussen, T.B., and Givskov, M. (2006). Quorum sensing inhibitors: a bargain of effects. Microbiology *152*, 895–904.

Rasmussen, T.B., Bjarnsholt, T., Skindersoe, M.E., Hentzer, M., Kristoffersen, P., Köte, M., Nielsen, J., Eberl, L., and Givskov, M. (2005a). Screening for quorum-sensing inhibitors (QSI) by use of a novel genetic system, the QSI selector. J. Bacteriol. *187*, 1799–1814.

Rasmussen, T.B., Skindersoe, M.E., Bjarnsholt, T., Phipps, R.K., Christensen, K.B., Jensen, P.O., Andersen, J.B., Koch, B., Larsen, T.O., Hentzer, M., *et al.* (2005b). Identity and effects of quorum-sensing inhibitors produced by *Penicillium* species. Microbiology *151*, 1325–1340.

Reimmann, C., Ginet, N., Michel, L., Keel, C., Michaux, P., Krishnapillai, V., Zala, M., Heurlier, K., Triandafillu, K., Harms, H., *et al.* (2002). Genetically programmed autoinducer destruction reduces virulence gene expression and swarming motility in *Pseudomonas aeruginosa* PAO1. Microbiology *148*, 923–932.

Riedel, K., and Eberl, L. (2007). Social behavior of members of the genus *Burkholderia*: quorum sensing and biofilms. In *Burkholderia*. Molecular Microbiology and Genomics, Coenye, T., and Vandamme, P., eds. (Horizon Bioscience), pp. 251–282.

Riedel, K., Hentzer, M., Geisenberger, O., Huber, B., Steidle, A., Wu, H., Høiby, N., Givskov, M., Molin, S., and Eberl, L. (2001). *N*-acylhomoserine-lactone-mediated communication between *Pseudomonas aeruginosa* and *Burkholderia cepacia* in mixed biofilms. Microbiology *147*, 3249–3262.

Riedel, K., Arevalo-Ferro, C., Reil, G., Görg, A., Lottspeich, F., and Eberl, L. (2003). Analysis of the quorum-sensing regulon of the opportunistic pathogen *Burkholderia cepacia* H111 by proteomics. Electrophoresis *24*, 740–750.

Riedel, K., Köthe, M., Kramer, B., Saeb, W., Gotschlich, A., Ammendola, A., and Eberl, L. (2006). Computer-aided design of agents that inhibit the cep quorum-sensing system of *Burkholderia cenocepacia*. Antimicrob. Agents Chemother. *50*, 318–323.

Romero, M., Diggle, S.P., Heeb, S., Cámara, M., and Otero, A. (2008). Quorum quenching activity in *Anabaena* sp. PCC 7120: identification of AiiC, a novel AHL-acylase. FEMS Microbiol. Lett. *280*, 73–80.

Ryan, R.P., and Dow, J.M. (2010). Intermolecular interactions between HD-GYP and GGDEF domain proteins mediate virulence-related signal transduction in *Xanthomonas campestris*. Virulence *1*, 404–408.

Ryan, R.P., McCarthy, Y., Watt, S.A., Niehaus, K., and Dow, J.M. (2009). Intraspecies signaling involving the diffusible signal factor BDSF (cis-2-dodecenoic acid) influences virulence in *Burkholderia cenocepacia*. J. Bacteriol. *191*, 5013–5019.

Sajjan, U., Liu, L., Lu, A., Spilker, T., Forstner, J., and LiPuma, J.J. (2002). Lack of cable pili expression by *cblA*-containing *Burkholderia cepacia* complex. Microbiology *148*, 3477–3484.

Schaber, J.A., Triffo, W.J., Suh, S.J., Oliver, J.W., Hastert, M.C., Griswold, J.A., Auer, M., Hamood, A.N., and Rumbaugh, K.P. (2007). *Pseudomonas aeruginosa* forms biofilms in acute infection independent of cell-to-cell signaling. Infect. Immun. *75*, 3715–3721.

Schmid, N., Pessi, G., Deng, Y., Aguilar, C., Carlier, A.L., Grunau, A., Omasits, U., Zhang, L.-H., Ahrens, C.H., and Eberl, L. (2012). The AHL- and BDSF-Dependent Quorum Sensing Systems Control Specific and Overlapping Sets of Genes in *Burkholderia cenocepacia* H111. PLoS ONE 7, e49966.

Schmidt, S., Blom, J.F., Pernthaler, J., Berg, G., Baldwin, A., Mahenthiralingam, E., and Eberl, L. (2009). Production of the antifungal compound pyrrolnitrin is quorum sensing-regulated in members of the *Burkholderia cepacia* complex. Environ. Microbiol. *11*, 1422–1437.

Seyedsayamdost, M.R., Chandler, J.R., Blodgett, J.A.V., Lima, P.S., Duerkop, B.A., Oinuma, K.-I., Greenberg, E.P., and Clardy, J. (2010). Quorum-sensing-regulated bactobolin production by *Burkholderia thailandensis* E264. Org. Lett. *12*, 716–719.

Shepherd, R.W., and Lindow, S.E. (2009). Two dissimilar *N*-acyl-homoserine lactone acylases of *Pseudomonas syringae* influence colony and biofilm morphology. Appl. Environ. Microbiol. *75*, 45–53.

Sio, C.F., Otten, L.G., Cool, R.H., Diggle, S.P., Braun, P.G., Bos, R., Daykin, M., Cámara, M., Williams, P., and Quax, W.J. (2006). Quorum quenching by an *N*-acyl-homoserine lactone acylase from *Pseudomonas aeruginosa* PAO1. Infect. Immun. *74*, 1673–1682.

Sokol, P.A. (2003). The CepIR quorum-sensing system contributes to the virulence of *Burkholderia cenocepacia* respiratory infections. Microbiology *149*, 3649–3658.

Sokol, P.A., Malott, R.J., Riedel, K., and Eberl, L. (2007). Communication systems in the genus *Burkholderia*: global regulators and targets for novel antipathogenic drugs. Future Microbiol. *2*, 555–563.

Solis, R., Bertani, I., Degrassi, G., Devescovi, G., and Venturi, V. (2006). Involvement of quorum sensing and RpoS in rice seedling blight caused by *Burkholderia plantarii*. FEMS Microbiol. Lett. *259*, 106–112.

Song, Y., Xie, C., Ong, Y.-M., Gan, Y.-H., and Chua, K.-L. (2005). The BpsIR quorum-sensing system of *Burkholderia pseudomallei*. J. Bacteriol. *187*, 785–790.

Sousa, S.A., Ulrich, M., Bragonzi, A., Burke, M., Worlitzsch, D., Leitão, J.H., Meisner, C., Eberl, L., Sá-Correia, I., and Döring, G. (2007). Virulence of *Burkholderia cepacia* complex strains in gp91$^{phox-/-}$ mice. Cell Microbiol. *9*, 2817–2825.

Stoodley, P., Jørgensen, F., Williams, P., and Lappin-Scott, H.M. (1999). The role of hydrodynamics and AHL signalling molecules as determinants of the structure of *Pseudomonas aeruginosa* biofilms. In Biofilms: The Good, the Bad, and the Ugly, Bayston, R., Brading, M., Gilbert, P., Walker, J., and Wimpenny, J.W., eds. (Cardiff, UK: BioLine), pp. 323–330.

Subramoni, S., and Venturi, V. (2009). LuxR-family 'solos': bachelor sensors/regulators of signalling molecules. Microbiology *155*, 1377–1385.

Sulák, O., Cioci, G., Delia, M., Lahmann, M., Varrot, A., Imberty, A., and Wimmerová, M. (2010). A TNF-like trimeric lectin domain from *Burkholderia cenocepacia* with specificity for fucosylated human histo-blood group antigens. Structure *18*, 59–72.

Sulák, O., Cioci, G., Lameignère, E., Balloy, V., Round, A., Gutsche, I., Malinovská, L., Chignard, M., Kosma, P., Aubert, D.F., *et al.* (2011). *Burkholderia cenocepacia* BC2L-C is a super lectin with dual specificity and proinflammatory activity. PLoS Pathog. 7, e1002238.

Suárez-Moreno, Z.R., Caballero-Mellado, J., and Venturi, V. (2008). The new group of non-pathogenic plant-associated nitrogen-fixing *Burkholderia* spp. shares a conserved quorum-sensing system, which is tightly regulated by the RsaL repressor. Microbiology *154*, 2048–2059.

Suárez-Moreno, Z.R., Devescovi, G., Myers, M., Hallack, L., Mendonça-Previato, L., Caballero-Mellado, J., and Venturi, V. (2010). Commonalities and differences in regulation of *N*-acyl homoserine lactone quorum sensing in the beneficial plant-associated *Burkholderia* species cluster. Appl. Environ. Microbiol. *76*, 4302–4317.

Tielker, D., Hacker, S., Loris, R., Strathmann, M., Wingender, J., Wilhelm, S., Rosenau, F., and Jaeger, K.-E. (2005). *Pseudomonas aeruginosa* lectin LecB is located in the outer membrane and is involved in biofilm formation. Microbiology *151*, 1313–1323.

Toledo-Arana, A., Valle, J., Solano, C., Arrizubieta, M.J., Cucarella, C., Lamata, M., Amorena, B., Leiva, J., Penadés, J.R., and Lasa, I. (2001). The enterococcal surface protein, Esp, is involved in *Enterococcus faecalis* biofilm formation. Appl. Environ. Microbiol. *67*, 4538–4545.

Tomlin, K.L., Malott, R.J., Ramage, G., Storey, D.G., Sokol, P.A., and Ceri, H. (2005). Quorum-sensing mutations affect attachment and stability of *Burkholderia cenocepacia* biofilms. Appl. Environ. Microbiol. *71*, 5208–5218.

Twomey, K.B., O'Connell, O.J., McCarthy, Y., Dow, J.M., O'Toole, G.A., Plant, B.J., and Ryan, R.P. (2012). Bacterial cis-2-unsaturated fatty acids found in the cystic fibrosis airway modulate virulence and persistence of *Pseudomonas aeruginosa*. ISME J. *6*, 939–950.

Uehlinger, S., Schwager, S., Bernier, S.P., Riedel, K., Nguyen, D.T., Sokol, P.A., and Eberl, L. (2009). Identification of specific and universal virulence factors in *Burkholderia cenocepacia* strains by using multiple infection hosts. Infect. Immun. *77*, 4102–4110.

Ulrich, R.L. (2004). Quorum quenching: enzymatic disruption of *N*-acylhomoserine lactone-mediated bacterial communication in *Burkholderia thailandensis*. Appl. Environ. Microbiol. *70*, 6173–6180.

Ulrich, R.L., Deshazer, D., Brueggemann, E.E., Hines, H.B., Oyston, P.C., and Jeddeloh, J.A. (2004a). Role of quorum sensing in the pathogenicity of *Burkholderia pseudomallei*. J. Med. Microbiol. *53*, 1053–1064.

Ulrich, R.L., Deshazer, D., Hines, H.B., and Jeddeloh, J.A. (2004b). Quorum sensing: a transcriptional regulatory system involved in the pathogenicity of *Burkholderia mallei*. Infect. Immun. *72*, 6589–6596.

Ulrich, R.L., Hines, H.B., Parthasarathy, N., and Jeddeloh, J.A. (2004c). Mutational analysis and biochemical characterization of the *Burkholderia thailandensis* DW503 quorum-sensing network. J. Bacteriol. *186*, 4350–4360.

Uroz, S., Chhabra, S.R., Cámara, M., Williams, P., Oger, P., and Dessaux, Y. (2005). *N*-Acylhomoserine lactone quorum-sensing molecules are modified and degraded by *Rhodococcus erythropolis* W2 by both amidolytic and novel oxidoreductase activities. Microbiology *151*, 3313–3322.

Valade, E., Thibault, F.M., Gauthier, Y.P., Palencia, M., Popoff, M.Y., and Vidal, D.R. (2004). The PmlI-PmlR quorum-sensing system in *Burkholderia pseudomallei* plays a key role in virulence and modulates production of the MprA protease. J. Bacteriol. *186*, 2288–2294.

Venturi, V., Friscina, A., Bertani, I., Devescovi, G., and Aguilar, C. (2004). Quorum sensing in the *Burkholderia cepacia* complex. Res. Microbiol. *155*, 238–244.

Vergunst, A.C., Meijer, A.H., Renshaw, S.A., and O'Callaghan, D. (2010). *Burkholderia cenocepacia* creates an intramacrophage replication niche in zebrafish embryos, followed by bacterial dissemination and establishment of systemic infection. Infect. Immun. *78*, 1495–1508.

Vial, L., Lépine, F., Milot, S., Groleau, M.-C., Dekimpe, V., Woods, D.E., and Déziel, E. (2008). *Burkholderia pseudomallei, B. thailandensis,* and *B. ambifaria* produce 4-hydroxy-2-alkylquinoline analogues with a methyl group at the 3 position that is required for quorum-sensing regulation. J. Bacteriol. *190*, 5339–5352.

Wahjudi, M., Papaioannou, E., Hendrawati, O., Van Assen, A.H.G., Van Merkerk, R., Cool, R.H., Poelarends, G.J., and Quax, W.J. (2011). PA0305 of *Pseudomonas aeruginosa* is a quorum quenching acylhomoserine lactone acylase belonging to the Ntn hydrolase superfamily. Microbiology *157*, 2042–2055.

Wang, L.-H., He, Y., Gao, Y., Wu, J.E., Dong, Y.-H., He, C., Wang, S.X., Weng, L.-X., Xu, J.-L., Tay, L., *et al.* (2003). A bacterial cell–cell communication signal with cross-kingdom structural analogues. Mol. Microbiol. *51*, 903–912.

Watnick, P., and Kolter, R. (2000). Biofilm, city of microbes. J. Bacteriol. *182*, 2675–2679.

Wei, Y., Ryan, G.T., Flores-Mireles, A.L., Costa, E.D., Schneider, D.J., and Winans, S.C. (2011). Saturation mutagenesis of a CepR binding site as a means to identify new quorum-regulated promoters in *Burkholderia cenocepacia*. Mol. Microbiol. *79*, 616–632.

Whiteley, M., Bangera, M.G., Bumgarner, R.E., Parsek, M.R., Teitzel, G.M., Lory, S., and Greenberg, E.P. (2001). Gene expression in *Pseudomonas aeruginosa* biofilms. Nature *413*, 860–864.

Williams, P. (2007). Quorum sensing, communication and cross-kingdom signalling in the bacterial world. Microbiology *153*, 3923–3938.

Winsor, G.L., Khaira, B., Van Rossum, T., Lo, R., Whiteside, M.D., and Brinkman, F.S.L. (2008). The Burkholderia Genome Database: facilitating flexible queries and comparative analyses. Bioinformatics *24*, 2803–2804.

Winzer, K., Falconer, C., Garber, N.C., Diggle, S.P., Camara, M., and Williams, P. (2000). The *Pseudomonas aeruginosa* lectins PA-IL and PA-IIL are controlled by quorum sensing and by RpoS. J. Bacteriol. *182*, 6401–6411.

Wongtrakoongate, P., Tumapa, S., and Tungpradabkul, S. (2012). Regulation of a quorum sensing system by stationary phase sigma factor RpoS and their co-regulation of target genes in *Burkholderia pseudomallei*. Microbiol. Immunol. *56*, 281–294.

Wopperer, J., Cardona, S.T., Huber, B., Jacobi, C.A., Valvano, M.A., and Eberl, L. (2006). A quorum-quenching approach to investigate the conservation of quorum-sensing-regulated functions within the *Burkholderia cepacia* complex. Appl. Environ. Microbiol. *72*, 1579–1587.

Wu, H., Song, Z., Hentzer, M., Andersen, J.B., Molin, S., Givskov, M., and Høiby, N. (2004). Synthetic furanones inhibit quorum-sensing and enhance bacterial clearance in *Pseudomonas aeruginosa* lung infection in mice. J. Antimicrob. Chemother. *53*, 1054–1061.

Yang, L., Nilsson, M., Gjermansen, M., Givskov, M., and Tolker-Nielsen, T. (2009). Pyoverdine and PQS mediated subpopulation interactions involved in *Pseudomonas aeruginosa* biofilm formation. Mol. Microbiol. *74*, 1380–1392.

Yang, L., Barken, K.B., Skindersoe, M.E., Christensen, A.B., Givskov, M., and Tolker-Nielsen, T. (2007). Effects of iron on DNA release and biofilm development by *Pseudomonas aeruginosa*. Microbiology *153*, 1318–1328.

Zhang, H.-B., Wang, L.-H., and Zhang, L.-H. (2002). Genetic control of quorum-sensing signal turnover in *Agrobacterium tumefaciens*. Proc. Natl. Acad. Sci. U.S.A. *99*, 4638–4643.

Zlosnik, J.E.A., Hird, T.J., Fraenkel, M.C., Moreira, L.M., Henry, D.A., and Speert, D.P. (2008). Differential mucoid exopolysaccharide production by members of the *Burkholderia cepacia* complex. J. Clin. Microbiol. *46*, 1470–1473.

Genomics of *Burkholderia* Phages

11

Karlene H. Lynch and Jonathan J. Dennis

Abstract

Sequenced phages that infect and/or lysogenize strains of *Burkholderia pseudomallei*, *Burkholderia mallei*, *Burkholderia thailandensis*, and the *Burkholderia cepacia* complex (Bcc) belong to the tailed phage order *Caudovirales* and families *Myoviridae*, *Podoviridae* and *Siphoviridae*. Within these families, phages can be classified as part of at least ten distinct genera based on CoreGenes analysis of protein relatedness. Members of these genera show similarity to myoviruses P2, Bcep781, PB1, and BcepMu, podoviruses Bcep22 and BPP-1, and siphoviruses PA73, BcepNazgul, ϕE125, and BcepGomr. Prophage islands have also been identified in the sequenced genomes of *B. pseudomallei*, *B. thailandensis*, *Burkholderia cenocepacia*, and *Burkholderia xenovorans*, but not *B. mallei*. Similar to phages infecting other organisms, *Burkholderia* phage genomes are diverse, unique, and modular. In contrast to many other bacterial pathogens, characterized *Burkholderia* virulence factors do not appear to be phage-encoded. However, several putative lysogenic conversion genes have been identified that require functional characterization during the development of phage therapy protocols for this genus.

Introduction

As discussed throughout this book, the genus *Burkholderia* is a diverse and unique group of organisms. Although all bacteria in this genus share certain characteristics, distinct traits can be identified at the isolate, strain, and species level. While this statement describes *Burkholderia* bacteria, the same holds true for the bacteriophages of this genus. Although *Malleomyces pseudomallei* (now *Burkholderia pseudomallei*) phages had been identified in the English literature as early as the mid-1950s (Wolochow and Green, 1956; Smith and Cherry, 1957), genomic characterization of *Burkholderia* phages has only begun in the last decade. Thirty complete *Burkholderia* phage genome sequences are now available in GenBank (Table 11.1) and thanks to the availability of next-generation sequencing, many more are now in the pipeline. These characterized viruses infect and/or lysogenize strains of *B. pseudomallei*, *Burkholderia mallei*, *Burkholderia thailandensis*, and several species of the *Burkholderia cepacia* complex (Bcc).

Previous reviews on phages of *Burkholderia* (Summer *et al.*, 2007; Semler *et al.*, 2012; Lynch and Dennis, 2012) focus on the Bcc and the relevance of these viruses with respect to pathogenicity and phage therapy. Here, we present a comprehensive review of *Burkholderia* phage genomics (encompassing all sequences currently available in GenBank) with a focus on comparative genomics, functional characterization, and the potential for lysogenic conversion. The latter category is not restricted to virulence factors, but instead encompasses all proteins that may alter the phenotype of a lysogenized host. In addition, we provide an overview of partially sequenced phages and of prophage islands identified in fully sequenced *Burkholderia* genomes. Similar to their hosts, these phages are part of several related groups and have numerous similarities and differences both within and between genera.

Table 11.1 Overview of *Burkholderia* phages with complete genome sequences in GenBank

Family	Genus	Phage	Source	Size (bp)	GC (%)	Genes	CoreGenes 3.1 correlation (%)	RefSeq accession	INSDC accession	Reference
Myoviridae	P2-like	ΦE202	*B. thailandensis* E202	35,741	65.4	48	55.81	NC_009234.1	CP000623.1	Ronning *et al*. (2010)
		Φ52237	*B. pseudomallei* Pasteur 52237	37,639	64.8	47	51.16	NC_007145.2	DQ087285.2	Ronning *et al*. (2010)
		ΦE12–2	*B. pseudomallei* E12–2	36,690	64.6	50	48.84	NC_009236.1	CP000624.1	Ronning *et al*. (2010)
		ΦK96243	*B. pseudomallei* K96243	36,286	65.4	49	55.81	NC_006350.1	BX571965.1	Holden *et al*. (2004)
		KL3	*B. cenocepacia* CEP511	40,555	63.2	52	53.49	NC_015266.1	GU911304.1	Lynch *et al*. (2010b)
		KS14	Soil	32,317	62.3	44	53.49	NC_015273.1	HM461982.1	Lynch *et al*. (2010b)
		KS5	Soil and *B. multivorans* ATCC 17616	37,236	63.7	46	51.16	NC_015265.1	GU911303.1	Lynch *et al*. (2010b)
	Bcep781-like	Bcep781	Soil	48,247	63.3	66	100	NC_004333.2	AF543311.2	Summer *et al*. (2006)
		Bcep43	Soil	48,024	63.4	65	96.97	NC_005342.2	AY368235.2	Summer *et al*. (2006)
		Bcep1	Soil	48,177	63.6	71	92.42	NC_005263.2	AY369265.2	Summer *et al*. (2006)
		BcepNY3	Soil	47,382	63.6	70	92.42	NC_009604.1	EF602154.1	None
	PB1-like	BcepB1A	Soil	47,399	54.5	73	24.73	NC_005886.2	AY616033.2	Summer *et al*. (2006)
		BcepF1	Soil	72,415	55.9	127	43.01	NC_009015.1	EF153632.1	None
	BcepMu-like	BcepMu	*B. cenocepacia* J2315	36,748	62.9	53	100	NC_005882.1	AY539836.1	Summer *et al*. (2004)
		ΦE255	*B. thailandensis* E255	37,446	63.1	55	86.79	NC_009237.1	CP000622.1	Ronning *et al*. (2010)
		KS10	*B. cenocepacia* K56–2	37,635	62.9	49	30.19	NC_011216.1	EU822883.1	Goudie *et al*. (2008)

Podoviridae	Bcep22-like	Bcep22	Soil	63,882	65.3	82	100	NC_005262.2	AY349011.3	Gill *et al*. (2011)
		BcepMigl	Soil	62,952	65.5	76	79.22	NC_019917.1	JX104231.1	None
		BcepIL02	Soil	62,715	66.2	76	70.13	NC_012743.2	FJ937737.2	Gill *et al*. (2011)
		DC1	Soil	61,847	66.2	74	67.53	NC_018452.1	JN662425.1	Lynch *et al*. (2012a)
	BPP-1-like	BcepC6B	Soil	42,415	65.2	46	38.78	NC_005887.1	AY605181.1	None
Siphoviridae	PA73-like	KL1	Sewage	42,832	54.6	55	80.77	NC_018278.1	JF939047.1	Lynch *et al*. (2012b)
	BcepNazgul-like	BcepNazgul	Soil	57,455	60.6	73	100	NC_005091.2	AY357582.2	None
		AH2	Soil	58,065	61.3	78	54.79	NC_018283.1	JN564907.1	Lynch *et al*. (2012b)
	ΦE125-like	ΦE125	*B. thailandensis* E125	53,373	61.2	71	100	NC_003309.1	AF447491.1	Woods *et al*., (2002)
		Φ1026b	*B. pseudomallei* 1026b	54,865	60.7	83	80.28	NC_005284.1	AY453853.1	DeShazer (2004)
		Φ644–2	*B. pseudomallei* 644	48,674	60.4	71	73.24	NC_009235.2	CP000625.2	Ronning *et al*. (2010)
		Bcep176	*B. multivorans* ATCC 17616	44,856	61.5	81	38.03	NC_007497.1	DQ203855.1	None
		KS9	*B. pyrrocinia* LMG 21824	39,896	60.7	50	35.21	NC_013055.1	FJ982340.1	Lynch *et al*. (2010a)
	BcepGomr-like	BcepGomr	Soil	52,414	56.3	75	100	NC_009447.1	EF523948.1	None

Where possible, taxonomic categories are based on assignments by Lavigne *et al.* (2008, 2009). Size is based on the most recent RefSeq or INSDC record. GC content and genes are taken from Genome Annotation Reports in NCBI Genome (excluding ΦK96243). CoreGenes 3.1 correlation scores (Mahadevan *et al.*, 2009) were calculated from http://binf.gmu.edu:8080/CoreGenes3.1 using the RefSeq accession (excluding Bcep22 and BcepMigl, for which the INSDC accession was used). The ΦK96243 data assume coverage from 145,410–181,695 of the *B. pseudomallei* K96243 chromosome 1 sequence (including the putative *attL* but not *attR*).

Principles of bacteriophage comparative genomics

Although a large proportion of currently available phage genomes are from those that infect enterobacteria, dairy bacteria, or mycobacteria (Brüssow, 2001; Casjens, 2005; Hatfull, 2010), the general principles that apply to these genomes also apply to those infecting *Burkholderia* species. Three common principles in the field of bacteriophage genomics are that phage genomes are incredibly varied, contain many unique sequences, and have a mosaic organization (Hatfull, 2008; Hatfull and Hendrix, 2011). These observations stem from the study of an ever-increasing number of sequenced phage genomes, starting with that of MS2 in 1976 and ϕX174 in 1977 (Fiers *et al.*, 1976; Sanger *et al.*, 1977).

There is an exceptional degree of variety observed in characterized phages with respect to nucleic acid type, virion morphology (and by extension virion morphogenesis genes), and genome size, organization, and content. Phages can be single- or double-stranded DNA or RNA and may have modifications such as hydroxymethylation as found in enterobacteria phage T4 (Ackermann, 2003; Miller *et al.*, 2003). Most characterized phages (and all characterized *Burkholderia* phages) have dsDNA genomes and belong to the order *Caudovirales*. This order is comprised of tailed phages and is divided into three families: the long contractile tailed *Myoviridae*, the short non-contractile tailed *Podoviridae*, and the long non-contractile tailed *Siphoviridae* (Ackermann, 2003). Each of these families requires a unique set of morphogenesis proteins for virion assembly that, although it may carry out similar functions, is divergent among individual phages (Lynch and Dennis, 2012). Within the *Caudovirales*, genome sizes vary by over an order of magnitude, from the 11,660 bp *Mycoplasma* phage P1 (Tu *et al.*, 2001) to the 497,513 bp *Bacillus megaterium* phage G. As discussed in detail below for the *Burkholderia* phages (and as illustrated in Figs. 11.1–11.9), variations in genome organization and especially genome content may be observed both within and among different phage genera.

In many phage genomes, including those of *Burkholderia*, many of the gene sequences encode hypothetical proteins with uncharacterized functions. For example, following sequencing and annotation of 18 *Pseudomonas aeruginosa* phages, 1562 of 1894 proteins (82%) could not be assigned a putative function (Kwan *et al.*, 2006). A similar result was seen in 27 *Staphylococcus aureus* phages for 1,402 of 2,170 proteins (65%) (Kwan *et al.*, 2005). Furthermore, in these studies 55% of *P. aeruginosa* phage proteins and 24% of *S. aureus* phage proteins showed no similarity to either phage or bacterial proteins in GenBank based on BLAST searches (Kwan *et al.*, 2005, 2006). Similar results are seen in environmental metagenomics surveys; in a study of marine viruses in California, 1352 of 1,934 (70%) sequences showed no matches via BLAST analysis (Breitbart *et al.*, 2002). These results serve to underscore the diversity of the phage population and suggest that uncharacterized sequences will continue to be found within newly isolated phages. As is true for *Burkholderia* phages, broad functional characterization studies are required in order to assess the functions of these novel proteins.

A conserved feature among phage genomes is that they are comprised of numerous smaller modules, resulting in a mosaic structure. These modules represent portions of genes, whole genes, or clusters of genes that likely contribute to a common function (such as virion morphogenesis, lysis, or integration) (Hendrix, 2003). These modules can move among phages by homologous or, perhaps more commonly, non-homologous recombination (Hendrix, 2003). Ample opportunities for recombination with either co-infecting phages or prophages arise during the phage life cycle. This statement may be particularly true for temperate phages because they do not degrade the bacterial DNA as many lytic phages do and can integrate into the bacterial chromosome for extended periods of time as a prophage (Hendrix, 2003). In the *Pseudomonas* and *Staphylococcus* phage studies discussed above (Kwan *et al.*, 2005 and 2006), proteins for which GenBank matches were identified tended to be similar to other *Pseudomonas* and *Staphylococcus* phage proteins, respectively, suggesting that recombination occurs readily within these phage populations. The same holds true for the *Burkholderia* phages discussed here, with some exceptions (such as *B. cenocepacia* phage KL1), as each phage is most closely related to another *Burkholderia* phage.

Applications of *Burkholderia* phage genomics

Phage therapy development

One of the main objectives of *Burkholderia* (particularly Bcc) phage genomics studies is to characterize possible candidates for clinical administration. Although this chapter does not discuss the history of and theory behind phage therapy, many excellent reviews discussing these subjects are available (see Merril *et al.*, 2003; Thiel, 2004; Levin and Bull, 2004; Kropinski, 2006). As *Burkholderia* spp. strains tend to be broadly antibiotic-resistant, the feasibility of treating infections with phages instead of antibiotics is currently being assessed for both plant and human disease. In recent years, this approach has been shown to be particularly successful for another Gram-negative opportunistic pathogen, *P. aeruginosa* (Harper and Enright, 2011). In recent preliminary studies, phages have been shown to be effective against ear infections in a canine model (Hawkins *et al.*, 2010), lung or systemic infections in a mouse model (Debarbieux *et al.*, 2010; Morello *et al.*, 2011; Tiwari *et al.*, 2011; Alemayehu *et al.*, 2012), and skin infections in an *ex vivo* human model (Vieira *et al.*, 2012). Most significantly, a *P. aeruginosa*-specific lytic phage cocktail (BioPhage-PA) was shown to be both safe and effective against ear infections in a UK Phase I/II trial (Wright *et al.*, 2009). The company that developed this cocktail, AmpliPhi Biosciences, is now working to develop BioPhage-PR for use in cystic fibrosis (CF) patients and is working towards Phase III trials for BioPhage-PA (AmpliPhi Biosciences, 2011).

Phage therapy against *Burkholderia* has thus far been shown to be effective in invertebrate, vertebrate, and plant models. The first three reports, published in 2009 and 2010, described the activity of phages against *B. cenocepacia*, a pathogenic and clinically prevalent Bcc strain in CF patients. The first report to show that phages are effective against *Burkholderia in vivo* was published by Seed and Dennis (2009). Greater wax moth larvae (*Galleria mellonella*) were used as an infection model with two CF epidemic *B. cenocepacia* strains, K56-2 and C6433. Myoviruses KS4-M, KS12, and KS14 were administered at various multiplicities of infection (MOIs) to larvae infected with a lethal dose of *B. cenocepacia*. For all three phages, administration immediately following infection resulted in a generally dose-dependent increase in survival. Even when treatment with KS4-M or KS12 was delayed by 6 or 12 h, a similar result was observed (Seed and Dennis, 2009).

The second study, published by Carmody *et al.* (2010), was the first to show successful phage therapy against *Burkholderia* spp. in a mammalian model. Mice were infected with *B. cenocepacia* AU0728 and treated with podovirus BcepIL02 at an MOI of 100, either intranasally or intraperitoneally. Intraperitoneal treatment was found to be the most effective under these conditions, resulting in decreased bacterial density, TNF-α cytokine levels, and MIP-2 chemokine levels (Carmody *et al.*, 2010). In the third study, published by Lynch *et al.* (2010a), a genetically engineered *B. cenocepacia*-specific phage was shown to be effective *in vivo*. As discussed below, the majority of characterized *Burkholderia* phages are temperate. To construct a lytic variant for therapeutic use in this proof-of-concept study, the siphovirus KS9 repressor gene was replaced with a trimethoprim-resistance cassette, resulting in the lytic variant KS9c. Unlike KS9, KS9c did not stably integrate into the chromosome of *Burkholderia pyrrocinia* LMG 21824 or *B. cenocepacia* K56-2. When KS9c was administered at an MOI of 50 or 100 immediately following infection of *G. mellonella* larvae with a lethal dose of K56-2, survival was comparable to that of uninfected controls (Lynch *et al.*, 2010a).

Most recently, Adachi *et al.* (2012) showed that *Burkholderia*-specific phage treatment might be effective both in plants and against species other than *B. cenocepacia*. Phage treatment is being considered for use against *Burkholderia glumae* (seedling rot) and *Burkholderia plantarii* (seedling blight) in rice because of the negative environmental and financial impact of conventional chemical treatments. In this study, rice seed was infected with a phage-susceptible or phage-resistant strain of *B. glumae* or a phage-susceptible strain of *B. plantarii*, treated with one of three phages (BGPP-Ar, BGPP-Sa, or BPP-Ya), and germinated. The proportion of rice seedlings with disease symptoms was significantly lower when treated with phages, but only if the *Burkholderia* strain was phage-susceptible. In this system, phage treatment was even shown to be more effective than chemical treatment using ipconazole/copper (II) hydroxide, indicating that phages may be a superior treatment for both diseases (Adachi *et al.*, 2012).

Progress has also been made with respect to the development of technologies for *Burkholderia* phage delivery to the lungs (reviewed in Semler *et al.*, 2012; Hoe *et al.*, 2013). Bcc phages have been successfully aerosolized using nebulization (KS4-M), following freeze drying/lyophilization with lactose/lactoferrin (KS4-M), and following spray drying with trehalose/leucine (KS4-M and KS14) (Golshahi *et al.*, 2008; Golshahi *et al.*, 2011; Matinkhoo *et al.*, 2011). Other pathogenic species such as *B. pseudomallei*, *B. mallei*, and *Burkholderia gladioli* can also cause lung infections (Kennedy *et al.*, 2007; Balder *et al.*, 2010) and these aerosolization techniques may be broadly applicable for clinical (and perhaps even agricultural) delivery of *Burkholderia* phages.

Virulence factor identification

A second objective of *Burkholderia* phage genomics studies is to identify genes that may contribute to host pathogenicity. The characterization of these genes is important both to further our understanding of *Burkholderia* virulence and to ensure that phages used therapeutically do not carry these or related genes (Lynch and Dennis, 2012). In many other pathogens, prophage genes can contribute substantially to virulence. Classical examples include cholera toxin of *Vibrio cholerae*, diphtheria toxin of *Corynebacterium diphtheriae*, and shiga toxin of enterohaemorrhagic *Escherichia coli* (reviewed in Boyd and Brüssow, 2002). Although these examples are solely toxins, phage-encoded virulence genes can have a variety of effects including adherence and immune evasion (Boyd and Brüssow, 2002). Numerous virulence factors have been identified in the Bcc (Loutet and Valvano, 2010) and in *B. pseudomallei* and *B. mallei* (Galyov *et al.*, 2010), but these proteins have not been found to be phage-encoded. Certain *Burkholderia* phage genes with possible links to pathogenicity have been identified (Summer *et al.*, 2007) and these sequences will be discussed throughout this chapter. However, it is important to note that the relationship of these genes to virulence is speculative, as none have been experimentally confirmed. In fact, it is thought that *Burkholderia* phages tend not to encode virulence factors, a trait that makes their development into therapeutic agents more feasible (Summer *et al.*, 2007; Lynch and Dennis, 2012). Because *Burkholderia* species thrive in several niches – including environmental, industrial, and clinical – selection may promote the retention of prophage genes that facilitate survival under multiple conditions (Summer *et al.*, 2007).

Overview of *Burkholderia* phages and prophages

Although the complete sequences of 30 *Burkholderia* phages are available in GenBank (Table 11.1), this number represents only a fraction of the total number of such phages isolated to date. Isolated but not yet sequenced Bcc phages include both temperate phages isolated from lysogens – CP1 (Cihlar *et al.*, 1978), CP75 (Matsumoto *et al.*, 1986), NS1 and NS2 (Nzula *et al.*, 2000), DK1–DK3 and MM1–MM10 (Langley *et al.*, 2003), KS8 and KS11 (Seed and Dennis, 2005), and BcP15 (Hens *et al.*, 2005) – and temperate or lytic phages isolated from environmental (largely rhizosphere, soil, and sediment) samples – JB1, JB3, JB5, RL1, and RL2 (Langley *et al.*, 2003), KS1–KS3, KS6, (Seed and Dennis, 2005), and KS12 (Seed and Dennis, 2009). These phages have been partially characterized with respect to traits such as host range, virion morphology, transduction efficiency, and *in vivo* efficacy. Isolated but not sequenced *B. thailandensis* and *B. pseudomallei* phages include A through H (Smith and Cherry, 1957), ϕC2, ϕC13, and ϕC32 (Sariya *et al.*, 2006), ST2, ST7, ST70, ST88, and ST96 (Yordpratum *et al.*, 2010). Numerous *B. thailandensis* and *B. pseudomallei* phages have also been reported in the French and Russian literature (Leclerc and Sureau, 1956; Denisov and Kapliev, 1991 and

1995; Grishkina and Merinova, 1993; Manzeniuk *et al.*, 1993 and 1994; Iliukhin *et al.*, 2011).

As discussed above, phages BGPP-Ar and BGPP-Sa can infect strains of both *B. glumae* and *B. plantarii*, while BPP-Ya infects only *B. plantarii* (Adachi *et al.*, 2012). *B. glumae* prophage Bglu421 and Bcc phages BcepBrny, BcepEtu, and BcepFife have been sequenced, but these data sets are not publicly available (Mera, 2006; Robinson *et al.*, 2006). *B. gladioli* phages have been described in a Russian report (Iliukhin *et al.*, 2011) and select Bcc phages have been found to infect strains of this species (Langley *et al.*, 2003; Seed and Dennis, 2005). Other *Burkholderia* species in the NCBI Taxonomy Browser (including *B. acidipaludis, B. andropogonis, B. bannensis, B. bryophila, B. caledonica, B. caribensis, B. caryophylli, B. endofungorum, B. ferrariae, B. fungorum, B. ginsengisoli, B. glathei, B. graminis, B. heleia, B. hospita, B. kururiensis, B. megapolitana, B. mimosarum, B. nodosa, B. oklahomensis, B. oxyphila, B. phenazinium, B. phenoliruptrix, B. phymatum, B. phytofirmans, B. rhizoxinica, B. sabiae, B. sacchari, B. sartisoli, B. sediminicola, B. silvatlantica, B. singaporensis, B. soli, B. sordidicola, B. terrae, B. terricola, B. tropica, B. tuberum, B. unamae* and *B. zhejiangensis*) and several more-recently named species within the Bcc (*B. arboris, B. contaminans, B. diffusa, B. lata, B. latens, B. metallica, B. seminalis* and *B. ubonensis*) have not yet been characterized with respect to their phage susceptibility or content.

As in many bacteria, prophages can be found in the genomes of most fully sequenced *Burkholderia* species (excluding *B. mallei*). Prophage genes have been identified in Bcc transcriptome studies and exhibit dynamic transcription within the host cell in the environment, including in response to stressors such as hydrogen peroxide (O'Grady *et al.*, 2009; Nishiyama *et al.*, 2010; Peeters *et al.*, 2010). Although phage gene content varies between strains, certain isolates (such as *Burkholderia multivorans* ATCC 17616 and *B. cenocepacia* J2315) have been found to be polylysogenic (Summer *et al.*, 2004; Mera, 2006; Goudie *et al.*, 2008; Lynch *et al.*, 2010b). Viable phages have been isolated and sequenced from *B. thailandensis, B. pseudomallei*, and Bcc lysogens (Woods *et al.*, 2002; DeShazer, 2004; Summer *et al.*, 2004; Goudie *et al.*, 2008; Ronning *et al.*, 2010; Lynch *et al.*, 2010a,b), while phage genes and phage-like genomic islands have been identified in the genomes of both these species and *B. xenovorans* (Holden *et al.*, 2004, 2009; Yu *et al.*, 2006; Chain *et al.*, 2006). Furthermore, the majority of the sequenced phages discussed below are temperate (confirmed or putatively). These phages were either isolated from *Burkholderia* lysogens or isolated from environmental samples and then found to carry an integration module.

Burkholderia phage genomes

Taxonomy

As discussed above, tailed phages are taxonomically classified as members of the order *Caudovirales* and families *Myoviridae, Podoviridae*, or *Siphoviridae* (Ackermann, 2003). However, at the lower levels of phage taxonomy, there has been some debate as to the most effective way to classify phages. The difficulty arises due to the mosaic nature of bacteriophage genomes, where different modules within the same genome may have been acquired from a variety of phage lineages (Hendrix *et al.*, 1999). As a result, classical taxonomy – for which the hierarchy relies on predominantly vertical gene transfer – is less appropriate for phages than it is for other biological groups (Lawrence *et al.*, 2002). Several alternative taxonomic approaches have been proposed (Lawrence *et al.*, 2002; Rohwer and Edwards 2002; Lavigne *et al.*, 2008, 2009). In this chapter, *Burkholderia* phages will be categorized based on the system proposed by Lavigne *et al.* (2008, 2009) for classification of fully sequenced *Podoviridae* and *Myoviridae*. This system uses the comparison program CoreGenes [or the updated program CoreGenesUniqueGenes (CGUG)] to identify sets of proteins shared (with a BLASTP score of at least 75) between a genome of interest and a reference genome (Mahadevan *et al.*, 2009). Two taxonomic levels may be identified in this analysis, either genus (if the genome of interest has at least 40% of its proteins in common with respect to the reference genome) or subfamily (20–39% of its proteins) (Lavigne *et al.*, 2008).

Using this classification system, *Burkholderia* myoviruses and podoviruses can be divided into the following groups (Table 11.1): P2-like myoviruses, Bcep781-like myoviruses, PB1-like myoviruses, BcepMu-like myoviruses, Bcep22-like podoviruses, and BPP-1-like podoviruses (Lavigne *et al.*, 2008 and 2009). Although this classification system has not yet officially been extended to the *Siphoviridae* (although there are plans to do so) (Mahadevan *et al.*, 2009), we can tentatively classify sequenced *Burkholderia* siphoviruses as PA73-like, BcepNazgul-like, ϕE125-like, and BcepGomr-like (Table 11.1). For simplicity, the phages described here with sub-family level relatedness (20–39% correlation for BcepB1A, KS10, BcepC6B, Bcep176, and KS9) are described with the genus to which they are the most similar (Table 11.1).

P2-like myoviruses

The P2-like myoviruses are the largest group of *Burkholderia* phages characterized to date, with at least seven members from *B. thailandensis, B. pseudomallei*, and the Bcc. Enterobacteria phage P2 is a characterized temperate myovirus (part of the *Peduovirinae* subfamily and P2-like viruses genus) (Lavigne *et al.*, 2009) that lysogenizes strains of *E. coli*. A review on P2-like phage evolution by Nilsson and Haggård-Ljungquist (2007) notes that comparative genomics of these phages reveals two commonalities. First, this group of phages tends to maintain the same modules and organization thereof. Second, phages infecting and/or lysogenizing the same organism tend to be more closely related, with structural proteins showing the strongest relatedness, followed by regulatory proteins. Genes may also be identified that are found in only one phage type (Nilsson and Haggård-Ljungquist, 2007). As discussed below, all of these commonalities are true for the P2-like *Burkholderia* phages.

A progressive Mauve comparison of *B. thailandensis* phage ϕE202, *B. pseudomallei* phages ϕ52237, ϕE12–2, and ϕK96243, and Bcc phages KL3, KS14, and KS5 is shown in Fig. 11.1. It is important to note that the major difference in gene organization between ϕE202/ϕ52237/ϕE12–2 (Fig. 11.1, top) and ϕK96243/KL3/KS14/KS5 (Fig. 11.1, bottom) – the placement of the head morphogenesis module at the left and right end of the genome, respectively – is due to the latter sequences representing the integrated prophage gene arrangement (or a syntenic arrangement for prophage KS14). These phages encode similar modules to P2 and maintain them in a similar order. The first gene in the ϕE202/ϕ52237/ϕE12–2 annotations is the small terminase subunit gene *M*. Hypothetically, if the *cos* sites were annealed in P2 to make the genome circular, key modules would be ordered as follows, starting at *M*: head morphogenesis-integration-tail morphogenesis-lysis (Nilsson and Haggård-Ljungquist, 2007). These modules (excluding integration for ϕE12–2 and KS14) are found in the same order in ϕE202, ϕ52237, ϕE12-2, and the KS14 prophage and in a cyclical permutation of this order in the ϕK96243, KL3, and KS5 prophages.

As predicted, the P2-like *Burkholderia* phages are most closely related to other P2-like *Burkholderia* phages and predicted prophage elements. For example, when the putative ϕK96243 sequence is analysed in a BLASTN search, the closest matches when sorted by query coverage (those over 50%) are the P2-like phages described here and uncharacterized sequences from *B. pseudomallei, B. glumae, B. thailandensis, B. cenocepacia*, and *B. rhizoxinica*. These results suggest that P2-like prophages and prophage elements are likely to be widely distributed throughout the entire *Burkholderia* genus.

CoreGenes analysis indicates that the P2-like *Burkholderia* phages encode proteins similar to approximately half (48.84–55.81%) of those in P2 (Table 11.1). These proteins are predominantly structural. CoreGenes comparison of P2, ϕE202, ϕ52237, ϕE12–2, and ϕK96243 indicates that these five phages share 19 homologues from three distinct modules (head morphogenesis, lysis, and tail morphogenesis), from left to right in P2: portal vertex protein Q, terminase large subunit P, capsid-scaffolding protein O, capsid protein N, terminase small subunit M, capsid completion protein L, tail protein X, lysis proteins LysB and LysC, baseplate assembly protein V, baseplate wedge subunit W, baseplate assembly proteins J and I, major tail sheath protein FI, major tail tube protein FII, tail protein E, tail tape measure protein T, and tail proteins U and D. The only P2 structural proteins not included here are the tail completion proteins R and S, tail fibre protein

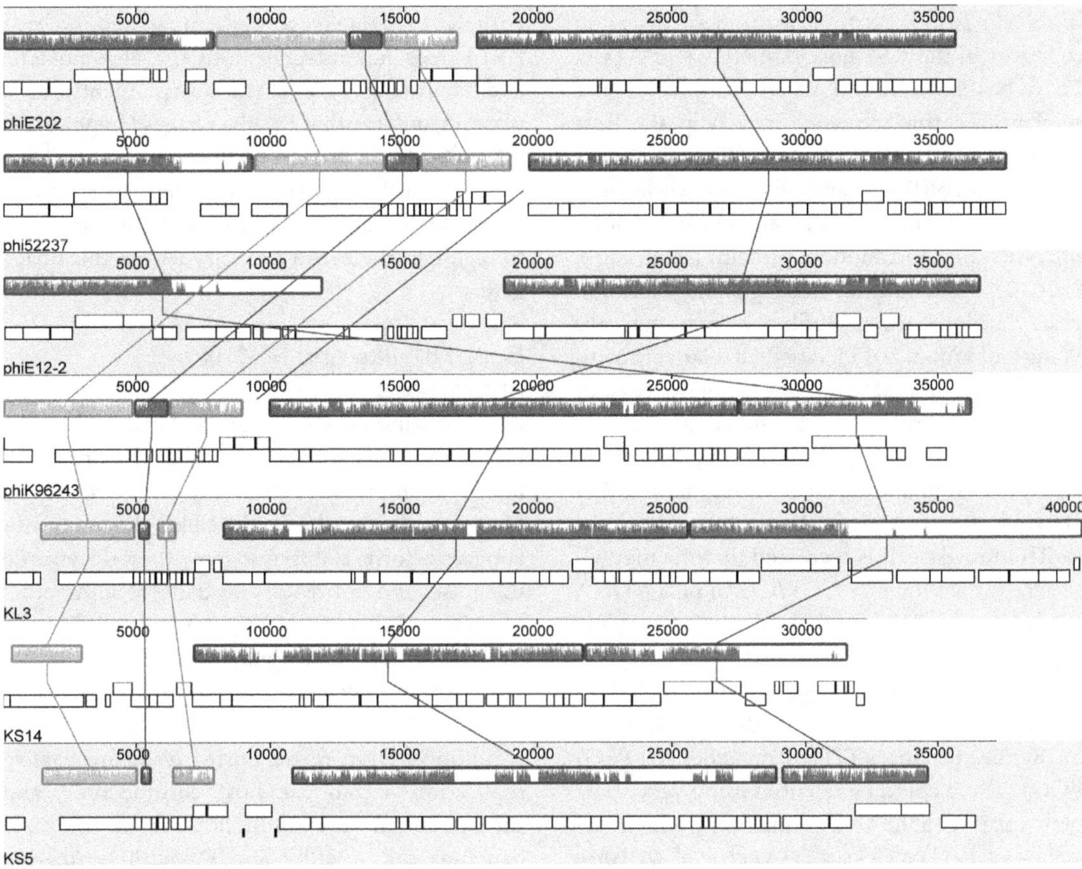

Figure 11.1 Whole genome alignment of P2-like *Burkholderia* myoviruses. Alignments were performed using default parameters for the Mauve plugin (progressive Mauve algorithm) in the Geneious package (Darling *et al.*, 2010; Drummond *et al.*, 2012). A colour version of this figure is located in the plate section at the back of the book.

H, tail fibre assembly protein G, and tail protein E + E' (R and E + E' are absent in ϕ52237, R and S are absent in ϕE12–2, and H and G are absent in all four phages). A similar analysis of P2, KL3, KS5, and KS14 indicates that these four phages share 20 homologues: the list is identical to the one above, except that the Bcc phages all encode R, S, and E + E′ homologues, KS14 encodes a G homologue, and KS5 lacks LysB and LysC homologues. The only structural proteins not shared within either of these groups (excluding KS14) are the H and G tail fibre-related proteins, which would be expected as these proteins recognize specific structures on the surface of the host cell which would be different between *E. coli* and various *Burkholderia* species (Lynch *et al.*, 2010b).

The comparisons above reveal that the *Burkholderia* phages encode homologues of P2 structural proteins, which must be maintained as a unit to preserve protein–protein interactions for virion morphogenesis, while encoding dissimilar accessory proteins (Lynch *et al.*, 2010b). Four examples of such accessory proteins (or sets thereof) have been identified in KS5, ϕK96243, ϕE12–2, and KL3. First, KS5 encodes a reverse transcriptase (Lynch *et al.*, 2010b). In P2-like prophage P2-EC30, a reverse transcriptase mediates exclusion of T5 infection (Odegrip *et al.*, 2006) and this protein may serve an analogous function within KS5 lysogens. Second, similar to ϕE125 and Bcep22, KS5 carries an insertion sequence (Woods *et al.*, 2002; Lynch *et al.*, 2010b;

Gill *et al.*, 2011). As transposition of insertion sequences in the KS5 host *B. multivorans* ATCC 17616 has been found to increase at higher temperatures, this sequence may help the host to adapt upon human infection (Ohtsubo *et al.*, 2005). Third, ϕK96243 and ϕE12–2 encode putative type II toxin–antitoxin addiction modules with both killer and antidote proteins (Ronning *et al.*, 2010), potentially facilitating phage exclusion or the development of biofilms or persister cells (Wang and Wood, 2011). Finally, a relatively complex module of putatively interacting accessory proteins was found in KL3 (Lynch *et al.*, 2010b). Here, we identified a module involved in phage DNA protection and repair encoding an EcoRII-C restriction endonuclease, Vsr endonuclease, and EcoRII methylase. It is predicted that the methylase acts at the sequence CCWGG in phage DNA (producing $CC^{m}WGG$, where C^{m} is 5-methylcytosine), protecting it from EcoRII–C cleavage of host or superinfecting phage DNA at CCWGG sites. Deamination of this 5-methylcytosine base to a thymine produces T/G mismatches (C[T/G]WGG). These could be repaired using very short patch repair with the Vsr endonuclease, which recognizes C(T/G)WGG sites (Lynch *et al.*, 2010b). An analogous module encoded by the unrelated siphovirus AH2 is discussed below.

Another feature distinguishing P2 and the P2-like *Burkholderia* phages are the sites used for prophage integration. P2 can use numerous *attB* sites in *E. coli*, none of which are reported to be in tRNA genes (Barreiro and Haggård-Ljungquist, 1992). In contrast, P2-like phages in *Burkholderia* almost exclusively use such sites. ϕE202, ϕ52237, and ϕK96243 are predicted to integrate at a phenylalanine tRNA gene (Holden *et al.*, 2004; Ronning *et al.*, 2010), while KL3 integrates at a threonine tRNA gene (Lynch *et al.*, 2010b). ϕE12–2 is predicted to integrate into a noncoding region, although the mechanism is unknown because the annotation for this phage does not include an integrase gene (Ronning *et al.*, 2010). KS5 integrates into an AMP nucleosidase gene, a previously uncharacterized prophage integration site (Lynch *et al.*, 2010b). Finally, KS14 is notable as it does not encode a tyrosine recombinase and its serine recombinase lacks features conserved in serine integrases. By performing plasmid isolation with KS14-lysogenized cells, it was found that KS14 does not integrate into the chromosome and is instead maintained independently as a plasmid, unlike other P2-like phages (Lynch *et al.*, 2010b). Although each of these phages shares a similar morphology, such differences in accessory gene content and mechanisms for lysogeny serve to highlight the broad diversity within this phage genus.

Bcep781-like and PB1-like myoviruses

A key distinguishing factor of the Bcep781-like myoviruses is that they are one of the only published examples of strictly lytic *Burkholderia* phages (although the Bcep22-like phages do not appear to form stable lysogens, they do encode integrase and repressor proteins) (Summer *et al.*, 2006; Gill *et al.*, 2011; Lynch *et al.*, 2012a). In general, temperate phages are thought to be more common because they enjoy the evolutionary benefits of protection from environmental conditions, DNA repair during lysogeny, passive replication within the host chromosome, and the potential for induction from multiple daughter cells (Miller and Day 2008; Abedon 2009). In *Streptococcus pneumoniae*, lytic phages are less commonly isolated and only two have been genomically characterized (Martin *et al.*, 1996; Sabri *et al.*, 2011). In an extreme case, all *Clostridium difficile* phages isolated to date have been found to be temperate (Meader *et al.*, 2010). The low prevalence of lytic *Burkholderia* phages (particularly those with a broad host range) has prompted the development of engineered lytic derivatives of temperate phages for potential clinical use (Lynch *et al.*, 2010a).

The Bcep781-like myovirus genus comprises Bcep781, Bcep43, Bcep1, BcepNY3, and the more distantly related *Xanthomonas* phage OP_2 (Lavigne *et al.*, 2009). Little information is available regarding BcepNY3 aside from its GenBank record and an undergraduate thesis (Orchard, 2008), but a detailed analysis of the other three Bcc phages is provided by Summer *et al.* (2006). Bcep781, Bcep43, and Bcep1 were all originally isolated from soil samples in different locations of New York state: muck soil in 1978, muck soil in 1999, and soil in 1999, respectively. Each of these phages has a narrow host range, infecting only one or two Bcc strains. Headful packaging is predicted to be used as the Bcep781 genome was found to be cyclically permuted and minimally terminally redundant.

The genomes of the Bcep781-like myoviruses (including BcepNY3) are very similar with respect to size, GC content, organization, and sequence. The sizes differ by less than one kbp (between 47,382 bp and 48,247 bp), while the GC content ranges between 63.3% and 63.6% (Table 11.1). These phages encode 65–71 proteins, and with the exception of Bcep1, one tRNA (serine for Bcep781, leucine for Bcep43, and tyrosine for BcepNY3). As can be seen in Fig. 11.2, the proteins are arranged in a similar order and with a high coding density. Two regions are transcribed from the forward strand and two from the reverse strand (one gene is also found on the opposite strand in the head morphogenesis module). Between these regions are two low-GC (~30%) segments containing multiple promoters, predicted to allow transcription from both strands (Summer *et al.*, 2006).

Bcep781-like phages are very closely related to one another (87.4–97.6% nucleotide identity for Bcep781, Bcep43, and Bcep1), but show minimal relatedness to other phages with four exceptions (Summer *et al.*, 2006). First, a shorter but related uncharacterized prophage element was identified in *Photorhabdus luminescens* TT01, evidence of module assortment both between different genera and between lytic and temperate phages. Second, proteins of the *Xanthomonas oryzae* pv. *oryzae* phage OP_2 and the *Acinetobacter baumannii* phage AB1 show similarity to several proteins from these phages (Inoue *et al.*, 2006; Li *et al.*, 2012). The fourth related phage, BcepB1A, is discussed below. Variations in the nucleotide sequence of Bcep781, Bcep43, and Bcep1 were found to be caused by single nucleotide polymorphisms (mostly synonymous in coding regions) and some short insertions and deletions. CoreGenes analysis of the Bcep781-like sequences including BcepNY3 indicates that homologues of the following Bcep781 proteins are found in all four Bcep phages: gp02–gp14, gp16–gp28, gp30–gp38, gp40–gp42, gp44–gp58, gp60, and gp62–gp66, a total of 59 proteins.

Each phage contains distinct modules for (from left to right) head morphogenesis, lysis, and tail morphogenesis (Summer *et al.*, 2006).

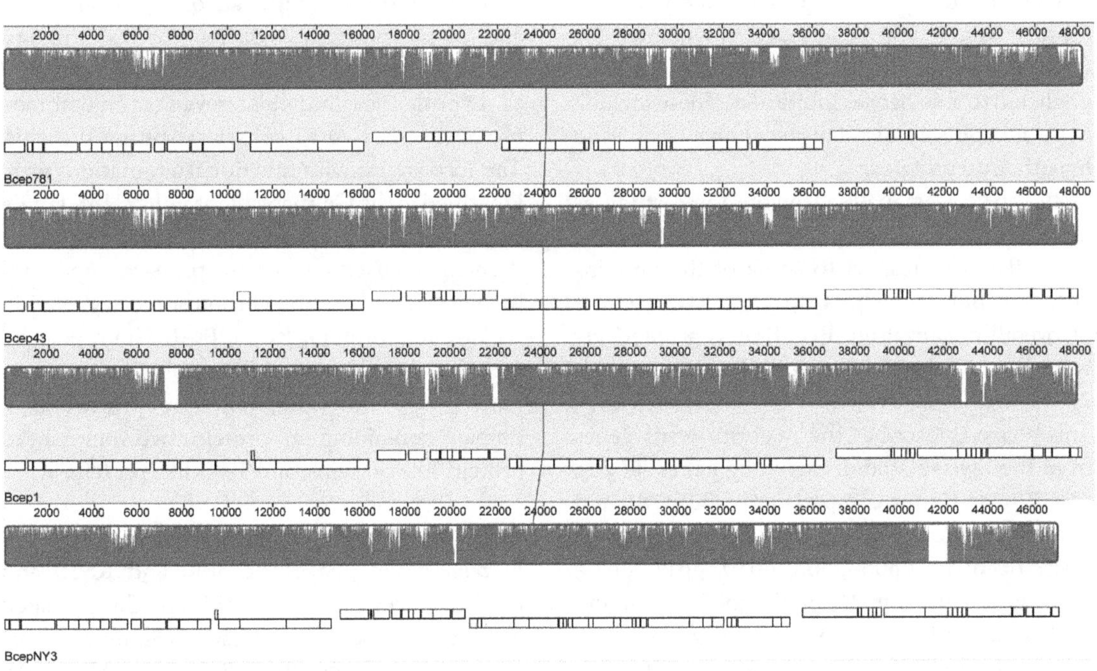

Figure 11.2 Whole genome alignment of Bcep781-like myoviruses.

In addition, DNA binding proteins – including a putative adenine methylase, DNA polymerase III β-clamp (truncated), resolvase, helicase, and DNA polymerase I – are dispersed throughout the genome. Although originally annotated as VirE, Bcep781 gp52 and similar proteins may instead be primases (Summer *et al.*, 2006 and 2007; Orchard, 2008). In the head morphogenesis module, the major capsid protein, decorator protein, scaffold protein/prohead protease, minor head proteins, and terminase large subunit were identified. The terminase small subunit is expected to be present, but could not be identified based on similarity searches. The lysis module is comprised of a holin (transcribed from the reverse strand), endolysin, Rz, and Rz1 (all transcribed from the forward strand), interspersed with two additional hypothetical proteins. To functionally characterize the Bcep781 and Bcep43 endolysins (which have the same sequence), the gene was expressed in *E. coli* and cell membranes were disrupted with chloroform. Absorbance decreased for the cells expressing the endolysin, but not for the vector control, indicating that this gene was correctly annotated and functional. In the tail morphogenesis module, transcribed from the reverse strand, most of the annotated proteins are distantly related to the tail proteins of myoviruses P2 and Mu. Using an interactive sequence tag (IST) library screen, ten Bcep781 proteins were predicted to associate as multimers. These include head morphogenesis, tail morphogenesis, and hypothetical proteins.

BcepB1A was analysed in the same study as the Bcep781-like Bcep phages and was found to be similar with respect to some of the proteins encoded, but not with respect to GC content or overall organization. BcepB1A is a putatively lytic phage isolated from soil (Robinson *et al.*, 2006) with a relatively low GC content, 54.5%. This phage transcribes the majority of its genes from the reverse strand, excluding genes 37 and 66–72. CoreGenes comparison of Bcep781 and BcepB1A indicates that they share fourteen proteins: BcepB1A gp06–gp08, gp10, gp11, gp14, gp17, gp21, gp23, gp26, gp27, gp63, gp65, and gp67 (Bcep781 gp49, gp32, gp33, gp36, gp38, gp44, gp04, gp08, gp12, gp17, gp16, gp59, gp58, and gp20, respectively). These proteins are not part of a single module, nor are they maintained in the same order as in Bcep781. Functionally, these proteins are predicted to be involved in head and tail morphogenesis and helicase activity. In addition to the Bcep781-like phages, BcepB1A also shows some similarity to the podoviruses Bcep22 and BcepC6B (Summer *et al.*, 2006) and to the *Agrobacterium* myovirus 7–7–1 (Kropinski *et al.*, 2012).

In the classification scheme of Lavigne *et al.* (2009), BcepB1A is not identified as a Bcep781-like phage, but as related to *Pseudomonas* phage PB1-like viruses, along with the lytic soil phage BcepF1 (Fig. 11.3) (Robinson *et al.*, 2006; Ceyssens *et al.*, 2009). BcepF1 is the largest *Burkholderia* phage sequenced to date at 72,415 bp. This genome sequence has not been formally described in a publication, but the GenBank sequence record indicates that BcepF1 infects *B. ambifaria* and encodes 127 proteins. Two regions are transcribed from the forward strand and two from the reverse strand. In CoreGenes analysis, BcepF1 shares 40 proteins with PB1, including head and tail morphogenesis, lysis, DNA binding, and hypothetical proteins. In contrast, 23 proteins are shared between either PB1 or BcepF1 and BcepB1A (two different sets of 23 proteins, but with most overlapping), all of which are head and tail morphogenesis or hypothetical proteins. Out of 127 BcepF1 proteins, 103 were annotated as hypothetical proteins, seven as membrane proteins, and two as conserved phage proteins. The 15 proteins with functional annotations were identified in the GenBank file as HflC/HflK family inner membrane lipoprotein, ParB-like nuclease domain, archaeo-eukaryotic primase superfamily helicase, eukaryotic chromosome segregation ATPase-like protein, RecD, PseT, Gram-positive type DNA polymerase III α subunit, two spanins, phage endolysin, tail fibre protein, SLT domain-containing tail protein, two minor head protein-like proteins, and TerL-like protein.

BcepMu-like myoviruses

B. cenocepacia phages BcepMu and KS10 and *B. thailandensis* phage ϕE255 are three related temperate *Burkholderia* phages (Summer *et al.*, 2004; Goudie *et al.*, 2008; Ronning *et al.*, 2010). Each of these phages has a different distribution

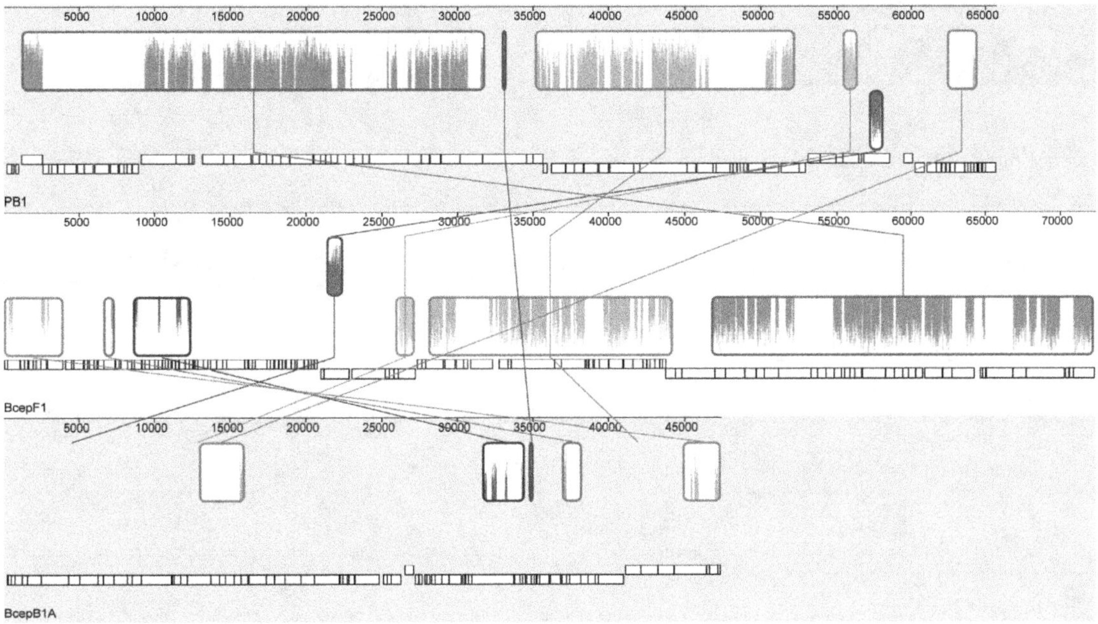

Figure 11.3 Whole genome alignment of PB1-like myoviruses. A colour version of this figure is located in the plate section at the back of the book.

in *Burkholderia,* with BcepMu lysogenizing and infecting ET12 *B. cenocepacia,* KS10 lysogenizing *B. cenocepacia* and infecting *B. cenocepacia, B. stabilis* and *B. ambifaria,* and ϕE255 lysogenizing *B. thailandensis* and infecting *B. mallei.* Although these phages are transposable and share some similar proteins with enterobacteria phage Mu, they are not strictly Mu-like based on CoreGenes analysis. Most of the homologues shared between Mu and either BcepMu or ϕE255 are the same: Mor, putative portal protein, virion morphogenesis late F open reading frame (ORF), putative virion morphogenesis protein, tail fibre assembly protein (in ϕE255 only), and five hypothetical proteins. In contrast, KS10 encodes 13 proteins similar to Mu, but only three in common with BcepMu and ϕE255 (virion morphogenesis late F ORF and two hypothetical proteins). Other KS10 proteins similar to those of Mu include transposase, DNA transposition protein, major tail subunit, putative tape measure protein, putative DNA circulation protein, putative tail protein, putative baseplate assembly protein, and three hypothetical proteins. Unlike the P2-like *Burkholderia* phages that all shared conserved structural genes, BcepMu and ϕE255 encode similar morphogenesis proteins that are dissimilar to those of both Mu and KS10. Comparing BcepMu and ϕE255 with CoreGenes, these phages share 46 proteins (86.79% correlation) in modules for transposition, transcriptional regulation, lysis, head morphogenesis, and tail morphogenesis. Comparing BcepMu and KS10, these phages only share 16 proteins (30.19% correlation), most of which are hypothetical (excluding an HU DNA binding protein, endolysin, virion morphogenesis protein, capsid protein, and baseplate assembly protein).

In addition to differences in gene content, these phages also show differences in organization. Although the gene order of BcepMu and ϕE255 appears dissimilar in the progressive Mauve comparison (Fig. 11.4), these differences are an artefact of the ϕE255 sequence record in which the left and right genome ends are found near the middle of the sequence (Ronning *et al.,* 2010). In both phages, the transposition module is found near the left end, followed by the lysis, head, and tail morphogenesis modules. A similar organization is seen in the related prophage-like sequences of SalMu in *Salmonella typhi,* ChromoMu of *Chromobacterium violaceum,* and PhotoMu of *Photorhabdus luminescens* (Summer *et al.,* 2004).

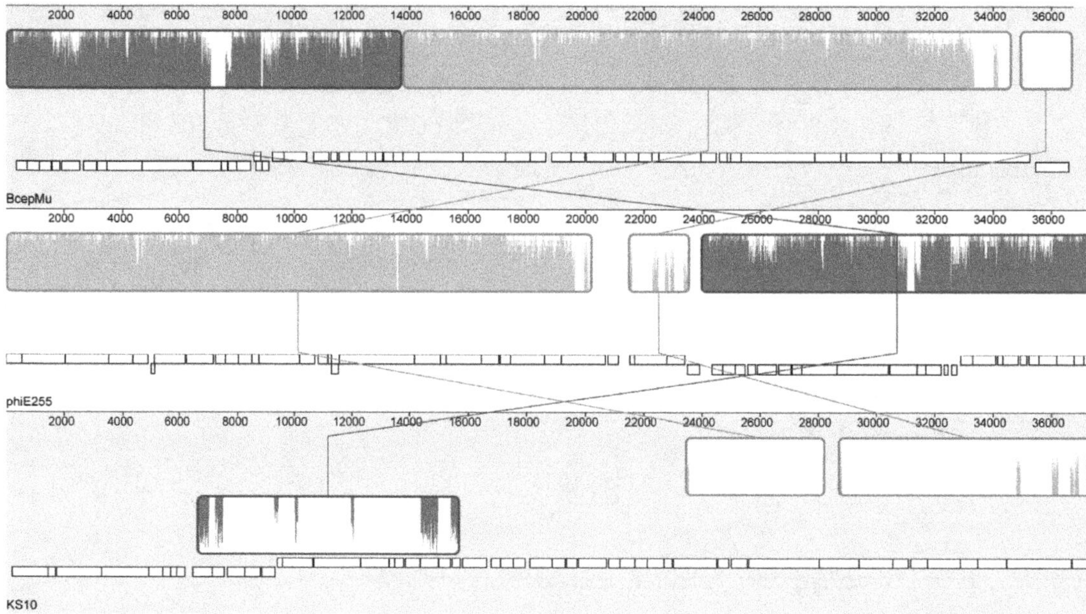

Figure 11.4 Whole genome alignment of BcepMu-like myoviruses. A colour version of this figure is located in the plate section at the back of the book.

In contrast, the organization of KS10 is genuinely dissimilar from that of BcepMu and ϕE255. Here, the head morphogenesis genes are separated into two distinct modules, one on the far left end and one near the centre. The lysis and transposition modules are located between the two head morphogenesis regions and, similar to BcepMu and ϕE255, the tail morphogenesis module is found at the right end. Because the head morphogenesis proteins interact closely with one another, it is extremely uncommon to find these genes spatially separated from each other in the sequence (where they would tend not to re-assort as a single unit) (Goudie *et al.*, 2008).

BcepMu-like phages encode four putative virulence factors, although the effect of these proteins on pathogenicity is currently unknown. KS10 lacks these genes and lysogeny by this phage does not increase *B. cenocepacia* virulence in the *G. mellonella* model (Goudie *et al.*, 2008). In BcepMu, a putative ExeA-like type II secretion protein (gp8) and lipopolysaccharide acyltransferase (gp53) were identified in the original annotation (Summer *et al.*, 2004). However, it was noted in a subsequent review that the putative secretion protein may instead be a transposase B subunit with no effect on pathogenicity (Summer *et al.*, 2007). Two additional proteins were identified by Ronning *et al.* (2010). In ϕE255, a region containing five genes not found in BcepMu encodes a putative ATP-binding cassette transporter protein, gp29. Although it is unknown if this protein acts as an importer or exporter, this class of protein can facilitate antibiotic resistance in other species (Michalet and Dijoux-Franca, 2009). Finally, a lytic transglycosylase has been identified in both BcepMu and ϕE255. Although lytic transglycosylases can be involved in processes such as secretion or colonization, analysis of the BcepMu sequence suggests that this protein acts as the endolysin and not as a virulence determinant (Summer *et al.*, 2004; Zahrl *et al.*, 2005; Stapleton *et al.*, 2007). Although KS4-M (a variant of BcepMu) has been shown to be effective *in vivo* (Seed and Dennis, 2009), it remains to be determined if any of these proteins may have a detrimental effect during clinical use.

Bcep22-like podoviruses

As is true for other species, podoviruses are the least commonly isolated *Burkholderia* phage morphotype (Ackermann, 2003). Most of these phages are Bcep22-like and are part of a newly characterized group of closely related Bcc-specific phages. These include Bcep22 (isolated from New York soil), BcepIL02 (isolated from Illinois maize soil), DC1 (isolated from Alberta *Dracaena* sp. soil), and BcepMigl (isolated from soil) (Gill *et al.*, 2011; Lynch *et al.*, 2012a). Characterization of the first three phages has been published, but information regarding BcepMigl is limited to its GenBank record. Bcep22 and BcepIL02 infect select *B. cenocepacia* strains, while DC1 infects strains of *B. cepacia*, *B. cenocepacia*, and *B. stabilis*. The relatively broader host range of DC1 could be beneficial if this group of phages is to be developed for therapeutic use. The lytic or temperate nature of these phages is currently undefined. All four phages encode putative repressors (capable of multimerization based on IST analysis in Bcep22) (Gill *et al.*, 2011) and integrases: tyrosine recombinases in BcepIL02, BcepMigl, and DC1 and a serine recombinase at the same locus in Bcep22. However, Bcep22, BcepIL02, and DC1 have not been found to stably lysogenize Bcc hosts, potentially resulting from a defect in integration (Gill *et al.*, 2011; Lynch *et al.*, 2012a).

Bcep22 and BcepIL02 are cyclically permuted and predicted to use headful packaging (Gill *et al.*, 2011), which is likely true for DC1 and BcepMigl as well. The genomes are arranged into modules involved in DNA replication, head morphogenesis, tail morphogenesis, and lysis, interspersed with many hypothetical proteins. These phages encode at least two potential lysogenic conversion genes, including phosphoadenosine phosphosulphate (PAPS) reductase (in all four phages) and PagP (in BcepIL02 and DC1) (Summer *et al.*, 2007; Gill *et al.*, 2011). PAPS reductase (also found in BcepB1A, ϕE125, and ϕ644-2) is predicted to facilitate survival in sulphur-poor conditions, as it is part of the pathway that converts sulphate to thiosulphate (Schiff, 1979; Woods *et al.*, 2002; Summer *et al.*, 2007; Ronning *et al.*, 2010). Iron uptake is also inhibited under these conditions, as pyochelin siderophore production decreases (Farmer and Thomas, 2004). PagP is a lipid A palmitoyltransferase that, in other species, has been associated with cationic antimicrobial peptide resistance and inhibition of TLR4 signalling (Guo *et al.*, 1998; Kawasaki *et al.*, 2004).

Similarities in organization and sequence of the Bcep22-like phages are represented in a progressive Mauve alignment, where almost the entirety of each genome is aligned in a single locally collinear block (LCB, Fig. 11.5). A short section at the beginning of the DC1 genome is linked to the end of the other three genomes due to a revision in the BcepIL02 start site annotation (DC1 bp 1 corresponds to BcepIL02 bp 1 in accession FJ937737.1 but bp 62,658 in the revised FJ937737.2). CoreGenes analysis of these four phages indicates that their ~ 62–64 kbp genomes have 48 – or approximately two-thirds – of their proteins in common. These ORFs include a number of interesting products such as a nuclease/RecT pair, carbon storage regulator, multiple (3–4) similar tail fibre proteins, acetyltransferase, PAPS reductase, and signal anchor release (SAR) endolysins, in addition to a serine tRNA gene.

One of the most interesting proteins found in all four Bcep22-like phages is an extremely large multidomain ORF encoded near the right end of the genome. This protein ranges between 4602 anf 4667 amino acids in length (gene length of 13,809 bp to 14,004 bp) and is virion-associated in Bcep22 and BcepIL02 (Gill *et al.*, 2011). Within this protein, several distinct domains can be identified including soluble lytic transglycosylase, methylase, and helicase. This protein is predicted to prevent bacterial restriction of phage DNA similar to enterobacteria phage P1 DarB, a virion-associated protein with a relatively large size and a similar domain organization (with respect to the methylase and helicase domains) (Iida *et al.*, 1987; Gill *et al.*, 2011).

Both physical and genomic characteristics of Bcep22-like phages make them excellent potential candidates for therapeutic use. These phages do not stably lysogenize Bcc strains (Gill *et al.*, 2011; Lynch *et al.*, 2012a), may have a broad host range in the case of DC1 (Lynch *et al.*, 2012a), and BcepIL02 has been shown to be active against *B. cenocepacia* in a mouse model of infection (Carmody *et al.*, 2010). Furthermore, the presence of a CsrA-like carbon storage regulator, previously shown to have anti-biofilm activity in *E. coli*, *Citrobacter freundii*, and *Salmonella enterica* sv. Typhimurium (Jackson *et al.*, 2002), may allow these phages to be more effective against

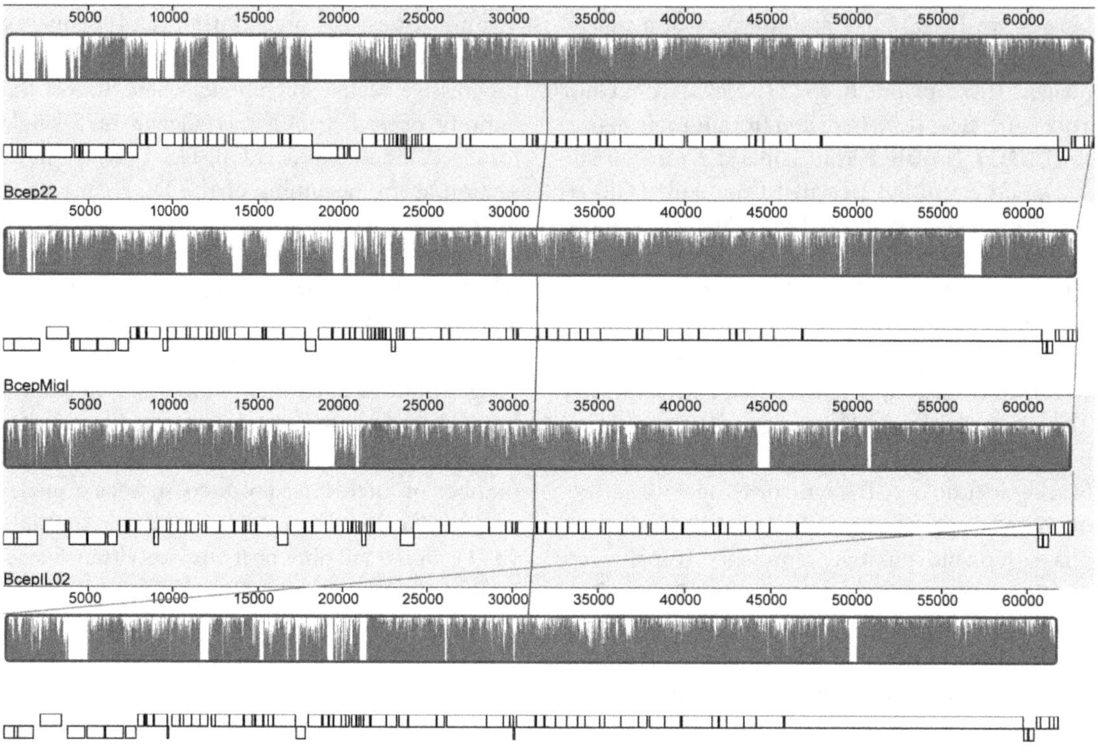

Figure 11.5 Whole genome alignment of Bcep22-like podoviruses.

Bcc biofilms in the CF lung (Lynch *et al.*, 2012a). Although the lipid A palmitoyltransferase is a potential virulence factor, the effect of this protein in *Burkholderia* has not been studied and the absence of stable lysogeny makes it less likely that expression from a prophage could occur.

BPP-1 like podoviruses

BcepC6B is the only *Burkholderia* podovirus characterized to date with similarity to *Bordetella* phage BPP-1. The most notable trait of BPP-1 is its ability to modify its host range using a reverse transcriptase. The BPP-1 tail fibre tip gene *mtd* (major tropism determinant) contains a 134 bp region called the variable repeat (Liu *et al.*, 2002; Dai *et al.*, 2010). An almost identical region called the template repeat is transcribed into mRNA, reverse transcribed into cDNA, and altered at adenine bases to generate a novel sequence that can be inserted at the variable repeat to generate unique tail fibre tip proteins (Liu *et al.*, 2002; Doulatov *et al.*, 2004). Based on CoreGenes analysis, BcepC6B encodes a similar tail fibre protein, but not the reverse transcriptase or Mtd. The homologues found in both phages encode head and tail morphogenesis proteins, DNA binding proteins (including a repressor, primase/helicase, helicase, and DNA polymerase), and hypothetical proteins. The organization of these two sequences is different in the progressive Mauve alignments because the BPP-1 sequence begins at the *att* site (Fig. 11.6). Both phages also show limited similarity to ε15, a *S. enterica* podovirus in the ε15-like phage genus (Kropinski *et al.*, 2007; Lavigne *et al.*, 2008). Walmagh *et al.* (2013) functionally characterized the gp22 endolysin of BcepC6B. This protein was predicted to be a lytic transglycosylase based on BLASTP analysis. In an analogous fashion to the assay described above for the Bcep781/Bcep43 endolysin, gp22 expressed and purified from *E. coli* was able to decrease the absorbance of *P. aeruginosa*, *S. enterica* sv. Typhimurium, and *E. coli* cells with chloroform-disrupted membranes (Walmagh *et al.*, 2013).

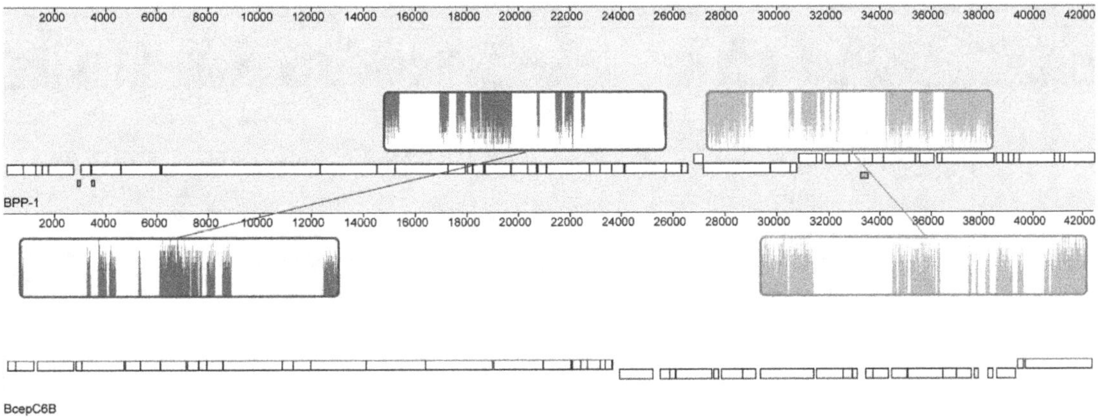

Figure 11.6 Whole genome alignment of BPP-1-like podoviruses. A colour version of this figure is located in the plate section at the back of the book.

PA73-like and BcepNazgul-like siphoviruses

Although the PA73-like and BcepNazgul-like siphoviruses are genomically distinct, they have been grouped together here because of the phenotypic similarities of two characterized representatives. These two phages, the PA73-like KL1 and the BcepNazgul-like AH2, were isolated from sewage plated on *B. cenocepacia* K56–2 and soil from the plant species *Nandina* plated on *B. cenocepacia* C6433, respectively (Lynch *et al.*, 2012b). Both of these siphoviruses infect only *B. cenocepacia* strains and at low titres show delayed plaque formation at 30°C (> 16 h for the formation of turbid plaques) and no plaque formation at 37°C. This phenotype had not been observed for phages that we had previously characterized and so the genomes were initially predicted to be similar. However, restriction fragment length polymorphism analysis and subsequent whole genome sequencing indicated that the two genomes showed almost no relatedness (Lynch *et al.*, 2012b).

KL1 is closely related to PA73 (*Pseudomonas* phage 73) and to the recently characterized vB_Pae-Kakheti25 (Kwan *et al.*, 2006; Karumidze *et al.*, 2012), but not to any other sequenced phage. A progressive Mauve comparison shows that the PA73 and KL1 genomes are very similar with respect to length and organization and are aligned in a single LCB (Fig. 11.7). Both sequences share a similar length (~43 kbp) and low GC content (54.6% in KL1 and 53.6% in PA73). A *cos* site was not identified in KL1 (Lynch *et al.*, 2012b), so it likely uses a headful packaging mechanism similar to the cyclically permuted vB_Pae-Kakheti25 (Karumidze *et al.*, 2012). CoreGenes analysis indicates that PA73 and KL1 share 42 homologues. Proteins unique to PA73 include a putative peptidyl-tRNA hydrolase (peptide chain release factor) and hypothetical proteins, while proteins unique to KL1 include two proteins in the lysis module and hypothetical proteins. These phages share a modular organization, with regions containing (from left to right) lysis, head morphogenesis, tail morphogenesis, and (with a more diffuse arrangement) DNA binding proteins. These DNA binding proteins include DNA polymerase, DNA polymerase III β subunit, superfamily II helicase/restriction enzyme, exonuclease/recombinase (similar to BcepGomr), transcriptional regulator, primase, Vsr endonuclease, and dCMP deaminase.

AH2 is related in a mosaic fashion to BcepNazgul, a soil phage that infects *B. ambifaria*. Both the length (~ 57–58 kbp) and GC content (61.3% in AH2 and 60.6% in BcepNazgul) of these phages are similar and their *cos* site sequences differ at only one position. A progressive Mauve comparison shows that the sequences of these phages are aligned in a single LCB (Fig. 11.8), but unlike the Bcep22-like (Fig. 11.5) and PA73-like (Fig. 11.7) phages, the regions of similarity within the LCB are discontinuous. CoreGenes analysis shows that

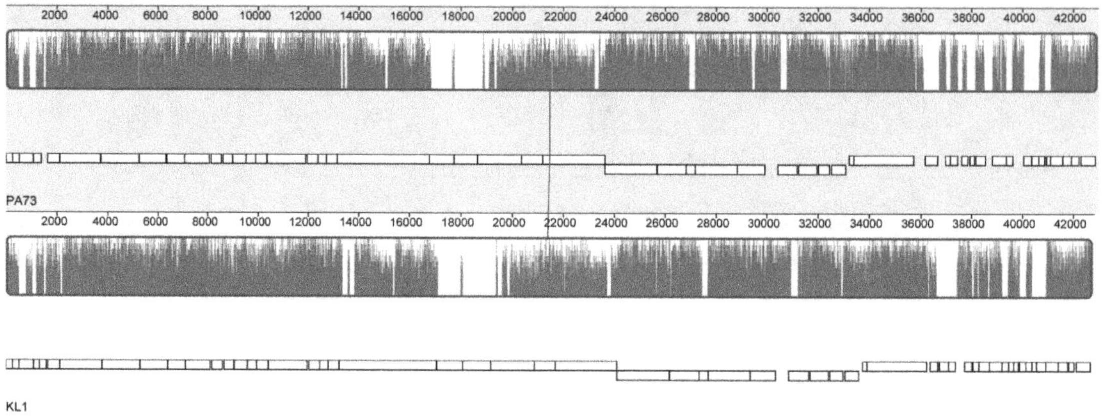

Figure 11.7 Whole genome alignment of PA73-like siphoviruses.

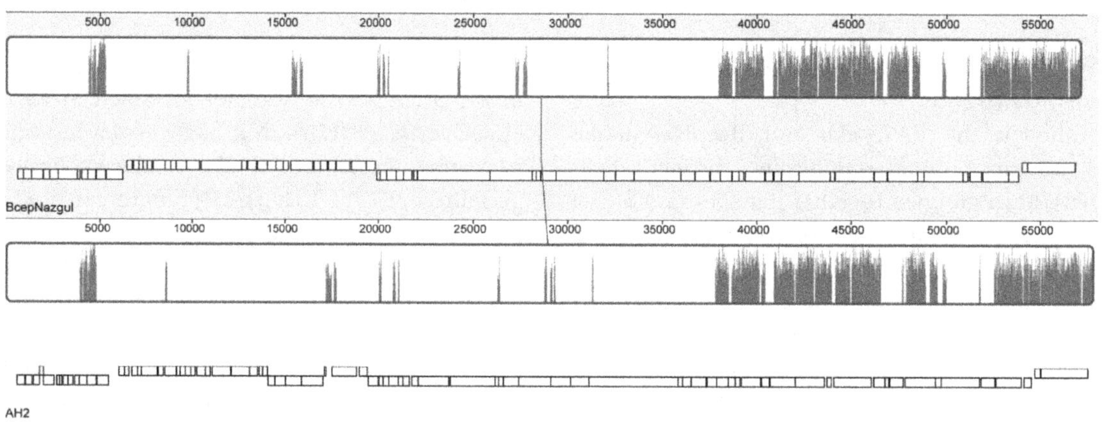

Figure 11.8 Whole genome alignment of BcepNazgul-like siphoviruses.

these phages have 40 proteins in common, over half of their ORFs. These proteins include those in the tail morphogenesis, head morphogenesis, lysis (excluding the endolysin), and DNA binding protein modules, plus several hypothetical proteins. While the putatively temperate AH2 encodes both an integrase and repressor, the putatively lytic BcepNazgul (Robinson *et al.*, 2006) is a mosaic that lacks similar proteins.

An interesting module encoded by AH2 but not BcepNazgul is predicted to be involved in DNA protection and repair. Genes 32–35 encode a Vsr endonuclease, UvrC excinuclease, restriction enzyme, and methylase, respectively (Lynch *et al.*, 2012b). As discussed above for KL3, we predict that AH2 DNA is methylated by the gp35 methylase and that non-AH2 DNA (superinfecting phage or bacterial) is digested by the gp34 restriction enzyme. Methylation-inducted mismatches can then be repaired with the gp32 Vsr endonuclease, while the gp33 UvrC excinuclease may be involved in nucleotide excision repair (Lynch *et al.*, 2012b). Proteins encoded by BcepNazgul but not by AH2 include the endolysin, three conserved tail assembly proteins, Nu3, DNA polymerase I, T4 NrdA.1-like protein, and hypothetical proteins.

The genomes of KL1 and AH2 are predicted to have evolved in a convergent fashion, as they are unrelated but produce nearly identical phage phenotypes (Lynch *et al.*, 2012b). The PA73-like vB_Pae-Kakheti25 is not identified as having this phenotype, as it readily forms clear, turbid, or haloed plaques on a wide range of *P. aeruginosa* strains (Karumidze *et al.*, 2012), while the plaque morphologies of PA73 and BcepNazgul are unknown. The unique delayed plaque development of KL1 and AH2 may be due to virion interaction with a host receptor expressed in stationary phase, extended latent periods, lysis inhibition, or the production of small plaques from which lytic products diffuse (Lynch *et al.*, 2012b). These assumptions were neither confirmed nor refuted following one-step growth curve analysis, as neither phage propagated under the conditions tested (Lynch *et al.*, 2012b). Despite sampling bias that would hamper the isolation of phages with such specific requirements for plaque development, future isolation of similar *Burkholderia* phages may allow us to determine the mechanisms causing this unique phenotype.

ΦE125-like siphoviruses

As a formal CoreGenes classification manuscript has not yet been published for the *Siphoviridae*, the *Burkholderia* phages described here are tentatively identified as the ϕE125-like siphoviruses. Although these phages – including *B. thailandensis* phage ϕE125, *B. pseudomallei* phages ϕ1026b and ϕ644-2, and the less closely related Bcc phages Bcep176 and KS9 – are described as lambdoid (Summer *et al.*, 2007), they cannot be classified as λ-like based on a strict CoreGenes definition as each phage shows less than 15% correlation with λ. When compared as a unit, only six proteins in a single module are shared between the ϕE125-like siphoviruses and λ: the tail proteins H, M, L, K, I, and J. Instead, when the *Burkholderia* phages are compared using ϕE125 as the reference genome, the CoreGenes correlation is greater than 70% for the *B. pseudomallei* phages and ~35–40% for the Bcc phages (Table 11.1). When all five genomes are compared, 18 common proteins are identified belonging to at least three modules: capsid morphogenesis (including the terminase large subunit, portal protein, major capsid protein, and head-tail adaptor), tail morphogenesis (including the major tail subunit, tail assembly chaperone, tail tape measure, two minor tail proteins, two tail component proteins, and tail tip fibre protein), and hypothetical proteins. Although the major capsid proteins of ϕE125 and ϕ1026b are similar

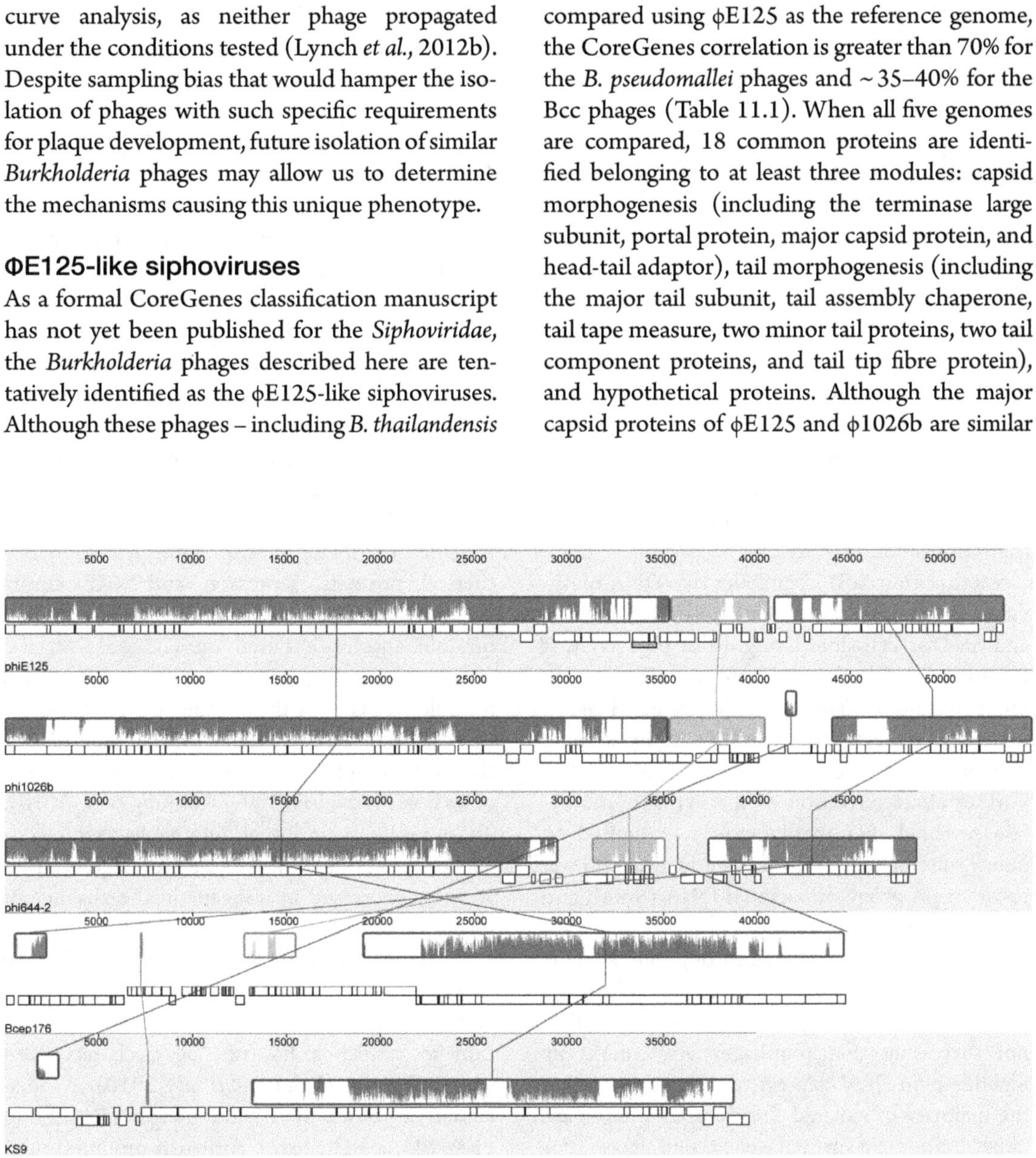

Figure 11.9 Whole genome alignment of ϕE125-like siphoviruses. A colour version of this figure is located in the plate section at the back of the book.

based on CoreGenes analysis, they are sufficiently dissimilar that ϕE125-specific antibodies do not bind to the ϕ1026b capsid (DeShazer, 2004).

As shown in the progressive Mauve comparison, each phage has a similar organization with ϕE125, ϕ1026b, and ϕ644–2 being the most closely related (Fig. 11.9). The red boxes present at the left ends of ϕE125, ϕ1026b, and ϕ644–2 are at the right ends of both Bcep176 (because the sequence is reverse complemented compared to ϕE125) and KS9 (because the gene order represents the KS9 prophage sequence). Based on BLASTN analysis, these phages are not closely related to any previously characterized non-*Burkholderia* phages. Similar to the P2-like phages, the ϕE125-like siphoviruses largely integrate at tRNA genes: proline tRNA for ϕE125 and ϕ1026b and serine tRNA for ϕ644-2 (Woods *et al.*, 2002; DeShazer, 2004; Ronning *et al.*, 2010). However, these integration sites are not conserved for the Bcc phages as Bcep176 and KS9 integrate into an inactivated lysine exporter family pseudogene and a GTP cyclohydrolase II gene, respectively (Lynch *et al.*, 2010a).

Several functional characterization studies have been performed for the ϕE125-like phages, focusing on methylation, repression, prophage transcription, and virulence.

Methylation: ϕE125 encodes two DNA methylases in two separate modules, one for adenine and one for cytosine. Using a dot blot assay, it was found that both virion and *B. mallei* lysogen DNA (but not *B. thailandensis* lysogen or *B. mallei* non-lysogen DNA) contained N^6-methyladenine and N^4-methylcytosine (Woods *et al.*, 2002). Further characterization of the cytosine methylase revealed that methylation was limited to unintegrated phage DNA post-induction and not to either prophage or bacterial DNA (Smith and Jeddeloh, 2005).

Repression: As discussed throughout, specific adaptive benefits are conferred to temperate phages because of their lifestyle. It is therefore not surprising that prophages are commonly identified in *Burkholderia* genomes and that the majority of isolated *Burkholderia* phages are either temperate or putatively temperate. For phage therapy purposes, it is generally accepted that temperate phages are suboptimal because of the potential for lysogeny, superinfection immunity, lysogenic conversion, and specialized transduction (reviewed in Lynch *et al.*, 2010a). To functionally characterize a *Burkholderia* phage repressor and to develop a proof-of-principle that temperate phages can be converted to a lytic lifestyle through genetic modification, we identified and knocked out the putative repressor gene in KS9 (Lynch *et al.*, 2010a). Following replacement of this gene with a trimethoprim-resistance cassette, it could no longer stably lysogenize either *B. pyrrocinia* LMG 21824 or *B. cenocepacia* K56–2. As discussed above, treatment with this mutant was able to rescue *G. mellonella* larvae infected with twice the 50% lethal dose (LD_{50}) of K56–2 (Lynch *et al.*, 2010a).

Prophage transcription: The only *Burkholderia* phage with a characterized transcriptome is ϕ1026b. Prophage gene expression was assayed using RNAseq of ϕ1026b-lysogenized *B. pseudomallei* DD503 (Ronning *et al.*, 2010). As predicted, expression of structural and replication proteins was limited, in contrast to strong repressor protein expression. The ORFs with the highest expression levels were three hypothetical proteins, followed by a PAAR repeat-containing protein, VRR-NUC domain protein, addiction module toxin (RelE/StbE family), two hypothetical proteins, repressor, and XRE family putative transcriptional regulator (addiction module antidote). Based on GC content, the presence of promoters and/or terminators, and homology, many of these genes are predicted to be morons (DNA acquired by horizontal transfer that may provide an advantage to the phage and/or its host) (Abedon, 2009; Ronning *et al.*, 2010). These results suggest that putative lysogenic conversion genes in *Burkholderia* prophages can be strongly expressed in lysogens in the absence of phage induction.

Virulence: ϕE125 and ϕ644-2 (in addition to BcepB1A and the Bcep22-like phages) both encode PAPS reductases, but their contribution to virulence has not been characterized experimentally (Ronning *et al.*, 2010). As discussed above for the P2-like phages ϕE12-2 and ϕK96243, a RelE toxin–antitoxin module found in ϕE125, ϕ1026b, and ϕ644–2 may be related to phage exclusion, biofilms, or persister cells (especially considering that the RelE homologue is highly expressed in ϕ1026b-lysogenized DD503) (Ronning *et al.*, 2010; Wang and Wood, 2011). Two proteins whose annotations suggested that they might be virulence-related have been ruled out as having an impact on lysogen phenotype. In KS9, a protein containing the conserved domain 'COG3950: predicted ATP-binding protein involved in virulence' was identified, but there was no difference in mortality observed between *G. mellonella* infected with wild type *B. cenocepacia* K56–2 and KS9-lysogenized K56-2 (Lynch *et al.*, 2010a). In ϕ1026b, a protein was putatively identified as a major facilitator family permease transporter, which may facilitate acquisition of sugars, amino acids, or other substances (Pao *et al.*, 1998). Insertional mutagenesis was used to knock out the permease in *B. pseudomallei* 1026b and the effect of the knockout was determined using both phenotype microarrays and LD_{50} assays in hamsters (DeShazer, 2004). In each assay, there were no significant differences between wildtype 1026b and the phage protein knockout (DeShazer, 2004).

BcepGomr-like siphoviruses

BcepGomr is a lytic siphovirus isolated from soil (Robinson *et al.*, 2006) whose gene order is conserved with ϕE125-like phages (with virion morphogenesis genes on the far left end of the chromosome and lysis genes on the far right) (Summer *et al.*, 2007). However, BcepGomr is dissimilar from these phages as CoreGenes analysis indicates that ϕE125 and BcepGomr share only three similar proteins, encoding the tape measure protein, tail tip fibre protein, and a hypothetical protein (ϕE125 gp15, gp21, and gp22). Furthermore, BcepGomr shows minimal relatedness to any previously sequenced phages: in BLASTN analysis, portions of BcepGomr show similarity to sequences from phages such as YuA and MP1412 of *Pseudomonas* and HK639 of *Escherichia*, but query coverage reaches a maximum of 2% (for *Burkholderia* phage KL1). BcepGomr is also dissimilar in that it has one of the lowest GC contents of the *Burkholderia Siphoviridae* (56.3%, lower than all other siphoviruses except KL1 with a GC content of 54.6%) (Table 11.1).

Most of the BcepGomr proteins with assigned functions are involved in virion morphogenesis, DNA synthesis, or lysis. However, an interesting finding is that several BcepGomr proteins show similarity to those of mycobacteriophage Rosebush (Summer *et al.*, 2007). Out of seven Rosebush proteins identified as being unique with respect to previously identified phage proteins (Pedulla *et al.*, 2003), four are encoded by BcepGomr (gp33, gp35, gp37, and gp62), plus a fifth protein (gp34) that is predicted to have a similar function but a dissimilar sequence. Originally, the annotations for these proteins suggested that they might be involved in modulating the host immune response or other processes. Rosebush gp3 (BcepGomr gp35) is similar to *Mesorhizobium loti* ExsB and was predicted to function in regulation (Pedulla *et al.*, 2003). In *Rhizobium meliloti*, ExsB regulates exopolysaccharide production, decreasing succinoglycan synthesis (Becker *et al.*, 1995). ALU1-P, a similar protein in *Arthrobacter viscosus*, allows for growth in high concentrations of aluminium (Jo *et al.*, 1997). Rosebush gp5 (BcepGomr gp37) was predicted to be involved in biosynthesis of pyrroloquinoline quinone (Pedulla *et al.*, 2003). Rosebush gp4 and gp6 (BcepGomr gp34 and gp33, respectively) were predicted to function together in tetrahydrobiopterin biosynthesis as a 6-pyruvoyltetrahydrobiopterin synthase and GTP cyclohydrolase, respectively (Pedulla *et al.*, 2003). Because tetrahydrobiopterin is a nitric oxide synthase cofactor, the activity of these proteins may affect host immunity during infection (Alderton *et al.*, 2001; Pedulla *et al.*, 2003).

Recently, it has been proposed that these proteins are more likely involved in queuosine biosynthesis (Hatfull, 2012). Queuosine is a modified guanosine-like nucleoside found in tRNA at the wobble position for asparagine, aspartate, histidine, and tyrosine (Dineshkumar *et al.*, 2002). In this biosynthesis pathway, Que proteins convert GTP to 7-aminomethyl-7-deazaguanine (via several intermediates), which is then incorporated into the tRNA and finally converted into queuosine (reviewed in Bandarian, 2012). Rosebush gp2–gp6 (and by extension BcepGomr gp31, gp35, gp34, gp37, and gp33 [all in the same module, but arranged differently than in Rosebush]) have been recently re-annotated as queuine tRNA ribosyltransferase, QueC, QueD, QueE, and QueF proteins, respectively (Hatfull, 2012). Queuosine biosynthesis genes have been identified in several other phages, with the largest set thus far belonging to Dp-1, a lytic *S. pneumoniae* siphovirus encoding seven queuosine-related proteins (Sabri *et al.*, 2011). The effect of phage-mediated queuosine tRNA modification on translation is currently unknown (Hatfull, 2012).

Incomplete *Burkholderia* phage sequences

Currently, partial sequences can be found in GenBank for *B. pseudomallei* phages ϕBp-AMP1 and Bups ϕ1 (Elliman, 2006; Gatedee *et al.*, 2012). Sequences from the T4-type *B. cepacia* phage 42 or PC42 are in fact from an *Acinetobacter* phage with a previously misidentified host (Félix d'Hérelle Reference Centre for Bacterial Viruses, 2012). ϕBp-AMP1 is a podovirus isolated from rice field soil in Thailand with a *B. pseudomallei*-restricted host range. Only the *ttpB* gene from this phage has been sequenced, encoding the tail tubular protein B. In a phylogenetic comparison of this gene to other *ttpB* sequences, it was found to be most closely related to *Ralstonia* phage RSB1 and a *B. thailandensis* MSMB43 prophage (Gatedee *et al.*, 2012).

Bups ϕ1 is a 55.1 kbp myovirus isolated from *B. pseudomallei* NCTC 13178 (Elliman, 2006). Eleven contigs ranging in size from 376 bp to 19,870 bp have been sequenced and deposited in GenBank and are related to *Burkholderia* prophage and phage sequences. Similar sequences to contig 6 were identified in an olive leaf genome project, suggesting a potentially novel source of *Burkholderia* phages (Dundar and Suakar, 2010). Putative virulence genes identified in Bups ϕ1 included an anaerobic dehydrogenase, an oxidoreductase, and ADP-ribosyltransferases (Elliman, 2006). To assess if the prophage can affect *B. pseudomallei* virulence, the 50% infectious dose (ID_{50}) of wildtype and putatively lysogenized isolates was compared in a BALB/c mouse model. Although one trial showed a significant difference in ID_{50} between the isolates, this result was not reproducible and may have been due to changes in bacterial gene expression following prophage integration (Elliman, 2006).

Burkholderia prophage sequences

Numerous prophage or prophage-like sequences have been identified in *Burkholderia* genome sequencing projects. For *B. pseudomallei*, the complete genome sequence of strain K96243 was determined in 2004 (Holden *et al.*, 2004). Out of 16 genomic islands, three were identified as prophages (including ϕK96243/GI 2, discussed above) and five were identified as being prophage-like. The distribution of three of these islands (ϕK96243/GI 2 and the prophage-like GI 6 and GI 9) was assessed in a panel of sequenced strains and uncharacterized environmental and clinical isolates (Tumapa *et al.*, 2008). In the sequenced strains, GI 2 was identified as well as GI 2-like and GI 9-like islands. In the PCR-screened uncharacterized isolates, most were PCR-positive for GI 2 (72%) and less so for GI 6 and GI 9 (23% and 12%, respectively). Similar percentages were observed for both environmental and clinical isolates, suggesting that the presence of phage sequences did not enhance virulence (Tumapa *et al.*, 2008).

In *B. thailandensis* strain E264, 15 genomic islands have been identified and eight of these are prophage-like (Yu *et al.*, 2006). These islands are presumably the result of relatively recent horizontal transfer as they are absent in *B. pseudomallei* and *B. mallei*. Similar to the PCR screening described above for genomic islands in *B. pseudomallei* isolates, two phage islands were found only in the sequenced strain, five were found in between 10% and 50% of the *B. thailandensis* environmental isolates, and one was found in over half of the isolates (Yu *et al.*, 2006). Numerous other prophage-like sequences and prophages have since been identified in other sequenced strains of *B. pseudomallei* and *B. thailandensis* (Tuanyok *et al.*, 2008; Ronning *et al.*, 2010).

In the *B. cenocepacia* J2315 genome, 14 genomic islands were found, including five putative prophages and one prophage-like sequence (Holden *et al.*, 2009). The sequences of two of these islands are discussed above: KS10/BcenGI7 and BcepMu/BcenGI14 (Summer *et al.*, 2004; Goudie *et al.*, 2008). In *B. multivorans* strains CGD1, CGD2, and ATCC 17616, Ronning *et al.* (2010) identified 8 prophage-like sequences including the characterized ATCC 17616 prophages Bcep176 (PI 17616-3) and

KS5 (PI 17616-4). Although no *B. xenovorans* phages have been characterized independently, 21 genomic islands were identified in the LB400 genome and seven of these were prophage-like or contained phage genes (Chain *et al.*, 2006). The *B. xenovorans* phage-like island GI 7 was also identified by Ronning *et al.* (2010) as PI LB400-1. Seo *et al.* (2011) determined the genome sequence of *B. gladioli* strain BSR3, but its genomic islands have not yet been described. Although *B. mallei* is susceptible to infection by many *B. thailandensis* and *B. pseudomallei* phages – including ϕE202, ϕ52237, ϕE12–2, ϕK96243, ϕE255, ϕE125, ϕ1026b, and ϕ644-2 – its genome does not carry prophages (Woods *et al.*, 2002; Holden *et al.*, 2004; DeShazer, 2004; Ronning *et al.*, 2010). It has been suggested that, because *B. mallei* is a host-adapted organism with a reduced genome (having evolved from *B. pseudomallei*), it both lost *B. pseudomallei* prophage sequences and was not lysogenized by new prophages in its limited environment (Ronning *et al.*, 2010). For *Burkholderia* species excluding *B. mallei*, the availability of *in silico* phage-finding programs such as Prophage Finder, Phage_Finder, Prophinder, and PHAST will allow novel prophages to be identified as *Burkholderia* genome sequencing projects continue to be completed (Bose and Barber, 2006; Fouts, 2006; Lima-Mendez *et al.*, 2008; Zhou *et al.*, 2011).

Conclusions

As in other species, a number of key themes emerge from comparative genomics analysis of *Burkholderia* phages. First, these sequences are very diverse, having been classified as part of at least 10 distinct taxonomic groups (Table 11.1). Diversity is not only observed between groups but within them; for example, the P2-like myoviruses share half of their proteins (those involved in virion morphogenesis), but the other half of their proteomes are dissimilar (Ronning *et al.*, 2010; Lynch *et al.*, 2010b). Second, although this chapter has focused on those genes encoding proteins with a predicted function, it is important to note that a large proportion of *Burkholderia* phage proteins are hypothetical. The relative number of hypothetical proteins ranges from very low in KS14 (with a small genome and P2-like structural proteins) to very high in BcepF1 (with a large genome similar to PB1, which also encodes many hypothetical proteins) (Ceyssens *et al.*, 2009; Lynch *et al.*, 2010b). Finally, the *Burkholderia* phages have a mosaic organization built up of function-specific modules for virion morphogenesis, lysis, DNA binding, integration, and other functions. In the vast majority of cases, the genes within these modules are found at a single locus, but a different arrangement is found in KS10 where the head morphogenesis genes are divided into two clusters within the genome (Goudie *et al.*, 2008). As occurs with P2-like phages, a specific set of structural gene modules can re-assort with independent accessory gene modules to produce different phages with the same overall morphology (Ronning *et al.*, 2010; Lynch *et al.*, 2010b).

Burkholderia phage genera tend to be comprised of either temperate phages isolated from lysogens (P2-like, BcepMu-like, ϕE125-like), lytic or putatively lytic phages isolated from the environment (Bcep781-like, PB1-like, BcepGomr-like), and temperate or putatively temperate phages isolated from the environment (Bcep22-like, BPP-1-like, PA73-like) (Table 11.1). There is some overlap within the P2-like group, where KS14 and KS5 were both isolated from soil (although KS5 was later found to be a *B. multivorans* prophage) (Lynch *et al.*, 2010b). The BcepNazgul-like phages are an interesting case because, although both BcepNazgul and AH2 were isolated from soil, the former is thought to be lytic while the latter is thought to be temperate (Robinson *et al.*, 2006; Lynch *et al.*, 2012b). In this case, BcepNazgul may be a natural lytic variant of AH2 that lost the integrase and repressor genes, or conversely AH2 may be a temperate variant of BcepNazgul that acquired these genes. The lytic or putatively lytic Bcep781-like, PB1-like, and BcepGomr-like genera contain only Bcc phages, while the sequenced *B. thailandensis* and *B. pseudomallei* phages are strictly temperate (Table 11.1). As putatively lytic phages infecting *B. pseudomallei* have been identified (Yordpratum *et al.*, 2011), this result may stem from sampling bias whereby sequenced phages were solely isolated from lysogens (Woods *et al.*, 2002; Holden *et al.*, 2004; DeShazer, 2004; Ronning *et al.*, 2010).

As discussed throughout this chapter, many *Burkholderia* phages encode putative lysogenic conversion proteins, but the actual effect of these products on the host phenotype is unclear. In cases where the effect of a phage gene on host virulence was examined experimentally – such as in ϕ1026b and KS9 – no differences were observed (DeShazer, 2004; Lynch *et al.*, 2010a). *Burkholderia* is therefore dissimilar to organisms such as *V. cholerae*, for which virulence is clearly prophage-mediated (Waldor and Mekalanos, 1996). The functions of many of the putative lysogenic conversion genes described here suggest that they may be more likely to facilitate general adaptation than strict pathogenicity (Summer *et al.*, 2007). Although these findings do little to further our understanding of the mechanisms of *Burkholderia* virulence, they are advantageous with respect to clinical administration where it is critical that putative virulence factors are absent. Furthermore, some lysogenic conversion genes could potentially have a positive effect in phage therapy, such as the anti-biofilm CsrA protein encoded by the Bcep22-like podoviruses (Gill *et al.*, 2011; Lynch *et al.*, 2012a). Before these phages are developed for clinical use, more comprehensive experiments are needed to evaluate the relevance of lysogenic conversion genes in *Burkholderia* phages with respect to both pathogenicity and phage therapy.

There are many broad future directions for the study of *Burkholderia* phage genomics. As discussed above, the numerous phages that have been isolated but not characterized are a promising and untapped source of potentially novel sequence data. Further sampling of isolates, strains, and environmental samples may provide phage types not currently identified in *Burkholderia*, such as jumbo or filamentous phages (as has been putatively identified in *B. pseudomallei* 668) (Askora *et al.*, 2011) or phages infecting and lysogenizing the many lesser-characterized *Burkholderia* species. Metagenomics surveys can identify the prevalence, diversity, and gene content of *Burkholderia* phages in specific niches, both clinical and environmental. Finally, whole genome sequencing of *Burkholderia* strains can provide insight as to possible differences in prophage carriage between clinical and environmental strains and if prophages contribute to virulence in toxigenic *Burkholderia* such as *B. gladioli* pv. *cocovenenans* (Lynch and Dennis, 2009). Although much of the future work on *Burkholderia* phages is likely to be focused on functional characterization, there is still much to do with regard to the isolation and comparative genomics of this group of bacterial viruses.

Three additional *Burkholderia* phage genome sequences are now available in GenBank:

- The P2-like *B. pseudomallei* myovirus ϕX216 (JX681814.1); (2012) BMC Microbiol. 12, 289.
- The ϕKMV-like Bcc podovirus JG068 (KC853746.1); (2013) BMC Genomics 14, 574.
- The P2-like *B. pseudomallei* myovirus ST79 (NC_021343.1); unpublished.

References

Abedon, S.T. (2009). Phage evolution and ecology. Adv. Appl. Microbiol. *67*, 1–45.

Ackermann, H.-W. (2003). Bacteriophage observations and evolution. Res. Microbiol. *154*, 245–251.

Adachi, N., Tsukamoto, S., Inoue, Y., and Azegami, K. (2012). Control of bacterial seedling rot and seedling blight of rice by bacteriophage. Plant Dis. *96*, 1033–1036.

Alderton, W.K., Cooper, C.E., and Knowles, R.G. (2001). Nitric oxide synthases: structure, function and inhibition. Biochem. J. *357*, 593–615.

Alemayehu, D., Casey, P.G., McAuliffe, O., Guinane, C.M., Martin, J.G., Shanahan, F., Coffey, A., Ross, R.P., and Hill, C. (2012). Bacteriophages ϕMR299-2 and ϕNH-4 can eliminate *Pseudomonas aeruginosa* in the murine lung and on cystic fibrosis lung airway cells. mBio 3, e00029-12.

AmpliPhi Biosciences. (2011). Product Pipeline. Available at: http://www.ampliphibio.com/index.php/pipeline/product_pipeline.

Askora, A., Kawasaki, T., Fujie, M., and Yamada, T. (2011). Resolvase-like serine recombinase mediates integration/excision in the bacteriophage ϕRSM. J. Biosci. Bioeng. *111*, 109–116.

Balder, R., Lipski, S., Lazarus, J.J., Grose, W., Wooten, R.M., Hogan, R.J., Woods, D.E., and Lafontaine, E.R. (2010). Identification of *Burkholderia mallei* and *Burkholderia pseudomallei* adhesins for human respiratory epithelial cells. BMC Microbiol. *10*, 250.

Bandarian, V. (2012). Radical SAM enzymes involved in the biosynthesis of purine-based natural products. Biochim. Biophys. Acta *1824*, 1245–1253.

Barreiro, V., and Haggård-Ljungquist, E. (1992). Attachment sites for bacteriophage P2 on the

Escherichia coli chromosome: DNA sequences, localization on the physical map, and detection of a P2-like remnant in *E. coli* K-12 derivatives. J. Bacteriol. *174*, 4086–4093.

Becker, A., Kuster, H., Niehaus, K., and Puhler, A. (1995). Extension of the *Rhizobium meliloti* succinoglycan biosynthesis gene cluster: identification of the *exsA* gene encoding an ABC transporter protein, and the *exsB* gene which probably codes for a regulator of succinoglycan biosynthesis. Mol. Gen. Genet. *249*, 487–497.

Bose, M., and Barber, R.D. (2006). Prophage Finder: a prophage loci prediction tool for prokaryotic genome sequences. *In Silico* Biol. *6*, 223–227.

Boyd, E.F., and Brüssow, H. (2002). Common themes among bacteriophage-encoded virulence factors and diversity among the bacteriophages involved. Trends Microbiol. *10*, 521–529.

Breitbart, M., Salamon, P., Andresen, B., Mahaffy, J.M., Segall, A.M., Mead, D., Azam, F., and Rohwer, F. (2002). Genomic analysis of uncultured marine viral communities. Proc. Natl. Acad. Sci. U.S.A. *99*, 14250–14255.

Brüssow, H. (2001). Phages of dairy bacteria. Annu. Rev. Microbiol. *55*, 283–303.

Carmody, L.A., Gill, J.J., Summer, E.J., Sajjan, U.S., Gonzalez, C.F., Young, R.F., and LiPuma, J.J. (2010). Efficacy of bacteriophage therapy in a model of *Burkholderia cenocepacia* pulmonary infection. J. Infect. Dis. *201*, 264–271.

Casjens, S.R. (2005). Comparative genomics and evolution of the tailed-bacteriophages. Curr. Opin. Microbiol. *8*, 451–458.

Ceyssens, P.J., Miroshnikov, K., Mattheus, W., Krylov, V., Robben, J., Noben, J.P., Vanderschraeghe, S., Sykilinda, N., Kropinski, A.M., Volckaert, *et al.* (2009). Comparative analysis of the widespread and conserved PB1-like viruses infecting *Pseudomonas aeruginosa*. Environ. Microbiol. *11*, 2874–2883.

Chain, P.S., Denef, V.J., Konstantinidis, K.T., Vergez, L.M., Agullo, L., Reyes, V.L., Hauser, L., Cordova, M., Gomez, L., Gonzalez, M., *et al.* (2006). *Burkholderia xenovorans* LB400 harbors a multi-replicon, 9.73-Mbp genome shaped for versatility. Proc. Natl. Acad. Sci. U.S.A. *103*, 15280–15287.

Cihlar, R.L., Lessie, T.G., and Holt, S.C. (1978). Characterization of bacteriophage CP1, an organic solvent sensitive phage associated with *Pseudomonas cepacia*. Can. J. Microbiol. *24*, 1404–1412.

Dai, W., Hodes, A., Hui, W.H., Gingery, M., Miller, J.F., and Zhou, Z.H. (2010). Three-dimensional structure of tropism-switching *Bordetella* bacteriophage. Proc. Natl. Acad. Sci. U.S.A. *107*, 4347–4352.

Darling, A.E., Mau, B., and Perna, N.T. (2010). progressiveMauve: multiple genome alignment with gene gain, loss and rearrangement. PLoS ONE *5*, e11147.

Debarbieux, L., Leduc, D., Maura, D., Morello, E., Criscuolo, A., Grossi, O., Balloy, V., and Touqui, L. (2010). Bacteriophages can treat and prevent *Pseudomonas aeruginosa* lung infections. J. Infect. Dis. *201*, 1096–1104.

Denisov, I.I., and Kapliev, V.I. (1991). The level of spontaneous phage production and sensitivity to melioidosis phages of museum cultures of *Pseudomonas pseudomallei*. Mikrobiol. Zh. *53*, 66–70.

Denisov, I.I., and Kapliev, V.I. (1995). The isolation and characteristics of cloned strains of *Pseudomonas pseudomallei* phages. Mikrobiol. Z. *57*, 53–56.

DeShazer, D. (2004). Genomic diversity of *Burkholderia pseudomallei* clinical isolates: subtractive hybridization reveals a *Burkholderia mallei*-specific prophage in *B. pseudomallei* 1026b. J. Bacteriol. *186*, 3938–3950.

Dineshkumar, T.K., Thanedar, S., Subbulakshmi, C., and Varshney, U. (2002). An unexpected absence of queuosine modification in the tRNAs of an *Escherichia coli* B strain. Microbiology *148*, 3779–3787.

Doulatov, S., Hodes, A., Dal, L., Mandhana, N., Liu, M., Deora, R., Simons, R.W., Zimmerly, S., and Miller, J.F. (2004). Tropism switching in *Bordetella* bacteriophage defines a family of diversity-generating retroelements. Nature *431*, 476–481.

Drummond, A.J., Ashton, B., Buxton, S., Cheung, M., Cooper, A., Duran, C., Heled, J., Kearse, M., Markowitz, S., Moir, R., *et al.* (2012). Geneious v5.6. http://www.geneious.com.

Dundar, E., and Suakar, O. (2010). Mining olive genome through library sequencing and bioinformatics: Novel sequences and new microsatellites. Afr. J. Biotechnol. *9*, 2499–2509.

Elliman, J. (2006). Bacteriophage of *Burkholderia pseudomallei*; friend or foe? PhD thesis, James Cook University.

Farmer, K.L., and Thomas, M.S. (2004). Isolation and characterization of *Burkholderia cenocepacia* mutants deficient in pyochelin production: pyochelin biosynthesis is sensitive to sulfur availability. J. Bacteriol. *186*, 270–277.

Félix d'Hérelle Reference Center for Bacterial Viruses. (2012). Host strain HER 1422. http://www.phage.ulaval.ca/fileadmin/php/pages/en/souche.php?crePdf=1.

Fiers, W., Contreras, R., Duerinck, F., Haegeman, G., Iserentant, D., Merregaert, J., Min Jou, W., Molemans, F., Raeymaekers, A., Van den Berghe, A., Volckaert, G., and Ysebaert, M. (1976). Complete nucleotide sequence of bacteriophage MS2 RNA: primary and secondary structure of the replicase gene. Nature *260*, 500–507.

Fouts, D.E. (2006). Phage_Finder: automated identification and classification of prophage regions in complete bacterial genome sequences. Nucleic Acids Res. *34*, 5839–5851.

Galyov, E.E., Brett, P.J., and DeShazer, D. (2010). Molecular insights into *Burkholderia pseudomallei* and *Burkholderia mallei* pathogenesis. Annu. Rev. Microbiol. *64*, 495–517.

Gatedee, J., Kritsiriwuthinan, K., Galyov, E.E., Shan, J., Dubinina, E., Intarak, N., Clokie, M.R., and Korbsrisate, S. (2011). Isolation and characterization of a novel podovirus which infects *Burkholderia pseudomallei*. Virol. J. *8*, 366.

Gill, J.J., Summer, E.J., Russell, W.K., Cologna, S.M., Carlile, T.M., Fuller, A.C., Kitsopoulos, K., Mebane, L.M., Parkinson, B.N., Sullivan, D., *et al.* (2011). Genomes and characterization of phages Bcep22 and BcepIL02, founders of a novel phage type in *Burkholderia cenocepacia*. J. Bacteriol. *193*, 5300–5313.

Golshahi, L., Seed, K.D., Dennis, J.J., and Finlay, W.H. (2008). Toward modern inhalational bacteriophage therapy: Nebulization of bacteriophages of *Burkholderia cepacia* complex. J. Aerosol Med. Pulm. Drug. Deliv. *21*, 351–359.

Golshahi, L., Lynch, K.H., Dennis, J.J., and Finlay, W.H. (2011). *In vitro* lung delivery of bacteriophages KS4-M and ΦKZ using dry powder inhalers for treatment of *Burkholderia cepacia* complex and *Pseudomonas aeruginosa* infections in cystic fibrosis. J. Appl. Microbiol. *110*, 106–117.

Goudie, A.D., Lynch, K.H., Seed, K.D., Stothard, P., Shrivastava, S., Wishart, D.S., and Dennis, J.J. (2008). Genomic sequence and activity of KS10, a transposable phage of the *Burkholderia cepacia* complex. BMC Genomics *9*, 615.

Grishkina, T.A., and Merinova, L.K. (1993). Spontaneous phage production in *Pseudomonas pseudomallei* and in a range of hosts of melioidosis phages among representatives in the genus *Pseudomonas*. Mikrobiol. Z. *55*, 43–47.

Guo, L., Lim, K.B., Poduje, C.M., Daniel, M., Gunn, J.S., Hackett, M., and Miller, S.I. (1998). Lipid A acylation and bacterial resistance against vertebrate antimicrobial peptides. Cell *95*, 189–198.

Harper, D.R., and Enright, M.C. (2011). Bacteriophages for the treatment of *Pseudomonas aeruginosa* infections. J. Appl. Microbiol. *111*, 1–7.

Hatfull, G.F. (2008). Bacteriophage genomics. Curr. Opin. Microbiol. *11*, 447–453.

Hatfull, G.F. (2010). Mycobacteriophages: Genes and genomes. Annu. Rev. Microbiol. *64*, 331–356.

Hatfull, G.F. (2012). The secret lives of mycobacteriophages. Adv. Virus Res. *82*, 179–288.

Hatfull, G.F., and Hendrix, R.W. (2011). Bacteriophages and their genomes. Curr. Opin. Virol. *1*, 298–303.

Hawkins, C., Harper, D., Burch, D., Anggard, E., and Soothill, J. (2010). Topical treatment of *Pseudomonas aeruginosa* otitis of dogs with a bacteriophage mixture: a before/after clinical trial. Vet. Microbiol. *146*, 309–313.

Hendrix, R.W. (2003). Bacteriophage genomics. Curr. Opin. Microbiol. *6*, 506–511.

Hendrix, R.W., Smith, M.C., Burns, R.N., Ford, M.E., and Hatfull, G.F. (1999). Evolutionary relationships among diverse bacteriophages and prophages: all the world's a phage. Proc. Natl. Acad. Sci. U.S.A. *96*, 2192–2197.

Hens, D.K., Ghosh, A.N., and Kumar, R. (2005). A new small temperate DNA phage BcP15 isolated from *Burkholderia cepacia* DR11. Arch. Virol. *150*, 2421–2428.

Hoe, S., Semler, D.D., Goudie, A.D., Lynch, K.H., Matinkhoo, S., Finlay, W.H., Dennis, J.J., and Vehring, R. (2013). Respirable bacteriophages for the treatment of bacterial lung infections. J. Aerosol Med. Pulm. Drug Deliv. Online ahead of print: doi:10.1089/jamp.2012.1001.

Holden, M.T.G., Titball, R.W., Peacock, S.J., Cerdeño-Tárraga, A.M., Atkins, T., Crossman, L.C., Pitt, T., Churcher, C., Mungall, K., Bentley, S.D., *et al.* (2004). Genomic plasticity of the causative agent of melioidosis, *Burkholderia pseudomallei*. Proc. Natl. Acad. Sci. U.S.A. *101*, 14240–14245.

Holden, M.T.G., Seth-Smith, H.M.B., Crossman, L.C., Sebaihia, M., Bentley, S.D., Cerdeño-Tárraga, A.M., Thomson, N.R., Bason, N., Quail, M.A., Sharp, S., *et al.* (2009). The genome of *Burkholderia cenocepacia* J2315, an epidemic pathogen of cystic fibrosis patients. J. Bacteriol. *91*, 261–277.

Iida, S., Streiff, M.B., Bickle, T.A., and Arber, W. (1987). Two DNA antirestriction systems of bacteriophage P1, *darA*, and *darB*: characterization of *darA*$^-$ phages. Virology *157*, 156–166.

Iliukhin, V.I., Merinova, L.K., Ageeva, N.P., and Senina, T.V. (2011). Bacteriocins and bacteriophages of pathogenic *Burkholderia*. Zh. Mikrobiol. Epidemiol. Immunobiol. *1*, 12–19.

Inoue, Y., Matsuura, T., Ohara, T., and Azegami, K. (2006). Sequence analysis of the genome of OP_2, a lytic bacteriophage of *Xanthomonas oryzae* pv. *oryzae*. J. Gen. Plant Pathol. 72, 104–110.

Jackson, D.W., Suzuki, K., Oakford, L., Simecka, J.W., Hart, M.E., and Romeo, T. (2002). Biofilm formation and dispersal under the influence of the global regulator CsrA of *Escherichia coli*. J. Bacteriol. *184*, 290–301.

Jo, J., Jang, Y.S., Kim, K.Y., Kim, M.H., Kim, I.J., and Chung, W.I. (1997). Isolation of *ALU1-P* gene encoding a protein with aluminum tolerance activity from *Arthrobacter viscosus*. Biochem. Biophys. Res. Commun. *239*, 835–839.

Karumidze, N., Thomas, J.A., Kvatadze, N., Goderdzishvili, M., Hakala, K.W., Weintraub, S.T., Alavidze, Z., and Hardies, S.C. (2012). Characterization of lytic *Pseudomonas aeruginosa* bacteriophages via biological properties and genomic sequences. Appl. Microbiol. Biotechnol. *94*, 1609–1617.

Kawasaki, K., Ernst, R.K., and Miller, S.I. (2004). Deacylation and palmitoylation of lipid A by Salmonellae outer membrane enzymes modulate host signaling through Toll-like receptor 4. J. Endotoxin Res. *10*, 439–444.

Kennedy, M.P., Coakley, R.D., Donaldson, S.H., Aris, R.M., Hohneker, K., Wedd, J.P., Knowles, M.R., Gilligan, P.H., and Yankaskas, J.R. (2007). *Burkholderia gladioli*: Five year experience in a cystic fibrosis and lung transplantation center. J. Cyst. Fibros. *6*, 267–273.

Kropinski, A.M. (2006). Phage therapy – Everything old is new again. Can. J. Infect. Dis. Med. Microbiol. *17*, 297–306.

Kropinski, A.M., Kovalyova, I.V., Billington, S.J., Patrick, A.N., Butts, B.D., Guichard, J.A., Pitcher, T.J., Guthrie, C.C., Sydlaske, A.D., Barnhill, L.M., *et al.* (2007).

The genome of ε15, a serotype-converting, Group E1 *Salmonella enterica*-specific bacteriophage. Virology *369*, 234–244.

Kropinski, A.M., Van den Bossche, A., Lavigne, R., Noben, J.P., Babinger, P., and Schmitt, R. (2012). Genome and proteome analysis of 7-7-1, a flagellotropic phage infecting *Agrobacterium* sp. H13-3. Virol. J. *9*, 102.

Kwan, T., Liu, J., DuBow, M., Gros, P., and Pelletier, J. (2005). The complete genomes and proteomes of 27 *Staphylococcus aureus* bacteriophages. Proc. Natl. Acad. Sci. U.S.A. *102*, 5174–5179.

Kwan, T., Liu, J., DuBow, M., Gros, P., and Pelletier, J. (2006). Comparative genomic analysis of 18 *Pseudomonas aeruginosa* bacteriophages. J. Bacteriol. *188*, 1184–1187.

Langley, R., Kenna, D.T., Vandamme, P., Ure, R., and Govan, J.R.W. (2003). Lysogeny and bacteriophage host range within the *Burkholderia cepacia* complex. J. Med. Microbiol. *52*, 483–490.

Lavigne, R., Seto, D., Mahadevan, P., Ackermann, H.-W., and Kropinski, A.M. (2008). Unifying classical and molecular taxonomic classification: analysis of the *Podoviridae* using BLASTP-based tools. Res. Microbiol. *159*, 406–414.

Lavigne, R., Darius, P., Summer, E.J., Seto, D., Mahadevan, P., Nilsson, A.S., Ackermann, H.-W., and Kropinski, A.M. (2009). Classification of *Myoviridae* bacteriophages using protein sequence similarity. BMC Microbiol. *9*, 224.

Lawrence, J.G., Hatfull, G.F., and Hendrix, R.W. (2002). Imbroglios of viral taxonomy: genetic exchange and failings of phenetic approaches. J. Bacteriol. *184*, 4891–4905.

Leclerc, H., and Sureau, P. (1956). Research on bacteriophages of the Whitmore bacillus in stagnant waters of Hanoi. Bull. Soc. Pathol. Exot. Filiales *49*, 874–882.

Levin, B.R., and Bull, J.J. (2004). Population and evolutionary dynamics of phage therapy. Nat. Rev. Microbiol. *2*, 166–173.

Li, P., Chen, B., Song, Z., Song, Y., Yang, Y., Ma, P., Wang, H., Ying, J., Ren, P., Yang, L., *et al.* (2012). Bioinformatic analysis of the *Acinetobacter baumannii* phage AB1 genome. Gene *507*, 125–134.

Lima-Mendez, G., Van Helden, J., Toussaint, A., and Leplae, R. (2008). Prophinder: a computational tool for prophage prediction in prokaryotic genomes. Bioinformatics *24*, 863–865.

Liu, M., Deora, R., Doulatov, S.R., Gingery, M., Eiserling, F.A., Preston, A., Maskell, D.J., Simons, R.W., Cotter, P.A., Parkhill, J., and Miller, J.F. (2002). Reverse transcriptase-mediated tropism switching in *Bordetella* bacteriophage. Science *295*, 2091–2094.

Loutet, S.A., and Valvano, M.A. (2010). A decade of *Burkholderia cenocepacia* virulence determinant research. Infect. Immun. *78*, 4088–4100.

Lynch, K.H., and Dennis, J.J. (2009). *Burkholderia*. In Molecular Detection of Foodborne Pathogens, Liu, D. ed., (Boca Raton, FL: CRC Press) pp. 331–343.

Lynch, K.H., and Dennis, J.J. (2012). Cangene gold medal award lecture – genomic analysis and modification of *Burkholderia cepacia* complex bacteriophages. Can. J. Microbiol. *58*, 221–235.

Lynch, K.H., Seed, K.D., Stothard, P., and Dennis, J.J. (2010a). Inactivation of *Burkholderia cepacia* complex phage KS9 gp41 identifies the phage repressor and generates lytic virions. J. Virol. *84*, 1276–1288.

Lynch, K.H., Stothard, P., and Dennis, J.J. (2010b). Genomic analysis and relatedness of P2-like phages of the *Burkholderia cepacia* complex. BMC Genomics *11*, 599.

Lynch, K.H., Stothard, P., and Dennis, J.J. (2012a). Characterization of DC1, a broad-host-range Bcep22-like podovirus. Appl. Environ. Microbiol. *78*, 889–891.

Lynch, K.H., Stothard, P., and Dennis, J.J. (2012b). Comparative analysis of two phenotypically similar but genomically distinct *Burkholderia cenocepacia*-specific bacteriophages. BMC Genomics *13*, 223.

Mahadevan, P., King, J.F., and Seto, D. (2009). CGUG: *in silico* proteome and genome parsing tool for the determination of 'core' and unique genes in the analysis of genomes up to *ca.* 1.9 Mb. BMC Res. Notes *2*, 168.

Manzeniuk, O.I., Volozhantsev, N.V., Astashkin, E.I., and Svetoch, E.A. (1993). Transduction of *Pseudomonas mallei* bacteria. Mol. Gen. Mikrobiol. Virusol. *4*, 37–40.

Manzeniuk, O.I., Volozhantsev, N.V., and Svetoch, E.A. (1994). Identification of the bacterium *Pseudomonas mallei* using *Pseudomonas pseudomallei* bacteriophages. Mikrobiologiia *63*, 537–544.

Martin, A.C., Lopez, R., and Garcia, P. (1996). Analysis of the complete nucleotide sequence and functional organization of the genome of *Streptococcus pneumoniae* bacteriophage Cp-1. J. Virol. *70*, 3678–3687.

Matinkhoo, S., Lynch, K.H., Dennis, J.J., Finlay, W.H., and Vehring, R. (2011). Spray-dried respirable powders containing bacteriophages for the treatment of pulmonary infections. J. Pharm. Sci. *100*, 5197–5205.

Matsumoto, H., Itoh, Y., Ohta, S., and Terawaki, Y. (1986). A generalized transducing phage of *Pseudomonas cepacia*. J. Gen. Microbiol. *132*, 2583–2586.

Meader, E., Mayer, M.J., Gasson, M.J., Steverding, D., Carding, S.R., and Narbad, A. (2010). Bacteriophage treatment significantly reduces viable *Clostridium difficile* and prevents toxin production in an *in vitro* model system. Anaerobe *16*, 549–554.

Mera, L. (2006). Bcep176 and Bglu421 – two novel phages contributing to the understanding of pathogenicity and diversity in *Burkholderiacae*. Undergraduate thesis, Texas A&M University.

Merril, C.R., Scholl, D., and Adhya, S.L. (2003). The prospect for bacteriophage therapy in Western medicine. Nat. Rev. Drug Discov. *2*, 489–497.

Michalet, S., and Dijoux-Franca, M.-G. (2009). ABC transporters and resistance to antibiotics. In ABC Transporters and Multidrug Resistance, Boumendjel, A., Boutonnat, J., and Robert, J. eds., (Hoboken, NJ: John Wiley & Sons, Inc.) pp. 179–193.

Miller, E.S., Kutter, E., Mosig, G., Arisaka, F., Kunisawa, T., and Ruger, W. (2003). Bacteriophage T4 genome. Microbiol. Mol. Biol. Rev. *67*, 86–156.

Miller, R.V., and Day, M. (2008). Contribution of lysogeny, pseudolysogeny, and starvation to phage ecology. In

Bacteriophage Ecology, Abedon, S.T. ed., (Cambridge: Cambridge University Press) pp. 114–143.

Morello, E., Saussereau, E., Maura, D., Huerre, M., Touqui, L., and Debarbieux, L. (2011). Pulmonary bacteriophage therapy on *Pseudomonas aeruginosa* cystic fibrosis strains: First steps towards treatment and prevention. PLoS ONE *6*, e16963.

Nilsson, A.S., and Haggård-Ljungquist, E. (2007). Evolution of P2-like phages and their impact on bacterial evolution. Res. Microbiol. *158*, 311–317.

Nishiyama, E., Ohtsubo, Y., Nagata, Y., and Tsuda, M. (2010). Identification of *Burkholderia multivorans* ATCC 17616 genes induced in soil environment by *in vivo* expression technology. Environ. Microbiol. *12*, 2539–2558.

Nzula, S., Vandamme, P., and Govan, J.R.W. (2000). Sensitivity of the *Burkholderia cepacia* complex and *Pseudomonas aeruginosa* to transducing bacteriophages. FEMS Immunol. Med. Microbiol. *28*, 307–312.

Odegrip, R., Nilsson, A.S., and Haggård-Ljungquist, E. (2006). Identification of a gene encoding a functional reverse transcriptase within a highly variable locus in the P2-like coliphages. J. Bacteriol. *188*, 1643–1647.

O'Grady, E.P., Viteri, D.F., Malott, R.J., and Sokol, P.A. (2009). Reciprocal regulation by the CepIR and CciIR quorum sensing systems in *Burkholderia cenocepacia*. BMC Genomics *10*, 441.

Ohtsubo, Y., Genka, H., Komatsu, H., Nagata, Y., and Tsuda, M. (2005). High-temperature-induced transposition of insertion elements in *Burkholderia multivorans* ATCC 17616. Appl. Environ. Microbiol. *71*, 1822–1828.

Orchard, R.C., 2nd (2008). Genomic analysis of *Burkholderia* and *Rhodococcus equi* bacteriophages. Undergraduate thesis, Texas A&M University.

Pao, S.S., Paulsen, I.T., and Saier, M.H.,Jr. (1998). Major facilitator superfamily. Microbiol. Mol. Biol. Rev. *62*, 1–34.

Pedulla, M.L., Ford, M.E., Houtz, J.M., Karthikeyan, T., Wadsworth, C., Lewis, J.A., Jacobs-Sera, D., Falbo, J., Gross, J., Pannunzio, N.R., *et al.* (2003). Origins of highly mosaic mycobacteriophage genomes. Cell *113*, 171–182.

Peeters, E., Sass, A., Mahenthiralingam, E., Nelis, H., and Coenye, T. (2010). Transcriptional response of *Burkholderia cenocepacia* J2315 sessile cells to treatments with high doses of hydrogen peroxide and sodium hypochlorite. BMC Genomics *11*, 90.

Robinson, S.M., Hall, J., Das, A., Rosenbloom, A., Gonzalez, C., Young, R., and Summer, E.J. (2006). Genomic analysis of virulent soil phages of *Burkholderia*. Poster, American Society for Microbiology 106th General Meeting.

Rohwer, F., and Edwards, R. (2002). The Phage Proteomic Tree: a genome-based taxonomy for phage. J. Bacteriol. *184*, 4529–4535.

Ronning, C.M., Losada, L., Brinkac, L., Inman, J., Ulrich, R.L., Schell, M., Nierman, W.C., and DeShazer, D. (2010). Genetic and phenotypic diversity in *Burkholderia*: Contributions by prophage and phage-like elements. BMC Microbiol. *10*, 202.

Sabri, M., Hauser, R., Ouellette, M., Liu, J., Dehbi, M., Moeck, G., Garcia, E., Titz, B., Uetz, P., and Moineau, S. (2011). Genome annotation and intraviral interactome for the *Streptococcus pneumoniae* virulent phage Dp-1. J. Bacteriol. *193*, 551–562.

Sanger, F., Air, G.M., Barrell, B.G., Brown, N.L., Coulson, A.R., Fiddes, C.A., Hutchison, C.A., Slocombe, P.M., and Smith, M. (1977). Nucleotide sequence of bacteriophage ϕX174 DNA. Nature *265*, 687–695.

Sariya, L., Prempracha, N., Keelapang, P., and Chittasophon, N. (2006). Bacteriophage isolated from *Burkholderia pseudomallei* causes phenotypic changes in *Burkholderia thailandensis*. ScienceAsia *32*, 83–91.

Schiff, J.A. (1979). Pathways of assimilatory sulphate reduction in plants and microorganisms. Ciba Found. Symp. *72*, 49–69.

Seed, K.D., and Dennis, J.J. (2005). Isolation and characterization of bacteriophages of the *Burkholderia cepacia* complex. FEMS Microbiol. Lett. *251*, 273–280.

Seed, K.D., and Dennis, J.J. (2009). Experimental bacteriophage therapy increases survival of *Galleria mellonella* larvae infected with clinically relevant strains of the *Burkholderia cepacia* complex. Antimicrob. Agents Chemother. *53*, 2205–2208.

Semler, D.D., Lynch, K.H., and Dennis, J.J. (2012). The promise of bacteriophage therapy for *Burkholderia cepacia* complex respiratory infections. Front. Cell. Inf. Microbio. *1*, 27.

Seo, Y.S., Lim, J., Choi, B.S., Kim, H., Goo, E., Lee, B., Lim, J.S., Choi, I.Y., Moon, J.S., Kim, J., and Hwang, I. (2011). Complete genome sequence of *Burkholderia gladioli* BSR3. J. Bacteriol. *193*, 3149.

Smith, M.J., and Jeddeloh, J.A. (2005). DNA methylation in lysogens of pathogenic *Burkholderia* spp. requires prophage induction and is restricted to excised phage DNA. J. Bacteriol. *187*, 1196–1200.

Smith, P.B., and Cherry, W.B. (1957). Identification of *Malleomyces* by specific bacteriophages. J. Bacteriol. *74*, 668–672.

Stapleton, M.R., Horsburgh, M.J., Hayhurst, E.J., Wright, L., Jonsson, I.M., Tarkowski, A., Kokai-Kun, J.F., Mond, J.J., and Foster, S.J. (2007). Characterization of IsaA and SceD, two putative lytic transglycosylases of *Staphylococcus aureus*. J. Bacteriol. *189*, 7316–7325.

Summer, E.J., Gonzalez, C.F., Carlisle, T., Mebane, L.M., Cass, A.M., Savva, C.G., LiPuma, J., and Young, R. (2004). *Burkholderia cenocepacia* phage BcepMu and a family of Mu-like phages encoding potential pathogenesis factors. J. Mol. Biol. *340*, 49–65.

Summer, E.J., Gonzalez, C.F., Bomer, M., Carlile, T., Embry, A., Kucherka, A.M., Lee, J., Mebane, L., Morrison, W.C., Mark, L., *et al.* (2006). Divergence and mosaicism among virulent soil phages of the *Burkholderia cepacia* complex. J. Bacteriol. *188*, 255–268.

Summer, E.J., Gill, J.J., Upton, C., Gonzalez, C.F., and Young, R. (2007). Role of phages in the pathogenesis of *Burkholderia*, or 'Where are the toxin genes in *Burkholderia* phages?'. Curr. Opin. Microbiol. *10*, 410–417.

Thiel, K. (2004). Old dogma, new tricks – 21st Century phage therapy. Nat. Biotechnol. *22*, 31–36.

Tiwari, B.R., Kim, S., Rahman, M., and Kim, J. (2011). Antibacterial efficacy of lytic *Pseudomonas* bacteriophage in normal and neutropenic mice models. J. Microbiol. *49*, 994–999.

Tu, A.H., Voelker, L.L., Shen, X., and Dybvig, K. (2001). Complete nucleotide sequence of the *Mycoplasma* virus P1 genome. Plasmid *45*, 122–126.

Tuanyok, A., Leadem, B.R., Auerbach, R.K., Beckstrom-Sternberg, S.M., Beckstrom-Sternberg, J.S., Mayo, M., Wuthiekanun, V., Brettin, T.S., Nierman, W.C., Peacock, S.J., *et al.* (2008). Genomic islands from five strains of *Burkholderia pseudomallei*. BMC Genomics *9*, 566.

Tumapa, S., Holden, M.T., Vesaratchavest, M., Wuthiekanun, V., Limmathurotsakul, D., Chierakul, W., Feil, E.J., Currie, B.J., Day, N.P., Nierman, W.C., and Peacock, S.J. (2008). *Burkholderia pseudomallei* genome plasticity associated with genomic island variation. BMC Genomics *9*, 190.

Vieira, A., Silva, Y.J., Cunha, A., Gomes, N.C., Ackermann, H.-W., and Almeida, A. (2012). Phage therapy to control multidrug-resistant *Pseudomonas aeruginosa* skin infections: in vitro and ex vivo experiments. Eur. J. Clin. Microbiol. Infect. Dis. *31*, 3241–3249.

Waldor, M.K., and Mekalanos, J.J. (1996). Lysogenic conversion by a filamentous phage encoding cholera toxin. Science *272*, 1910–1913.

Walmagh, M., Boczkowska, B., Grymonprez, B., Briers, Y., Drulis-Kawa, Z., and Lavigne, R. (2013). Characterization of five novel endolysins from Gram-negative infecting bacteriophages. Appl. Microbiol. Biotechnol. *97*, 4369–4375.

Wang, X., and Wood, T.K. (2011). Toxin–antitoxin systems influence biofilm and persister cell formation and the general stress response. Appl. Environ. Microbiol. *77*, 5577–5583.

Wolochow, H., and Green, L.B. (1956). Isolation of a bacteriophage for *Malleomyces pseudomallei*. Bacteriol. Proc. *1956*, 60–61.

Woods, D.E., Jeddeloh, J.A., Fritz, D.L., and DeShazer, D. (2002). *Burkholderia thailandensis* E125 harbors a temperate bacteriophage specific for *Burkholderia mallei*. J. Bacteriol. *184*, 4003–4017.

Wright, A., Hawkins, C.H., Änggård, E.E., and Harper, D.R. (2009). A controlled clinical trial of a therapeutic bacteriophage preparation in chronic otitis due to antibiotic-resistant *Pseudomonas aeruginosa*; A preliminary report of efficacy. Clin. Otolaryngol. *34*, 349–357.

Yordpratum, U., Tattawasart, U., Wongratanacheewin, S., and Sermswan, R.W. (2011). Novel lytic bacteriophages from soil that lyse *Burkholderia pseudomallei*. FEMS Microbiol. Lett. *314*, 81–88.

Yu, Y., Kim, H.S., Chua, H.H., Lin, C.H., Sim, S.H., Lin, D., Derr, A., Engels, R., DeShazer, D., Birren, B., Nierman, W.C., and Tan, P. (2006). Genomic patterns of pathogen evolution revealed by comparison of *Burkholderia pseudomallei*, the causative agent of melioidosis, to avirulent *Burkholderia thailandensis*. BMC Microbiol. *6*, 46.

Zahrl, D., Wagner, M., Bischof, K., Bayer, M., Zavecz, B., Beranek, A., Ruckenstuhl, C., Zarfel, G.E., and Koraimann, G. (2005). Peptidoglycan degradation by specialized lytic transglycosylases associated with type III and type IV secretion systems. Microbiology *151*, 3455–3467.

Zhou, Y., Liang, Y., Lynch, K.H., Dennis, J.J., and Wishart, D.S. (2011). PHAST: a fast phage search tool. Nucleic Acids Res. *39*, W347–52.

Index

Current books of interest

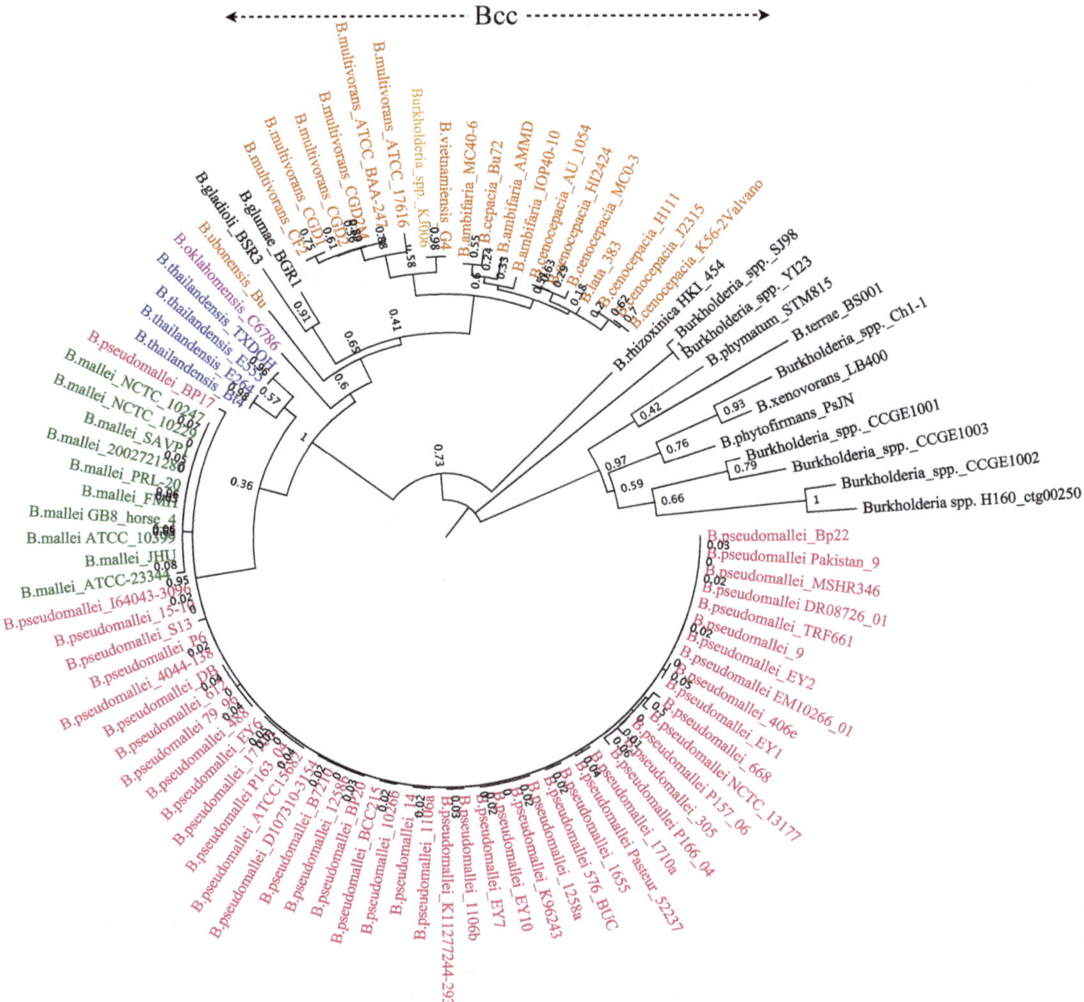

Figure 2.1 Phylogenetic analysis of the 16S rRNA gene from 96 *Burkholderia* genomes. Not all genomes listed in Table 2.1 are included here because the full length of their 16S rRNA gene could not be identified. The tree was generated by the maximum-likelihood method. The numbers next to the branches represent the bootstrap value, which was based on 1000iterations. Species names were labelled according to *Burkholderia* groups; *B. pseudomallei* (pink), *B. mallei* (green), *B. thailandensis* (blue), *B. oklahomensis* (purple), Bcc (orange), *Burkholderia* spp. and the rest are shaded with black.

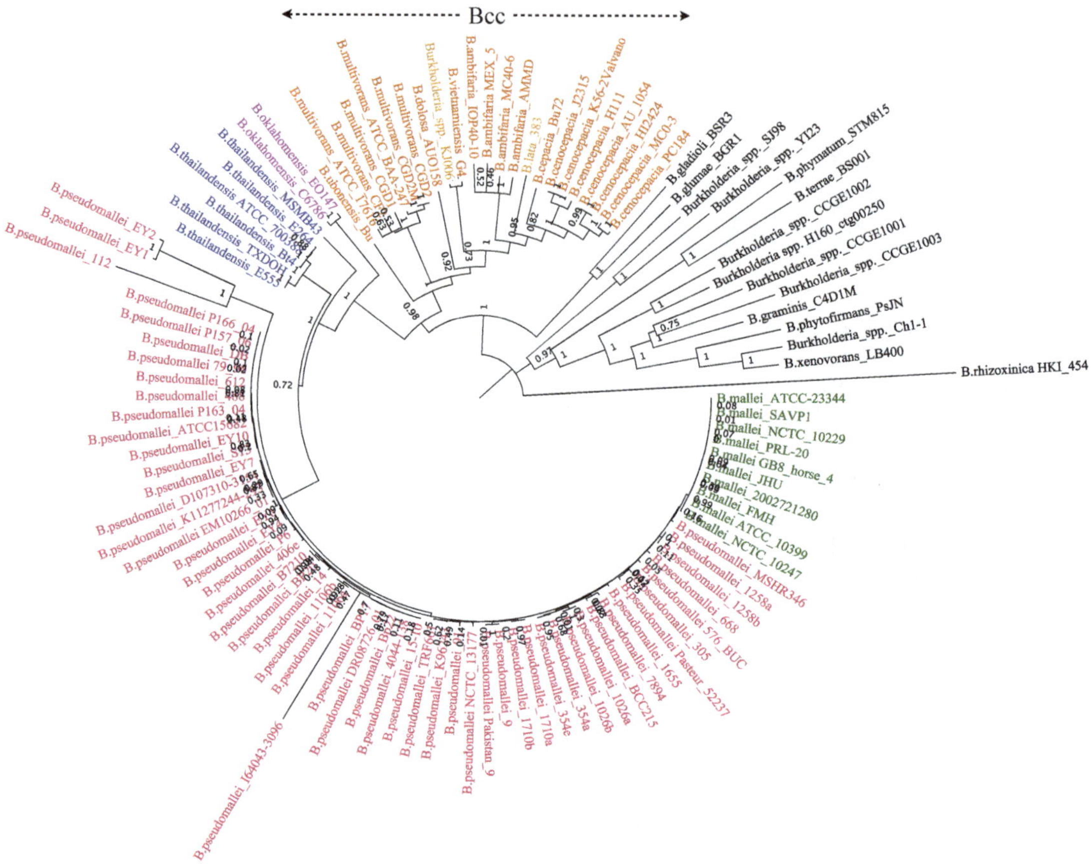

Figure 2.2 *In silico* MLST tree. This tree was constructed from the concatenated sequences of seven *housekeeping* genes used for MLST typing of *B. pseudomallei*. The tree was simulated by MEGA5, with the maximum-likelihood method using 1000iterations of bootstrapping. The numbers next to the branches represent the bootstrap value. The species groups are labelled similarly to Fig. 2.1.

Figure 2.3 (opposite page) A pan-genome family tree of 110 *Burkholderia* genomes. This is an alternative approach to construct a phylogenetic tree based on all the gene contents in the genome (pan-genome). The tree was generated from the presence or absence of gene families across all *Burkholderia* genomes. The red values next to the branches are bootstrap values.

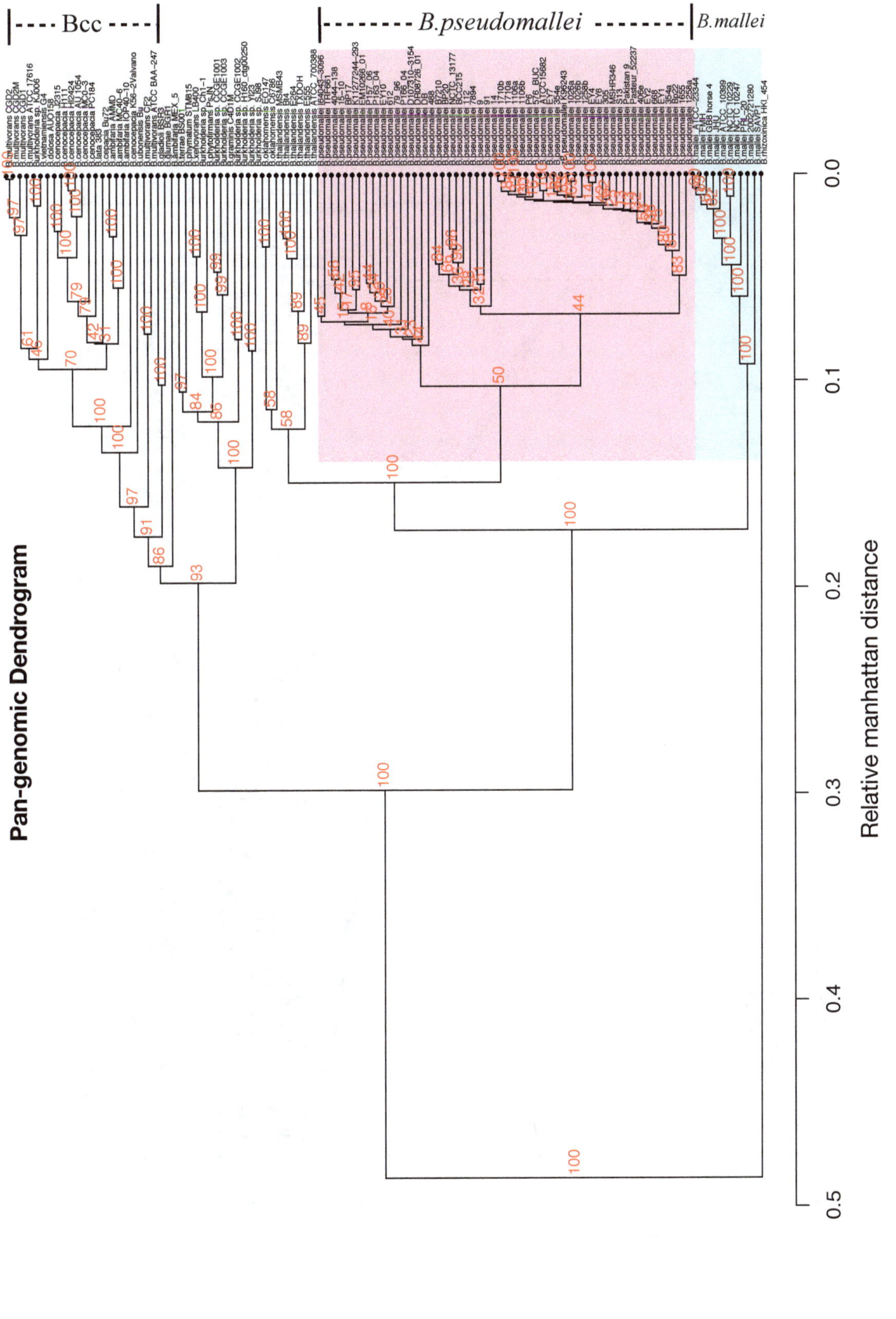
Pan-genomic Dendrogram
Bcc
B.pseudomallei
B.mallei
Relative manhattan distance
0.0
0.1
0.2
0.3
0.4
0.5

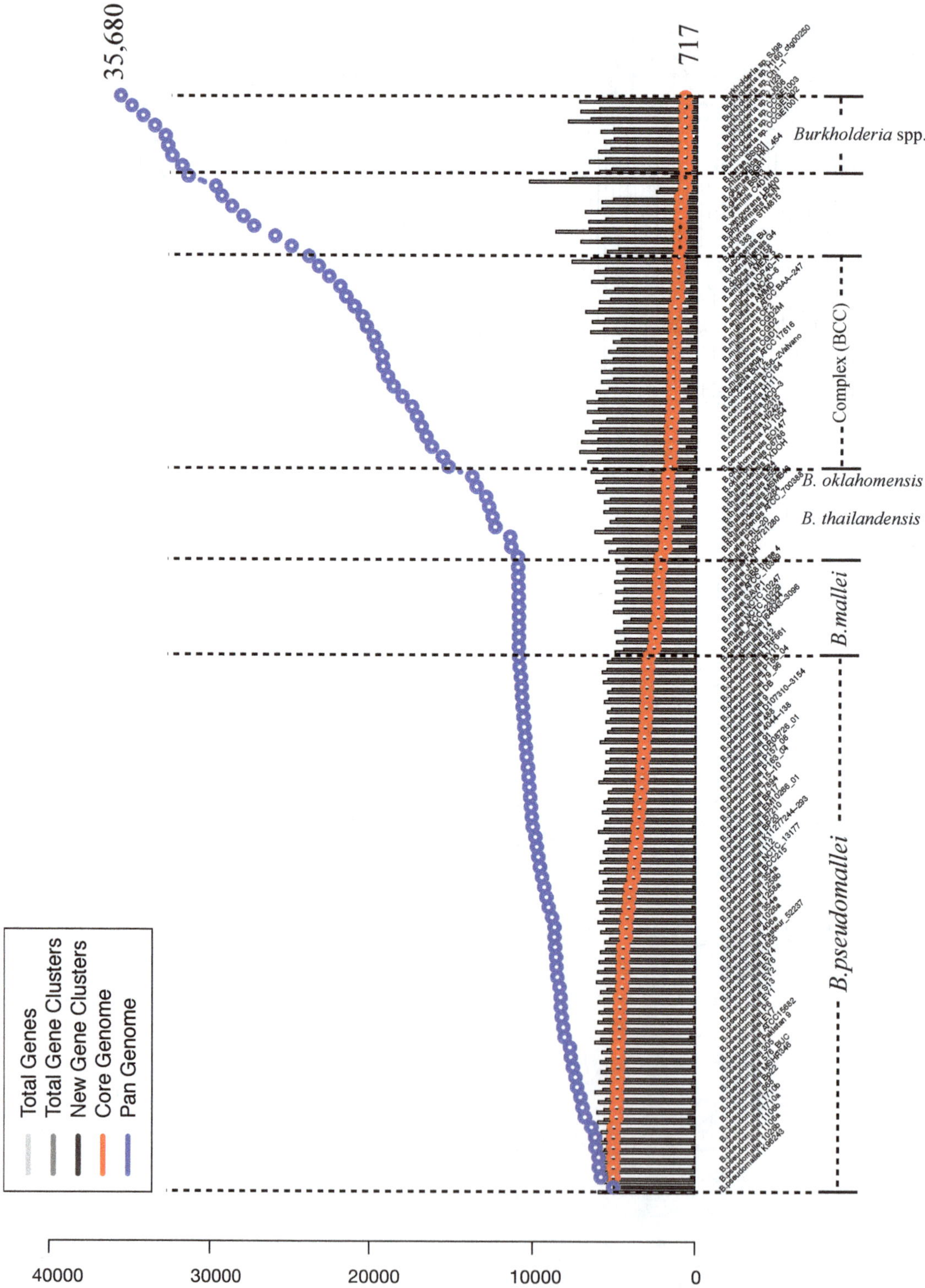
35,680
717
Burkholderia spp.
Complex (BCC)
B. oklahomensis
B. thailandensis
B.mallei
B.pseudomallei
Total Genes
Total Gene Clusters
New Gene Clusters
Core Genome
Pan Genome
40000
30000
20000
10000
0

Figure 2.4 (opposite page) Core/pan-genome plot of 110 *Burkholderia* genomes. The increasing number of genes in the pan-genome and core-genome are shown in blue and red dots, respectively. The last points express the total number of pan- (35,680 gene families) and core- genomes (717 gene families).

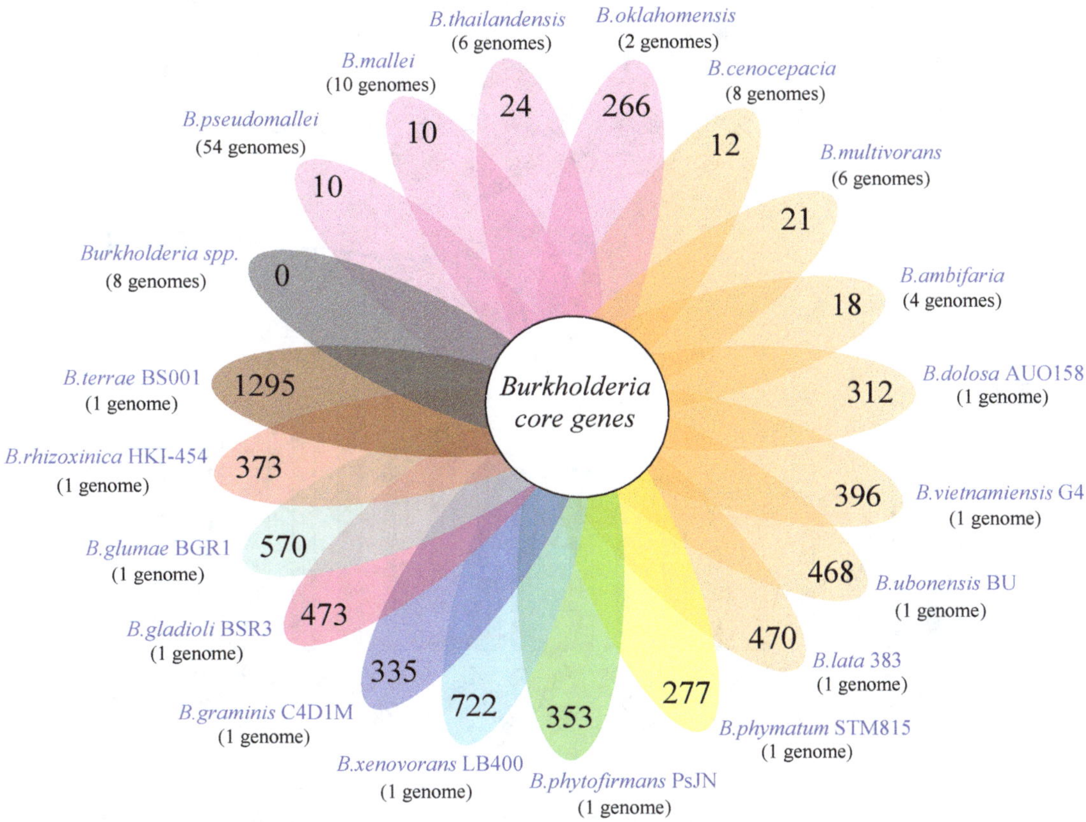

Figure 2.5 FlowerPlot of unique core gene families in each *Burkholderia* species. The figure represents the number of core gene families found in each genome as being unique to the *Burkholderia* species. The coloured petals represent different *Burkholderia* groups or species.

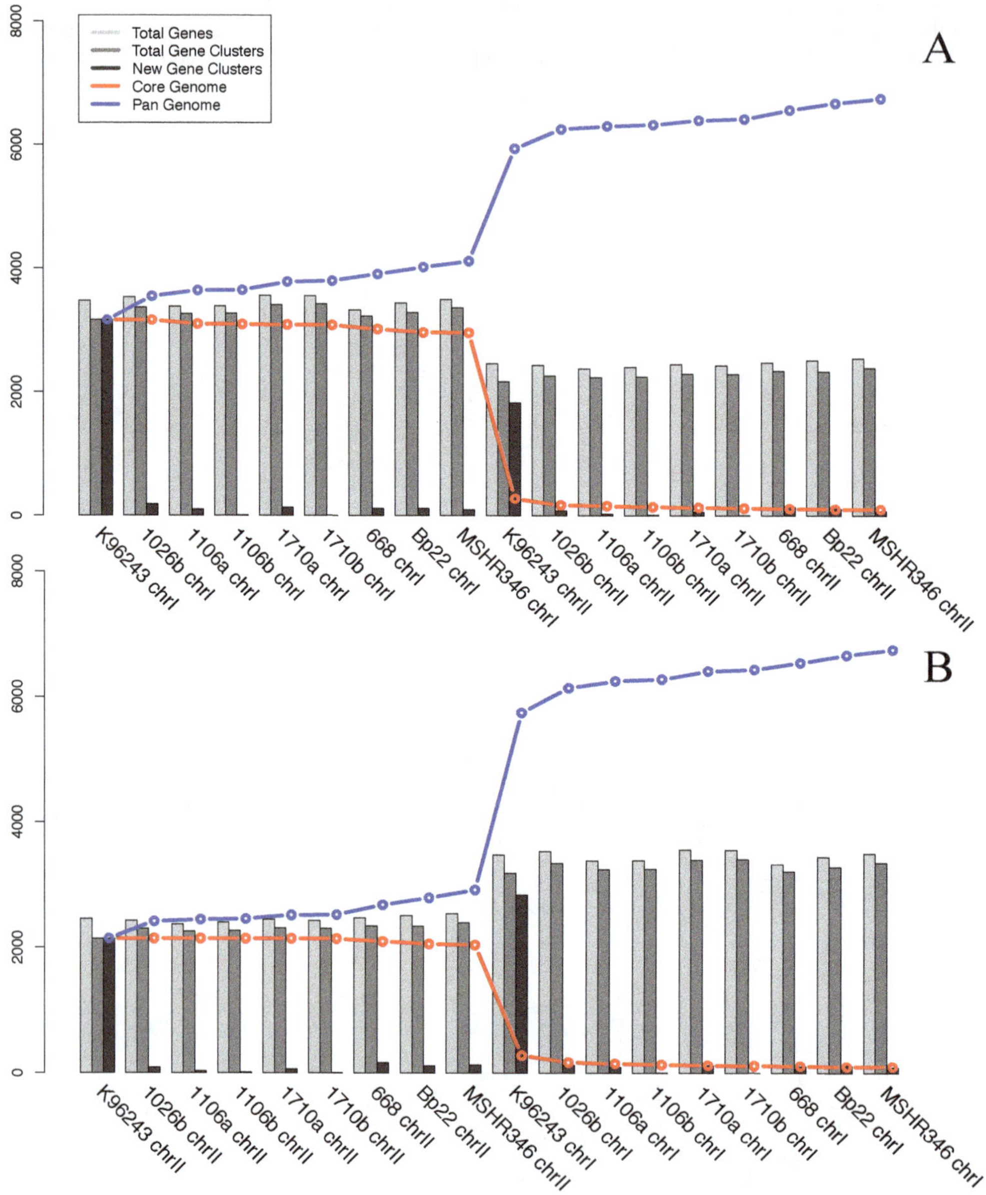

Figure 2.6 Core/pan-genome plot of nine *Burkholderia pseudomallei* genomes. (A) The plot starts with the first replicon. (B) The plot starts with the second chromosome. The increasing numbers of genes in the pan-genome and core-genome are shown in blue and red dots, respectively.

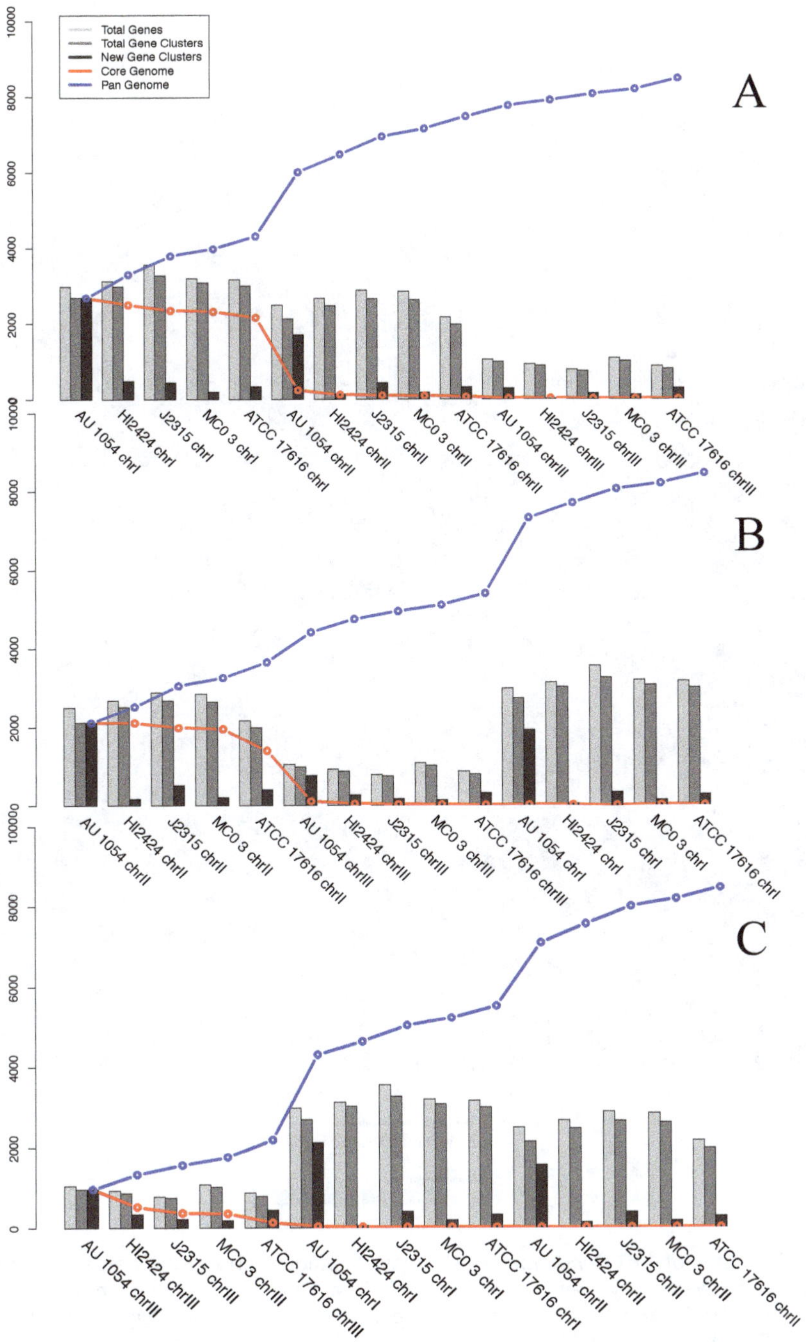

Figure 2.7 Core/pan-genome plot of four *Burkholderia cenocepacia* plus one *Burkholderia multivorans* genomes. (A) The plot starts from the first replicon. (B) The plot starts from the second chromosome, and (C) the plot starts from the third replicon. The increasing numbers of genes in the pan-genome and core-genome are shown in blue and red dots, respectively.

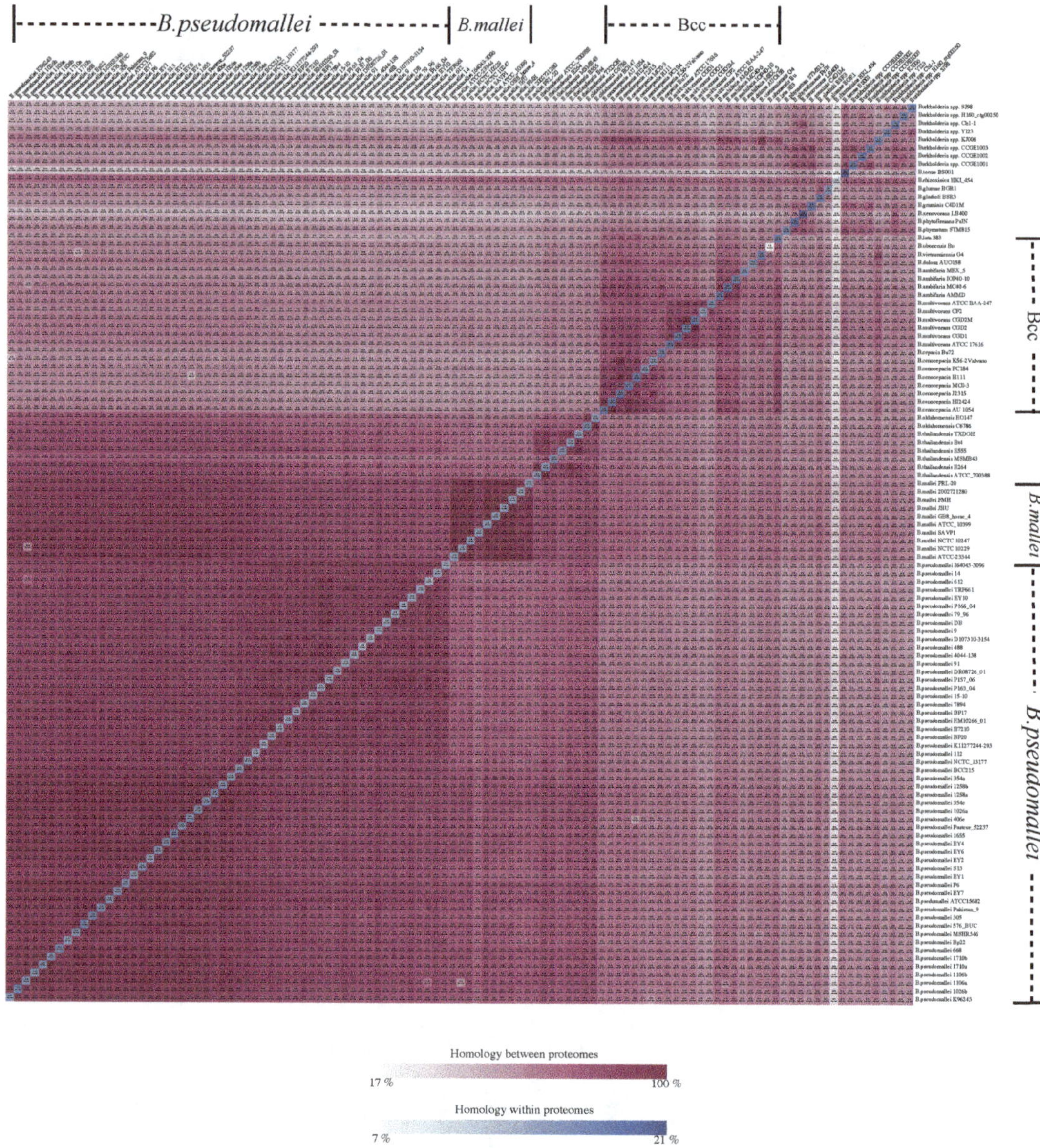

Figure 2.8 BLAST Matrix of 110 *Burkholderia* genomes. The figure shows the percentage of gene families found in common between any pair-genomes (pink). The blue colour represents the degree of gene duplication within each *Burkholderia* genome. The pairwise all-against-all alignment was performed by reciprocal BLASTP at the amino acid level.

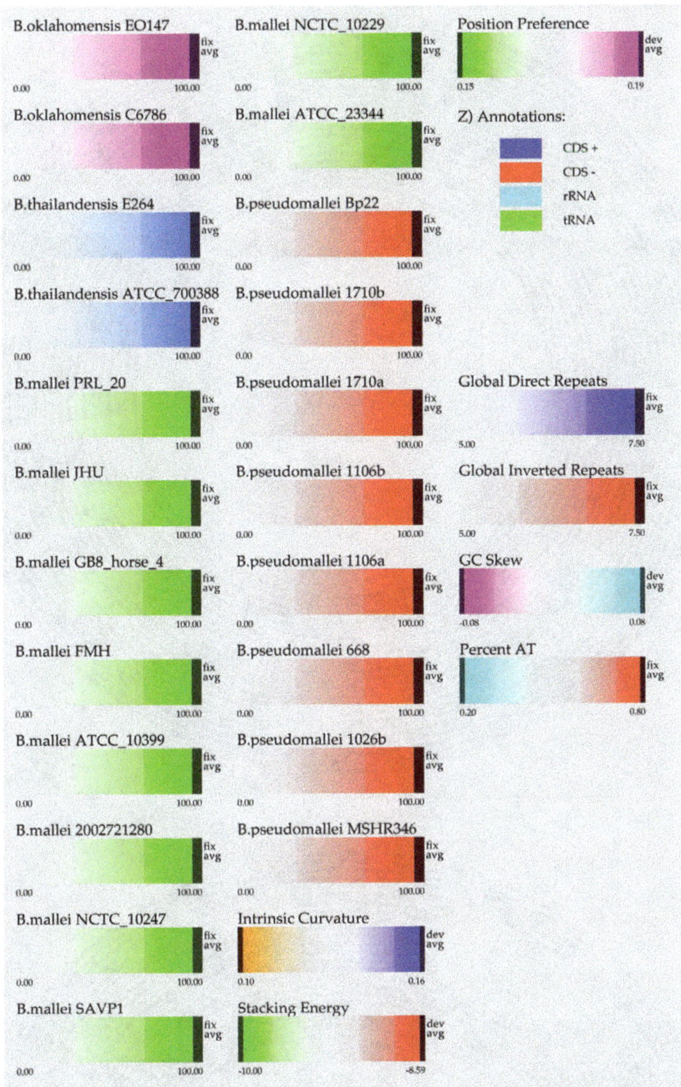

Figure 2.9 BLAST Atlas of eight *B. pseudomallei*, ten *B. mallei*, two *B. thailandensis* and two *B. oklahomensis*. The figure shows the BLAST atlas with B. pseudomallei K96243 chromosome I (top) and chromosome II (bottom) as references (continued on next page).

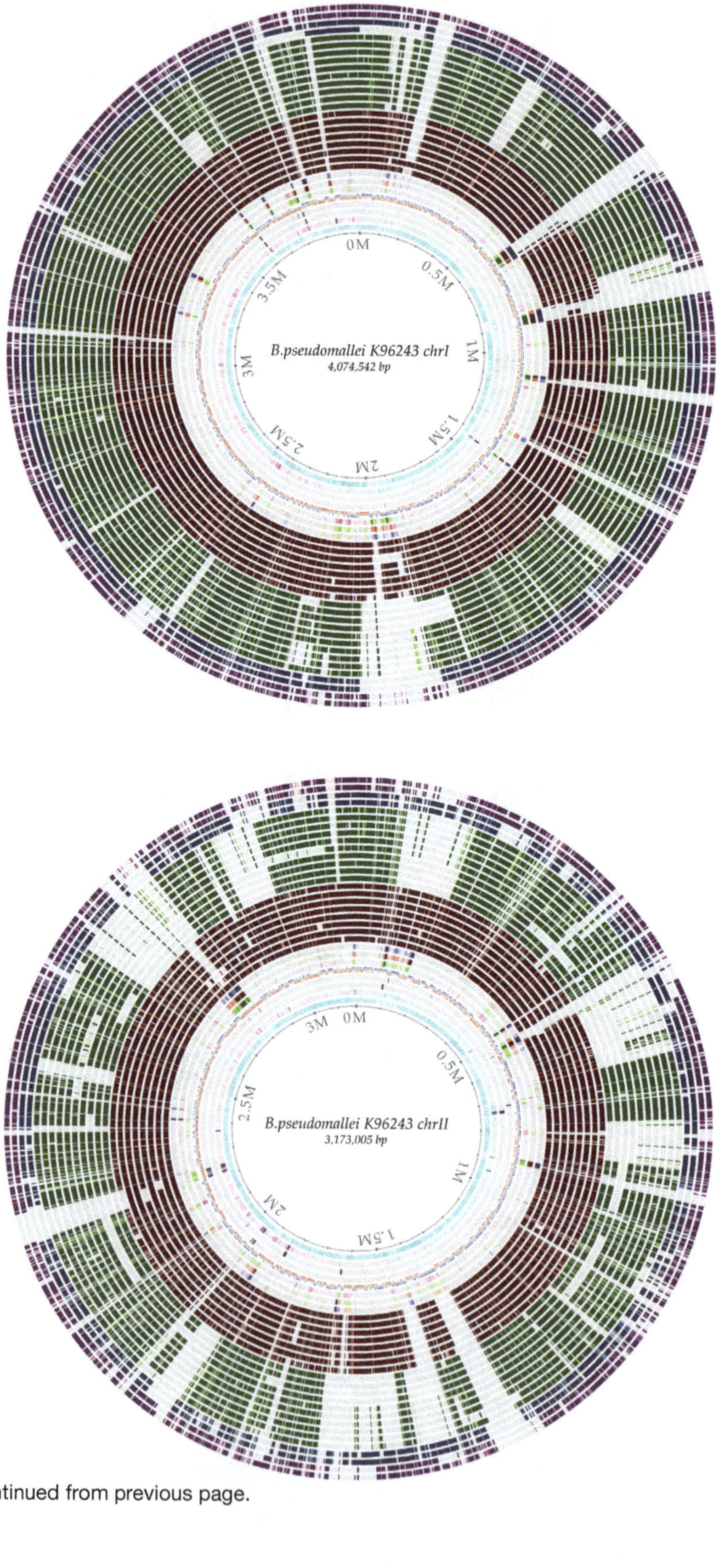

Figure 2.9 Continued from previous page.

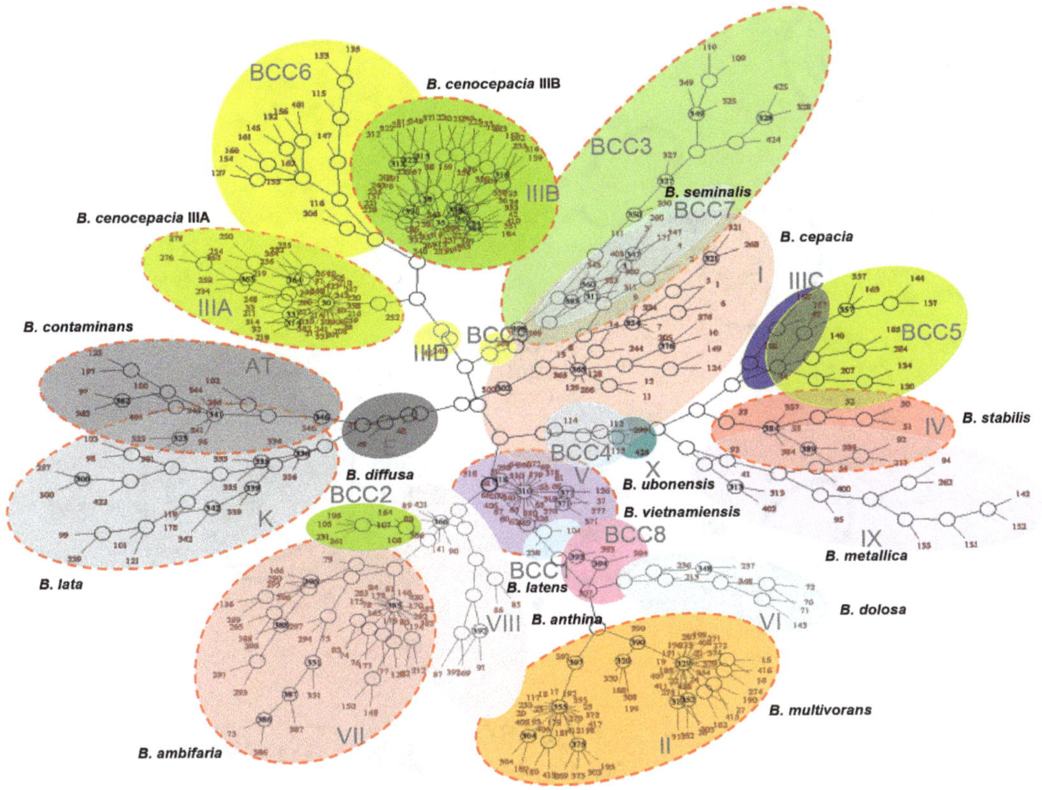

Key:

I - ***B. cepacia***	VI - ***B. dolosa***	BCC1 - ***B. latens***	AT - ***B. lata***
II - ***B. multivorans***	VII - ***B. ambifaria***	BCC2 - ***B. diffusa***	K - ***B. contaminans***
IIIA-D - ***B. cenocepacia***	VIII - ***B. anthina***	BCC3 - ***B. arboris***	
IV - ***B. stabilis***	IX - ***B. pyrrocinia***	BCC7 - ***B. seminalis***	
V - ***B. vietnamiensis***	X - ***B. ubonensis***	BCC9 - ***B. metallica***	**BCC 4, 5, 6 & 8 Novel groups**

Figure 3.3 Prediction of evolutionary relationships within the *B. cepacia* complex. Sequence data from the MLST database were examined using clonal frame analysis to predict the evolutionary relationships between each Bcc species and STs within them. The 17 currently named Bcc species as well as novel MLST groups are indicated adjacent to their respective clonal frame analysis clusters and also listed in the key. Data analysis was performed by Dr Adam Baldwin using the methods described by Didelot and Falush (2007).

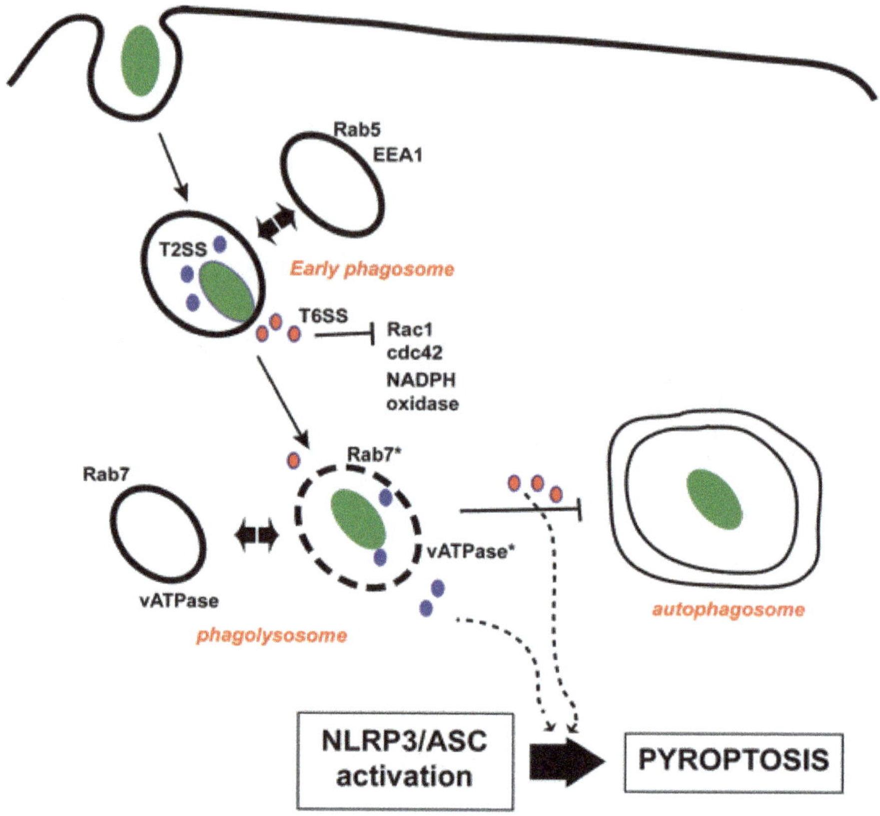

Figure 7.1 Diagram summarizing current knowledge of the intramacrophage survival mechanisms of *B. cenocepacia*. Upon engulfment the BcCV matures normally as an early phagosome by recruiting Rab5 and EEA1. However, further maturation is halted resulting in delayed acidification and recruitment of components of the NADPH oxidase and v-ATPase complexes. Delayed recruitment of the NADPH oxidase complex depends on a functional T6SS, which presumably secretes one or more effectors (in red) that inactivate Rac1 and Cdc42, also leading to defects in actin polymerization, phagocytosis, and cell migration. At later stages of maturation the BcCV accumulates Rab7 but this protein is inactivated by an unknown mechanism that most likely involves one or more effectors (in blue) secreted by the T2SS, which escape from the BcCV to the host cell's cytoplasm. The BcCV interacts with components of the autophagosome, but the full maturation of the autophagosome is also halted. BcCV membrane damage also triggers the assembly and activation of the NLRP3/ASC inflammasome resulting in macrophage cell death by pyroptosis. Asterisks indicate non-functional vATPase and Rab7, respectively.

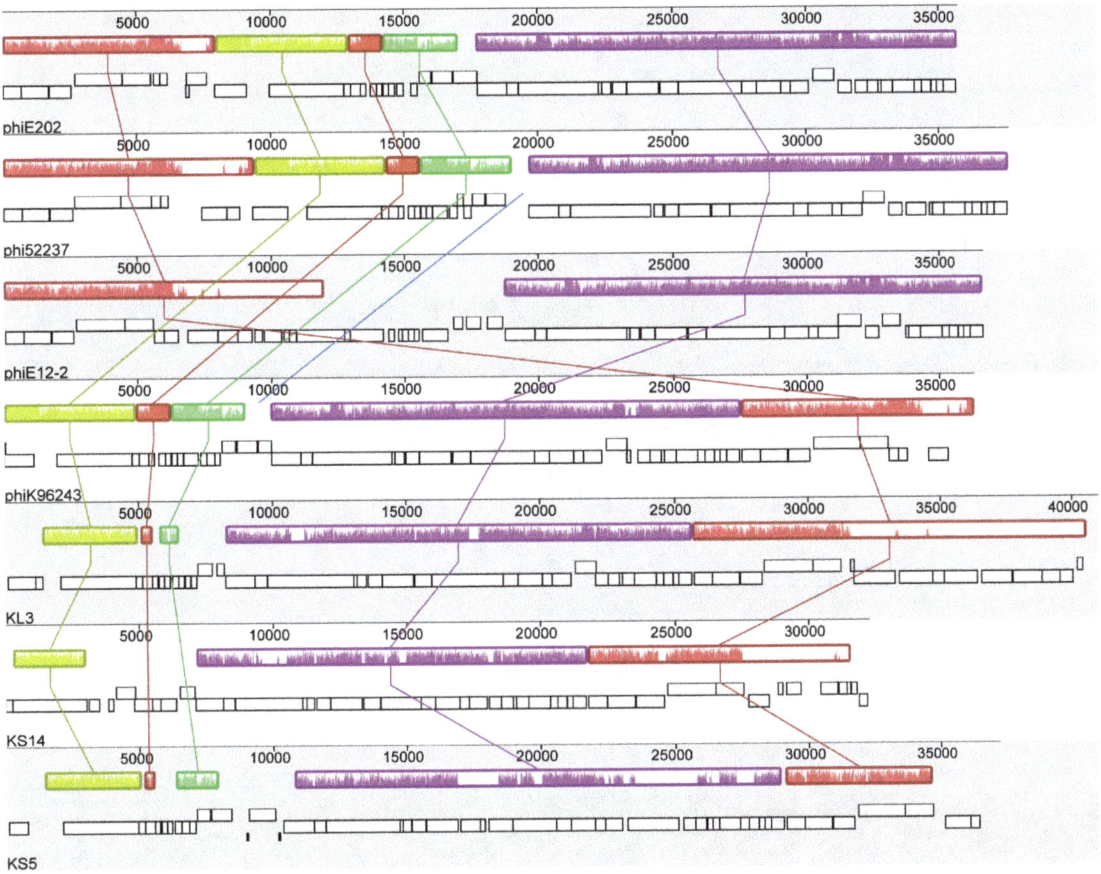

Figure 11.1 Whole genome alignment of P2-like *Burkholderia* myoviruses. Alignments were performed using default parameters for the Mauve plugin (progressive Mauve algorithm) in the Geneious package (Darling *et al.*, 2010; Drummond *et al.*, 2012).

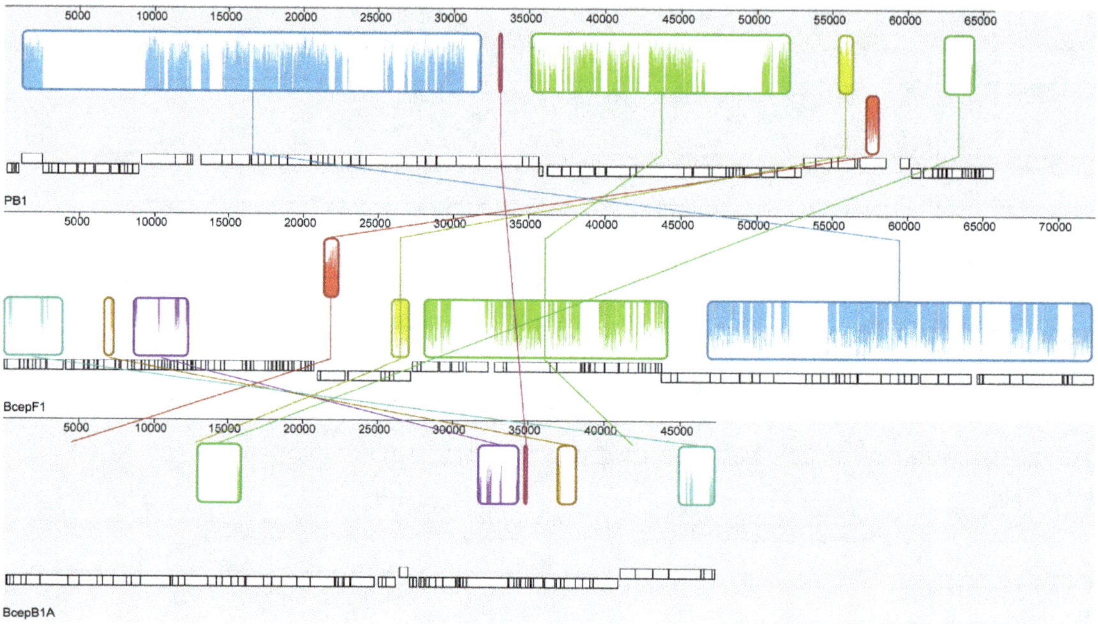

Figure 11.3 Whole genome alignment of PB1-like myoviruses.

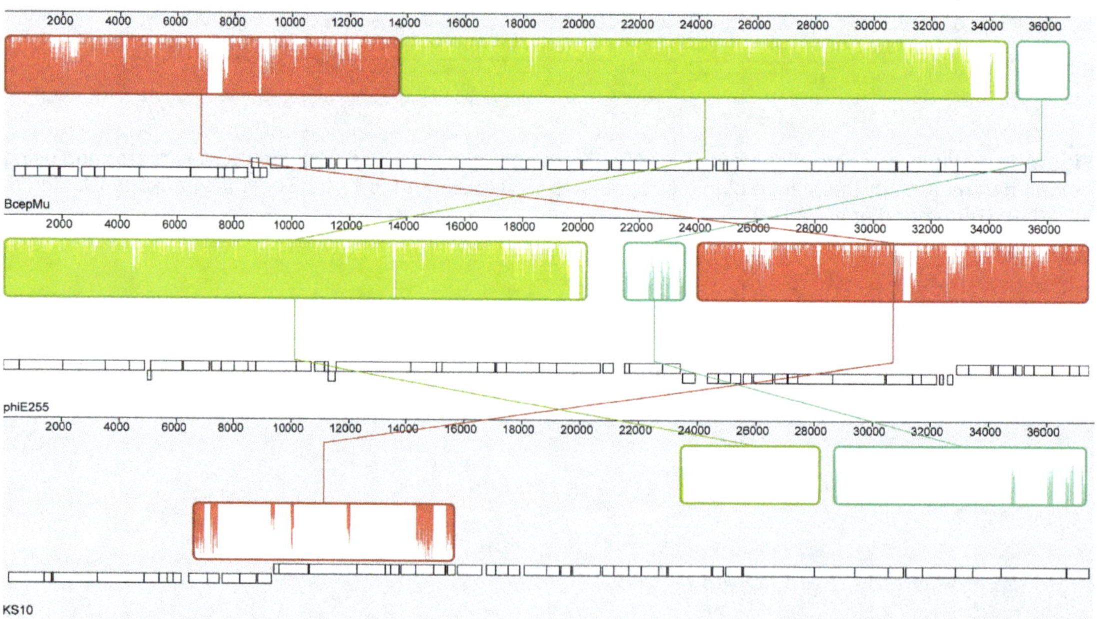

Figure 11.4 Whole genome alignment of BcepMu-like myoviruses.

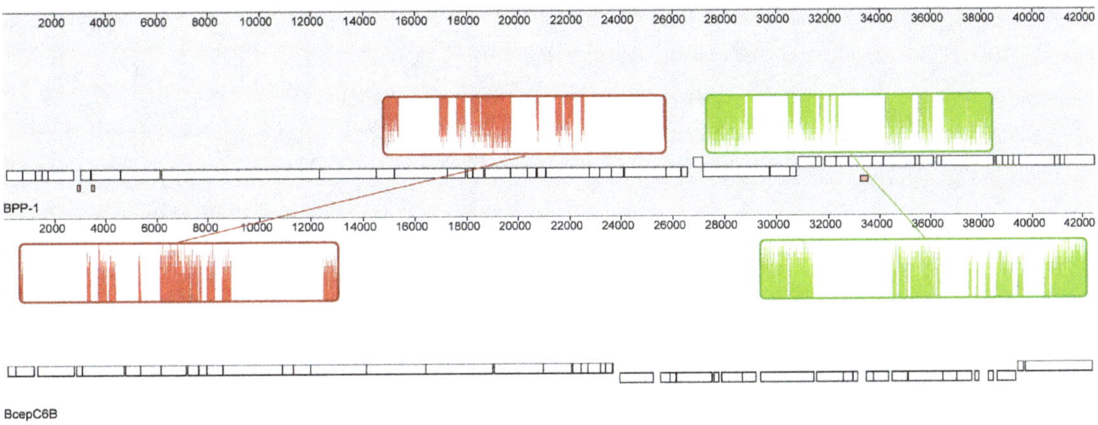

Figure 11.6 Whole genome alignment of BPP-1-like podoviruses.

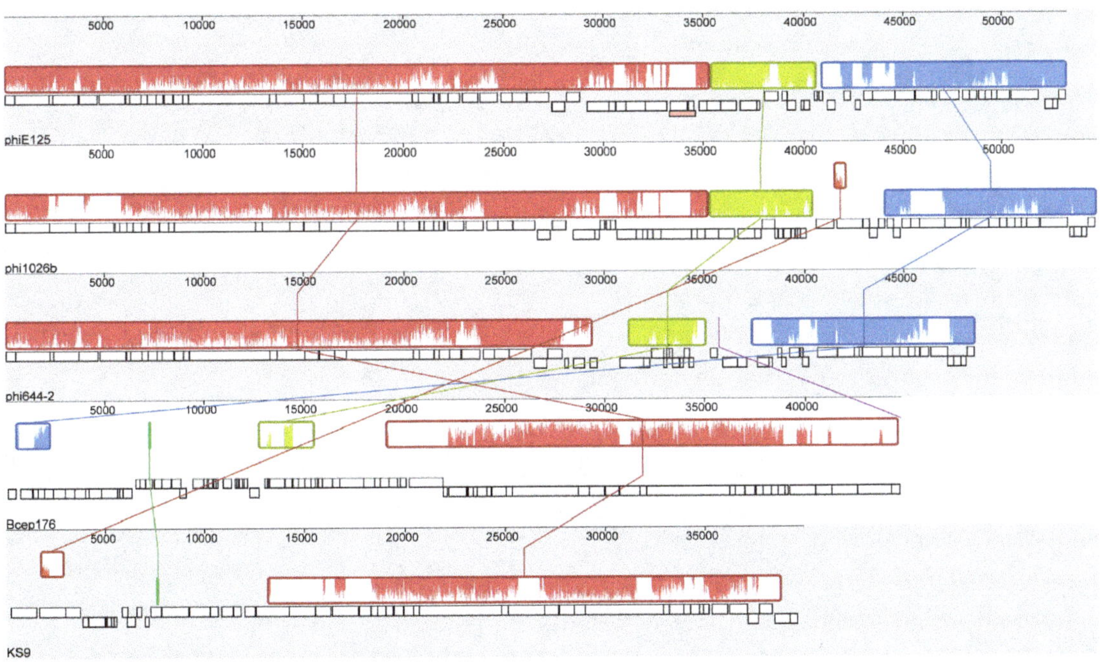

Figure 11.9 Whole genome alignment of ϕE125-like siphoviruses.

www.ingramcontent.com/pod-product-compliance
Lightning Source LLC
LaVergne TN
LVHW062227200926
848484LV00006B/1494
* 9 7 8 1 9 0 8 2 3 0 3 5 5 *